Computational Design and Robotic Fabrication

Series Editor

Philip F. Yuan, *Tongji University, Shanghai, China*

This open access book series includes compilations of selected papers from the International Conference on Computational Design and Robotic Fabrication. The books focus on novel techniques for computational design and robotic fabrication. It not only aims at the most recent research results from the key scholars in the computational design and robotic fabrication, but also offers an in-depth examination of intelligence in design and construction industry, referring to the invention and application of machine intelligence in architecture.The proceedings in this series are related with Sustainable Development Goals 11 (Sustainable Cities & Communities) and 9 (Industry Innovation & Infrastructure).

The contents make valuable contributions to academic researchers, designers, and engineers in the industry. As well, readers will encounter new ideas about understanding intelligence in architecture.

Indexed by EI Compendex and Scopus.

Hua Chai · Ding Wen Nic Bao · Zhe Guo ·
Philip F. Yuan

Editors

Symbiotic Intelligence

Proceedings of the 6th International Conference
on Computational Design and Robotic
Fabrication (CDRF 2024)

Editors
Hua Chai
Tongji University
Shanghai, China

Zhe Guo
Hefei University of Technology
Hefei, China

Ding Wen Nic Bao
RMIT University
Melbourne, VIC, Australia

Philip F. Yuan
Tongji University
Shanghai, China

ISSN 2731-9040   ISSN 2731-9059 (electronic)
Computational Design and Robotic Fabrication
ISBN 978-981-96-3432-3   ISBN 978-981-96-3433-0 (eBook)
https://doi.org/10.1007/978-981-96-3433-0

This Springer imprint is published by the registered company Springer Nature Singapore Pte Ltd.
The registered company address is: 152 Beach Road, #21-01/04 Gateway East, Singapore 189721, Singapore

If disposing of this product, please recycle the paper.

Preface

In 1960, American psychologist and computer scientist Joseph C. R. Licklider proposed the concept of "Man-Computer Symbiosis," creatively likening the mutual benefit seen in ecological systems between different species to the emerging relationship between humans and computers. Today, with advancements in technology reaching unprecedented levels of intelligence and autonomy, there is a renewed emphasis on establishing a positive and harmonious symbiotic relationship between humans and technology.

Dutch philosopher Peter-Paul Verbeek argues that the relationship between humans and technology can no longer be simply summarized by traditional mediated intentionality, where technology serves as a tool between humans and the world. The relationship between humans and machines is gradually manifesting diverse and complex attributes, notably hybrid intentionality and composite intentionality. In the field of architecture, the "hybrid" and "composite" of the designer and technological intentionality are reshaping the production paradigm and knowledge landscape of the discipline. The development of technologies such as machine learning, collaborative robots, and brain–computer interfaces is significantly enhancing the role of human–machine interaction in architectural design and production decision-making. At the same time, the unprecedented creativity demonstrated by artificial intelligence continually challenges and impacts the unique position and creative abilities of architects. Against the backdrop of increasingly prominent global challenges, the relationship between architects and technologies needs to evolve into a more intricate and mutually interdependent symbiotic state, collectively providing innovative solutions to issues including climate change, pandemic propagation, and resource and environmental crises.

As the architectural domain transitions into a new epoch characterized by the flourishing development of intelligent and autonomous technologies, architectural studies need to rethink and redefine the symbiotic patterns between human and technology. As the theme of CDRF 2024, "Symbiotic Intelligence" aims to promote global exchange and reflection on the transformation of architectural paradigms under the influence of intelligent technologies. This conference encourages the actively exploration of the rich possibilities of symbiotic intelligence in the field of architecture, collectively driving innovation and development in the architectural discipline.

Organization

Committees

Honorary Advisors

Philippe Block	ETH Zurich, Switzerland
Jane Burry	Swinburne University of Technology, Australia
Mark Burry	Swinburne University of Technology, Australia
Yiming Chen	Nanyang Technological University, Singapore
Lieyun Ding	Huazhong University of Science and Technology, China
Jian Gong	Shanghai Construction Group (SCG), China
Guoqiang Li	Tongji University, China
Jiaping Liu	Xi'an University of Architecture and Technology, China
Areti Markopoulou	Institute for Advanced Architecture of Catalonia, Spain
Achim Menges	University of Stuttgart, Germany
Antoine Picon	GSD, USA
Patrik Schumacher	Zaha Hadid Architects (ZHA), UK
Mette Ramsgaard Thomsen	Royal Danish Academy, Denmark
Zhiqiang Wu	Tongji University, China
Yimin (Mike) Xie	RMIT University, Australia
Xianzhong Zhao	Tongji University, China

Organization Committees

Philip F. Yuan	Tongji University, China
Neil Leach	Tongji University, China

Scientific Committees

Felix Amtsberg	University of Stuttgart, Germany
Alisa Andrasek	RMIT University, Australia
Ding Wen Nic Bao	RMIT University, Australia
Thomas Bock	Technical University of Munich, Germany

Serban Bodea	University of Stuttgart, Germany
Biayna Bogisian	Florida International University in Miami, USA
Daniel Bolojan	Florida Atlantic University, USA
Matias Del Campo	New York Institute of Technology, USA
Brad Cantrell	University of Virginia, USA
Tengwen Chang	National Yunlin University of Science and Technology, Taiwan, China
Kristof Crolla	University of Hong Kong, China
Benjamin Dillenburger	ETH Zurich, Switzerland
Marcus Farr	Tongji University, China
Melissa Goldman	University of Virginia, USA
Yunsong Han	Harbin Institute of Technology, China
Hua Hao	Southeast University, China
Wanyu He	Xkool, China
Tim Heath	University of Nottingham, UK
Alvin Huang	University of Southern California, USA
Weixin Huang	Tsinghua University, China
Guohua Ji	Nanjing University, China
Gene Ting-Chun Kao	ETH Zurich, Switzerland
Immanuel Koh	Singapore University of Technology and Design, Singapore
Neil Leach	Tongji University, China
Guan Lee	University College London, UK
Hyejin Lee	Tongji University, China
Biao Li	Southeast University, China
Linxue Li	Tongji University, China
Yujie Lu	Tongji University, China
Peng Luo	Tongji University, China
Andrea Macruz	Tongji University, China
Sandra Manninger	New York Institute of Technology, USA
Wes McGee	University of Michigan, USA
Xianchuan Meng	Nanjing University, China
Virginia Melnyk	Tongji University, China/Clemson University, USA
Kris Mun	University of Minnesota, USA
Guvenc Ozel	University of California, Los Angeles, USA
Gilles Retsin	University College London, UK
Klaas de Rycke	Bollinger + Grohmann, Germany
Bob Sheil	University College London, UK
Xing Shi	Tongji University, China
Miroslaw J. Skibniewski	University of Maryland, USA
Roland Snooks	RMIT University, Australia

Satoru Sugihara	Architectural Technology Laboratorial Venture
Chengyu Sun	Tongji University, China
Kostas Terzidis	Tongji University, China
Oliver Tessmann	Technische Universität Darmstadt, Germany
Kathy Velikov	University of Michigan, USA
Tomas Vivanco	Tongji University, China
Xiang Wang	Tongji University, China
Makoto Sei Watanabe	Tokyo City University, Japan
Dylan Wood	University of Stuttgart, Germany
Jing Wu	Southern University of Science and Technology, China
Leiqing Xu	Tongji University, China
Weiguo Xu	Tsinghua University, China
Michael Weinstock	The Architectural Association, UK
Chao Yan	Tongji University, China
Feng Yang	Tongji University, China
Jiawei Yao	Tongji University, China
Kaiho Yu	University of Applied Arts Vienna, Austria
Philip F. Yuan	Tongji University, China
Xu Zhen	Tianjin University, China

Contents

Artificial Intelligence in Design and Simulation

Digital Design Theory, Method and Education

Performance-based Design, Analytics and Optimization

VR, AR and Interactive Technology

Urban Analytics, Urban Modelling and Simulation

Robotic Fabrication and Additive Manufacturing

Towards Lightweight Structure: Coupling Topology Optimization with Non-planar 3D Concrete Printing

Yuxin Lin, Alireza Bayramvand, and Mania Aghaei Meibodi(✉)

DART Laboratory, Taubman College of Architecture and Urban Planning, University of Michigan, Ann Arbor, MI 48109, USA
meibodi@umich.edu

Abstract. This study explores the integration of Topology Optimization (TO) with non-planar 3D concrete printing (NP-3DCP) to address environmental and material efficiency challenges in construction. We present a novel approach leveraging robotic NP-3DCP for creating lightweight, structurally optimized architectural components, specifically focusing on load-bearing walls. The advantages of this approach were demonstrated through the manufacturing of two prototypes, Shell Wall and Branch Wall, showcasing significant material savings and reduced carbon footprint without compromising structural integrity. Our methodology encompasses the generation of a non-planar printpath, material extrusion control for variable layer height, and the integration of rebar and thermal insulation within the casting process. The results showcase the potential of this approach in producing complex geometries with improved environmental performance, suggesting a promising direction for the future of sustainable concrete construction.

Keywords: Lightweight structure · Topology optimization · Robotic 3D concrete printing · Non-planar printpath

1 Introduction

The increasing urban population intensifies the need for housing and infrastructure. Reinforced concrete is important to meeting this rising demand, due to its strength, durability, affordability, and capability to create long-span structures [1]. Yet, the inefficient practices in concrete construction have led to excessive use of concrete, consequently amplifying cement production, which intensifies the scarcity of sand and accounts for over 8% of the global CO_2 emissions [2]. There is a pressing need to adopt material optimization strategies aimed at developing lightweight load-bearing concrete structures that can minimize material consumption while ensuring structural integrity and consequently reduce the carbon footprint. Topology Optimization (TO) is a technique that can achieve lightweight designs by optimally distributing material within a given design space, systematically eliminating structurally inefficient areas [3]. TO methods often results in forms entailing complex geometries characterized by extreme overhangs, ribbed geometries, and branches with varying sections and angles. Producing

H. Chai et al. (Eds.): CDRF 2024, *Symbiotic Intelligence*, pp. 3–12, 2025.
https://doi.org/10.1007/978-981-96-3433-0_1

these intricate geometries presents a challenge when using traditional concrete production techniques, primarily due to the increased complexity of the required formwork and the costs associated with fabricating non-standard formwork, which can consume over half of the structure's total budget [4].

The combination of 3D concrete printing (3DCP) with TO is a promising field of research as it has the potential to address the material efficiency with reduced cost for fabrication. Because 3DCP precisely places material only where it is needed. Previous studies have shown that coupling 3D printing (3DP) with structural and material optimization strategies can significantly reduce environmental impact [5], and is estimated to minimize the environmental footprint of construction practice by about 50% [6]. Nonetheless, there are challenges that arise in the production of complex geometries using 3DCP [7]. The most prevalent technique in 3DCP involves the planar deposition of material in a linear, horizontal manner. This approach, when applied to the manufacturing of complex geometries, results in undesired "staircase effect," (Fig. 1) where each layer is only partially supported by the one beneath it, impacting the overall buildability [8].

Non-planar 3D concrete printing (NP-3DCP) is a promising method for manufacturing complex parts as it enables variations in orientation and layer height (Fig. 1) [9]. This method is critical for accurately contouring the profile of intricate designs enhancing the geometric fidelity of the 3D printed complex parts. NP-3DCP diverges from traditional layer-by-layer, horizontal deposition, by using non-planar printpath that follows the exact contour of the part. To deposit concrete along the contour of the part in a non-planar fashion, it is essential to use a 6-axis robotic arm.

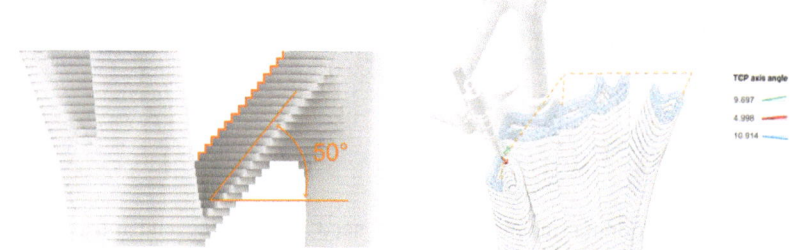

Fig. 1. The planar material deposition (left) results in "staircase effect" when printing overhanging structures which affects buildability. 3D simulation of the non-planar printpath and the material deposition process (Right)

2   Robotic Non-planar 3D Concrete Printing Strategies

Non-planar Printpath Required for Complex Geometries

Non-planar printpath (NP-printpath) guides the movement of the printhead in three dimensions within a single layer. Instead of strictly being confined to horizontal planar layers, the printhead is allowed a higher degree of freedom to vary in X, Y, and Z axes simultaneously while changing the Tool Center Point (TCP) orientation. This approach allows the creation of curved, sloped or angled surfaces when 3D printing, closely following the object's 3D geometry. This method effectively alleviates the "staircase effect" by depositing a new layer precisely on the previous one, facilitating the fabrication of a wide variety of topology-optimized forms.

The Need for Precise Material Deposition in NP-3DCP

A NP-printpath is accompanied by a variable layer height and orientation changes responding to the complex geometric features of the 3D printed object [10]. Thus, the precise control of variable material extrusion rate is essential to match the varying layer height [11, 12]. Similar to planar 3DCP, and perhaps even more crucially, achieving precision in NP-3DCP for complex geometry demands the deposition of an optimal amount of material at each layer. This not only impacts the buildability and structural integrity of the printed parts but also significantly enhances the surface finishing. The basic equation for 3DCP has been defined as:

$$H * W = F/V \tag{1.1}$$

Here H is layer height; W is extrudate material width; F is material flow rate based on pump speed; and V is nozzle travel speed [13]. The extruded material width (W) and the material flow rate (F) were kept constant in this paper; Thus, the variable layer height (H) depends on the nozzle travel speed (V). A smaller H requires a faster V, and vice versa. The method involves calculating the H by measuring each point on the current printpath and the nearest point on the preceding path, then subsequently using H to adjust the V during the printing process.

3 Topology Optimization and Printpath Generation Methods

Optimization Strategies and Geometric Post-processing Techniques

The main objectives in applying the TO method on concrete elements are reducing the weight of the component by optimizing material usage and enhancing performance capabilities [14]. Shell Wall was generated from two supports and two loads. Multiple iterations of optimization were followed by mesh reconstruction and simplification; the resulting structural ribs represent the wall's optimal material distribution. Subsequently, ribbed structures from TO were utilized as a frame to generate the minimal surface representing the in-between non-loading bearing area (Fig. 2). The Branch Wall design includes four support locations and six load-bearing points, resulting in a branched structure (Fig. 2). Further refinement was necessary through mesh relaxation and subdivision to enhance mesh quality. Additionally, the resulting form was modified to have a hollow structure with a wall thickness of 20 mm, equivalent to the filament width used in the printing process. This modification created voids within the structure, allowing it to serve as formwork for concrete casting. Additionally, these voids served as spaces to host steel rebars and insulation.

Fig. 2. Topology optimization process and geometric post-processing: Shell Wall (left) and Branch Wall (right)

NP-Printpath Generation in Relation to Complex Geometries

Methods of Generating NP-Printpath in Shell Wall:

The central axis of the structural ribs is discretized into segments with a uniform length. At each segmentation point, planes are orthogonally aligned along the central axis. The printpath for the ribs is determined by the intersection of these planes with the topology-optimized structural ribs. The printpath corresponding to each rib is then interconnected using planar contours, following the curvature of the minimal surface of the optimized geometry (Fig. 3). The printhead's orientation aligns tangent to the central axis and the planar contours for each corresponding layers.

Methods of Generating NP-Printpath in Branch Wall:

Branch Wall features four overhanging branches so it is critical to identify the branching moments within the geometry. First, we slice the TO generated geometry with a horizontal plane that moves from the bottom upwards. This process generated a series of intersected curves. For each curve, the central point was determined and sequentially connected vertically, forming the reeb graph that represents the geometry's topology. Through the reeb graph [15], critical points can be detected that are essential for segmenting the original geometry into distinct branches. Then we set paths at the local minima (gray mark) and local maxima (black mark), as well as the desired path in the middle on each branch. The final printpath was then generated by interpolating these pre-set paths (Fig. 3). The final step involves determining the orientation of the nozzle, which is accomplished by initially dividing each layer with division points every 50 mm. Subsequently, these division points are projected onto the previous contour using the shortest route. The vector originating at each division point and pointing towards its projection on the previous contour is regarded as the printing direction of the nozzle.

Bifurcation requires lead-in and lead-out strategies within the printpath design to address the challenge of navigating "islands"—disconnected sections within the branching areas that necessitate a robotic mechanism to navigate between them without depositing excessive material. The lead-in method starts the extruder 20 mm above each printpath, descends, and simultaneously extrudes material while avoiding collision with an already 3D-printed structure. The lead-out method stops material flow using a pinch valve at each printpath's endpoint and retraces a portion of the already printed path.

Then the printhead elevates and moves to the next island. This technique effectively eliminates any residual material on the nozzle, ensuring a more precise printing process.

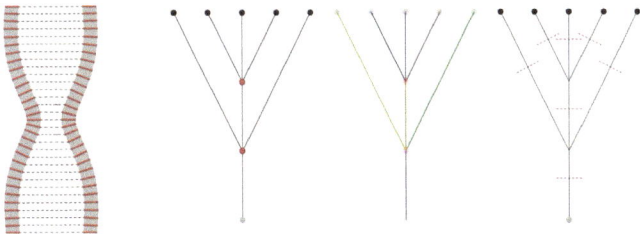

Fig. 3. The non-planar printpath generation process of Shell Wall (left) and Branch Wall (right)

4   Method for Remapping Extrusion Parameters to Complex 3D Geometry

Material Exploration for Pumpability and Extrudability to Achieve a Stable Flow Rate

Robotic 1-component (1k) system was utilized in this research and the mortar used in this study consists of fine sand, ordinary Portland cement, fly ash, water, and super-plasticizer. Components ratios were fine-tuned through testing various ratios of water-reducer (superplasticizer, in this context) to identify the optimal viscosity for 1) ensuring pumpability – a consistent material flow without separation or loss of consistency during pumping and 2) achieving extrudability – the ability of the mix to be extruded from the nozzle and maintain its shape upon deposition [16].

To guarantee the homogeneity of the material, the base mortar was thoroughly mixed for 12 min and then divided into four separate batches. Varying Superplasticizer-Binder Ratio ranging from 0.14% to 0.26%, were added to each batch, and followed by mixing the material for another 2 min. A funnel viscosity tool was used to visually assess the quality of the mixed concrete mortar. The exact mortar ratios for each batch as well as the measured results from the conducted tests and experiments are described in Table 1. Based on these findings, the optimal viscosity was concluded to be when the Superplasticizer-Binder Ratio was maintained at 0.18%.

Deposition Calibration Experiment for Variable Height Through Parallel Testing

The quality of the material deposition is directly influenced by the relation between the layer height and printing nozzle diameter (ND). Using layer heights smaller than the ND helps in preventing defects and enhancing the filament overlap, as it ensures layers are more effectively pressed. However, setting the layer height too small can lead to over-extrusion issues. Conversely, selecting layer heights larger than the nozzle diameter can lead to increased voids and suboptimal filament overlap [17]. The presented research utilized a circular nozzle with 20 mm in diameter and the average height per layer was

Table 1. Data of material explorations through different batches for pumpability, extrudability, and shape retainability after deposition

No	Water-cement ratio	Water-binder ratio	Superplasticizer-binder ratio	Pumpability	Extrudability	Shape retainability after deposition
01	0.553	0.25	0.14%	Hard	Separate	Good
02	0.553	0.25	0.18%	Good	Consistent	Good
03	0.553	0.25	0.22%	Good	Fair	Fair
04	0.553	0.25	0.26%	Fluid	Fluid	Bad

maintained at 0.5 time ND with maximum layer height below 0.83 times ND, which was showcased in the literature to have better layer bonding and buildability [17, 18].

12 experiments were conducted to evaluate the material behavior in relation to the robotic 3D printing settings, the results and setup for these experiments are described in Table 2. The printing nozzle diameter used measured 18 mm, two layer heights were tested 9 mm (50% ND) and 13.5 mm (75% ND), in combination with six printing speeds to identify the optimal speed for each layer height. An optimal nozzle travel speed has been visually assessed by evaluating the occurrence of material over-extrusion and under-extrusion. The results from the experiments demonstrated that 32 mm/s and 23 mm/s nozzle travel speeds performed well when combined with 9 mm and 13.5 mm layer heights, respectively.

Table 2. The top views of 12 printed samples with 2 different layer heights and corresponding 6 different nozzle travel speeds

H	9mm						13.5mm					
V (mm/s)	20	26	32	38	44	50	13	18	23	28	33	38
W (mm)	36	29	24.5	20	18.5	17.2	34.5	27.3	23.8	19	16	13
Visual Assessment												

Adjusting Nozzle Travel Speed Through Parameters Derived from NP-Printpath

The findings presented in Table 2 aligns with Eq. 1.1, demonstrating that when under optimal extrusion situations, the relationship between layer height and nozzle travel speed is an inverse relation, provided that the filament width and material flow rate remain constant. Speed can be remapped inversely proportional to different layer height through Eq. 1.1.

Manipulating the extrusion rate is achieved by altering the robot's nozzle travel speed that can be calculated in relation to the changing in layer height at each point. As described before, each layer of printpath has been uniformly sampled with points and a vector representing the shortest distance was used to set the printhead's orientation (Fig. 4). Here, layer height is determined through measuring the shortest distance. This value will serve as input H for Eq. 1.1 and the V will be calculated based on that.

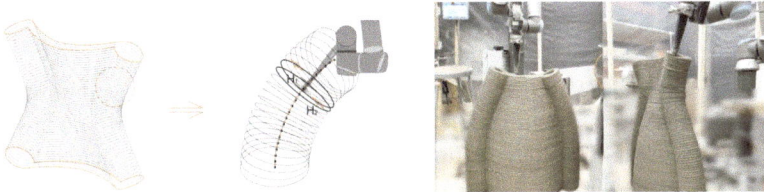

Fig. 4. Nozzle travel speed based on changing layer heights (left) and toolhead orientations (right)

5   Method for Rebar and Insulation Integration with Casting

Sequential Integration of Rebar and Casting Through Pauses: Shell Wall

Rebar integration along with NP-3DCP is essential for creating lightweight structures that possess the required strength to counter tensile loads. The ribbed geometry in Shell Wall, a common characteristic of structurally optimized forms, facilitates the effective incorporation of longitudinal reinforcement. The ribs are post rationalized to have a hollow cross section that hosts the rebars. Due to the complex geometries the rebars needed to be discretized into smaller sections that were sequentially manually integrated into the form, which was followed by the manual casting of concrete. This manual part of the process required the halting of the 3DP process at various stages of the manufacturing process as shown in Fig. 5.

Simultaneous 3DCP and Set-on-Demand Casting for Overhangs: Branch Wall

A similar approach was applied to construct the first half of the Branch Wall in which ribs and membrane areas are connected. Nonetheless, challenges arose for the top part when dealing with overhanging branches—particularly at angles greater than 75 degrees. The gravitational forces would often cause the freshly printed parts to slump, leading to collapse immediately after the deposition of the freshly printed material. To address this, a method was introduced to manually cast fast-setting concrete after the rebar reinforcement is successfully integrated (Fig. 5). The casting intervals were set every 25 to 50 mm of freshly 3D printed material based on the angle of the overhang.

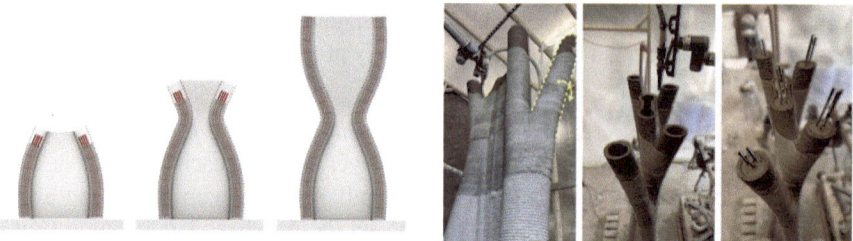

Fig. 5. Sequential integration of rebar and insulation with concrete casting (left). The collapse due to self-weight requires simultaneous 3DCP and set-on-demand casting (right)

6   Results and Discussion

Shell Wall and Branch Wall (Fig. 6) can be actualized for a wide range of topologically complex features in construction scale, the successful fabrication them demonstrates the potential of coupling Topology Optimization and non-planar 3D concrete printing to create ultra-lightweight walls. Shell Wall, having a width ranging from 490 mm at the ends to 140 mm in the middle and a height of 1.5 m, is a ribbed structure with a doubly curved membrane/surface in between. It weighs only 160 kg, showcasing the potential for zero waste production since no additional materials, such as scaffolding, were utilized. Meanwhile, Branch Wall, measuring 1.2 m in width and 2 m in height, features more geometric complexity such as multiple intersecting branches, overhanging structural ribs, and ultra-thin membranes. It weighs 367 kg, demonstrating a 68% reduction in material use compared to a standard load-bearing concrete wall with similar dimensions and load-bearing capacity.

Variable material extrusion rate achieved by adjusting the nozzle travel speed, was critical for the accurate fabrication complex geometrical features of the presented prototypes. The nozzle travel speed was determined based on changing layer height. Nevertheless, fluctuating flow rates were observed due to the increase of material viscosity through time during printing, necessitating manual adjustments to the nozzle speed based on visual inspections. Future research should focus on the development of a closed loop system capable of providing real-time feedback for immediate and accurate process adjustments.

Rebar reinforcement was successfully incorporated within the 150 mm tubular rib structural components of the Shell Wall, followed by concrete casting. For Branch Wall, the use of fast-setting concrete enabled the printing of branches with a diameter of 300 mm and lengths of up to 790 mm, while supporting overhangs of up to 70 degrees. Moreover, the integration of 50 mm thick XPS insulation into the double-layered sections improved the walls' thermal performance. Challenges such as shape deformation during printing and casting were observed due to the material setting time and hydrostatic pressure exerted by the casted concrete. Further investigation is needed in the geometric fidelity of the proposed method through live data collection.

Fig. 6. 3D concrete printed topology-optimized walls: Shell Wall (left) and Branch Wall (right)

7 Conclusion

The research provided evidence that the integration of TO with robotic NP-3DCP presents a viable solution for constructing lightweight, and structurally optimized architectural elements. Non-planar printpath and variable material deposition are necessary for successful 3DCP of complex geometries. The fabrication of the Shell Wall and Branch Wall prototypes illustrates the potential of this method in reducing material usage by up to 68% compared to traditional construction methods, thereby demonstrating the potential of this method in achieving a lower carbon footprint, offering a more sustainable alternative to traditional approaches. Further research needs to address challenges such as fluctuating flow rates, material deformation, real-time printpath adjustment and live 3DCP data synchronization to ensure geometric fidelity.

References

1. Belaïd, F.: How does concrete and cement industry transformation contribute to mitigating climate change challenges? Resour. Conserv. Recycl. Adv. [Internet]. **15**(200084), 200084 (2022). https://doi.org/10.1016/j.rcradv.2022.200084
2. Van Damme, H.: Concrete material science: past, present, and future innovations. Cem Concr Re [Internet]. **112**, 5–24 (2018). https://doi.org/10.1016/j.cemconres.2018.05.002
3. Bi, M., Tran, P., Xia, L., Ma, G., Xie, Y.M.: Topology optimization for 3D concrete printing with various manufacturing constraints. Addit. Manuf. [Internet]. **57**(102982), 102982 (2022). https://doi.org/10.1016/j.addma.2022.102982
4. García de Soto, B., Agustí-Juan, I., Hunhevicz, J., Joss, S., Graser, K., Habert, G., et al.: Productivity of digital fabrication in construction: cost and time analysis of a robotically built wall. Autom. Constr. [Internet]. **92**, 297–311 (2018). https://doi.org/10.1016/j.autcon.2018.04.004
5. Vantyghem, G., De Corte, W., Shakour, E., Amir, O.: 3D printing of a post-tensioned concrete girder designed by topology optimization. Autom. Constr. [Internet]. **112**(103084), 103084 (2020). https://doi.org/10.1016/j.autcon.2020.103084
6. Biernacki, J.J., Bullard, J.W., Sant, G., Banthia, N., Brown, K., Glasser, F.P., et al.: Cements in the 21st century: challenges, perspectives, and opportunities. J. Am. Ceram. Soc. [Internet]. **100**(7), 2746–2773 (2017). https://doi.org/10.1111/jace.14948

7. Kamhawi, A., Aghaei Meibodi, M.: Techniques and strategies in extrusion based 3D concrete printing of complex components to prevent premature failure. Autom. Constr. **168**, 105768 (2024). https://doi.org/10.1016/j.autcon.2024.105768

8. Lao, W., Li, M., Tjahjowidodo, T.: Variable-geometry nozzle for surface quality enhancement in 3D concrete printing. Addit. Manuf. [Internet]. **37**(101638), 101638 (2021). https://doi.org/10.1016/j.addma.2020.101638

9. Aghaei Meibodi, A., Lin, Y., Chen, H.: Hybrid approaches towards 3D concrete printing for lightweight reinforced concrete structures. Digital Concrete **2024** (2024). https://doi.org/10.24355/DBBS.084-202408140649-0

10. Mitropoulou, I., Bernhard, M., Dillenburger, B.: Print paths key-framing: design for non-planar layered robotic FDM printing. In: Symposium on Computational Fabrication. ACM, New York, NY, USA (2020)

11. Anton, A., Yoo, A., Bedarf, P., Reiter, L., Wangler, T., Dillenburger, B.: Vertical modulations. In: ACADIA Proceedings. ACADIA (2019)

12. Carneau, P., Mesnil, R., Roussel, N., Baverel, O.: Additive manufacturing of cantilever - From masonry to concrete 3D printing. Autom. Constr. [Internet]. **116**(103184), 103184 (2020). https://doi.org/10.1016/j.autcon.2020.103184

13. Anton, A., Reiter, L., Wangler, T., Frangez, V., Flatt, R.J., Dillenburger, B.: A 3D concrete printing prefabrication platform for bespoke columns. Autom. Constr. [Internet]. **122**(103467), 103467 (2021). https://doi.org/10.1016/j.autcon.2020.103467

14. Liu, D., Zhang, Z., Zhang, X., Chen, Z.: 3D printing concrete structures: state of the art, challenges, and opportunities. Constr. Build. Mater. [Internet]. **405**(133364), 133364 (2023). https://doi.org/10.1016/j.conbuildmat.2023.133364

15. Biasotti, S., Giorgi, D., Spagnuolo, M., Falcidieno, B.: Reeb graphs for shape analysis and applications. Theor. Comput. Sci. [Internet]. **392**(1–3), 5–22 (2008). https://doi.org/10.1016/j.tcs.2007.10.018

16. Yuan, P.F., Zhan, Q., Wu, H., Beh, H.S., Zhang, L.: Real-time toolpath planning and extrusion control (RTPEC) method for variable-width 3D concrete printing. J. Build. Eng. [Internet]. **46**(103716), 103716 (2022). https://doi.org/10.1016/j.jobe.2021.103716

17. Nair, S.A.O., Tripathi, A., Neithalath, N.: Examining layer height effects on the flexural and fracture response of plain and fiber-reinforced 3D-printed beams. Cem Concr Compos [Internet]. **124**(104254), 104254 (2021)

18. Lee, H., Kim, J.-H.J., Moon, J.-H., Kim, W.-W., Seo, E.-A.: Evaluation of the mechanical properties of a 3D-printed mortar. Materials (Basel) [Internet]. **12**(24), 4104 (2019). https://doi.org/10.3390/ma12244104

Robotic Micro-house – Experience with 3D Concrete Printing for Housing Construction

Deborah Benros[1](✉), Arman Hashemi[1](✉), Su Yunsheng[2](✉), Zhong[2](✉),
and Carl Callaghan[1](✉)

[1] University of East London, University Way, London 16 2RD, UK
{ddorosario,A.Hashemi,c.g.callaghan}@uel.ac.uk
[2] Tongji University, No. 281 Fuxin Road, Yangpu District, Shanghai, China
suyunsheng@tongji.edu.cn, 3288462166@qq.com

Abstract. This study introduces a digital manufacturing approach for mini and micro-housing design, deploying 3D concrete printing. Combining shape grammar with robotic construction methods, it aims to revolutionize architectural practice by enabling mass customization while ensuring creativity and feasibility. Traditional construction methods often hinder affordability and design diversity, necessitating innovative approaches. Shape grammars, rooted in design language principles, facilitate architectural design exploration. This research focuses on developing a novel generative system and harnessing automated printing for efficiency, reduction of embodied carbon and waste reduction. The study proposes leveraging digital manufacturing and 3D concrete printing. Its methodology involves developing a design grammar, a computer implementation, and fabricating prototypes. The approach, currently undergoing physical prototyping, demonstrates customizable housing solutions, advocating for a streamlined approach to housing design and construction while addressing affordability and customization.

Keywords: Generative system · shape grammar · robotic construction · mass-customization

1 Introduction and State of the Art

The paper explores the potential of automation and robotics in reducing construction costs, improving built quality, increasing diversity, and facilitating customization in housing units. It acknowledges the rich history of efforts in optimizing post-war housing, particularly in Europe, where strategies like prefab repetition aimed to optimize resources. However, these efforts often resulted in monotonous designs lacking personalization. John Habraken's expert design system, outlined in "Support Systems," sought to address this by introducing a directing grid for customization without compromising efficiency or costs [1].

© The Author(s) 2025
H. Chai et al. (Eds.): CDRF 2024, *Symbiotic Intelligence*, pp. 13–22, 2025.
https://doi.org/10.1007/978-981-96-3433-0_2

The introduction highlights the role of shape grammars, generative systems rooted in linguistic principles, in architectural design exploration. Many were the grammars applied to housing, some describing the Palladian villas system based on Palladio's design system expressed in his architectural treaty [2, 3], the Lloyd Wright Prairie houses illustrating shape rules three-dimensionally [4] or the first computer implementation of a three-dimensional shape grammar the Malagueira houses [5]. The above-mentioned housing grammars used the corpus of existing designs to post-rationalise the design system, infer design rules and recreate not only a set of existing designs but also a new corpus. While past studies have primarily focused on analysing existing design languages, there is a call for the development of novel generative systems capable of creating new architectural languages. Additionally, the introduction discusses the evolution of digital tools for housing design and construction, including robotic construction and automated prefabrication methods. It traces the history of prefab housing, noting its prominence in North America via the early twentieth century Sears house catalogue [6]. The flat-pack was however re-introduced recently with the Muji house and Wiki house, which offer customizable, affordable solutions [7, 8]. The latest is a zero-carbon, modular, open-source design system that enables self-build. All pieces are CNC machined and delivered ready to be easily assembled, allowing the choice of finishes and foundations. Major limitations are size, number of storeys (limited to 3 storeys) and availability of main material.

Moreover, the paper examines rule-based systems integrated with architectural design, such as the Actar and Manuel Gauza 'ABC' housing system [9]. These systems allowed for mass customization without additional costs, facilitating real-time modelling and automated production. Several notable projects highlight the potential of robotic construction in revolutionizing the building industry. One such project is the Icon gantry-operated concrete nozzle, which has been used to construct multiple high-quality houses in the US [10]. These houses were extruded on-site, showcasing impressive speed compared to traditional construction methods, with a single-story house completed within months from conception to completion. Additionally, experiments at ETH Zurich have explored mixed strategies for concrete extrusion, incorporating reinforcement into the fabrication process [11]. Similarly, Hanaa in Ithaca utilized Cornell's robotic lab to rapidly prototype 3D extruded concrete panels for a micro-home, with on-site assembly techniques [12]. However, challenges such as embodied carbon and transportation limitations must be addressed. Foster and Partners, in collaboration with the European Space Agency, proposed a Lunar Habitat concept that utilizes 3D printing extrusion nozzles and robots to create habitats on the lunar surface, addressing these challenges creatively [13, 14]. Similarly, Penn State University proposed a solution for Martian habitats using self-supporting conic structures and local aggregate mixed with concrete, demonstrating the potential of robotic construction in inhospitable environments [15, 16].

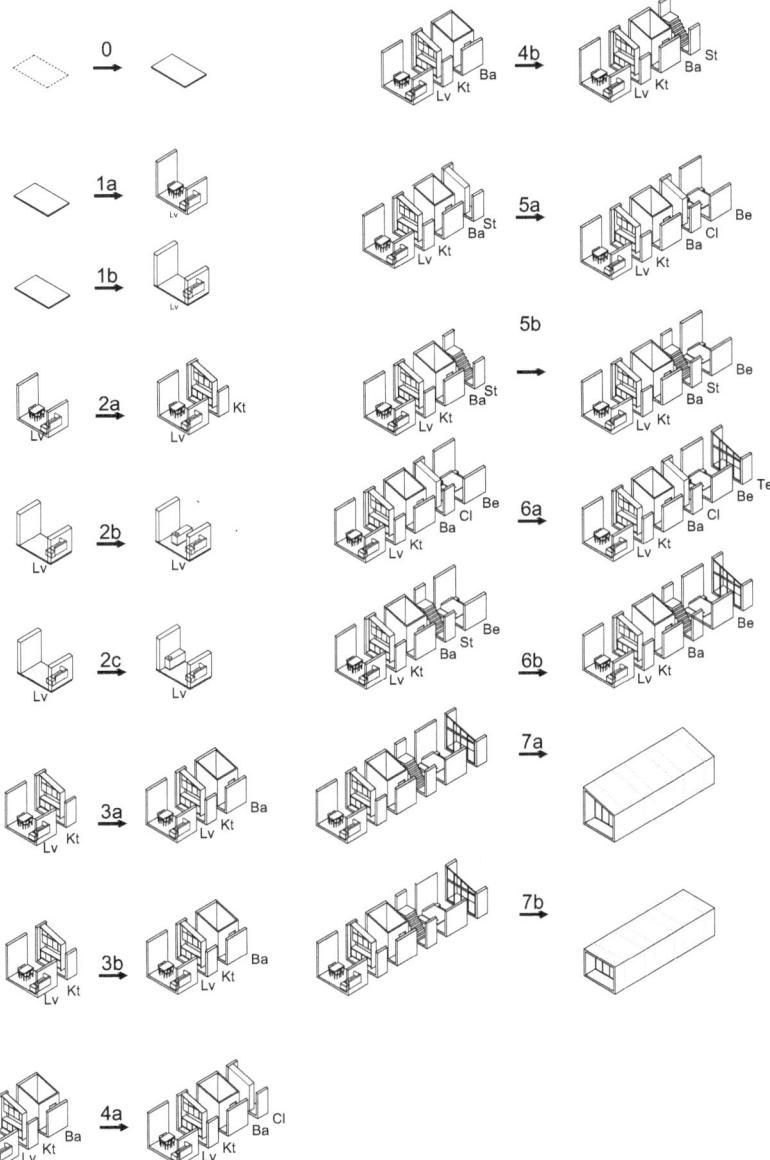

Fig. 1. Mini and Micro housing generative grammar share rules system. The length of the module varies from 1.2 m for a basic kitchenette or stair unit to a more generous modular multiple of 1.2 such as a minimum bedroom of 2.40 m.

These experiments underscore the importance of optimizing construction processes through robotic technology, particularly in remote or challenging environments. By leveraging robotic construction methods, construction can be expedited.

This exploration serves as a precursor to further delve into these themes in subsequent sections of the paper. Motivated by a British Council initiative, a team of academic researchers has been prompted to propose the adoption of robotic construction as a means to enhance the efficiency and sustainability of affordable, customizable housing.

Challenges such as embodied carbon and transportation limitations are identified as critical considerations in robotic construction endeavours. The potential of robotic construction in extreme environments is great. When allied with a shape grammar in its genesis and the implementation in a computer program the potential of the system is harnessed to allow even more flexibility. Many were the grammar implementations over the last years from the Malagueira [17] to Jowers system to study design emergency [18] to Economou [19] and Stouffs [20] many contributions as described in 50 years of grammars [21]. The current study does not offer this system for analysis and recreation but for the use in new designs.

2 Aims

The housing crisis demands innovative solutions that revolutionize construction process, from conception to completion, to address its complexity, length, and expense. To this end, our study outlines several objectives:

1. Development of a new design grammar for generative purposes.
2. Implementation of grammar using a design script, specifically Rhino Script.
3. Construction of a prototype. This prototype will serve as demonstration of the application of our generative design system and robotic construction methods.
4. Integration of novel methods of robotic construction, specifically 3D printing.
5. Examination of optimal structural components for 3D printing, investigate load-bearing walls, foundations, and roofs to determine the structural integrity.
6. Proposal of a net-zero carbon emission focused method for 3D printing.
7. Evaluation and improvement of sustainable materials for 3D printing

3 Methodology

Our methodology integrates a generative shape grammar with robotic construction, and is fourfold:

1. Shape grammar as a generative system: The generative system utilizes a three-dimensional grammar to produce innovative designs, addressing the complexity and cost constraints of the project. The grammar's multilingual nature enables the design of both compact affordable houses and micro-houses, serving as a proof of concept. Modular designs are created bottom-up, progressing from social areas to private spaces, with parametric rules as shown in Fig. 1. The derivation and generation of a solution is illustrated in Fig. 2.

2. Implementation of the grammar into a design script: The grammar is implemented in Rhino using RhinoScript in VBA to harness the modelling capabilities of Rhino and control outcomes. The script follows a structured process involving several stages, including defining dimensions, functional units, and spatial arrangements. The micro-house grammar structure is shown in Fig. 3.

3. Creation of a chosen design and modelling in real-time using Rhino. His iterative process allows for the exploration of various design options while ensuring adherence to the grammar's rules and principles. A modelled example is shown in Fig. 4, while Fig. 5 shows prototype examples printed at 1:100 scale.

4. The integration of a robotic arm on an ABB robot and a concrete printing nozzle for construction. The construction process involves printing external walls using additive methods of concrete extrusion, ensuring monolithic construction and reducing material waste. The real-scale micro-house will be constructed on-site in the digital fabrication lab, showcasing the integration of generative design and robotic construction.

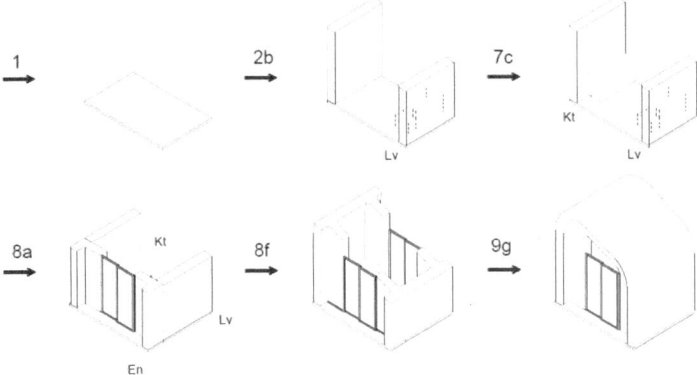

Fig. 2. Design derivation bottom-up approach with applications of rules 1, 2b. 7c, 8f and 9g a

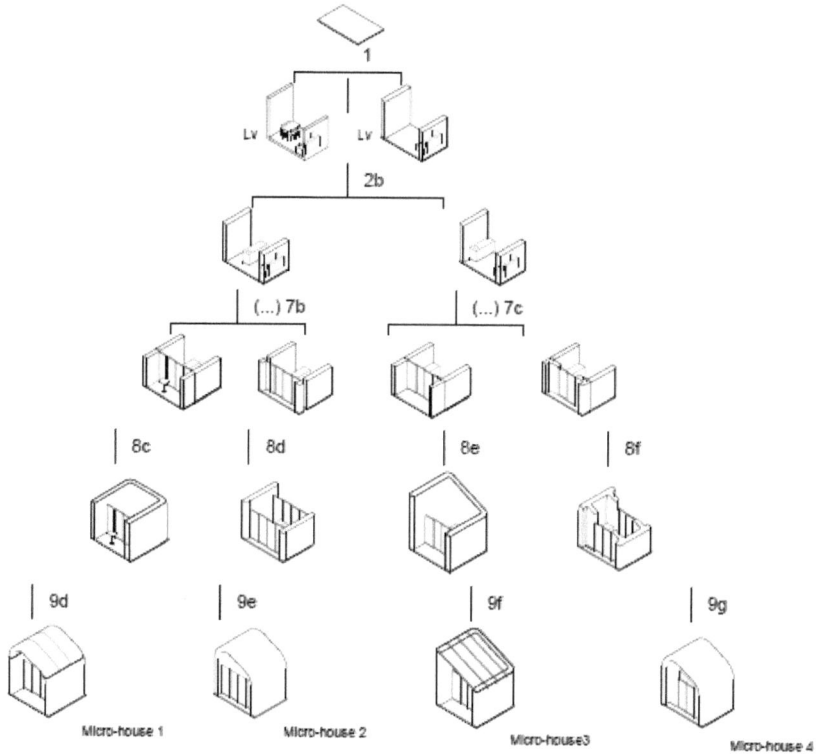

Fig. 3. Sub-grammar for micro-housing substructure with 4 different design solutions

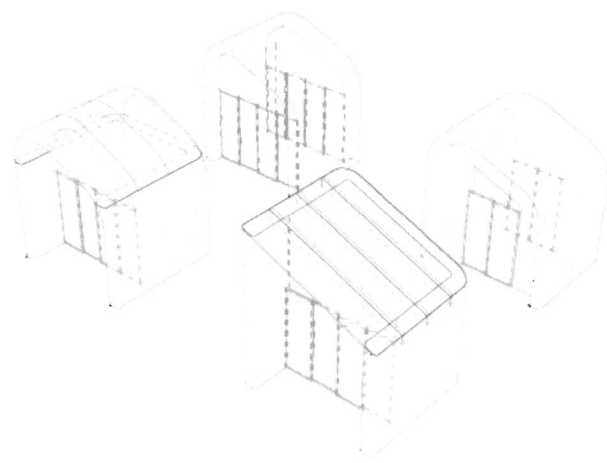

Fig. 4. Four design solutions generated by the micro-housing grammar 3D wireframe view

Fig. 5. Four design solutions generated by the micro-housing grammar perspective and four models 3D printed at 1:100 scale, complete and with the roof disassembled.

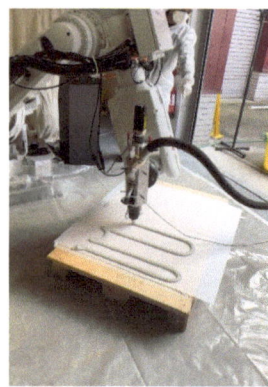

Fig. 6. Images of the dFuel lab preparing for prototyping using the concrete extrusion nozzle applied on the ABB robotic arm.

4 Conclusion

This study represents an effort to address the multifaceted challenges of affordable housing through the integration of generative design systems and robotic construction methods. By leveraging a novel approach that combines a generative shape grammar with advanced digital tools and additive manufacturing technologies, we have demonstrated the potential to revolutionize the conceptualization, design, and realization of housing solutions.

The physical prototyping of the micro-home is scheduled for May/June 2024, with initial tests conducted in the DFuel digital manufacturing lab (Fig. 6). Thus far, the paper has focused on discussing the generative system and its computer implementation. Our objectives included tackling the complex housing crisis. Through the development of a new design grammar for generative systems, we aimed to enable mass customization of housing while ensuring affordability. The implementation of this grammar into a design script facilitated the creation of diverse and highly customizable housing solutions, ranging from compact affordable houses to micro-homes. Furthermore, our methodology incorporated the use of robotic construction techniques, particularly concrete extrusion, to expedite the building process, minimize material waste, and enhance construction efficiency.

Despite the limited physical prototyping conducted thus far, our multilingual grammar has demonstrated its ability to generate a wide range of housing designs, adaptable to various construction methods and structural systems. The forthcoming prototyping experience will provide valuable insights into the feasibility and real-world applicability of our proposal, shedding light on any potential limitations and informing future iterations of the design process.

However, it is important to acknowledge the challenges and uncertainties inherent in pioneering endeavours such as ours. The use of robotic construction represents a paradigm shift in the construction industry, introducing new complexities and potential risks that must be carefully navigated. Additionally, while our study has focused primarily on addressing the housing crisis, the implications of our research extend beyond housing to other fields such as infrastructure development, disaster relief, and space exploration.

Looking ahead, future research should continue to explore the potential of generative design systems and robotic construction methods in addressing the diverse and complex challenges of the built environment. This includes further refining our generative grammar, optimizing robotic construction processes, and exploring new materials and construction techniques. Moreover, interdisciplinary collaboration and knowledge-sharing will be essential in driving innovation and overcoming the barriers to adoption of these transformative technologies.

This paper outlines a pioneering design methodology cantered on the application of a real-life grammar as a generative design system, aiming to transcend the boundaries of traditional housing design approaches. By harnessing the power of shape grammar and digital tools, coupled with automated construction using robotic means, our methodology offers a transformative approach to addressing the complex challenges of affordable housing.

Carbon emissions pose a significant threat to the environment, contributing to global warming and extreme weather events. The UK government's ambitious target of achieving net zero carbon emissions by 2050 underscores the urgency of adopting sustainable construction practices. The utilization of 3D printing technology in construction has emerged as a promising solution with the potential to significantly reduce carbon emissions. By carefully selecting low-carbon cement and optimizing the 3D printing process, we can make significant strides towards achieving this goal.

The contribution of this paper is threefold. Firstly, we have created and implemented a grammar-based generative system that empowers designers to create diverse and highly customizable housing solutions. Secondly, we have demonstrated the potential for mass customization of housing units, addressing the gap between highly specified, personalized designs and standardized, affordable housing. Lastly, we have streamlined the design process using robotic construction methods, leading to resource efficiency, waste reduction, and accelerated construction.

The implementation of our grammar-based generative system in a mainstream design tool like Rhino enables real-time modelling of generated solutions, allowing users to evaluate and assess their suitability. While previous implementations have focused on analytical post-rationalization of design, our approach prioritizes generative design, paving the way for the creation of new, innovative housing designs.

References

1. Habraken, N.J.: Variations: The Systematic Design of Supports. New edition, 216 p. MIT Press, Cambridge, MA (1976)
2. Stiny, G., Mitchell, W.J.: The Palladian grammar. Environ. Plan. B Plan. Des. **5**(1), 5–18 (1978)
3. Palladio, A.: The Four Books on Architecture, 472 p. The MIT Press, Cambridge, Mass. (2002)
4. Koning, H., Eizenberg, J.: The language of the prairie: Frank Lloyd Wright's prairie houses. Environ. Plan. B Plan. Des. **8**(3), 295–323 (1981)

5. Duarte, J.P.: Customizing mass housing : a discursive grammar for Siza's Malagueira houses [Internet] [Thesis]. Massachusetts Institute of Technology; 2001 [cited 2014 Jun 7]
6. Cooke, A.: Ahead of their time: the sears catalogue prefabricated houses. J. Des. Hist. (2001)
7. Granello, G., Reynolds, T., Prest, C.: Structural performance of composite WikiHouse beams from CNC-cut timber panels. Eng. Struct. 1(252), 113639 (2022)
8. WikiHouse [Internet]. [cited 2023 May 29]. Available from: https://www.wikihouse.cc/
9. Gausa, M., Salazar, J.: Housing + Singluar Housing. Spanish ed. Edition., 560 p. ActarD Inc (2003)
10. Dezeen, D.B.: 2022 [cited 2023 Jun 6]. ICON and Lake Flato build 3D-printed Available from: https://www.dezeen.com/2022/03/04/icon-lake-flato-3d-printed-house-zero-austin/
11. Graser, K., Baur, M., Hack, N., Apolinarska, A.: Dfab house A comprehensive demonstrator of Digital fabrication in architecture. In: Fabricate 2020 Proceedings. UCL Press, London
12. Zivkovic, S., Lok, L.: Making form work, experiments along the grain of concrete and timber. In: Fabricate 2020 Proceedings, p. 116–123. UCL Press, London (2020)
13. Fernandez-Galiano, L., Arquitectura Viva. 2013 [cited 2023 May 29]. AV Monografías 163–164 - Norman Foster In the 21st Century. Available from: https://arquitecturaviva.com/
14. Kestelier, X., Dini, E., Cesaretti, G., Colla, V.: The design of a lunar outpost: 3D printing regolith as a construction technique for environmental shielding on the moon. In: Fabricate 2014 Proceedings, pp. 200–5. UCL Press, London (2014)
15. Duarte, J.P., Duarte, J.P.: How can 3D Printed Homes for Mars address the housing crisis on Earth? | TED Talk [Internet]. [cited 2023 Jun 6]. Available from: https://www.ted.com/talks/jose_pinto_duarte_how_can_3d_printed_homes_for_mars_address_the_housing_crisis_on_earth
16. Xu, W., Huang, S., Han, D., Zhang, Z., Gao, Y., Feng, P., et al.: Toward automated construction: the design-to-printing workflow for a robotic in-situ 3D printed house. Case Stud. Constr. Mater. 1(17), e01442 (2022)
17. Duarte, J.P.: Towards the mass customization of housing: the grammar of Siza's houses at Malagueira. Environ. Plan. B Plan. Des. 32(3), 347–380 (2005)
18. Jowers, I., Prats, M., Eissa, H., Lee, J.H.: A study of emergence in the generation of Islamic geometric patterns. 39–48 (2010)
19. Hong, T.C.K., Economou, A.: Implementation of shape embedding in 2D CAD systems. Autom. Constr. 1(146), 104640 (2023)
20. Stouffs, R.: Where associative and rule-based approaches meet. A shape grammar plug-in for grasshopper. In: CAADRIA 2018 Proceedings. Hong Kong: Association for Computer-Aided Architectural Design Research in Asia (2018)
21. Haakonsen, S.M., Rønnquist, A., Labonnote, N.: Fifty years of shape grammars: a systematic mapping of its application in engineering and architecture. Int. J. Archit. Comput. (2023)

A Design-Fabrication Method for Thin-Vaulted Green Roof Through Integrated Hybrid Formwork with Clay Printing

Chenxi Jin, Chenhan Xu, and Weishun Xu[✉]

College of Civil Engineering and Architecture, Zhejiang University, Hangzhou 310058, China
xuweishun@zju.edu.cn

Abstract. As a critical element in urban sustainability efforts, green roofs (GRs) can reduce building energy use and enhance ecological performance via additional growing medium and plant layers. However, the extra building components result in increased load and structural thickness as well as construction complexity. This paper proposes a design-fabrication method for a thin-vaulted green roof prototype with a compressive-only surface and upstand ribs along unevenly distributed stress lines for a lightweight and material-efficient structure. The structural efficiency of the proposed roof has been proven high compared to conventional flat roof through simulation. To realize such a non-standard and multifunctional structure, a hybrid formwork system is presented to deal with geometric complexity and functional integration by treating stay-in-place formwork as a function part. In particular, 3D clay printing (3DCP) is selected as eco-friendly formwork staying in the GRs to integrate with plant growth substrate, leading to a more simplified and sustainable fabrication. An empirical construction experiment is conducted to validate the proposed method on a 1:5 scale.

Keywords: Green roof · Thin-vaulted slab · Hybrid formwork · Clay printing · Uneven Printing

1 Introduction

Green roofs (GRs) play a key role in reducing urban heat island and improving ecological performance (Shafique et al. 2018). Through the growing medium and plant layers, GRs decrease building energy use by providing additional thermal resistance (Susca 2019). Such extra building components increase structural weight and construction procedures (Bianchini and Hewage 2012), resulting in the challenge of high construction costs and carbon footprint.

Thin-vaulted floors have brought new perspectives in building strong and stiff structures with low mass by initiating internal compressive stresses (Liew et al. 2017). With the compressive-only surface carrying evenly distributed loads and a set of upstand ribs resisting asymmetric loading, the structure can be very thin without bending (Block et al. 2010). After a series of prototype investigations by the BRG, the thin-vaulted floor has been applied in a real building project, the EST-HiLo, proving a 70% ~ 80% reduction

© The Author(s) 2025
H. Chai et al. (Eds.): CDRF 2024, *Symbiotic Intelligence*, pp. 23–32, 2025.
https://doi.org/10.1007/978-981-96-3433-0_3

in weight compared to the flat slab (Ranaudo et al. 2021). Such unreinforced structures with concrete cast only in the compression zones largely reduce the use of material not only in the floor itself but also the whole structure, significantly mitigating the carbon emissions (Bhooshan et al. 2023).

As a pivotal method to integrate multifunctionality within building components (Gosselin et al. 2016), 3D printing has been applied to formwork fabrication, presenting new possibilities for streamlining customized construction process with non-standard geometries. 3D printed formwork (3DPF) can be categorized as removable and stay-in-place, the latter of which leads to interlocking geometry features between the formwork and building components (Jipa and Dillenburger 2022), creating high geometric complexity with potential to host cantilevers, undercuts, voids, and inner chambers (Anton et al. 2020). Such geometric complexity allows functional integration in the negative space of the structural components through 3DPF. Moreover, eco-friendly formwork should be given priority as a side product to provide ecological solutions (Kemper 2022) and expand the area for plant growth (Crawford et al. 2022).

This study proposes to use 3D clay printing (3DCP) for the formwork fabrication due to its ability to cultivate plants (S. Barnes et al. 2022), as well as its plasticity and dissolvability facilitating the integration with irrigation and drainage equipment. The existing research has explored the feasibility to treat stay-in-place clay mold as the growing media in green walls (Wang et al. 2020), yet comparable investigations regarding GRs are absent. Moreover, 3DCP has been proven feasible to print layers with varying heights (Motamedi et al. 2020) from a double-curved surface to a plane (Xu and Huang 2020), offering empirical backing for the fabrication of GRs with thin-vaulted structure.

Moreover, further investigation is required at two stages to implement thin-vaulted structures and 3DCP formwork for GRs. At the stage of prototype design, the efficiency of the thin-vaulted structure with integral soil substrate and clay molds should be demonstrated. At the stage of fabrication, multi-functions of GRs need to be integrated into hybrid formwork considering the geometric complexity of the structural components and the negative space in between.

This paper presents a design-fabrication method for thin-vaulted GRs to reduce the weight and thickness of the roof structure while integrating fabrication procedures, realized by hybrid formwork mainly consists of 3DCP molds. The method was validated through a case-based empirical experiment, which included constructing a 1:5 scale prototype of one-quarter of the designed roof.

2 Method

2.1 Thin-Vaulted Green Roof Prototype

The thin vaulted flooring system is based on a compressive-only surface, with a shallow arching action to induce internal compressive stress, resulting in a lightweight yet strong structure. To reduce the use of concrete and thus dead load of the component, the vaulted roof is supported by upstand dense ribs along stress lines minimizing the structural thickness. Such an uneven distribution of the upstand ribs defines serial geometries of voids between the structure, providing negative space for functional integration. While the earlier tests of a similar prototype focused on thermal performance (Bedarf et al. 2023; Jipa et al. 2019), the application to GRs can lead to a more comprehensive utilization of hollow space integrated with growing media, embedding irrigation and drainage system, as described in Fig. 1. Four sloping walls in corners with curved boundaries are set as the restraint for the horizontal thrust forces to achieve better lateral stability as well as form open spaces underneath.

The in-between spaces of ribs are filled with substrate, of which the boundaries are defined by 3DCP molds with thickening path allowing not only the fast fabrication of varied geometry but also the mix with soil as time goes by. Beyond serving as the substrate for plant growth, clay molds also hold promise for combining with pipes for irrigation and drainage due to material removability. Each plant media volume is connected to at least four holes, including two for the irrigation pipe passing through and the others for self-draining due to the arched surface, in which way the water can be concentrated to the circular boundaries for the feasibility of drainage through walls. Notably, these volumes vary in section profiles across the arched surface, enabling a broader selection of plant species. On the other hand, the fill of clay molds provides a well-distributed weight to the surface to better resist asymmetric loading in compression.

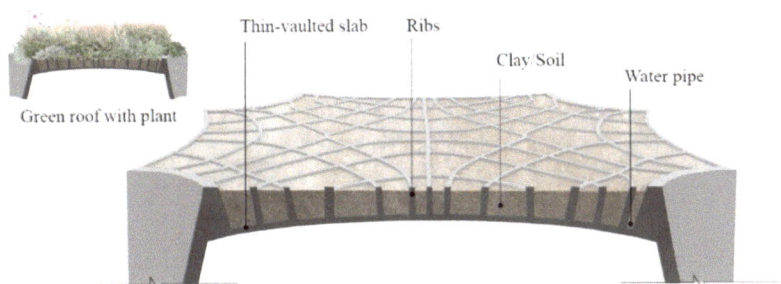

Fig. 1. Section of thin-vaulted green roof prototype

The geometry of the roofs was determined through a combination of form finding using reversed funicular surface simulation with Kangaroo and stress line analysis using Karamba. While the roof prototype has been designed with 10 m in length and 8 m in width, the target substrate depth was set to vary from 100 mm to 500 mm for plant growing, with an additional 50 mm for the thickness of the concrete vault. Placing materials on the stress line leads to a more effective distribution of loads, enhancing structural stability and reducing deflection and deformation of the floor (Halpern et al. 2013). Accordingly, the rib pattern corresponded to the stress line with a minimum distance of 100 mm. Moreover, the width of the ribs diminishes from bottom to top, providing the benefits of reducing self-weight while enhancing structural strength. Figure 2 illustrates the process of form-finding.

The structural behavior was then analyzed in the finite element software Karamba. The loads were applied according to the Chinese design code for GRs in public buildings as follows: the structure self-weight, a superimposed dead load of 1.2 to 6.5 kN/m^2 including substrate and plant, and a uniformly distributed live load of 3 kN/m^2. For ease

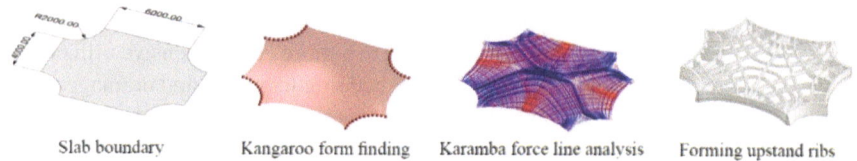

Slab boundary   Kangaroo form finding   Karamba force line analysis   Forming upstand ribs

Fig. 2. Form-finding process

of construction, we chose a 6 cm rib width based on Fig. 3(a), where a width of 2 cm conforms to safety standards for maximum displacement. Compared to the flat roof with cruciform ribs (Fig. 3(b)), the thin-vaulted roof performs better in structural stability (the maximum displacement is 0.039 cm) with a much lower self-weight (25670 kg, 37.6% of the flat roof). Furthermore, the thin-vaulted roof allows varying heights from 10 cm to 50 cm, while the flat roof has only 25 cm of soil depth. This highlights the higher structural efficiency of the thin-vaulted green roof prototype, as it achieves better performance with lower weight and height, as described in Table 1.

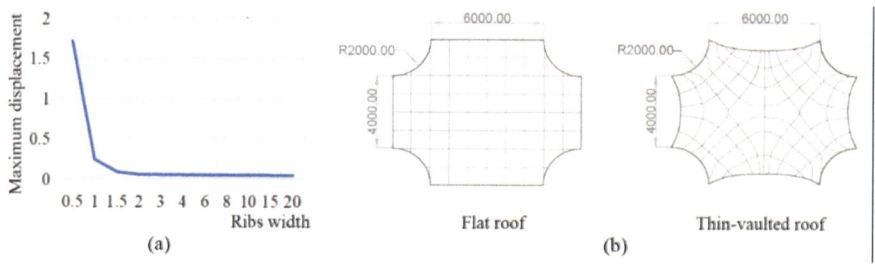

Fig. 3. (a) The change trend of maximum displacement with the increase of width of ribs. (b) Two types of roof slabs: flat roof and thin vaulted roof

Table 1. Performance of three proposed GR cases

Roof slab type	Material	Maximum displacement [cm]	Self-weight [Kg]	Soil depth [cm]	Total height [cm]	Concrete usage ratio
Flat roof	C30/37	0.131	68209	25	95	37.6%
Thin-vaulted roof	C30/37	0.039	25670	10–50	71	

2.2 Integrating Structural Components Through Hybrid Formwork

To deal with the geometric complexity resulting from the multifunctionality of GRs, a hybrid formwork system is proposed in this study to satisfy the requirements for shaping different geometries of building components and further integration of subsystem functions into the negative space. Moreover, the hybrid formwork can reduce cumulative error produced by clay molds due to unavoidable local deformation.

The hybrid formwork is divided into four parts: 3DCP molds, PVC flexible pipe, CNC-milled foam, and a wood frame (Fig. 4(a)). As the stay-in-place formwork, 3DCP components shaping the ribs fill the negative spaces in between, passed through by the flexible pipe for irrigation and drainage along the non-linear network. To define the global shape, CNC-milled foam provides a non-linear boundary and double-curved surface, while a wood frame facilitates lateral resisting and error control, both of which are reusable.

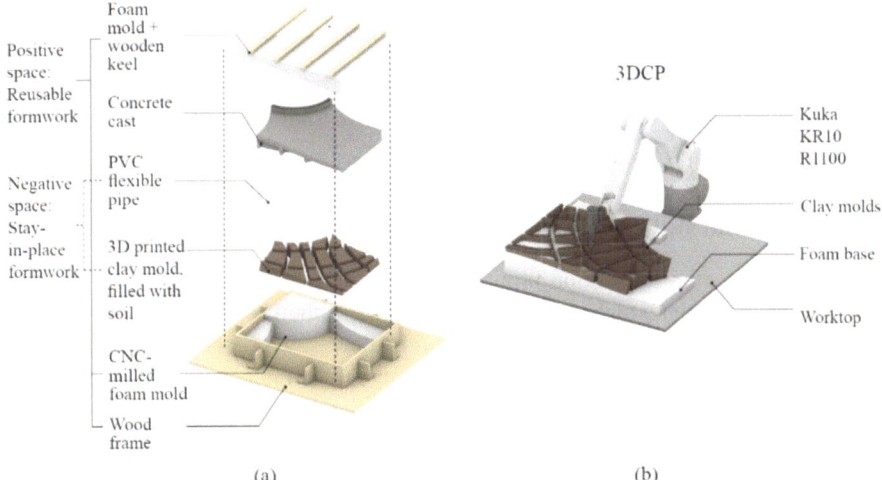

Fig. 4. (a) Hybrid formwork system. (b) Printing on foam base

Such 3DCP components need to be printed on the double-curved CNC milled foam surface (Fig. 4(b)) with varying layer heights to ensure continuously variable depth of the ribs, as illustrated in Table 2. Compared to the cutting-edge fabrication workflow of thin-vaulted slab (Bedarf et al. 2023), this method eliminates the step of labeling geometrically similar components by directly positioning on the foam. Furthermore, to achieve the reusability of the foam in most instances, the structure of the roof slabs with different boundary types has been optimized. The properties of the roof slabs including maximum displacement, structural weight, and concrete usage are shown in Fig. 5.

Table 2. Structural efficiency verification

	Slab (a)	Slab (b)	Slab (c)
Maximum displacement [cm]	0.032	0.088	0.072
Structural weight [Kg]	23932	22496	19657
Substrate weight [Kg]	18347	16771	12486
Area [m^2]	56.29	52.78	49.29

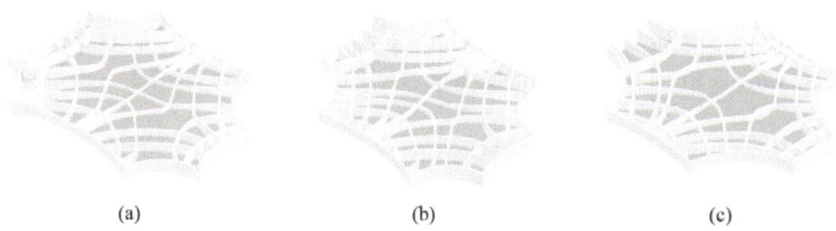

(a)   (b)   (c)

Fig. 5. Slabs of different boundary types

3 Fabrication Experiment

3.1 Design Context: Renovation of Residential Area

We used a residential renovation project with low floor height and demand for public green space to contextualize and take advantage of the proposed roof prototype in our empirical fabrication experiment. The thin-vaulted green roofs are placed between two residential buildings, creating a shared and functional integrated space to stimulate communication among residents. As the support of the roofs, the walls form a series of courtyards as well as functional spaces including shared kitchens, dining areas, lounge spaces, gym, etc. contained in an open space under the roofs (Fig. 6).

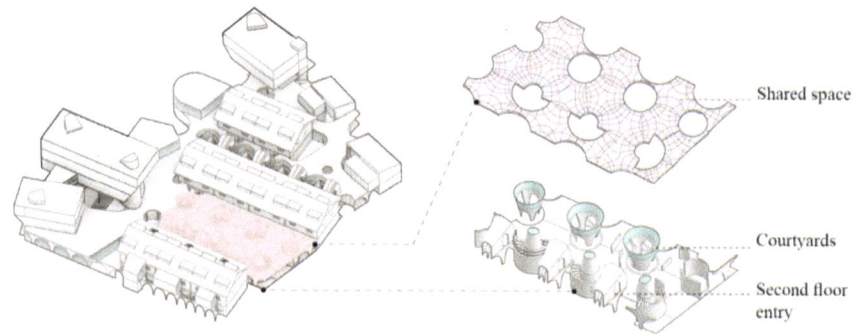

Fig. 6. The design scheme

3.2 Fabrication

To validate the feasibility of our methodology in a lab environment, 1/4 of the designed slab prototype was fabricated in 1:5 scale. Considering the ease of construction, the width of the ribs and the thickness of the vault were enlarged to 2 cm and 5 cm minimum. The construction process consists of three main steps: 3DCP, formwork assembly, and casting. In this feasibility test, only a portion of the pipes were installed and placed along the edges of the floor slab.

We employed an industrial robot, specifically the Kuka KR10 R1100, to carry out the 3D printing task considering the size of fabrication and the amount of clay material used in formwork. To ensure precise printing on the foam base, the foam base was re-mapped to the digital model for calibration by moving the 3DCP extruder to each feature point under manual operation mode (Xu and Huang 2020). Such a printing strategy with varying path height from a double-curved surface to a plane requires the continuous spatial change of extruder to keep vertical to the printing path (Fig. 7(b)), which effectively eliminates cumulative errors in the structure's height with the lace pressing (Motamedi et al. 2020). In this case, the sequence of printing was decided from the highest to lowest side of the foam base depending on the distance between contours and the maximum inclination of the extruder during printing. To reduce clay deformation caused by the static pressure of concrete during casting, we implemented a triple-layered path design shown in Fig. 7(c), leaving the remaining space to be filled with soil. The clay moisture level was set at 45%, based on industry standards. A total of 21 3DCP components were successfully printed in 12 h, spreading over 3 days.

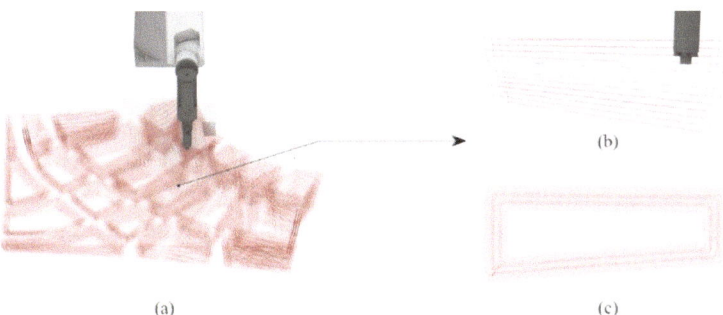

(a)   (b)

                                           (c)

Fig. 7. (a) Shapes of 3DCP component and toolpath planning. (b) Printing from a double-curved surface to a plane. (c) Triple-layered path

After printing, the 3DCP molds were cured for 72 h to be moved from the foam base and enough strength for concrete casting. The structure was reversed during the installation to leave a parallel space between the foam and 3DCP molds forming the thin vault. For precise assembly, a laser-cut paper showing the boundary of the ribs was placed on the wood frame, which was removed after positioning all the 3DCP molds. To facilitate subsequent demolding, the surface of non-clay components was coated with a thin layer of Vaseline, serving as the demolding agent. A customized highly fluid self-compacting cement was used for casting. During the casting process, when the concrete reached the line marked by the highest point of the foam, we capped it with foam and proceeded to pour the concrete through the gaps left by the curved walls. The whole casting procedure took less than 30 min.

The concrete was cured for 24 h before demolding. The foam and wood frame could be easily removed, and the 3DCP molds and flexible pipes were left in the slabs. It was found that there was no visible air bubble on the finished surface and the size of all ribs stayed very close to the designed profile. The shrubs and ground cover common in South China were planted. All materials de-molded were recycled or re-used. Figure 8 demonstrates the fabrication process.

Clay printing   Formwork assembly   Casting & Demolding   Planting

Fig. 8. Fabrication process and result

3.3 Test Results

The construction test yielded a thin-vaulted green roof prototype that faithfully adhered to the original design with controllable tolerance. Construction errors primarily manifested in acceptable width differences of the ribs, attributed to the lateral pressing and the shrinkage of 3DCP molds. The fabrication process spanned 7 days in total, underscoring its potential application in mass customization.

4   Conclusion

This study introduces a design-fabrication methodology for thin-vaulted green roofs, aimed at reducing the self-weight and thickness while streamlining fabrication procedures. The efficiency of the proposed roof structure is proved by finite element analysis compared with the common flat roof. More specifically, the reduction of weight reaches 62.4% while conforming to China's building code, which is consistent with the earlier research (Ranaudo et al. 2021). To deal with the geometric complexity due to the multi-functionality of GRs, the hybrid formwork is applied for functional integration, primarily comprising 3DCP molds as the boundary of the substrate to expand the plant growth area. Empirical experiment validates the method to fabricate the proposed thin-vaulted GRs prototype, transforming formwork in negative space into a part of the permanent building structural components, with the integration of soil layers, irrigation, and drainage, effectively reducing the carbon footprint while enhancing geometric complexity.

However, as an early-stage research, future work needs to further explore the integration of more complex and detailed tectonic layers of GRs, ensuring waterproofing and thermal continuity across multiple architectural materials and forms. Moreover, the structural analysis with concentrated load should be conducted considering the various use scenarios of GRs, including the placement of large equipment or pavilions.

Secondly, clay is not the only choice for substrate formwork. A comparable experiment of different eco-friendly formwork is worth conducting to investigate their mixability with soil and plant growth conditions. In addition, more factors of the formwork should be taken into account for plant growth, including fertility, water storage capacity, and pH value.

Moreover, further studies should be conducted on the installation of a full-scale slab to further determine its connection details for real-world construction. In most existing works, the slabs are supported by reinforced steel cables, leading to the separation between the slab and vertical structural components. This study offers new possibilities in connecting thin-vaulted slabs with walls through post-cast strip, providing functional continuity including drainage and lighting. Therefore, the design and fabrication strategies for the details connecting the slabs and supports need further study.

Acknowledgements. This research is supported by the National Natural Science Foundation of China under Grant 52208036 and Center for Balance Architecture at Zhejiang University.

References

Liew, A., López López, D., Van Mele, T., Block, P.: Design, fabrication and testing of a prototype, thin-vaulted, unreinforced concrete floor. Eng. Struct. **137**, 323–335 (2017). https://doi.org/10.1016/j.engstruct.2017.01.075

Anton, A., Bedarf, P., Yoo, A., Dillenburger, B., Reiter, L., Wangler, T., Flatt, R.J., Burry, J., Sabin, J., Sheil, B., Skavara, M.: Concrete choreography: prefabrication of 3d-printed columns. In: Fabricate 2020, Making Resilient Architecture, , pp. 286–293. UCL Press (2020)

Bedarf, P., Calvo-Barentin, C., Schulte, D.M., Şenol, A., Jeoffroy, E., Dillenburger, B.: Mineral composites: stay-in-place formwork for concrete using foam 3D printing. Arch. Struct. Constr. **3**, 251–262 (2023). https://doi.org/10.1007/s44150-023-00084-x

Bianchini, F., Hewage, K.: How "green" are the green roofs? Lifecycle analysis of green roof materials. Build. Environ. **48**, 57–65 (2012). https://doi.org/10.1016/j.buildenv.2011.08.019

Block, P., DeJong, M., Davis, L., Ochsendorf, J.: Tile vaulted systems for low-cost construction in Africa 7 (2010)

Crawford, A., In-na, P., Caldwell, G., Armstrong, R., Bridgens, B.: Clay 3D printing as a bio-design research tool: development of photosynthetic living building components. Archit. Sci. Rev. **65**, 185–195 (2022). https://doi.org/10.1080/00038628.2022.2058908

Gosselin, C., Duballet, R., Roux, P., Gaudillière, N., Dirrenberger, J., Morel, P.: Large-scale 3D printing of ultra-high performance concrete – a new processing route for architects and builders. Mater. Des. **100**, 102–109 (2016). https://doi.org/10.1016/j.matdes.2016.03.097

Halpern, A.B., Billington, D.P., Adriaenssens, S.: The ribbed floor slab systems of Pier Luigi Nervi. Proceedings of IASS Annual Symposia **2013**, 1–7 (2013)

Jipa, A., Barentin, C.C., Lydon, G., Rippmann, M., Lomaglio, M., Schlüter, A., Block, P.: 3D-Printed Formwork for Integrated Funicular Concrete Slabs (2019)

Jipa, A., Dillenburger, B.: 3D printed formwork for concrete: state-of-the-art, opportunities, challenges, and applications. 3d Print. Addit. Manuf. **9**, 84–107 (2022). https://doi.org/10.1089/3dp.2021.0024

Motamedi, M., Mesnil, R., Oval, R., Charier, M., Baverel, O.: Scaffold-free robotic 3D printing of a double-layer clay shell. In: Proceedings of IASS Annual Symposia. International Association for Shell and Spatial Structures (IASS), pp. 1–13 (2020)

Ranaudo, F., Van Mele, T., Block, P.: A low-carbon, funicular concrete floor system: design and engineering of the HiLo floors. Presented at the IABSE Congress, Ghent 2021: Structural Engineering for Future Societal Needs, Ghent, Belgium, pp. 2016–2024 (2021). https://doi.org/10.2749/ghent.2021.2016

Barnes, S., Kirssin, L., Needham, E., Baharlou, E., Carr, D.E., Ma, J.: 3D printing of ecologically active soil structures. Addit. Manuf. **52**, 102670 (2022). https://doi.org/10.1016/j.addma.2022.102670

Shafique, M., Kim, R., Rafiq, M.: Green roof benefits, opportunities and challenges – a review. Renew. Sustain. Energy Rev. **90**, 757–773 (2018). https://doi.org/10.1016/j.rser.2018.04.006

Susca, T.: Green roofs to reduce building energy use? A review on key structural factors of green roofs and their effects on urban climate. Build. Environ. **162**, 106273 (2019). https://doi.org/10.1016/j.buildenv.2019.106273

Wang, S., Liu, C., Zhang, G.L., Luo, Q.H., Xu, W., Raspall, F.: Digital Planting - Fabrication of Integrated Concrete Green Wall via Additive Manufacturing. Presented at the CAADRIA 2020: RE:Anthropocene, Bangkok, Thailand, pp. 145–151 (2020). https://doi.org/10.52842/conf.caadria.2020.1.145

Xu, W., Huang, Z.: Robotic fabrication of sustainable hybrid formwork with clay and foam for concrete casting. In: Blucher Design Proceedings. Presented at the Congreso SIGraDi 2020, Editora Blucher, Medellín, Colombia, pp. 377–383 (2020). https://doi.org/10.5151/sigradi2020-52

Bhooshan, S., Bhooshan, V., Megens, J., Casucci, T., Van Mele, T., Block, P.: Print-path design for inclined-plane robotic 3D printing of unreinforced concrete. In: Gengnagel, C., Baverel, O., Betti, G., Popescu, M., Thomsen, M.R., Wurm, J. (eds.) Towards Radical Regeneration. DMS 2022. Springer, Cham (2023). https://doi.org/10.1007/978-3-031-13249-0_16

Kemper, B.N.: Bio-formwork: large scale 3D deposition of thermoplastic starch in architecture. In: Open Conference Proceedings, vol. 2, pp. 65–70 (2022). https://doi.org/10.52825/ocp.v2i.130

Mobile Construction Positioning Method Based on the Robotic Arm and Laser-Camera Method

Xiaofan Gao, Hao Wu, Xingjie Xie, Yifan Zhou, and Philip F. Yuan[✉]

College of Architecture and Urban Planning, Tongji University, Shanghai, China
philipyuan007@tongji.edu.cn

Abstract. This paper introduces a positioning method based on the robot arm and laser-camera method, which is mainly oriented to the on-site additive construction and blockwork using mobile robots, and ensures the positioning accuracy of about 5 mm. This method provides a location scheme that is cheaper than directly using total stations and has higher accuracy than LiDAR or VSLAM. In addition, the method overcomes the limitation of the arm span on the positioning process: it is usually located by scanning the ArUco marker on the site by the camera at the end of the robot arm, which means that intensive positioning markers are required. The main equipment involved in the positioning process includes a laser light mounted at the end of the robot and a laser receiving panel as a marking point. The receiving panel is observed by an RGB camera to ensure that the light spot is projected in the center of the panel. The whole process is as follows: in advance, a total station is used to arrange several receiving panels at an interval of 5 to 10 m on the site, and then the mobile robot will shoot the laser at nearby receiving panels in turn, and push back its position through each axis angle to complete the positioning. The accuracy of this method mainly depends on the accuracy of the 4–6 axis of the robot. Therefore, this paper also introduces a calibration method to measure the mechanical error of the robot arm itself, to improve the positioning accuracy as much as possible. With the help of this method, many construction processes will achieve mobile construction and can bring some innovation to the construction process.

Keywords: Mobile construction · Site positioning · Laser-camera method

1 Introduction

In full-scale construction scenarios, construction robots are typically designed to prioritize small size and flexibility. Research by Zhang X. et al. [1] highlights scalability issues in 3D printing processes, especially problematic in construction where printers should not exceed the building size. Distributed mobile construction offers numerous advantages but also presents substantial challenges compared to fixed prefabrication. It struggles to maintain similar accuracy, requires sensors with a broader range to accommodate larger building scales, and must manage the costs of equipping multiple robots with these sensors.

© The Author(s) 2025
H. Chai et al. (Eds.): CDRF 2024, *Symbiotic Intelligence*, pp. 33–45, 2025.
https://doi.org/10.1007/978-981-96-3433-0_4

Precise positioning is essential for capitalizing on the benefits of digital construction. The positioning phase in mobile construction currently grapples with challenges related to scale, accuracy, and cost, with existing technologies often unable to meet all three demands concurrently.

On-site construction demands a larger positioning scale than fixed prefabrication, often involving the use of total stations. The constantly changing building environment and the potential for obstructions from newly constructed components necessitate the use of additional measurement devices or markers. An alternative on-site positioning method typically utilizes an end-effector camera on the robotic arm to scan visual markers. However, the limited reach of these arms, particularly in lightweight robots, requires a dense placement of markers throughout potential construction sites.

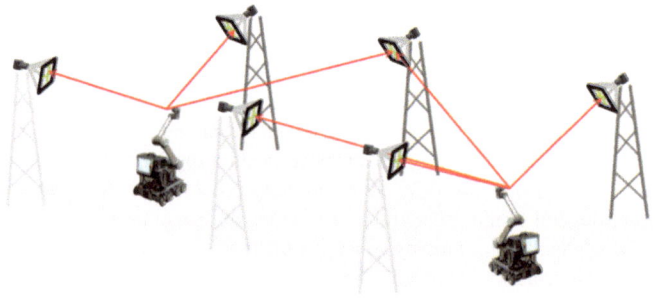

Fig. 1. Application scenario of mobile positioning based on laser-camera method

This paper presents a positioning method based on robotic arms and a laser-camera approach, mainly aimed at on-site additive manufacturing and masonry processes with mobile robots. This study focuses on two innovative points to expand the possibilities in the positioning phase of mobile construction. First, while the robotic arm joint itself is a measuring instrument with a certain degree of accuracy, it typically only functions within its reach. If utilized, it can replace the function of a total station within a smaller range (<10 m). Second, by combining the cost advantages of cameras with the accuracy of lasers, it reduces the cost of measurement in large-scale scenarios, facilitating mass deployment and offering potential solutions for distributed robots. This method provides a cheaper positioning solution than directly using total stations and higher accuracy than LIDAR or visual SLAM. Therefore, this method will have the following advantages:

1. Utilizing the precision of the robotic arm for measurements beyond its reach.
2. Low-cost measurement, facilitating mass deployment, and overcoming obstruction issues to some extent.

2 Related Works

2.1 Positioning Based on Visual Markers

Monocular cameras offer a cost-effective solution for short-range (<0.5 m) measurements, with ArUco markers being a popular choice for determining the position of robots in construction. These markers are typically established using traditional measuring tools like tape measures or total stations, and the robot's end-effector camera identifies them to calculate its location.

Several studies have capitalized on this approach for various construction applications. Zhang et al. [1, 2] investigated continuous printing by placing QR codes on the ground, achieving 2.2 mm in average accuracy via a camera on a vehicle. Giftthaler et al. [3] explored the mobile construction of curved steel walls by scanning QR codes at wall bases, reaching 3 mm in accuracy. Furthermore, Hua Chai et al. [4] examined the construction of wooden structures by positioning QR codes on ceilings; the end-effector camera then scans several codes to reposition the robot with an average accuracy of 3 mm.

However, the drawback of these methods is their reliance on short-range measurements, limiting their effectiveness to the robotic arm's reach and complicating large-area construction due to the dense placement of markers and associated operational challenges.

2.2 Laser-Camera Method

RGB cameras, as a type of inexpensive sensor, are increasingly widely used. RGB cameras can achieve sub-millimeter resolution in measurements at distances less than 0.5 m. Lasers, on the other hand, address the issue of large-scale measurements. When combined, they can complement each other's advantages.

In recent years, a method has gradually been developed that involves using an RGB camera to observe the spot of a laser on a whiteboard for measurement, capable of measuring slight directions or displacements at a considerably low cost and in a simple manner. This method is often referred to as the Laser-Camera (LaC) method. Yang et al. [5] used this method to correct laser angles. Yitian Han et al. [6] combined this method with an accelerometer to detect slight vibrations occurring in bridges.

Although this method cannot directly position mobile robots, it can significantly reduce measurement costs when cleverly combined with other methods. Tsuruta et al. [7] mounted two projection panels and two cameras on a cart to observe the spots projected by a total station, obtaining orientation information. The global accuracy of this method is 3 mm.

2.3 Total Stations and Resection

Laser-based measurement techniques are particularly effective for large-scale projects due to their precise distance and angle measurements, enhancing positioning method automation and efficiency. Early implementations date back to the last century; for instance, Beliveau et al. tracked an AGV cart using three total station-like devices in real-time, offering position and orientation data. In 2012, Inoue et al. managed to achieve a global positioning accuracy of 2 mm by intermittently tracking two prisms on an AGV with a single station, initially using a laser distance meter for rough positioning to achieve 2 cm accuracy.

Recent construction robots like FastBrick's Hadrian X utilize three automatic total stations for end-effector tracking and adjustment, similar to Dusty Robotics' Field Printer. Alternative approaches involve resection algorithms with sensors on movable bases to determine position by observing preset markers, benefiting large-area 3D scanning. Dingliang et al.'s use of laser trackers, total stations, and 3D scanners for tunnel point cloud stitching exemplifies this.

These methods secure accurate mobile robot positioning in extensive settings; however, their high equipment cost and susceptibility to obstruction during construction limit their application. The challenge is to reduce sensor costs without sacrificing the flexibility and feasibility of mobile construction scenarios.

3 Methodology

The positioning method primarily involves equipment including a laser mounted on a robotic arm, and several laser receivers serving as the markers. As shown in Fig. 1, receivers are pre-arranged on the construction site at intervals of 5–10 m, with center positions measured by a total station. The mobile robot directs the laser towards these receivers sequentially, with the direction determined by the robotic arm's joint angles to pinpoint its position.

Each receiver comprises an RGB camera and a projection panel. The camera captures the laser spot when illuminated, guiding precise adjustments to align the spot with the panel center.

Figure 2 depicts the feedback loop with these components. The mobile robot, equipped with a UR10 collaborative arm, is controlled via the UR RTDE (Real-time Data Exchange) protocol, focusing on adjusting the 5th and 6th joints for precise laser direction.

The receiver's core is an ESP32 microcontroller coupled with an OV2640 camera, facilitating wireless communication with the control computer. The receiver analyzes a 128×128 YUV422 image to identify the laser spot's pixels and compute their average coordinates, transmitting this data back to the computer at 10 Hz to adjust the laser's orientation in real-time.

A key element in ensuring measurement accuracy is the absolute precision of the robotic arm's joints, given that they are used as angle measuring instruments. For this reason, we conducted a feasibility assessment of accuracy: collaborative robots like the UR10 use servo motors with dual encoders and harmonic gear reducers, with their precision mainly dependent on the encoder's accuracy. The absolute accuracy of such encoders is around ± 2 arc min, with a repeatability of not less than ± 5 arc sec. In terms of laser measurement, this can roughly translate to an absolute deviation of less than ± 3 mm for a light spot projected 5 m away; and an absolute error of less than 6 mm for a spot projected 10 m away.

Apart from encoder accuracy, the assembly error of the robotic arm itself also needs to be considered. Since the robot's assembly precision only guarantees accuracy within

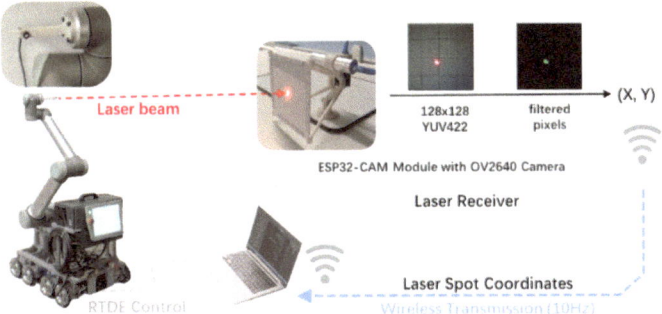

Fig. 2. Communication and information processing mechanism

its span range, this study also introduces methods for calibrating the errors of the 5th and 6th joints of the arm and the laser mounting errors to enhance positioning accuracy as much as possible.

4 Laser Receiver

In this method, the laser receiver serves as the marker point and guides the laser to aim directly at the center, thereby ensuring accuracy in positioning. During the positioning process, many receivers will be arranged across the site at intervals of 5–10 m. As shown in Fig. 3, a laser receiver is primarily composed of a projection panel, an ESP32-CAM module, OV2640 camera, power supply, and adjustable support components. The base of the receiver is equipped with magnets for easy attachment to metal construction elements.

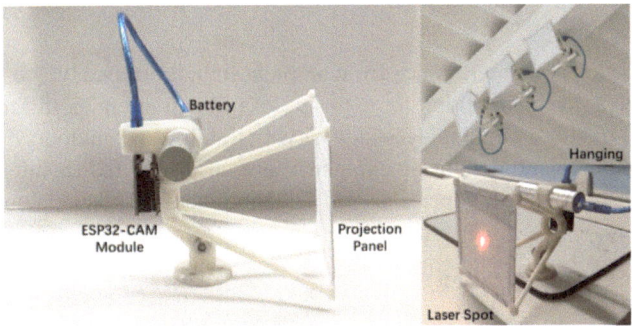

Fig. 3. A demo of laser receiver

The ESP32-CAM is a cost-effective microcontroller module for the IoT industry, boasting a 240 MHz dual-core CPU with some image processing capabilities and wireless communication functions. The camera utilizes an OV2640 (2-megapixel resolution) lens with a 120-degree field of view. During use, cropping and scaling are performed on the original 800 × 600 image to obtain a 128 × 128 image that only includes the projection panel part.

4.1 Extraction of the Laser Spot

Based on the 128 × 128 YUV422 format image obtained from the OV2640 camera, we use a certain rule to individually judge the color of each pixel, which allows us to isolate pixels considered to be part of the spot. To determine this rule, we visualized the distribution of YUV colors in a single frame image. As shown in Fig. 4, the color of the spot has a significantly different distribution than that of the projection panel. A simple judgment rule minimizes the computational load on the ESP32 module. Accordingly, we defined the judgment rule as follows, which corresponds to the green area in the figure:

$$\begin{cases} Y + 2V > 496 \\ \\ 90 < U < 145 \end{cases} \tag{4.1}$$

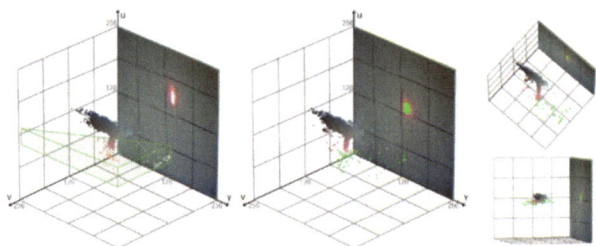

Fig. 4. Distribution of YUV colors in a single frame image and the corresponding judgment rule area

Another issue is the interference from ambient light. Fortunately, the OV2640 provides a manual adjustment option for exposure. As shown in Fig. 5, different exposure levels are required for optimal images indoors with lighting and outdoors on sunny days. Additionally, implementing light-blocking measures also helps reduce interference from ambient light, although this paper does not delve into detailed research on this aspect.

4.2 Wireless Communication

Compared to typical wireless image transmission modules, a crucial difference in this laser receiver is that it only transmits the coordinates of the spot, significantly reducing the load on wireless communication and increasing the distance for stable communication. As shown in Fig. 2, the process of obtaining images from the camera and filtering them pixel by pixel is completed entirely within the ESP32-CAM module. Therefore, we utilized a low-resolution 128 × 128 image to ensure the ESP32 could process the image information in real-time.

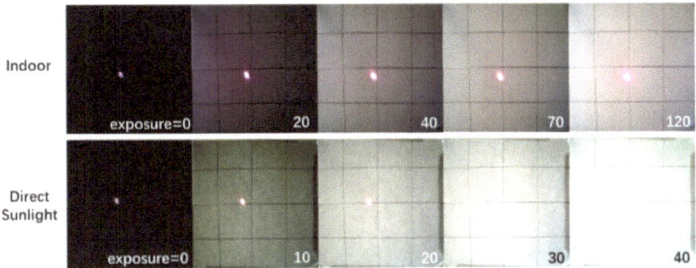

Fig. 5. Images of the OV2640 at different exposure levels

Regarding accuracy, the 128 × 128 image resolution is fully sufficient for positioning needs. The receiver panel measures 100 mm*100 mm, which roughly means 1 pixel reflects 1 mm of actual size. Since the coordinates of the spot are the average of multiple pixels (usually 20 to 100 pixels), such a resolution configuration is adequate to provide less than 0.1 mm resolution.

4.3 Guide the Laser to the Center

The task of the receiver is to obtain the position of the spot in real-time and inform the robotic arm how to move to direct the laser to the center, forming a control feedback loop. However, the image update frequency is limited (10Hz), and inevitably, there is a delay (tested to be 0.3 s). Using laser coordinates with delay to directly adjust the laser direction can make it difficult to achieve convergence and may even lead to oscillation. The correct approach is to use the mechanical arm's state, delayed by the corresponding time, in calculations:

$$pointer_{aim} = pointer_{t-delay} + convert\_to\_3D(spot_t - center) * rate \qquad (4.1)$$

where $pointer_{aim}$ is the position where the laser should be pointed at the moment, and $pointer_{t-delay}$ is the laser's previous pointing position. $spot_t$ is the current spot coordinates, which actually represent the state 0.3 s ago due to the delay. **convert_to_3D** is a transformation that converts two-dimensional coordinates into three-dimensional space coordinates, dependent on the receiver's orientation, and this value does not need to be very accurate, as its error only slightly affects convergence speed. Additionally, center is not the center of the 128 × 128 image or the 100 mm*100 mm projection panel but is the pixel coordinates measured by the total station when the receiver was prearranged.

5 Laser Pointer Calibration

The laser pointer is mounted at the end of the mobile robot's robotic arm and directed towards the receiver by adjusting its orientation. As depicted in Fig. 6, the installation ensures that the laser beam intersects perpendicularly with Axis6. During the positioning process, only Axis5 and Axis6 are rotated while other joints remain stationary, primarily to avoid introducing errors and wobbling caused by significant shifts in the center of gravity. The calibration method is divided into three stages (Sects. 5.1 to 5.3): from coarse to precise, gradually achieving accurate placement of the laser.

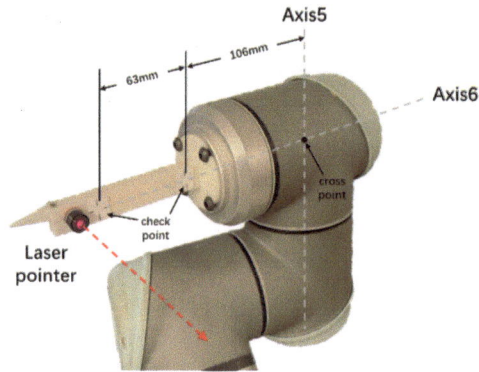

Fig. 6. Installing the laser pointer

Additionally, the laser extends about 100 mm from the flange to avoid occlusion by the mechanical arm during the calibration processes described in Sect. 5.3. For the validation tasks in Sect. 6, two checkpoints and a hypothetical cross point are preset on the plastic components. These are not necessary for practical use, meaning the laser can be relocated to other positions post-calibration.

5.1 Coarse Calibration of the Laser

Utilizing the receiver's guiding function, the laser can be conveniently directed to the same spot. With automated programs, the auto-calibration process is completed within 30 s, achieving a precision comparable to the 4-point TCP calibration method. However, this coarse calibration primarily adjusts the orientation roughly, as such angle accuracy does not meet the measurement needs for distances of 5–10 m. Nevertheless, this precision is sufficient for operations within arm's reach (Fig. 7).

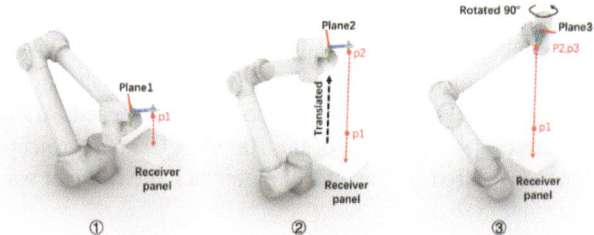

Fig. 7. Three steps of laser calibration

The calibration process is divided into three steps:

Step 1: The robotic arm holds the laser, directing it vertically downwards to a receiver below, recording the current flange plane as **Plane1**.
Step 2: The arm moves upwards by a suitable distance and adjusts horizontally to ensure the laser hits the original spot. The flange plane at this point is recorded as **Plane2**.

Plane2 is translated relative to **Plane1**, and their positional difference forms a vector **v**, indicating the direction of laser propagation.

Step 3: The flange is rotated around vector **v** by 90° (rotation center being irrelevant), and adjusted horizontally to strike the original position again, recording this as **Plane3**. **Plane3** maintains a rotational relationship with **Plane2**, defining the rotation axis that represents the laser's position and direction.

5.2   Precise Measurement of the Laser Direction

Section 5.1 provides a preliminary estimate of the laser's position. Subsequent methods refine the laser's directional accuracy. By positioning the robotic arm into the two poses depicted in Fig. 8, while keeping joints 1–4 static and adjusting only joints 5 and 6, the laser consistently targets the same distant point **P** across both poses. The distance **D** between the laser and **P**, approximately 8 m, is manually measured using a laser rangefinder with accuracy acceptable to within a centimeter.

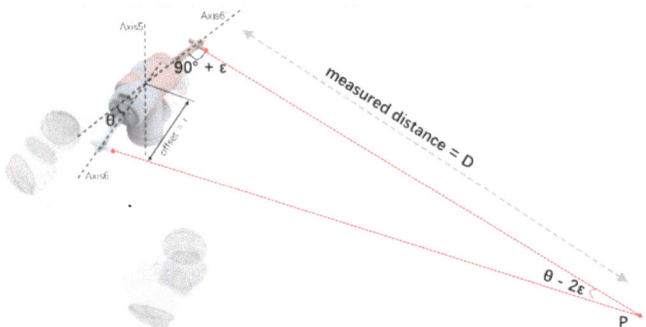

Fig. 8. Precise measurement of laser direction

Regarding the pitch angle relative to the vertical direction of Axis5, we need to calculate the corresponding angle φ for joint 6 when the pitch angle is zero (i.e., when the beam is perpendicular to Axis5). The angles of joint 6 in the two postures are denoted as A6 and A6', and the relationship is established as: $2\varphi = A6 + A6' + 180°$.

For the azimuth angle, which is determined by the angle of joint 5, we need to calculate the angle ε by which the beam deviates from being perpendicular to Axis6. The angles of joint 5 in the two postures are recorded as A5 and A5'. The difference between A5 and A5' is close to but not exactly 180° because of the offset distance **r** from Axis5 to the laser source, so we set $A5 - A5' = 180° + \theta$. Using the sine rule for triangles, the relationship: $r/\sin(\theta/2 - \varepsilon) = D / \sin(90° - \theta/2)$ allows for the calculation of ε based on known values of θ, **r**, and D.

With φ and ε determined, we complete calibrating the laser relative to the robotic arm.

5.3   Calibration of the Robotic Arm's 5–6 Joints

Apart from calibrating the installation position of the laser, it is essential to measure the mechanical errors of the robotic arm itself. The focus is on the angles α between Axis4 and Axis5, and β between Axis5 and Axis6, which due to the assembly process do not maintain the ideal 90° and can deviate by up to 0.2° for UR robot.

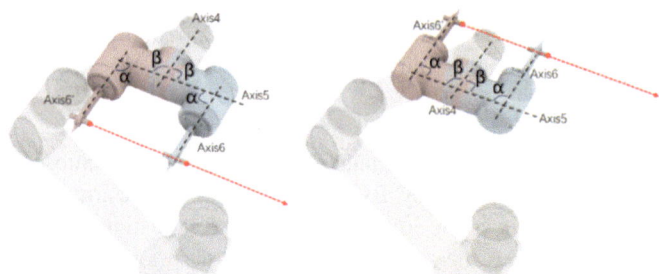

Fig. 9. Measurement steps of the robotic arm joint errors

As shown in Fig. 9, the measurement method includes four poses. In the first two poses displayed on the left, the laser beam would strike the same spot when α = β = 90° (assuming the laser is nearly perpendicular to Axis6). However, α and β are not ideally 90°, there is a horizontal deviation in the distant light spots, denoted as **x1**. With an distance **D** of approximate 8 m between the laser and the light spot, the relationship can be established as: **2(α + β) = x1/D**.

In the next two poses shown on the right, a similar horizontal deviation, noted as **x2**, is observed. This results in the relationship: **2(α − β) = x2/D**.

By combining these two equations, α and β can be precisely solved.

$$
\begin{cases}
2(\alpha + \beta) = x1/D \\
2(\alpha - \beta) = x2/D
\end{cases}
\tag{5.1}
$$

6 Field Measurement and Verification

Based on the preparatory work described previously, we conducted field measurements and validation using a total station to deploy receivers and verify the accuracy of positioning. As illustrated in Fig. 6, two checkpoints were measured using the total station to deduce the position of the cross point (Fig. 10).

Initially, four receivers were positioned at the corners of a rectangular space measuring 8.4 m*11.5 m. The centers of the receivers were measured using the total station, as shown in Table 1.

Furthermore, calibration of the laser and mechanical arm was performed using the method described in Chap. 5, as shown in Table 2.

The mobile robot remained stationary at specific locations, sequentially observing the direction of each receiver. Based on the resection method and the calibration parameters of the laser, it computed its position (Cross Point). Currently, manual control is used to direct the laser to the receiver points. Future integration with technologies like SLAM

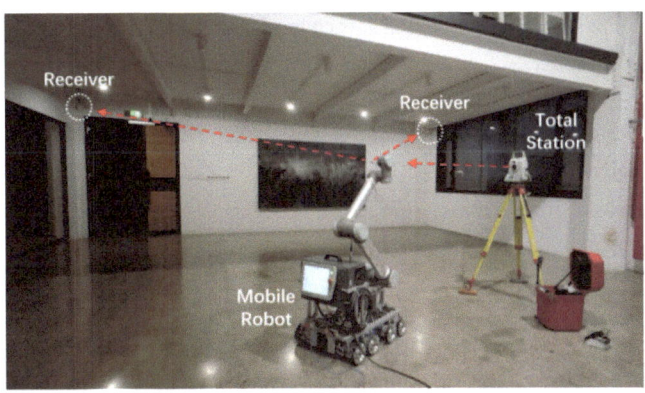

Fig. 10. Field positioning experiment: the mobile robot calculates its position by measuring the direction of four receivers, with the total station verifying the positioning accuracy

Table 1. Total station measurements of receiver centers

Position	X (mm)	Y (mm)	Z (mm)
Receiver 1	4719	239.6	2871.4
Receiver 2	4871.8	8736.7	2775.3
Receiver 3	−5611.8	−158	2379.5
Receiver 4	−6838.6	8588.1	2720.3

Table 2. Calibration results of the laser relative to the robotic arm flange

Calibration result	X (mm)	Y (mm)	Z (mm)	Direction X	Direction Y	Direction Z
Coarse measurement	96.2243	0.168819	−0.709096	−0.01396	0.99986	0.00787
Precise measurement	-	-	-	−0.00519	0.99997	0.00562

or AI vision may enable automatic receiver detection, but this is not the current research focus. Tables 3 and 4 present data from one of the positioning processes.

Experiments were conducted at multiple locations within an 8.4 m*11.5 m rectangular room, yielding an average positional error of 11.2 mm. Furthermore, to verify the effectiveness of the calibration method, results calculated using only the coarse calibration showed an average error of 15.5 mm.

Additional experiments were conducted in a smaller 5.2 m*5.6 m rectangular space, resulting in an average error of 4.6 mm; using only coarse calibration, the average error was 7.5 mm.

Table 3. Joint angles when pointing at each receiver center (A1-A4 remain constant)

UR robot joints	A1 (rad)	A2 (rad)	A3 (rad)	A4 (rad)	A5 (rad)	A6 (rad)
Pointing at Receiver 1	−1.5708	−0.785398	−1.5708	−0.785398	**0.429688**	**1.73303**
Pointing at Receiver 2	−1.5708	−0.785398	−1.5708	−0.785398	**1.55786**	**1.71497**
Pointing at Receiver 3	−1.5708	−0.785398	−1.5708	−0.785398	**5.08246**	**1.7325**
Pointing at Receiver 4	−1.5708	−0.785398	−1.5708	−0.785398	**3.49195**	**1.77293**

Table 4. Total station measurements of check points and error analysis of cross point

Position	X (mm)	Y (mm)	Z (mm)	Length (mm)
Check point 1	−1795.8	4710.7	1443	-
Check point 2	−1733.9	4725.3	1443.5	-
Cross point (ground truth)	−1629.75	4749.86	1444.34	-
Cross point (calculated)	−1625.52	4745.67	1454.18	-
Error	4.23221	−4.19842	9.83875	11.503

7   Conclusion and Future Work

The positioning method presented in this paper achieves an accuracy of 11.2 mm for receivers spaced 8.5 m*11.5 m apart and 4.6 mm for receivers 5.2 m*5.2 m apart. Furthermore, the precise calibration method introduced herein improved positioning accuracy by approximately 30% compared to the coarse calibration.

This paper serves as an interim research achievement and paves the way for future efforts. During the research process, it became evident that static measurements do not meet real-time requirements. The method requires the mobile robot to sequentially measure four surrounding receivers while stationary, thus it is a static positioning technique unsuitable for processes requiring continuous motion, such as 3D printing. Additionally, there is a lack of a mechanism to autonomously locate receivers. Currently, the identification of positioning receivers is manually performed, which significantly disrupts the automation process and urgently requires enhancement. Future work may explore the use of linear laser scanning or AI vision to address these challenges.

Acknowledgements. This work is supported by the Shanghai Science and Technology Committee (Grant No. 21DZ1204500).

References

1. Zhang, X., et al.: Large-scale 3D printing by a team of mobile robots. Autom. Constr. **95**, 98–106 (2018). https://doi.org/10.1016/j.autcon.2018.08.004

2. Tiryaki, M.E., Zhang, X., Pham, Q.-C.: Printing-while-moving: a new paradigm for large-scale robotic 3D Printing. In: 2019 IEEE/RSJ International Conference on Intelligent Robots and Systems (IROS), pp. 2286–2291 (2019). https://doi.org/10.1109/IROS40897.2019.8967524

3. Giftthaler, M., et al.: Mobile robotic fabrication at 1:1 scale: the in situ fabricator: system, experiences and current developments. Constr. Robot. **1**(1–4), 3–14 (2017). https://doi.org/10.1007/s41693-017-0003-5

4. Sustarevas, J., Kanoulas, D., Julier, S.: Autonomous mobile 3D printing of large-scale trajectories. In: 2022 IEEE/RSJ International Conference on Intelligent Robots and Systems (IROS), pp. 6561–6568 (2022). https://doi.org/10.1109/IROS47612.2022.9982274

5. Yang, Y.: Beam orientation of EAST visible optical diagnostic using a robot-camera system. Fusion Eng. Des. (2021)

6. Han, Y., Wu, G., Feng, D.: Structural modal identification using a portable laser-and-camera measurement system. Measurement **214**, 112768 (2023)

7. Tsuruta, T., Miura, K., Miyaguchi, M.: Mobile robot for marking free access floors at construction sites. Autom. Constr. **107**, 102912 (2019)

8. Beliveau, Y.J., Fithian, J.E., Deisenroth, M.P.: Autonomous vehicle navigation with real-time 3D laser based positioning for construction. Autom. Constr. **5**(4), 261–272 (1996). https://doi.org/10.1016/S0926-5805(96)00140-9

9. Inoue, F., Ohmoto, E.: Development of high accuracy position marking system in construction site applying automated mark robot (n.d.)

10. 杨丁亮, 邹进贵. 长大隧道点云的绝对定位配准方法. 测绘通报 **2022**(S2), 179–184 (2002). https://doi.org/10.13474/j.cnki.11-2246.2022.0583.er

A Symbiotic Approach for Developing Shoreline Infrastructure

Sara Pezeshk$^{(\boxtimes)}$ and Shahin Vassigh$^{(\boxtimes)}$

Florida International University, Miami, USA
{spezeshk,svassigh}@fiu.edu

Abstract. This research project presents an alternative approach to addressing the complex challenges of sustainability of the coastlines by integrating advanced technology solutions with ecological conservation principles. The paper introduces the Ecoblox, a modular infrastructure system consisting of interlocking blocks devised for attachment to seawalls to improve marine biodiversity at the water edge. The design of the Ecoblox system employs environmental data, data analytics, AI-powered generative algorithms, and digital fabrication to produce blocks with complex shapes and textures suitable for bio-marine habitats.

The project is executed in two phases. This paper describes the initial phase, encompassing the prototyping process, construction, testing, and analysis of various Ecoblox versions. The primary objective of this phase is to assess multiple designs and evaluate their effects on biodiversity. Building upon the insights gained from the initial phase, Phase II of the study focuses on developing data-driven strategies and applying robotic 3D printing to refine the system's design and construction.

Keywords: Symbiotic Intelligence · Coastal Architecture · AI-enhanced Design · Modular Shoreline Infrastructure · Ecoblox · Biodiversity Promotion · Climate Resilience · Ecosystem-Centric design · Environmental Sustainability · Robotic 3D Printing

1 Introduction

Climate change is the century's defining challenge (Rogelj et al. 2021; Masson-Delmotte et al. 2018), and the coastal regions are experiencing full impact. Rising sea levels are causing severe coastal erosion, excessive sedimentation, water contamination, frequent floods, hurricanes, severe storms, and storm surges (Azevedo de Almeida and Mostafavi 2016). In addition, rising ocean temperatures, one of the major consequences of climate change, poses a significant threat to marine ecosystems, leading to loss of biodiversity (Breitburg et al. 2018).

The widespread use of hard or gray infrastructures like concrete seawalls and barriers to mitigate coastal erosion has been detrimental to coastal ecosystems (Pilkey and Cooper 2012; European Environment Agency 2018). A NOAA study predicts that a third of the US coastline could be covered in concrete by 2100 (Dilling and Lemos 2011). Gray infrastructures damage natural habitats, exacerbating biodiversity loss in marshes,

© The Author(s) 2025
H. Chai et al. (Eds.): CDRF 2024, *Symbiotic Intelligence*, pp. 46–54, 2025.
https://doi.org/10.1007/978-981-96-3433-0_5

mangroves, and coral reefs (Worm et al. 2006; Arkema et al. 2013; Bulleri and Chapman 2010; United Nations Environment Program 2019, p. 45). Seawalls have reduced suitable nesting habitats for marine wildlife, decreasing their survival rates (Bulleri and Chapman 2010).

To address the problematic approach of using gray infrastructure, several innovative strategies have been implemented along the shorelines. Hybrid infrastructures, which integrate gray and green infrastructure, are gaining significant attention and are being implemented in several projects across the country (Maltby and Waldon 2018, Othman and Shaari 2021). Hybrid infrastructures sequester carbon while providing the same protection as seawalls and levees by combining plants and dunes. In addition, studies have shown that they can be an effective means of coastal protection while promoting biodiversity and ecosystem services (Melián-González et al. 2020; Kirshen et al. 2018).

A few successful projects in recent years have highlighted the potential benefits of hybrid infrastructures. For example, the Living Seawall project in Australia resulted in a 36% increase in biodiversity upon retrofitting seawalls with habitat panels (Dafforn et al. 2015; Campbell et al. 2019; Browne et al. 2019; Firth et al. 2016). Similarly, the Seattle Waterfront project used light-penetrating surfaces and biological stormwater management, allowing marine life like juvenile salmon to return to the shoreline (Dyson and Yocom 2014; Beekman 2019). A pilot study conducted during the project's early phase revealed that textured treatment of the seawall promoted microhabitats (Dyson and Yocom 2014).

While innovative coastal management approaches are essential, projects like the Living Seawall and Seattle Waterfront, which focus on creating complex textures conducive to marine habitat, represent limited strategies for enhancing biodiversity and sustainability of the shorelines. Emerging technologies such as networked sensors, big data analytics, parametric design algorithms, and robotic fabrication are introducing new possibilities for hybrid infrastructure design and construction. For instance, parametric design facilitates the creation of complex geometries based on optimization rules using AI, enabling the integration of various parameters such as site conditions and performance criteria (Wassim Jabi and Ozel 2017). Furthermore, robotic construction methods can be integrated with parametric design tools to generate complex, customized, and optimized geometries and forms.

This project builds on these recent technological innovations to create "Ecoblox," a modular interlocking system of blocks engineered for seawall attachment. Designed with specific geometries and textures, Ecoblox aims to foster ecological functionality by developing microhabitats that promote the colonization and growth of diverse marine organisms. The project's first stage, Phase I, which is elaborated upon in this paper, aims to establish a proof of concept for Ecoblox. Following this, Phase II is dedicated to refining the system through a data-driven methodology (Fig. 1).

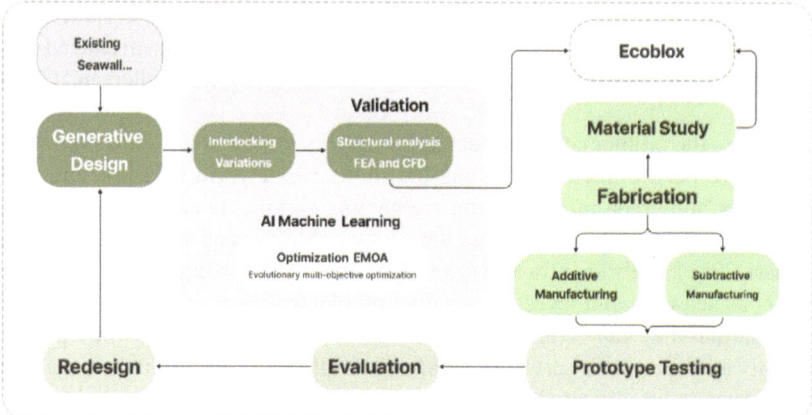

Fig. 1. Diagram describing the project development process.

2   Prototype Development: Phase I

In Phase 1 of the project, the key objectives were: 1) Conducting comprehensive research on factors influencing seawall and hybrid infrastructure design, including structural integrity, durability, and impact on biodiversity; 2) Collecting and analyzing data to inform the design and fabrication of textured surface prototypes. 3) Applying AI algorithms for optimization, enabling efficient and adaptive design iterations of the textured geometries; 4) Constructing and testing these prototypes in coastal waters to evaluate their performance in promoting biodiversity; 5) Conducting sustainability evaluations to assess the ecological impacts and biodiversity enhancement provided by the prototypes.

2.1   Fabrication

The fabrication process in this phase involved an iterative design workflow combining data analysis, parametric modeling, and digital fabrication techniques. This allowed us to explore various designs and rapidly produce physical prototypes for in-situ testing. A key focus of the fabrication approach was developing various surface textures to evaluate their influence on biodiversity. We hypothesized that creating distinctive biomimetic textures inspired by natural marine habitats like coral structures, Ecoblox surfaces would provide favorable microhabitats and enhance biodiversity.

An experimental arrangement was created to evaluate this hypothesis. We designed prototypes inspired by two natural patterns. The Voronoi pattern emulated the intricate structure of honeycomb coral, and the reaction-diffusion pattern drew from the complex folds of brain coral. The patterns designed for the blocks were digitally decoded using Grasshopper, a visual programming tool, and then precisely transferred onto the block surfaces using a CNC milling machine. This process detailed the patterns onto a flexible foam, which acted as molds for the casting. In addition, we experimented with various concrete mixes with lower cement content to decrease the carbon footprint of blocks.

The subsequent pouring of concrete mixes into molds resulted in prototypes measuring $1' \times 2' \times 2''$, displaying three unique designs: flat, honeycomb coral, and brain coral.

Five concrete prototypes featuring two distinct surface textures and three different concrete mixtures were placed in an intertidal zone. Additionally, a control prototype block with no surface treatment was produced to serve as a benchmark for comparative analysis with the textured prototypes. The prototypes were left undisturbed for two years, allowing for an extended data collection period. This duration was crucial for a thorough analysis of the tiles' effectiveness (Fig. 2).

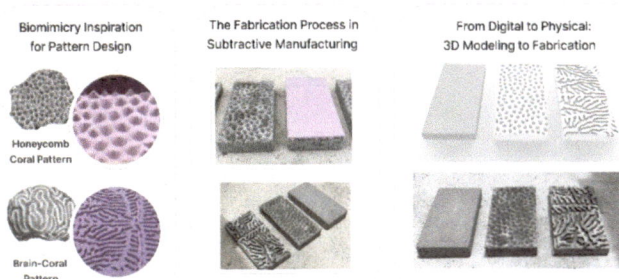

Fig. 2. The sequence of images showing the stages of creating prototypes: digital designs for honeycomb and brain coral patterns, their CNC-milled foam molds, and the final concrete prototypes, including a flat control tile.

2.2 Data Collection and Analysis

The ecological dynamics of the prototype were evaluated using a comprehensive data collection process, which included counting attached invertebrates and photographing and scanning the blocks to assess algae coverage. This analysis revealed the growth of

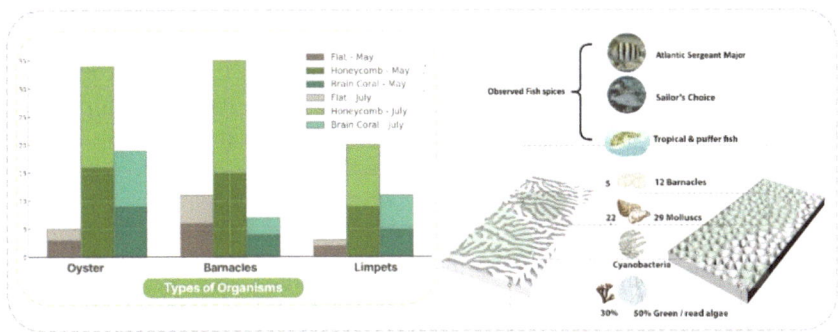

Fig. 3. Bar chart and the diagram showing the invertebrates count in 3 months.

three species of oysters, barnacles, and limpets on the textured blocks, whereas the flat control block exhibited negligible growth. The bar chart in Fig. 3 provides a detailed account of the results recorded over three months.

The analysis results demonstrated that surfaces featuring enhanced texture and rugosity significantly boosted natural growth, setting the stage for the project's next phase (Fig. 4).

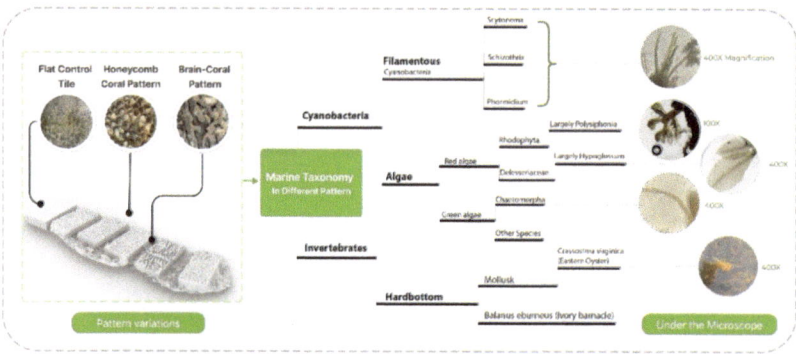

Fig. 4. Marine taxonomy for growth on Ecoblox

3   Prototype Optimization: Phase II

In Phase 2, which is currently underway, the project will focus on leveraging robotic 3D printing to construct the interlocking system for the blocks in a sustainable way. In addition, AI-driven generative optimization, utilizing environmental data sets, will enhance the blocks' texture and improve their structural integrity. Incorporating local data, such as solar radiation and air and water temperatures from the block's deployment site, will inform the design of new types of patterns and textures. These data-driven patterns could offer significant thermal stress relief for marine invertebrates and more accurately designed habitats for marine species.

Using robotic 3D printing to construct the blocks presents multiple benefits. First, it facilitates the fabrication of complex geometries with control over material distribution. Second, it allows the development of the Ecoblox without the formwork required for casting, a potentially waste-free process. Lastly, it will enable the exploration of various concrete mixes and sustainable alternative materials in the printing process (Fig. 5).

3.1   Interlocking System

To develop the interlocking system for the blocks, the project builds on a methodology introduced by Estrin et al. in 2021. This methodology explores a design principle that focuses on structuring material into interlocked elements without needing connectors or binders, purely based on their geometry and mutual arrangement. This approach

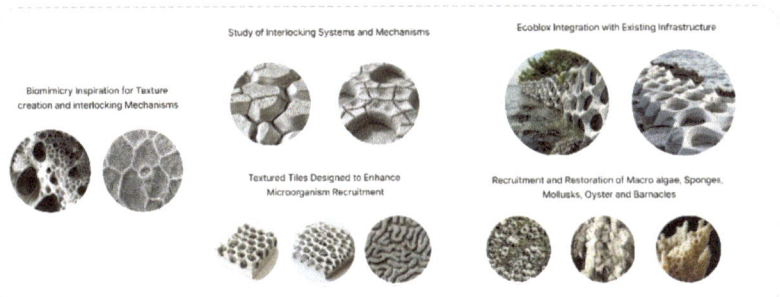

Fig. 5. Proposed design process in Phase II

draws inspiration from natural growth processes, mirroring the complex yet orderly development observed in biological and geological formations (Estrin et al. 2021).

To develop the interlocking system, we will employ lofting algorithmic operations. Specifically, our approach integrates a growth-based design algorithm inspired by Voronoi tessellation to design space-efficient, structurally robust shapes exhibiting superior directional interlocking. This innovative design methodology facilitates the creation of components that offer enhanced flexibility and strength, which is particularly advantageous in withstanding multidirectional or unpredictable loads, which is typical at the water's edge.

We plan to use Finite Element Analysis (FEA) software to model the edge geometries, enabling a detailed examination of how different edge conditions affect the overall strength and stability of the interlocking system. FEA will allow us to incorporate the material properties of concrete for various types of 3D printed infills inside the blocks. The analysis results will provide insights into the concrete's behavior under simulated loads, including cracking and other failure mechanisms (Fig. 6).

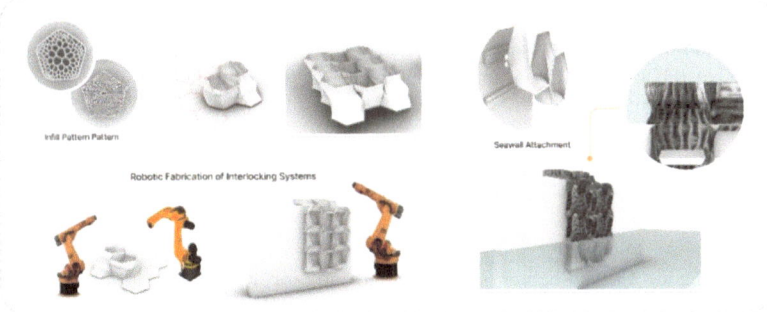

Fig. 6. Illustration of interlocking systems for seawall attachment

3.2 Testing and Evaluation

The Ecoblox prototypes' testing and biodiversity evaluation will involve their placement across different site locations, each characterized by unique water conditions, mainly varying salinity levels. This critical task aims to assess the prototypes' effectiveness and identify any areas of failure. The findings from these tests will inform necessary adjustments and enhancements to the prototype designs.

We will monitor and measure the growth of benthic communities, including macroalgae, sponges, mollusks, and barnacles on the prototype surfaces. These organisms obtain their nutrition by filtering and extracting organic particles from the water in their surroundings. This evaluation will ascertain the blocks' effectiveness in attracting marine life and enable us to gauge their impact on mitigating water pollution in the surrounding area.

4 Closing Remarks

As climate change and coastal erosion continue to present formidable challenges to global ecosystems, developing innovative approaches to coastal development and seawall construction is imperative. Advanced technologies, including Artificial Intelligence (AI), sensor technologies, big data analytics, and robotic manufacturing, are bringing new possibilities for data-driven approaches to the built environment. Through informed design, these approaches can help minimize ecosystem harm and promote sustainable development.

The Ecoblox system described here aims to offer a new perspective on coastal design, emphasizing support for biodiversity, enhancing climate resilience, and promoting environmental sustainability. This system's successful implementation and validation promise to catalyze a shift toward ecosystem-centric infrastructure design.

Acknowledgement. This work is supported by the National Science Foundation (NSF) Award No. 2329345: CREST-PRP. Any opinions, findings, conclusions, or recommendations expressed herein are those of the authors and do not necessarily reflect the views of the National Science Foundation. Additionally, this project is supported by the Environmental Protection Agency (EPA) under Grant No. 03D09124. This is publication #1781 from the Institute of Environment at Florida International University.

References

Arkema, K.K., et al.: Coastal habitats shield people and property from sea-level rise and storms. Nat. Clim. Chang. **3**(10), 913–918 (2013). https://doi.org/10.1038/nclimate1944

Azevedo de Almeida, B., Mostafavi, A.: Resilience of infrastructure systems to sea-level rise in coastal areas: impacts, adaptation measures, and implementation challenges. Sustainability **8**(11), Article 1115 (2016)

Beekman, G.: See the new $7.1 million shortcut to Seattle's waterfront. The Seattle Times (2019). Retrieved from https://www.seattletimes.com/seattle-news/politics/see-the-new-7-1-million-shortcut-to-seattles-waterfront/

Breitburg, D., et al.: Declining oxygen in the global ocean and coastal waters. Science **359**(6371), eaam7240 (2018). https://doi.org/10.1126/science.aam7240

Browne, N., Smith, J., Jones, A., Green, D.: Enhancing biodiversity through living seawalls: a case study of habitat panels in Australia. J. Mar. Biol. Conserv. **15**(3), 245–256 (2019)

Bulleri, F., Chapman, M.G.: The impact of artificial structures on marine ecosystems. Mar. Environ. Res. **70**(5), 365–377 (2010). https://doi.org/10.1016/j.marenvres.2010.05.002

Bulleri, F., Chapman, M.G.: The introduction of coastal infrastructure as a driver of change in marine environments. J. Appl. Ecol. **70**(5), 365–377 (2010). https://doi.org/10.1111/j.1365-2664.2009.01751.x

Campbell, M.L., Bishop, M.J., Mayer-Pinto, M.: Living seawalls: Biologically active artificial structures can enhance the biodiversity and ecological functioning of urban seascapes. Front. Ecol. Evol. **7**, 1–13 (2019). https://doi.org/10.3389/fevo.2019.00234

Dafforn, K.A., Glasby, T.M., Airoldi, L., Rivero, N.K., Mayer-Pinto, M., Johnston, E.L.: Marine urbanization: an ecological framework for designing multifunctional artificial structures. Front. Ecol. Environ. **13**(2), 82–90 (2015). https://doi.org/10.1890/140050

Dilling, L., Lemos, M.C.: Creating usable science: opportunities and constraints for climate knowledge use and their implications for science policy. Glob. Environ. Chang. **21**(2), 680–689 (2011). https://doi.org/10.1016/j.gloenvcha.2010.11.006

Dyson, K., Yocom, K.: Ecological design for urban waterfronts. Urban Ecosyst. **18**(1), 189–208 (2014). https://doi.org/10.1007/s11252-014-0385-9

European Environment Agency.: Coastal erosion and climate change in Europe (2018). Retrieved from https://www.eea.europa.eu/publications/coastal-erosion-and-climate-change-in-europe

Estrin, Y., Krishnamurthy, V.R., Akleman, E.: Design of architectured materials based on topological and geometrical interlocking. J. Market. Res. **15**, 1165–1178 (2021)

Firth, L.B., et al.: Ocean sprawl: challenges and opportunities for biodiversity management in a changing world. Oceanogr. Mar. Biol. Annu. Rev. **54**, 189–262 (2016)

Kirshen, P., Ruth, M., Carbone, G.: Hybrid infrastructure and its role in improving resilience to climate change. Environ Sci Policy **86**, 76–84 (2018)

Masson-Delmotte, V., Zhai, P., Pörtner, H.O., Roberts, D., Skea, J., Shukla, P.R., Waterfield, T.: Global warming of 1.5 C. An IPCC Special Report on the impacts of global warming. **1**, 93–174 (2019)

Maltby, E., Waldon, M.G.: Integrating gray and green infrastructure for water management: a review of concepts and research. J. Environ. Manage. **221**, 324–334 (2018). https://doi.org/10.1016/j.jenvman.2018.05.066

Melián-González, S., Mendez, F.J., Losada, I.J., Torres-Navarro, G.: Hybrid coastal protection: a review of the main concepts, designs, and ecological effects. J. Clean. Prod. **260**, 121131 (2020)

Othman, N., Shaari, N.: An overview of green-gray infrastructure integration as an urban flood management solution. Environ. Sci. Pollut. Res. **28**(2), 1197–1215 (2021). https://doi.org/10.1007/s11356-020-10832-9

Pilkey, O.H., Cooper, J.A.G.: Society's choices: mitigating the impacts of climate change on the coasts. National Academies Press, Washington, DC (2012)

Rogelj, J., et al.: Mitigation pathways compatible with 1.5°C in the context of sustainable development. In: IPCC (Ed.), Global Warming of 1.5°C. Cambridge University Press (2021)

United Nations Environment Programme.: Global environment outlook: Summary for policy-makers (2019). Retrieved from https://www.unep.org/resources/report/global-environment-outlook-6-summary-policy-makers

Wassim, J.E., Ozel, F.: Robotic Fabrication in Architecture, Art, and Design 2016. Springer International Publishing, Cham, Switzerland (2017). https://doi.org/10.1007/978-3-319-58487-9

Worm, B., et al.: Impacts of biodiversity loss on ocean ecosystem services. Science **314**(5800), 787–790 (2006). https://doi.org/10.1126/science.1132294

Digital Fabrication and Construction Method

Enhancing Seagrass Habitat Restoration: 3D Scanning and FGF for Tetrapod Prosthesis

You Sub Bang, Seok Won Choi, and Jungwon Yoon[✉]

University of Seoul, Seoulsiripdae-ro 163, Seoul 02504, Korea
jwyoon@uos.ac.kr

Abstract. This study proposes a novel solution for addressing coastal erosion and ecosystem degradation worldwide. By attaching a specially designed prosthesis to the curved surface of tetrapods, common marine structures in Korea, we aim to restore vital seaweed forest ecosystems within coastal areas. Leveraging 3D scanning, printing technologies, and robotic arms, we develop a customized attachment to seamlessly integrate with tetrapods, overcoming construction area limitations. Our approach utilizes Fused Granular Fabrication, employing biodegradable plastic made from recycled seaweed waste, promoting sustainable material use and disposal in marine environments. Method of attaching to the surface of the tetrapod was divided into three methods: Foldable 3D printing, On-surface 3D printing, and Barnacle 3D printing, which improves the two preceding methods. This innovative solution offers promise for sustainable coastal management and ecosystem restoration.

Keywords: Seagrass · Restoration · Tetrapod · 3D Scanning · FGF · Biomaterial

1 Introduction

1.1 Background

Advanced 3D printing (3DP) and 3D scanning (3DS) can make significant contributions to restoring damaged coastal ecosystems caused by human impacts as well as changes of the nature and climate. South Korea, surrounded by the sea on three sides, faces diverse marine environments. Furthermore, global coastal eco-systems have been severely damaged by landfill, coastal construction, and eutrophication, leading to the desolation of rocky ecosystems due to phenomena such as bleaching [1]. With increasing awareness of marine pollution, various protective measures are being taken to address wave erosion and coastal erosion [2]. The UN General Assembly declared the period from 2021 to 2030 as the "UN Decade on Ecosystem Restoration," highlighting the restoration of ecosystem structure and function as one of the major challenges in marine ecology [3].

Seagrass beds, primary producers in coastal ecosystems, play crucial roles as habitats and refuges for marine organisms, supporting marine biodiversity as breeding grounds. Seagrass beds are among the highest primary productivity areas in marine ecosystems, ranking second in net primary productivity after mangrove forests [4].

H. Chai et al. (Eds.): CDRF 2024, *Symbiotic Intelligence*, pp. 57–66, 2025.
https://doi.org/10.1007/978-981-96-3433-0_6

Table 1 shows that various countries are making efforts to restore seagrass, and recently, there are efforts to preserve the marine ecosystem by creating artificial coral reefs with 3DP tiles [5]. However, most seagrass restoration efforts have targeted species inhabiting weak wave action zones, overlooking those in highly exposed open waters where waves are stronger. The lack of research into restoring seagrass habitats in high-wave environments is attributed to the challenges associated with conventional restoration methods.

Table 1. Seagrass restoration project in various countries itself

Country	Year	Type	Water environment
Indonesia	2015	Transplantation, Monitoring	Weak wave
U.S.A	2019		
U.K	2021		
China	2022	Seed ball burial	

1.2 Scopes and Objectives

The purpose of this study is to evaluate the feasibility to introduce 3DP and 3DS technology in the technique for restoration of seagrass forests in the coastal waters of South Korea, focusing on commonly installed tetrapods (TTP). Typically, TTP is manufactured in a modular form through factories, resulting in minimal variations among individual units. Additionally, the faces of TTP exhibit a consistent curvature gradient. These characteristics of TTP offer significant advantages in the general application of prosthesis to be 3D printed and in the output on curved surfaces. Key aspects include material development using recycled seaweed and biodegradable compounds like PLA and PBAT for Fused Granular Fabrication (FGF) with industrial robot ABB IRB 1200. Prosthesis tailored to TTP surface are designed and conceptual experimented through FGF, using 3DS. This study aims to improve seagrass forest restoration through human intervention

Fig. 1. TTP, seaweed waste, and TTP on a prosthesis

in both pre-production and production stages, ensuring optimal prosthesis generation via FGF by design and adjustment processes (Fig. 1).

2   Methodology

2.1   Technical Setup

The technical setup for our experiment is organized with hardware and software as shown in Table 2. The main equipment for hardware is the robotic 3DP system, and the main software is set for G-Code generation for 3DP.

– **Robotic 3DP.** The Hardware is made up with an industrial robot arm, ABB IRB 1200, a large-scale pellet extruder manufactured by Dyze Design, Pulsar Beta version and a heating bed, as illustrated in Table 2.

Table 2. Production setup with hardware and software

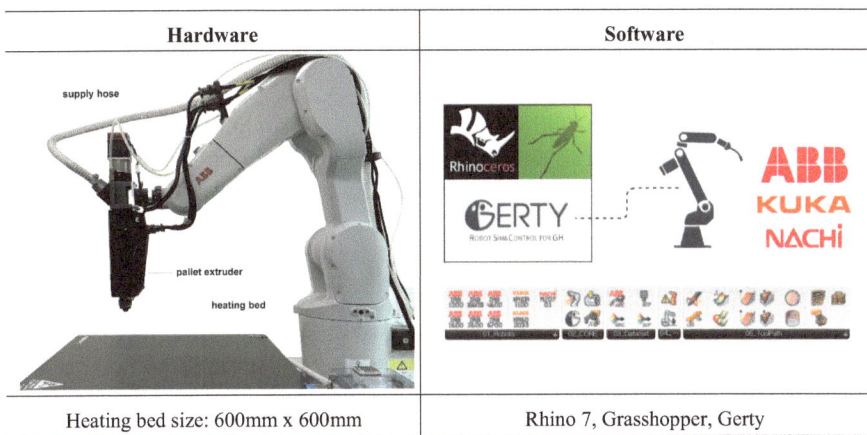

Hardware	Software
Heating bed size: 600mm x 600mm	Rhino 7, Grasshopper, Gerty

– **G-Code Generation.** We use Rhinoceros and Grasshopper scripts for design and FGF. A plug-in tool for Grasshopper has been developed by BAT Partners, Inc. in Korea, to integrate slicing, 3DP simulation and G-code file generation.

2.2   Design Workflow

The research process can be divided into four main processes of 3DS, Design, FGF, and Material Mix (Fig. 2).

● **3DS**. Utilized scanning tools to capture models, with Artec Eva offering precision scanning capabilities. Facilitated the digitalization of reference objects through image processing for 3D implementation.

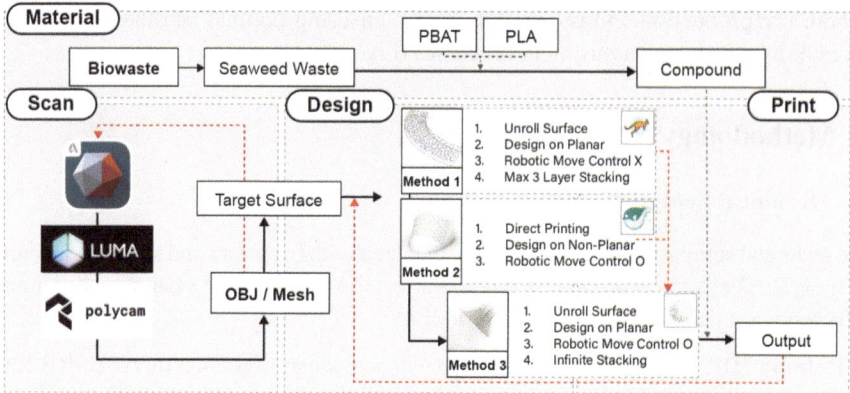

Fig. 2. Design workflow

- **Design**. Involved three methods: designing on an unfolded plane using the Unroll Surface technique, designing linearly by extracting specific points and curves from curved surfaces, and designing volumetric shapes from curved designs using Anemone. Each method has its pros and cons, with a hierarchy established based on the final output goals.
- **FGF**. Complemented the various design methods. Traditional layer-based 3DP was used for unfolded plane outputs, while designs requiring printing on inclined planes were adjusted either through iterative design modifications or considering the conditions of robotic arms.
- **Material Mix**. The materials used in FGF are combinations of Sea Polymer (SP) 1000, PLA and PBAT as below.

  1) **SP (Sea polymer) 1000**. Derived from PBAT-based bioplastics, SP1000, developed and supplied by Marine and Bio Company incorporates components extracted from seaweed byproducts (20% ratio) to enhance biodegradability.
  2) **750D**. A compound based on PLA, 750D exhibits excellent strength when used in 3DP and experiences minimal material supply bottleneck errors due to its pelletized form.
  3) **BG1000**. Based on PBAT, BG1000 offers low shrinkage rates and enhances flexibility in printed materials when used in 3DP applications.

2.3   Scanning Process

Due to the challenging nature of scanning TTPs at actual scale, in the experiments, scaled-down models of TTP (1:10) and partial TTP models (1:3) were produced using a 3D printer and then scanned. The scanning of the models was carried out using three different devices and software, as shown in the following Table 3.

In the experiment, precise and appropriately sized 3DS data from Artec Eva were utilized. To facilitate the smooth utilization of the scan data, mesh data in Rhino 7 was simplified using 'Quadremesh' to reduce the complexity of the mesh.

Table 3. Scan data with three different equipment and software

Artec Eva	Polycam	LUMA
1 Mesh	1 Mesh	45 Mesh
Scale: origin	Scale: none	Scale: none

3 Experiment

Experiment started by dividing into Method 1 and Method 2 and proceeded with Method 3 which improved the pros and cons of the two methods. Method 1 used Folding 3DP, and Method 2 applied on-surface 3DP. Method 3, which mixed the design of Method 1 and the printing way of Method 2, focused on creating Prosthesis that can be attached directly to TTP using Barnacle's principle.

3.1 Method 1: Folding 3DP

– **Design.** Method 1 involves printing on a planar surface followed by additional processing. Designing on a flat surface offers higher degrees of freedom compared to printing on uneven surfaces. The design involved unfolding the information obtained through scanning the target plane and implementing a porous design where seagrass roots could be fixed and allowed to grow within the area of the unrolled surface. Furthermore, the focus was on designing to maximize porosity within the limits of the maximum allowable thickness for post-printing processing. For design validation through various simulations, stepslover and spherecollide of 'kangaroo 2' for porosity between seagrasses were employed (Table 4).

Table 4. Design process in method 1

Simple ⟵   ⟶ Complex

– **Printing.** During the process of outputting the designed G-code, the printing considerations required for Method 1 are the same as when printing on a conventional flat surface. This includes considerations such as the speed of the robotic arm, the feed rate of the material, and the temperature within the extruder. When the output design retains heat, applying force again to the unfolded target for cooling allows obtaining prosthesis designed for curved surfaces. However, it was noted that additional flexibility needs to be ensured and sensitive consideration should be given to design points where deformation occurs, due to doubts about the altered bonding state of polymers during the deformation process and the possibility of brittle fracture beyond the limit deformation. These points were identified as limitations (Table 5).

Table 5. Production process in method 1

| FGF | Post-processing | Result |

3.2   Method 2: On-Surface 3DP

– **Design.** Method 2 involves extracting points, lines, and surfaces from confirmed irregular planes and designing based on the extracted information on curved surfaces. Since the designed G-code needs to be directly printed on the surface of the curved surface, care must be taken in the design process to avoid collisions between the nozzle and the paths of the layers. In Method 2, a mesh-like design was planned to fix the seagrass using the simplest method, 'Net on Surface'. Despite using 'Net on Surface', it was planned to stack a minimum of 3 layers to create voids between the layers. Some sections were intentionally induced to free-fall to create space even between the z-axis below, resembling the shape of net and quilting (Fig. 3).

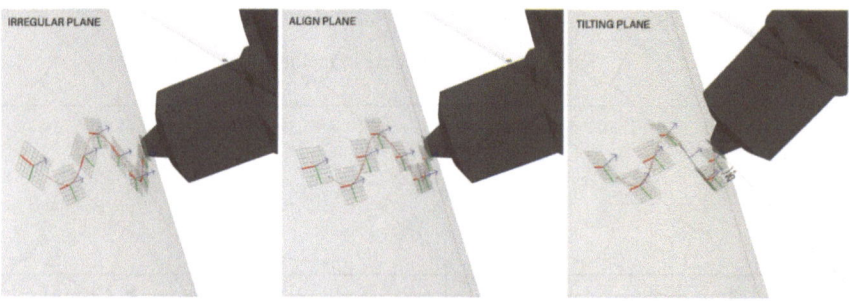

Fig. 3. Plane adjustment process in method 2

– **Printing.** Unlike 'Method 1', where a primary output was necessary to print on irregular planes, the output served as the bed and was printed using FDM (Fused Deposition Modeling) 3DP with PLA filament. Additionally, to prevent excessive tilting during printing for the stability of robot and extruder movements, the normal vector of each point on the plane was uniformly tilted by 25 degrees to enhance stability and efficiency. Initially, segmented planes on irregular surfaces were used during the initial printing process. However, it was found that this caused significant delays in the robot's movement due to the constantly changing xy-axis direction, resulting in slower movement than expected in dramatic curved sections. In other words, there was inconsistency in the extrusion speed, leading to the phenomenon where the printing speed was not consistent. Consequently, to minimize delays caused by the robot's movement on the fourth axis, the xy-values of the irregular planes were uniformly standardized while maintaining the normal vector values. By utilizing the stability values obtained, various designs became feasible on irregular planes, forming the basis for the advancement into Method 3 (Table 6).

Table 6. Production process in method 2

Printing and post-processing

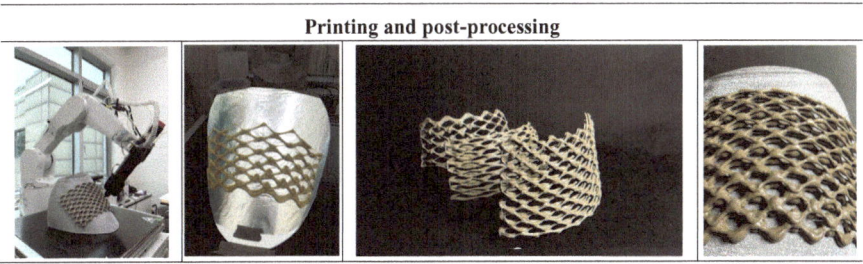

3.3 Method 3: Barnacle 3DP

– **Design.** Method 3 combines the advantages of the design freedom of Method 1 and the capabilities acquired through Method 2, which utilizes the robotic arm for printing

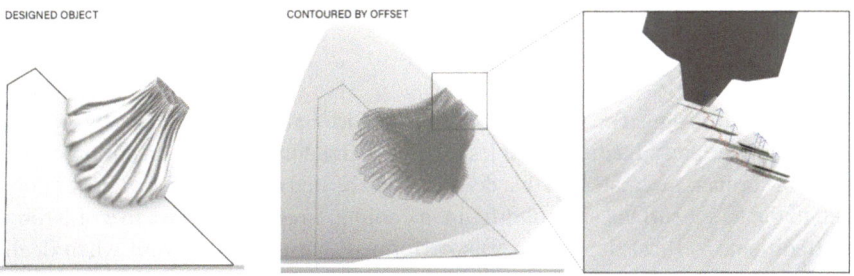

Fig. 4. Design process in method 3

on inclined surfaces. Starting with the first layer designed in Method 1 on the surface of the curved surface, Method 3 creates an artificial seaweed forest resembling growth on the surface using "anemone". The designed object is divided into each layer based on the offset value of the original first output instead of the z-axis. The advantage lies in maximizing the surface area at the bottom for easy adhesion to the opposite surface, while the numerous creases formed inside provide ideal spaces for each seaweed to root and be protected. Protecting the seaweed roots is crucial for their establishment in rough waves, a condition our design satisfies (Fig. 4).

– **Printing.** The complexity of the design and the large number of control points per layer imply the existence of more planes for the 'TCP' to traverse. Consequently, adjustments were made to the direction of progression of each plane based on average values, a consideration not addressed in Method 2 due to the smaller number of control points. This modification is essential for the limitations of the robotic arm's joints and efficient operation. During actual printing, the offset value of the original plane was used instead of the z-axis direction of gravity. Therefore, our output, affected by gravity, needed to be rapidly cooled to solidify, significantly impacting reducing errors in our planned design. Our output fits precisely on irregular surfaces, providing significant stability upon assembly (Fig. 5).

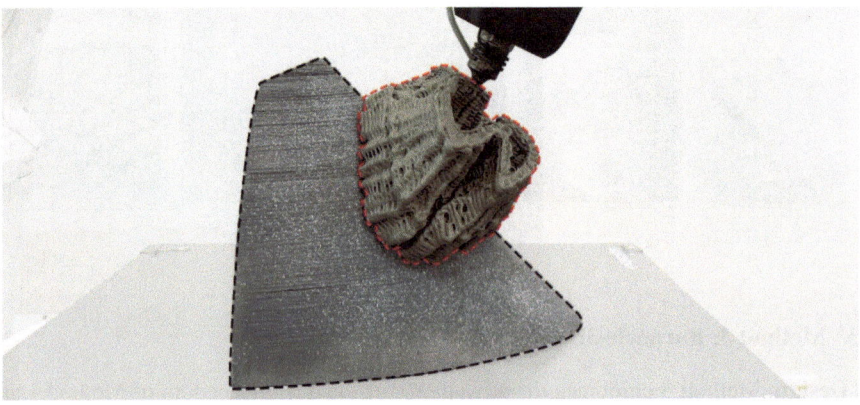

Fig. 5. Production process in method 3

3.4 Comparative Discussion of Methods

In method 1, we designed porous prosthesis on the surface of a curved object by unfolding the target surface and adjusting variables to achieve freedom in design. In this process, we utilized the property of materials that deform when heated above a certain temperature to attach to the target surface and solidify upon cooling. From the perspective of a robotic arm, traditional vertical deposition was performed, but problems arose when dealing with surfaces that unfolded into double curves or became more complex, resulting in significant errors during unfolding (Table 7).

Table 7. Comparison of three different methods

Process	Method 1	Method 2	Method 3
Design	Porosity design, Simulation	Simple net design	Layering of freely configured design
3DP	Planar 3DP	Non-planar 3DP	
Attachment	Bending and wrapping on surface	Direct 3DP on surface	

On the other hand, Method 2 addressed issues arising from 3DP on curved surfaces using a limited design approach called "net on surface." By adopting a non-planar deposition method instead of solely vertical stacking, it resolved issues related to the tilting angle of the extruder affecting material supply. Additionally, by determining an appropriate material feeding speed when the material adheres to the curved surface during the intersection of the surface and 3DP, Method 2 enabled 3DP on surfaces with double curves or complex shapes, provided that the nozzle size of the extruder is suitable for the surface.

Building upon the freedom of design in Method 1 and the non-planar 3DP results in Method 2, Method 3 executed freeform design stacking on curved surfaces and further adjusted the data to match the rotation radius of axes 4 and 6 of the robotic arm. Consequently, it became evident that 3DP of prosthesis is feasible not only for objects with simple curved surfaces like TTP but also for objects with various curved surface profiles.

Fig. 6. Prosthesis installed, in Tongyeong, South Korea

4   Conclusion

This study contributes to the coastal erosion and restoration of coastal ecosystems, particularly seagrass forests in areas prone to wave action, by creating prosthesis based on TTP commonly found in South Korea's coastal areas. In this process, recycling of waste seaweed, design mimicry of seagrass forests, and direct 3DP technology on curved surfaces were employed. Through three different prosthetic fabrication methods, we confirmed the degree of deformability of materials at a certain temperature when processed to fit existing curved surfaces, up to a certain thickness. We also assessed the freedom of design by unfolding the target curved surfaces. Furthermore, when directly fabricating on the target surface using a robotic arm, we examined how to limit the movement of the robotic arm concerning material supply and proposed ways to minimize the load on the robotic arm during buildup on the curved surface. These results offer inspiration for creating a variety of prosthesis through 3DP. Of course, empirical research is needed to determine whether the produced outputs can form seagrass forests and sustainably protect them. However, this innovation solution offers promise for sustainable coastal management and ecosystem restoration (Fig. 6).

References

1. Islam, Md.S., Tanaka, M.: Impacts of pollution on coastal and marine ecosystems including coastal and marine fisheries and approach for management: a review and synthesis. Mar. Poll. Bull. **48**(7–8), 624–649 (2004)
2. Spalding, M.D., et al.: The role of ecosystems in coastal protection: adapting to climate change and coastal hazards. Ocean Coast. Manag. **90**, 50–57 (2014)
3. Borja, A.: Grand challenges in marine ecosystems ecology. Front. Mar. Sci. **1** (2014)
4. Duarte, C.M., et al.: Assessing the capacity of seagrass meadows for carbon burial: current limitations and future strategies. Ocean Coast. Manag. **83**, 32–38 (2013)
5. Lange, C., Ratoi, L., Co, D.L.: Reformative coral habitats. In: RE: Anthropocene, Design in the Age of Humans-Proceedings of the 25th International Conference on Computer-Aided Architectural Design Research in Asia, CAADRIA 2020, vol. 2, pp. 465–474 ((2020))

Developing a Computational Design to Fabrication Method for 3D Knitted Stay-in-Place Moulds for Building Envelopes Tiles

Yoav Sterman[1]([✉]), Yasha Jacob Grobman[1], and Yiska Goldfeld[2]

[1] Faculty of Architecture and Town Planning, Technion, Haifa, Israel
stermnan.yoav@technion.ac.il
[2] Faculty of Civil Engineering, Technion, Haifa, Israel

Abstract. Contemporary building envelopes primarily rely on repetitive elements that fulfill mainly a singular purpose - a barrier between interior and exterior spaces. Implementing building envelopes featuring intricate geometries and non-repetitive tiles can significantly enhance the environmental performance of the structure. However, the current manufacturing processes for the required moulds are plagued by high costs, time consumption, labor-intensiveness, and mainly by using non-recyclable moulds. To address these challenges, the paper presents an innovative solution for designing and fabricating building envelope tiles with complex geometries by employing stay-in-place 3D knitted moulds. The moulds are digitally fabricated using innovative knitting procedures implemented on an industrial flat double-bed machine. The paper presents preliminary research outcomes, including a new digital design methodology for creating knitted moulds and a new fabrication method for buildings' envelope tiles.

Keywords: Computerized knitting · Parametric design · Building envelope · Non-repetitive tiles · Complex geometry · Spring-based simulation · Stitch length

1 Background and State of the Art

Contemporary building envelopes are required to meet increasing demands for environmental performance and sustainability, as well as fast and economical production feasibilities that minimize the impact on the environment. Recent advancements in digital design computation and fabrication, as well as the development of new materials and construction technologies, have enabled new approaches to building envelopes. These approaches, which focused on the potential of complex geometries to improve environmental performance, offered additional functionalities, such as green walls, thereby benefiting various stakeholders (Hershkovich et al. 2021; Grobman et al. 2023).

However, the fabrication of complex envelope forms remains a significant challenge. Current fabrication technologies for complex building envelopes rely on costly and non-recyclable moulds, such as those produced by CNC-based mould fabrication, or highly

H. Chai et al. (Eds.): CDRF 2024, *Symbiotic Intelligence*, pp. 67–76, 2025.
https://doi.org/10.1007/978-981-96-3433-0_7

time-consuming processes, such as 3D printing. Consequently, in existing solutions, the cost of moulds can account for up to 60% of the tile's total cost, and has a negative impact on the environment (Nawy 2008). To address these limitations, the current study aims to develop a fabrication method that employs hybrid knitted moulds constructed with advanced computational processes.

Knitting is a common technology for producing textiles for various applications. Most knitted textiles are used for apparel. In recent years, knitting technology has been integrated into the production process of reinforcement systems for the construction industry (Gries et al. 2016). By composing high-strength and resistance to corrosion yarns, and fine-grained and high-strength cementitious mixtures, it is possible to construct thin-walled composite elements (Brennan et al. 2013; Peled et al. 2017). The obtained textile-reinforced concrete (TRC) structures are light, durable, and structurally efficient. As a result, optimal 2D and 3D structural elements can be constructed (e.g., Hegger et al. 2018; Tysmans et al. 2011; Shames et al. 2014). In recent years, the TRC technology has been implemented in various applications, such as tanks, pipelines (Lieboldt et al. 2006; Perry and Goldfeld 2021), noise barriers (Funke et al. 2015), precast elements (Papantoniou et al. 2008), sandwich elements (e.g. Shames et al. 2014), lost formwork for slabs (Banholzer and Brameshuber 2005), beams, columns and for strengthening of existing concrete structures (e.g. Koutas et al. 2019). However, the geometry of those structures is limited to 2D or relatively simple 3D elements, such as cylinders (Perry and Goldfeld 2021), and special moulds are designed to facilitate their production.

In recent years, computer graphics researchers have explored the use of knitting technology to produce complex 3D shapes using algorithms that can be executed by knitting machines. McCann et al. (2016) developed a compiler that can transform a 3D CAD model into knitting instructions to create textiles in a desired 3D shape. Other researchers have developed methods for producing textured knits (Hofmann et al. 2019) and simulating the deformation of knitted fabrics that mix different compression structures (Karmon et al. 2018). These studies have demonstrated the potential of knitting technology to create intricate and complex 3D knitted structures in the field of textiles.

Recently, some projects have explored the integration of knitting technology for moulds or formwork in cement-based materials (Waimer and Knippers 2015; Popescu 2019; Popescu et al. 2021). However, no solution has been developed to create 3D thin-walled tiles with complex geometries. Furthermore, existing applications have focused on large-scale elements that require additional support or complex frameworks in the casting process (*P Wall | Concreteworks*, n.d.) To address these limitations, this study proposes creating a self-supporting textile mould system that eliminates the need for external support during the casting process.

2   Research Objectives and Specific Aims

The research aims to formulate a computational methodology tailored toward the development and fabrication of 3D knitted stay-in-place moulds intended for building envelope tiles. The tiles will be characterized by parametric design, facilitating individualized configurations to accommodate diverse functional requirements while addressing the imperative of streamlined production processes. This entails the development of a computational framework to generate bespoke tile geometries and the formulation of knitting machine instructions. Concurrently, a novel approach will be explored for casting cementitious materials into the knitted textiles, accounting for the material's adaptability and gravitational influences during the pouring process.

3 Methodology

The 3D knitted stay-in-place moulds are produced using a Karl Mayer Stoll ADF 530 double-bed industrial knitting machine. Double needle bed configuration is utilized to produce moulds as tubular sleeve-shaped fabrics, which are sealed on all edges except for an opening for pouring the concrete. The fabric's tension between the mould's front face and rear face was adjusted by varying the stitch length parameter, which controls the size of the knitted loops. Different stitch length values were assigned for knitting the front face, which is knitted on the front needle bed, and for the back face, which is knitted on the rear needle bed. The back face is flat and tight, and the front face is loose, allowing the cementitious material to fill the mould. The concrete inflates the mould, stretching the textile in the process. The differentiation in the tiles' design is generated by locating stitches that connect the front and back side of the moulds, preventing the inflation of the mould around these stitches. The tile's final shape depends on the location and amount of the connecting stitches. The input file for the knitting machine to produce the tiles is a color-coded bitmap image. Each pixel in the bitmap image corresponds to a knitting stitch. White pixels correspond to plain knit stitches, and gray pixels correspond to tuck stitches that connect both sides of the mould. Since the input design of the tile is a 2D image, it is a challenging task to predict the final appearance of the tiles after the concrete inflates them. Hence, a physics engine was incorporated which includes a simulation and visualization of the inflated shape of the tile. This allows for an interactive and iterative design workflow and inspection of different design options.

An automated workflow for designing the tiles was developed, demonstrated in Fig. 1. This workflow consists of three main stages: a parametric simulation tool, knitting the moulds, and a casting setup. The next sections describe in detail these three stages.

3.1 Parametric Simulation Tool

The design tool is implemented in a Rhino/Grasshopper environment (https://www.grasshopper3d.com). The simulation of the inflated shapes is embedded in the design tool using Kangaroo - a live physics engine for interactive simulation, form-finding, optimization, and constraint solving [http://kangaroo3d.com].

The input of the simulation tool is a two-color bitmap image, where each pixel may be white or gray. The size of the image in pixels defines the size of the tile. The white pixels define plain knit stitches, and the gray defines the location of connection stitches. The tool generates a colored mesh based on the input image. The physics engine simulation is based on dividing that mesh into edges and simulating each edge as a spring. Refereeing the knit stitches as spring is a known method for simulating knitting structures (Karmon et al. 2018; Oghazian 2022). The spring value defines the maximal stretch of the loop.

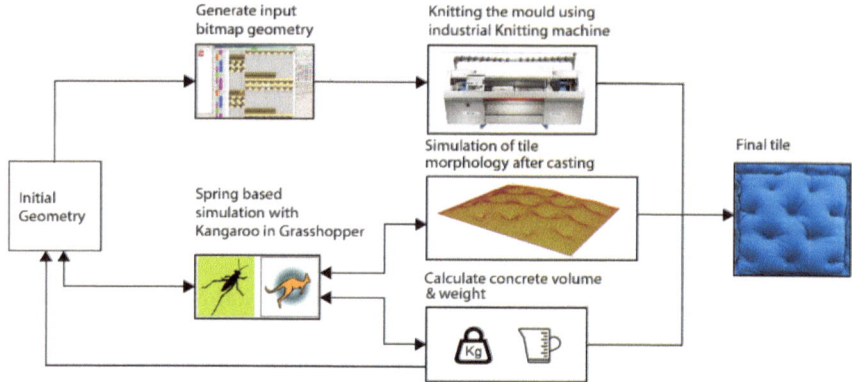

Fig. 1. Workflow diagram

It is determined by the loop length parameter defined in the fabrication process and by the specific yarn being used. The vertices of the mesh that are colored in gray are anchor points in the simulation and, thus, will not be affected. During the simulation, a global pressure force is applied to the mesh, lifting the vertices and constructing the inflated geometry.

Apart from a visual preview of the inflating process, the simulation may also be used to approximate the volume of the inflated moulds. To calculate the volume, we use the Z-value of the center point of the mesh's face boundaries and multiply it by the area of the projected boundaries on the XY plane. We then mass-add the results to calculate the estimated volume. The estimation value may be used to calculate the amount of cementitious matrix needed to fill the mould. This is especially valuable if the moulding is done on an industrial scale, when covering the façade of a building. The estimation may also be used to calculate the weight of the cast tile, by multiplying the volume by the specific gravity of the matrix used.

3.2   Knitting the Mould

The same bitmap used as input for the simulation is also used to generate the knitting instructions for the knitting machine. An automatic workflow in M1+, Stoll's knitting design program, was formulated, allowing the loading of any bitmap and automatically processing it to generate the knitting instructions. This workflow includes the definition of the Color Arrangement (CA) module, which seeks and assigns *plain knit* stitches to the white pixels and *front tuck* stitches to the gray pixels (Fig. 2). While the machine knits the mould's back face, every three knitting rows, it transfers a loop to the front needle bed and forms the tuck stitch, which lays the loop on a front needle. This action connects the front and back faces of the fabric in this specific location. The CA module also assigns different stitch tensions to the front and rear needle beds. The rear is assigned with a stitch length value of 10. For the front, different values may be assigned to control the magnitude of the inflation.

The moulds were knitted using nylon yarn. Since the moulds will be used as the tile's skin and will not removed after casting, the yarn must withstand outdoor conditions.

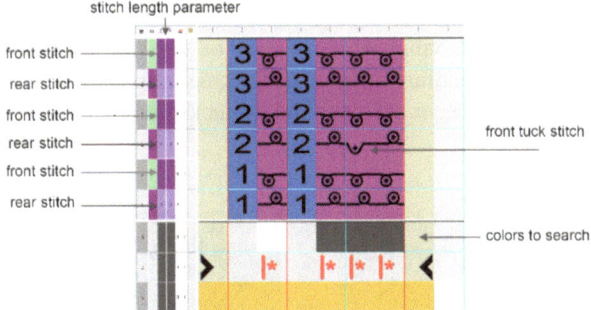

Fig. 2. The color arrangement for seeking and assigning stitches.

Therefore, in the future, we plan to replace the nylon yarn with UV-stabilized yarn that can withstand UV radiation.

After knitting, the difference between the fabric on the front face and the rear is evident (Fig. 3). The rear is flat and tight, and the front is loose with extra fabric that will be inflated when the mould is filled with concrete. The tuck stitches are only visible at the rear face (Fig. 3c).

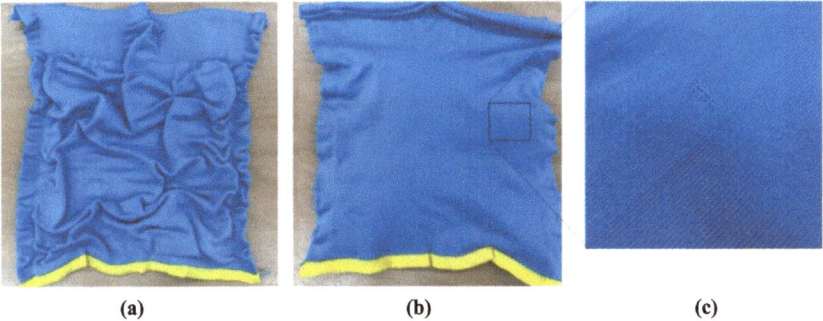

(a)   (b)   (c)

Fig. 3. The front (a) and back (b) fabric faces, (c) a detail showing the connecting tuck stitches on the back side of the fabric.

3.3 Casting Setup

In this stage, a designated method for casting cementitious material into the knitted moulds to produce the tiles was developed. Since the textile itself is soft, stabilizing the form before casting is required. We developed a frame (jig) for holding the knitted moulds while casting and allowing the mould to be filled with cementitious matrix using gravity. The jig is composed of a base part with an array of nails around its edges and a frame with holes (Fig. 4a). The knitted fabric is impaled on the base jig (Fig. 4b). The frame is clamped on top, locking the knitted mould in place. A gap in the frame allows the insertion of a tube threaded into an opening in the fabric (Fig. 4c). The frame with the fabric lays horizontally while the matrix flows through the tube, filling the mould with the force of gravity. A vibration table may be used to improve the flow of the matrix into the mould, depending on the chosen cementitious material. The material is distributed evenly, inflating the moulds around the tuck stitches, thus generating the final tile geometry (Fig. 4d).

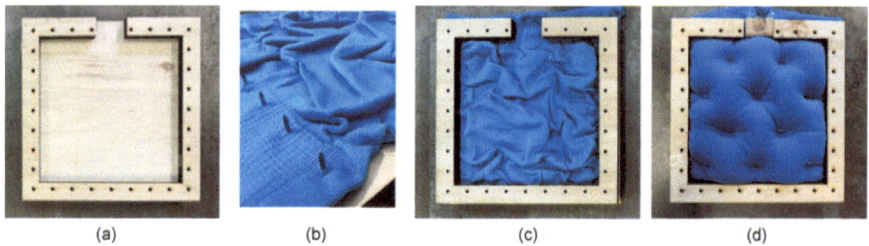

(a)   (b)   (c)   (d)

Fig. 4. The casting setup: the frame (jig) (a). The knitted mould is impaled on the frame (b). The frame top frame is clamped, locking the fabric in place (c). The mould after casting (d).

The thickness of the yarn that is being used determined the infiltration of the wet cementitious material through the small gaps between the knitting loops. Thicker yarn will prevent infiltration, but if a thinner yarn is used, tiny drops of material will infiltrate, creating a texture of droplets on the tile's surface (Fig. 5).

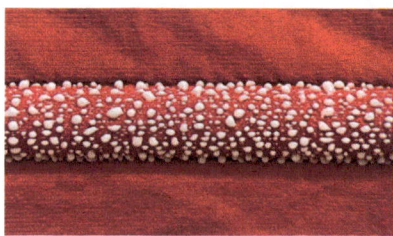

Fig. 5. Droplet texture on the fabric surface

4 Results

Various configurations of 3D fabrics were created, featuring distinct morphological characteristics such as linear or pointed cavities, protrusions, and perforations (see Fig. 5). An iterative computational simulation process tested the visual expression of the fabric after casting (see examples in Fig. 6).

In this preliminary stage, the selected tiles were fabricated with gypsum as the primary material under controlled conditions. A definitive formulation of a cementitious material mixture will be developed at a later stage of the research. The initial tiles were cased using a vibrating platform to ensure uniform material distribution. Subsequent

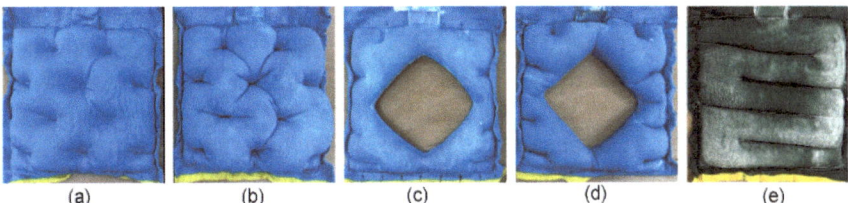

(a)   (b)   (c)   (d)   (e)

Fig. 6. Examples of knitted tiles

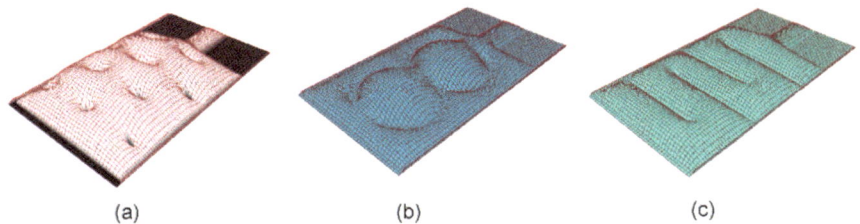

(a)   (b)   (c)

Fig. 7. Examples of physical simulation of 3D knitted tiles

casting procedures revealed that all cavities were filled without the need for vibration assistance.

To validate the simulation tool, the actual cast moulds were compared to the digital simulations (Fig. 7). The comparison shows that in terms of volume, the tiles without the holes (Fig. 6a, b, e) yield similar volume parameters to the simulation, while the tiles with the holes (Fig. 6c, d), further research is needed to understand the significant difference between the simulation and the actual tile (Table 1).

Table 1. A comparison between simulation estimation and actual dimensions of tiles

Type	Figure	Calculated volume (cm^3)	Measured volume (cm^3)	Difference in %
Cavities (stitch length value 11)	5.a	3270	3240	0.9
Cavities (stitch length value 12)	5.b	2500	1990	25.6
Diamond Hole (stitch length value 11)	5.c	790	1820	56.6
Diamond Hole (stitch length value 12)	5.d	630	1710	63.1
Linear bumps	5.e	1730	1710	1.2

A chosen sample (Fig. 6a) was utilized to assess the three-dimensional (3D) impact of various modifications in knitting techniques. Altering the stitch length parameter was observed to regulate the degree of inflation within the fabric. The results indicate that the fabric protruded beyond the mould boundary at a model height of approximately 5 cm (refer to Fig. 8c). Notably, the existing physical simulation framework needs to be revised to account for this phenomenon, as it primarily relies on gravitational forces and examines the model in a conventional orientation. Additional experimentation is required to ascertain whether casting in an inverted orientation yields improved congruence with computational simulations.

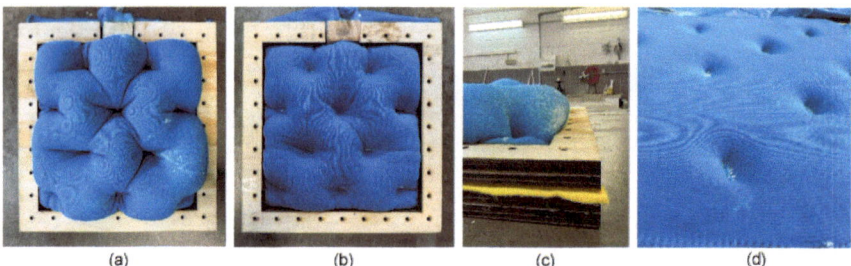

Fig. 8. Two knitted moulds with the same design but with different stitch length values. (a) assigned with a stitch length value of 12; (b) a stitch length value of 11; (c, d) the protrusion beyond the mould boundary in 12 stitch length.

Two additional intriguing phenomena were observed in the 3D tiles depicted in Fig. 8. Firstly, the proximity of cavity stitches to the edge is observed to induce a constrained flow of material, consequently reducing the model's height in the affected regions. Secondly, it is noted that the model's height diminishes towards the termination of a casting line despite a consistent amount of material being present, as evidenced by the model in Fig. 8a. Future research needs to examine ways to compensate for this phenomenon in order to achieve a constant width of material. Furthermore, an unexpected occurrence is evident on the backside of all the models, originally intended to remain flat but exhibiting small cavities in the physical model (refer to Fig. 8d). This occurrence can be attributed to the tension exerted by the stitches due to the material's presence in the protrusions. Subsequent investigations are needed to refine the simulation model to encapsulate these phenomena more accurately.

5   Conclusion

The study introduces a novel methodology for the design, simulation, and fabrication of three-dimensional (3D) 'stay-in-play' knitted moulds intended for building envelope tiles. The research findings underscore the viability of conceiving, knitting, and producing intricate 3D geometries. Various configurations of 3D tile geometries incorporating features such as cavities, bumps, and apertures were subjected to empirical testing, with initial attention directed toward exploring the relationship between stitches and the knitted material. However, an advanced comprehensive investigation is needed to systematically scrutinize the interdependencies between diverse stitch types and the resultant geometry they engender.

Regarding the simulation of 3D geometries, a bespoke workflow was devised utilizing the Kangaroo plugin for Rhinoceros. While this workflow demonstrates proficiency in accurately simulating bumps and protrusions, it encounters challenges in replicating precise outcomes when confronted with models featuring apertures. Therefore, further inquiry is essential to enhance the validation of various facets pertaining to the correspondence between the simulated form and the physical model.

References

Banholzer, B., Brameshuber, W.: Lost formwork elements made of textile reinforced concrete. In: FIB Symposium Keep Concrete Attractive, pp. 351–356 (2005)

Brennan, J., Walker, P., Pedreschi, R., Ansell, M.: The potential of advanced textiles for fabric formwork. Proc. ICE – Constr. Mater. **166**(4), 229–237 (2013)

Hegger, J., Curbach, M., Stark, A., Wilhelm, S., Farwig, K.: Innovative design concepts: application of textile reinforced concrete to shell structures. Struct. Concr. **19**, 637–646 (2018)

Gries, T., Raina, M., Quadflieg, T., Stolyarov, O.: Manufacturing of textiles for civil engineering applications. In: Textile Fibre Composites in Civil Engineering, pp. 3–24 (2016)

Hofmann, M., Albaugh, L., Sethapakadi, T., Hodgins, J., Hudson, S.E., McCann, J., Mankoff, J.: KnitPicking textures: programming and modifying complex knitted textures for machine and hand knitting. In: Proceedings of the 32nd Annual ACM Symposium on User Interface Software and Technology, pp. 5–16 (2019)

Karmon, A., Sterman, Y., Shaked, T., Sheffer, E., Nir, S.: KNITIT: a computational tool for design, simulation, and fabrication of multiple structured knits. In: Proceedings of the 2nd ACM Symposium on Computational Fabrication, pp. 1–10 (2018)

Koutas, L.N., Tetta, Z., Bournas, D.A., Traintafillou, T.C.: Strengthening of concrete structures with textile reinforced mortars: state-of-the-art review. J. Compos. Constr. **23**(1), 03118001 (2019)

Lieboldt, M., Helbig, U., Engler, T.: Textile reinforced concrete multilayer composite pipes. In: Proceeding of the ICTRC' 2006–1st International RILEM Conference on Textile Reinforced Concrete, pp. 369–378 (2006)

McCann, J., et al.: A compiler for 3D machine knitting. ACM Trans. Graphics (TOG) **35**(4), 1–11 (2016)

Nawy, E.G.: Concrete construction engineering handbook. CRC Press (2008)

Oghazian, F.: Calibrating a formfinding algorithm for simulation of tensioned knitted textile architectural models. In: van Ameijde, J., Gardner, N., Hyun, K.H., Luo, D., Sheth, U. (Eds.), POST-CARBON - Proceedings of the 27th CAADRIA Conference, Sydney, 9–15 April 2022, pp. 111–120. CUMINCAD (2022)

Wall, P.: Concreteworks. (n.d.). Retrieved March 21, 2024, from https://www.concreteworks.com/projects/p-wall

Peled, A., Bentur, A., Mobasher, B.: In: Textile Reinforced Concrete. Modern Concrete Technology 19. CRC Press Taylor & Francis Group (2017)

Perry, G., Goldfeld, Y.: Design methodology for TRC pipes: experimental and analytical investigations. Mater. Struct. **54**, 181(1–20) (2021)

Popescu, M.A.: KnitCrete: Stay-in-place knitted formworks for complex concrete structures. ETH Thesis (2019)

Popescu, M., Rippmann, M., Liew, A., Reiter, L., Flatt, R.J., Van Mele, T., Block, P.: Structural design, digital fabrication and construction of the cable-net and knitted formwork of the KnitCandela concrete shell. Structures **31**, 1287–1299 (2021)

Shames, A., Horstmann, M., Hegger, J.: Experimental investigations on textile-reinforced concrete (TRC) sandwich sections. Compos. Struct. **118**, 643–653 (2014)

Tysmans, T., Adriaenssens, S., Wasriels, J.: Form finding methodology for force-modelled anti-clastic shells in glass fibre textile reinforced cement composites. Eng. Struct. **33**, 2603–2611 (2011)

Waimer, F., Knippers, J.: A novel hybrid composite construction for complex concrete shells in architecture. In: 'Proceedings of IASS Annual Symposia', Vol. 28, International Association for Shell and Spatial Structures (IASS), pp. 1–11. June 8, 2022, 11:10 PM (2015)

Leveraging Motion Capture System for High Accuracy AR-Assisted Assembly

Hanning Liu, Xingjie Xie, Yujiao Li, Xiaofan Gao, Honglei Wu, Yao Zhang, and Philip F. Yuan(✉)

College of Architecture and Urban Planning (CAUP), Tongji University, Shanghai, China
philipyuan007@tongji.edu.cn

Abstract. Augmented Reality (AR) allows workers to construct buildings accurately and intuitively without the need for traditional tools like 2-D drawings and rulers. However, accurately tracking worker's pose remains a significant challenge in existing experiments due to their continuous and irregular movement. This research discusses a series of methods using cameras and algorithms to achieve the 6-DoF pose tracking function and reveal the relationship between each method and corresponding tracking accuracy in order to figure out a robust approach of AR-assisted assembly. This paper begins with a consideration of the possible limitations of existing methods including the image drift associated with visual SLAM and the time-consuming nature of fiducial markers. Next, the entire hardware and software framework was introduced, which elaborates on how the motion capture system is integrated into the AR-assisted assembly system. Then, some experiments have been carried out to demonstrate the connection between the system set up and pose tracking accuracy. This research shows the possibility to easily finish assembly task based on AR technology by integrating motion capture system.

Keywords: Augmented Reality · Pose tracking · AR-assisted assembly · Motion capture system · Freeform steel structure

1 Introduction

The Architecture, Engineering, and Construction (AEC) industry is embracing augmented reality (AR) technologies to speed up the design process, improve the construction quality and ensure the safety of construction workers [1]. Specifically, in the field of AR-assisted construction processes, workers are allowed to construct buildings accurately and intuitively without the need for traditional tools like 2-D drawings and rulers [2]. This not only enhances convenience for workers but also facilitates the observation of streamlined or parametrical designs that may be challenging to articulate through orthographic projection and dimensional annotation [3].

Moreover, there have already been numerous finished art installation practices worldwide utilizing AR technology, featuring variant materials and tectonic approaches such as a bent steel pipe pavilion [4], a lightweight timber structure [6], a collaboratively crafted bamboo weaving sculpture [9], etc. These practices are made possible by overlaying

© The Author(s) 2025
H. Chai et al. (Eds.): CDRF 2024, *Symbiotic Intelligence*, pp. 77–90, 2025.
https://doi.org/10.1007/978-981-96-3433-0_8

digital models onto real-world construction sites, facilitated by 6-DoF pose tracking advancements.

Robust 6-DoF pose tracking is a long-standing and well-established area of AR research [10]. From some of the earliest work on computer vision algorithms such as Direct Linear Transform (DLT) and Perspective-n-Point (PnP) [7, 8], to more recent work on multi-sensor data fusion approaches, integrating various types of data including those from inertial measurement units (IMUs) and simultaneous localization and mapping (SLAM) systems [11].

Despite much technical progress, however, the majority of AR-assisted construction projects nowadays still face a pose tracking issue and are only feasible for scenarios involving building materials and tectonics with significant tolerance allowances. In this paper, we investigate how to effectively improve the accuracy between digital overlays and the physical entities by integrating a 6-DoF motion capture pose tracking system into the assembly construction process. A handle-like AR device equipped with a power unit, computing unit, display unit and camera unit with highly reflective markers attached is introduced. This research shows that it outperforms more commonly used pose tracking methods in terms of accuracy, pose tracking scope, and assembly time consuming.

2   Related Works

In the 6-DoF pose tracking research field, two primary strategies are employed to tackle this challenge: marker-less and marker-based. The marker-less strategy commonly utilizes visual SLAM and multi-sensor data fusion technology [12]. The marker-based strategy relies on highly reflective markers or fiducial markers. Fiducial markers typically involve ArUco Marker and AprilTags [13, 14].

Visual SLAM algorithm tracks 6-DoF pose by analyzing how feature points within the environment change over time [15]. However, construction sites are often open and empty, present repeating elements and dynamic changes, which pose significant challenges to accurate pose tracking. Thus, existing research indicates that marker-less tracking systems are susceptible to error accumulation over extended periods [16].

Compared to visual SLAM algorithm, fiducial marker detection and pose estimation typically require fewer computational resources, making them suitable for real-time applications on mobile devices. Besides it is much cheaper than expensive pose tracking device [18]. Highly reflective markers, once attached to the device requiring pose tracking, remain in place and require no further adjustments, offering convenience and saving considerable time [17].

The subsequent section presents a detailed discussion on the feasibility of three distinct marker-based strategies: fiducial markers only, visual SLAM combined with fiducial markers and motion capture systems combined with highly reflective markers.

2.1   Fiducial Markers Only

An experiment using a designer-friendly Grasshopper plugin is conducted to test the accuracy and stability of fiducial markers-only strategy. While it exhibits impressive result when the camera is close to fiducial markers, its performance suffers under real-world conditions. Camera shake or increased distance from markers leads to frequent loss of camera pose, hindering its effectiveness. Additionally, the system's dependence on markers within the camera's view makes it unusable in common assembly scenarios where markers not always be seen. So fiducial markers only strategy isn't robust enough to be used in AR-assisted assembly scenario.

2.2 Visual SLAM and Fiducial Markers

A method of combining multiple fiducial markers with multi-sensor data fusion approaches has been proposed to enhance 6-DoF pose tracking accuracy [5]. However, the localized property of fiducial markers makes them can only enhance pose tracking accuracy within the vicinity of where the markers are placed. As a result, after completing construction tasks in one area, the process of setting up a new set of markers for each subsequent assembly area is necessary to achieve the desired level of accuracy. So, the time-consuming process of setting up a large number of fiducial markers needs to be optimized.

Besides, an experiment using Fologram on a mobile phone is conducted to evaluate the accuracy of the visual SLAM combined with fiducial markers strategy. The structure in this experiment consists of many round pipes and a concrete base, standing 3 m tall. As fiducial markers can only be attached to flat planes, they are restricted to the ground and cannot be mounted on the round pipes, leading to reduced accuracy at high positions. Additionally, the rough surface of the concrete base further diminishes accuracy compared to an ideal scenario. So, the stringent requirements of fiducial markers limit the scope of 6-DoF pose tracking and introduce unforeseen deviations.

2.3 Motion Capture System and Highly Reflective Markers

The motion capture system provides 6-DoF pose tracking with deviations of less than 1mm, establishing it as a widely accepted benchmark in research papers for ground truth data comparison [17]. This system enables pose tracking across its entire field of view, unhindered by non-flat elements. Additionally, the highly reflective markers attached to the device eliminate the need for frequent setup. However, despite its prominence, there is currently no research integrating the motion capture system with an image overlay feature to enable AR-assisted assembly work.

3 Methods

The following experiments were conducted in an 8.0 * 7.6m site (Fig. 1), using 6-DoF motion capture system provided by NOKOV and a custom-designed AR device. The 6-DoF motion capture system comprises 8 high-speed infrared optical lenses, each equipped with numerous infrared LED lights mounted on their panels. These lights illuminate highly reflective markers, which then reflect light back to the lenses, enabling the calculation of the 2D coordinates of each marker. Finally, by integrating data from multiple lenses capturing different perspectives, the system computes the 3D coordinates of each highly reflective marker.

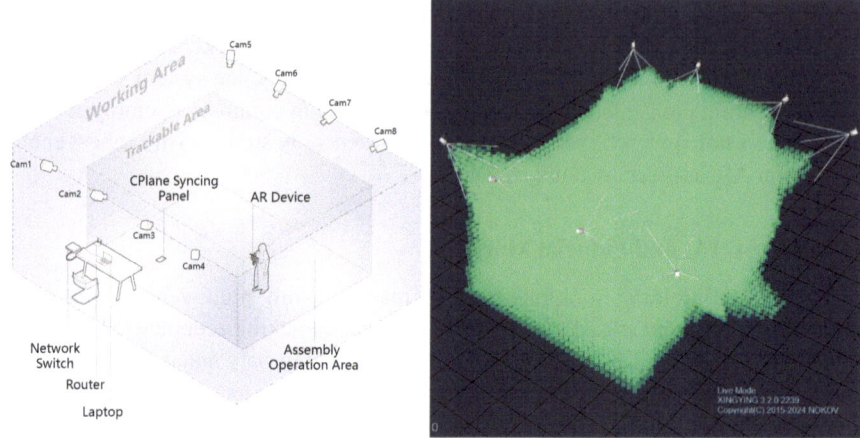

Fig. 1. AR-assisted assembly site setup (Left) and 6-DoF pose tracking area (Right)

With three or more markers, a rigid body can be defined within the motion capture system. Rigid bodies derive their pose from the average position of all markers and are resilient to the loss of individual markers, resulting in enhanced stability and reliability. The coordinate system of the motion capture system is established during the calibration phase and remains fixed thereafter, yet it becomes invisible in the real-world space post-calibration. To ensure synchronization between real-world and digital spaces, a rigid body comprising four markers acts as the reference coordinate system throughout the tracking process.

The custom-designed AR device is comprised of five essential units (Fig. 2): an image-capturing unit, a 6-DoF pose tracking unit, a computation unit, a display unit, and a power unit. The image-capturing unit features a 5-million-pixel camera tasked with capturing real-world imagery. Positioned in close proximity to the 6-DoF pose tracking unit, its captured images are overlaid with those captured by the digital camera. The 6-DoF pose tracking unit is equipped with five highly reflective markers. Three of these markers are strategically positioned near the image-capturing unit, facilitating straightforward calculation of their transform relations. The remaining two markers are placed on the left and right sides of the device to enhance visibility to the high-speed infrared optical lenses. The computation unit, a compact Intel NUC, serves as the core of the custom-designed AR device. It manages all tasks related to receiving 6-DoF pose data, rendering digital models, and overlaying images. The display unit is a screen with a resolution of 1024 * 768, perfectly matching the aspect ratio of the camera image sensor, enabling full-screen display of images. The power unit consists of a Li-Po battery with a capacity of 4000 mAh, providing continuous operation of the device for over 4 h.

This AR-assisted assembly system comprises two computers. The first computer connects to the network switch of the motion capture system via a cable, while the other serves as the core of the custom-designed AR device. Each computer runs a different software framework, together forming the system. The software framework of the first computer includes Rhino & Grasshopper and necessary tools to process data from

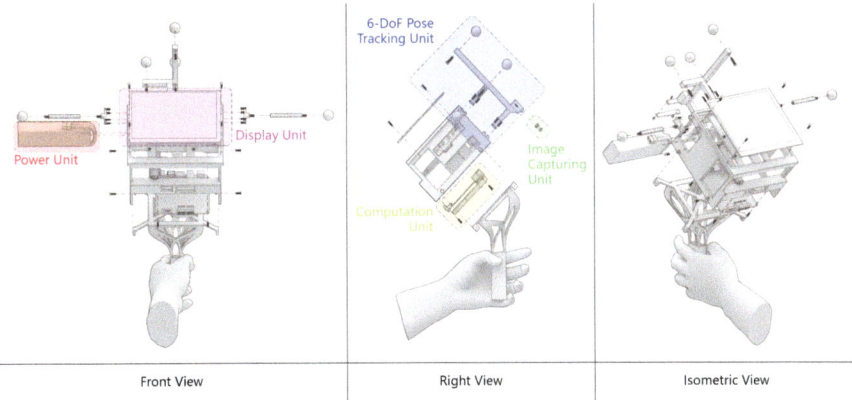

Fig. 2. Custom-designed AR device with five units

high-speed infrared optical lenses. A self-developed plugin for Grasshopper facilitates the transfer of 6-DoF pose data from the NOKOV SDK to the Rhino & Grasshopper platform. Subsequently, the data is transmitted to the AR device via UDP and local Wi-Fi. The software framework of the second computer revolves around Unity. Rhino & Grasshopper on this computer function as a plugin for Unity, with the assistance of the open-source Rhino Inside project. It receives pose data transmitted from the first computer via UDP, then utilizes the Callback method provided by Rhino Inside to establish the pose of the digital camera in Unity and synchronize digital model information.

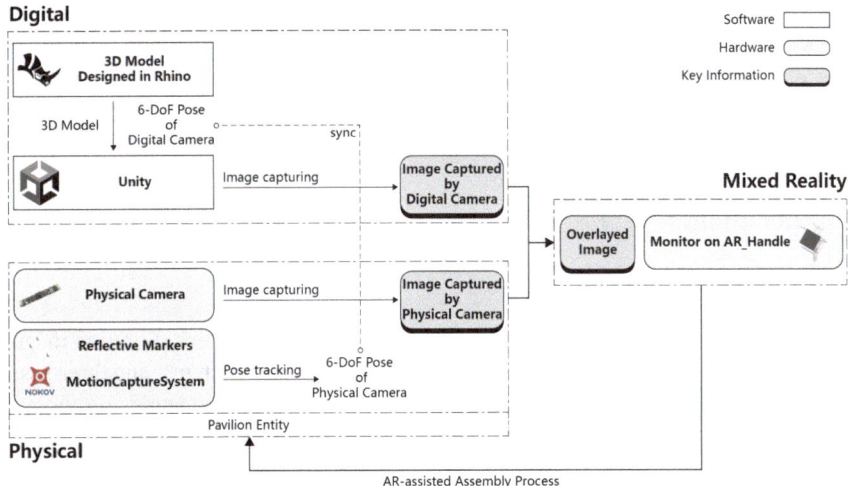

Fig. 3. AR-assisted assembly workflow

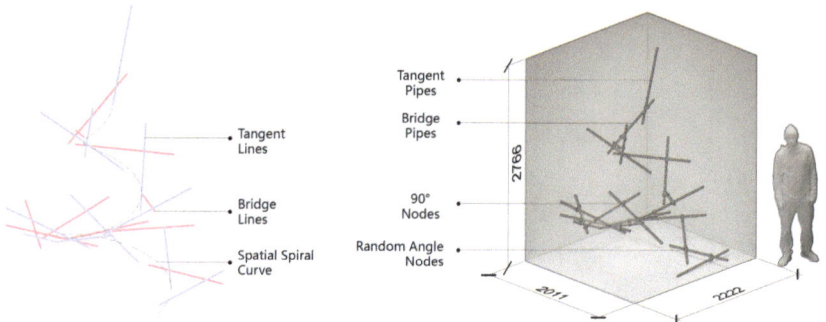

Fig. 4. Design diagram (Left), model in digital environment (Right)

The AR-assisted assembly workflow encompasses three environments (Fig. 3): the digital environment, the physical environment, and the mixed reality environment. Designers begin by completing their design tasks in the digital environment. They then synchronize the digital and physical environments using the motion capture system to create a mixed reality environment. With the aid of the mixed reality environment, designers can achieve results with greater accuracy and intuition than was previously possible.

The assembly experiment pavilion design in this research is tailored to suit the need for testing the accuracy of the AR-assisted assembly workflow (Fig. 4). It comprises 21 aluminum pipes, each 1 m in length, and 20 connection nodes capable of rotating 360 degrees. Among the 21 pipes, 11 form tangent lines of a spatial spiral curve, while the remaining 10 act as bridges to connect with pipes ahead and behind them separately. Locating the positions of these pipes without AR assistance is challenging, allowing for the assessment of our workflow's effectiveness. The height of the pavilion is set to about 2.8 m to test the 6-DoF pose tracking accuracy far from the ground. The connection nodes' angles are specially configured, with 10 set at 90 degrees and the remaining 10 at random angles, enabling the assembly process to be divided into multiple subprocesses.

4   Experiments

4.1   AR-Assisted Assembly Process

The experiment comprises two preparation steps: camera parameter calibration in Unity, synchronization between the real-world and digital coordinate systems; and three assembly steps: assembly of the orthogonal pipes, fixation of the rotation angles for the nodes with random angles, and the final assembly of the pavilion.

At the beginning of the experiment series, the first step involves calibrating the parameters of the physical camera to prevent image distortion and ensure seamless overlay with the digital camera in Unity (Fig. 5). This is achieved using a custom-developed Grasshopper plugin integrated with algorithms from the OpenCV library to estimate the camera's pose. Subsequently, Rhino Inside is employed to synchronize model information and camera pose with Unity. Finally, adjustments are made to the camera parameters of the digital camera in Unity to achieve perfect alignment between the images from the physical and digital cameras.

The second step is aligning the coordinate system between real-world and digital space (Fig. 6). A panel equipped with four highly reflective markers is initially defined as a rigid body. The motion capture system recognizes it and provides pose information as a reference construction plane in Rhino. Subsequently, digital models are oriented with respect to the reference construction plane before being sent to Unity.

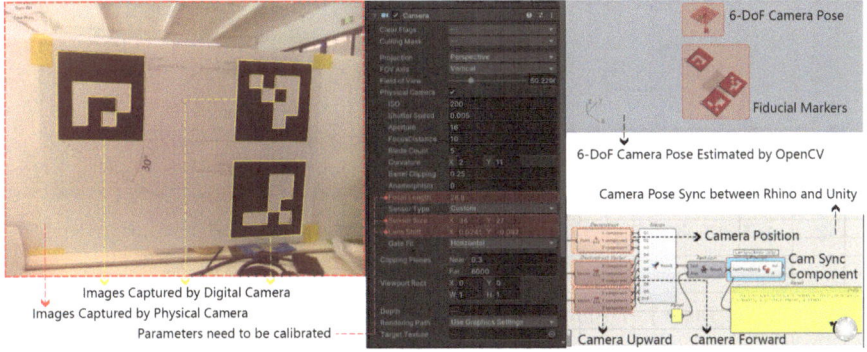

Fig. 5. Overlaid images in Unity (Left), camera parameters in Unity (Middle), 6-DoF pose estimate and sync (Right)

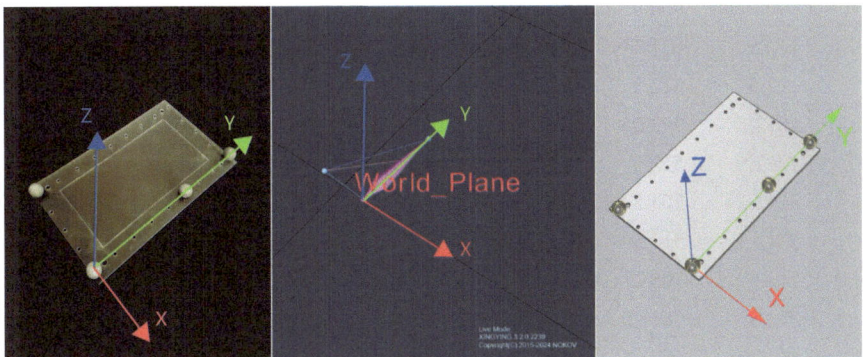

Fig. 6. Photo of the CPlane Syncing Panel (Left), panel tracked by motion capture system (Middle) and panel in Rhino space (Right)

In the first step of assembly process, the orthogonal pipes are assembled (Fig. 7). The pavilion comprises 21 pipes, with 11 aligned tangentially to a spiral curve, and 10 acting as bridge pipes connecting the tangential ones. Each tangential pipe is perpendicular to its corresponding bridge pipe, collectively forming an assembly element. The position of the connection node, represented by one-dimensional data—the distance between the node and the end of each pipe—is the only variable among these elements. Initially, all the orthogonal elements are arranged in the Rhino space, then a custom-designed AR device is used to precisely locate the connection node in physical space. After confirming the overlaid images, an additional deviation check is performed solely using the motion capture system to evaluate the deviation with ground truth. The results of the check indicate deviations of less than 3.2mm, with an average value of 1.835mm (Table 1).

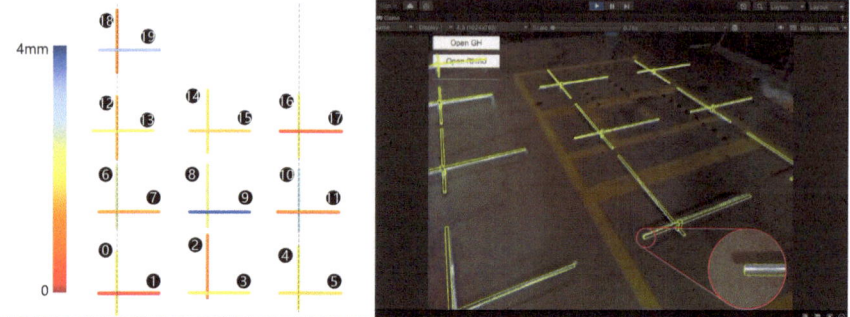

Fig. 7. Position deviation of each pipe (Left), deviation check of overlaid image (Right)

Table 1. Position deviation of the middle point of each pipe

Pipe index	X value of ideal middle point/mm	Y value of ideal middle point/mm	Z value of ideal middle point/mm	X value of real middle point/mm	Y value of real middle point/mm	Z value of real middle point/mm	Error/mm
0	112.5	671.34	20	113.4	672.6	21.07	1.89
1	312.5	512.5	65	313.08	512.98	64.87	0.76
2	1612.5	942.05	20	1612.75	941.74	18.68	1.38
3	1812.5	512.5	65	1812.42	511.41	66.48	1.84
4	3112.5	754.54	20	3112.09	752.66	20.28	1.94
5	3312.5	512.5	65	3311.76	513.84	64.08	1.78
6	112.5	2055.11	20	111.43	2055.66	21.89	2.24
7	312.5	1812.5	65	311.1	1812.27	65.69	1.58
8	1612.5	2071.66	20	1610.77	2070.64	19.49	2.07
9	1812.5	1812.5	65	1814.45	1810.69	63.29	3.15
10	3112.5	2007.19	20	3114.12	2008.6	21.1	2.41

(continued)

Table 1. (*continued*)

Pipe index	X value of ideal middle point/mm	Y value of ideal middle point/mm	Z value of ideal middle point/mm	X value of real middle point/mm	Y value of real middle point/mm	Z value of real middle point/mm	Error/mm
11	3271.5	1812.5	65	3272.79	1813.12	64.9	1.43
12	112.5	3167.43	20	113.46	3167.27	18.7	1.62
13	210.5	3112.5	65	211.13	3111.55	66.51	1.89
14	1612.5	3260.63	20	1612.8	3258.89	20.31	1.79
15	1812.5	3112.5	65	1812.47	3113.98	64.11	1.72
16	3112.5	3193.72	20	3112.14	3194.41	21.91	2.06
17	3312.5	3112.5	65	3311.81	3112.41	65.72	1.00
18	112.5	4550.93	20	111.49	4550.05	19.52	1.43
19	312.5	4412.5	65	311.16	4410.83	63.32	2.72
Avg	/	/	/	/	/	/	1.835

After accurately placing the orthogonal elements, the assembly process enters its second step, utilizing random rotation nodes to connect the elements sequentially (Fig. 8). During this stage, two random parameters are considered: the rotation angle and the length of the node between itself and the end of each pipe. The rotation parameter can be predetermined with the help of overlaid images provided by AR technology, adding convenience to the final assembly process. After confirming the overlaid images, an additional deviation check is also performed using the motion capture system to evaluate the deviation with ground truth. The angle deviations are less than 0.2 degree, with an average value of 0.116 degree (Table 2).

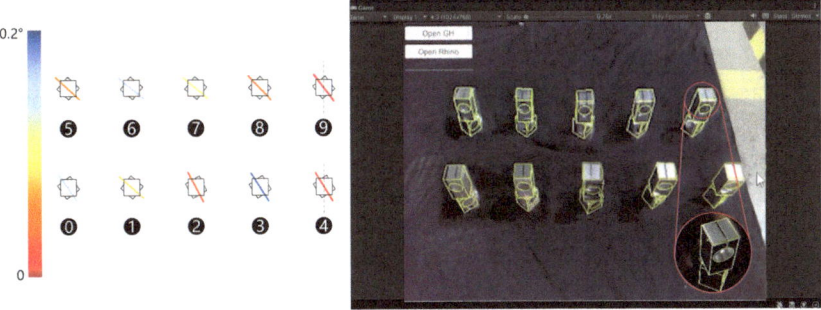

Fig. 8. Angle deviation of each random node (Left), deviation check of overlaid image (Right)

In the final step of the pavilion assembly (Fig. 9), the process is simplified similar to the first step. The only factor to consider is the distance between the node and the

Table 2. Angle deviation of the random nodes

Node index	Ideal angle/°	Real angle/°	Error/°
0	144.46	144.57	0.11
1	130.23	130.19	0.04
2	151.55	151.42	0.13
3	146.98	147.17	0.19
4	147.79	147.63	0.16
5	131.97	131.89	0.08
6	128.79	128.91	0.12
7	129.24	129.22	0.02
8	136.05	135.94	0.11
9	142.38	142.18	0.20
Avg	/	/	0.116

end of each pipe, allowing the pavilion to be easily built with the assistance of AR technology theoretically. Due to the pipes being randomly scattered in the air, and the lack of consideration and calculation of structural stability and deformation beforehand, the pipes are unable to remain stable in their ideal positions, even with the support of some extra vertical pipes. However, the pipes still maintain their correct orientation, experiencing a positional deviation of less than 22 mm and 0.90 degrees, with average values of 10.31 mm and 0.49 degrees (Table 3).

4.2 Deviation Analysis

A method employing the motion capture system is utilized to obtain the 3D coordinate data of the constructed pavilion's pipes and nodes (Fig. 10). Then the position and angle deviation data of the designed and constructed pavilion's pipes and nodes are compared to test the accuracy of the AR-assisted assembly workflow.

In steps 1 and 2, the pipes exhibit a position deviation of less than 3.2 mm, while the nodes demonstrate an angle deviation smaller than 0.2 degrees. Furthermore, the outlines of the digital models align perfectly with the models in the physical environment. The deviation check results underscore the high accuracy of integrating the motion capture system into the AR-assisted assembly workflow.

In step 3 of the assembly process, the position deviation is less than 22 mm, with the angle deviation smaller than 0.9 degrees. And some pipes are lower than the digital models in the overlaid images. The larger deviation observed in this step is primarily attributed to the deformation of the cantilever parts within the structure.

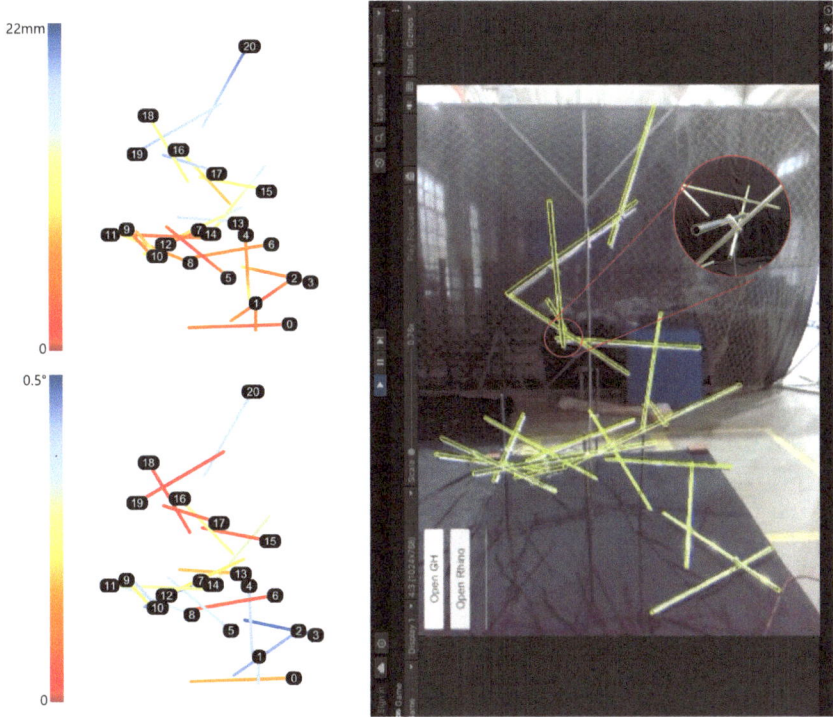

Fig. 9. Position and angle deviation of each pipe (Left), deviation check of overlaid image (Right)

Table 3. Position and angle deviation of all pipes

Pipe index	X value of ideal middle point/mm	Y value of ideal middle point/mm	Z value of ideal middle point/mm	X value of real middle point/mm	Y value of real middlepoint/mm	Z value of real middle point/mm	Position error/mm	Angle error/°
0	1339.3	179.77	3.62	1341.75	182.69	182.69	4.71	0.34
1	1410.39	434.67	44.89	1406.8	429.84	429.84	6.09	0.68
2	1359.58	765.08	121.63	1360.79	764.52	764.52	4.10	0.80
3	1308.16	1200.83	269.63	1309.08	1200.71	1200.71	6.86	0.90
4	921.51	1298.94	307.52	922.22	1299.91	1299.91	7.74	0.73
5	296.85	1244.92	419.48	296.41	1243.11	1243.11	1.86	0.64
6	649.51	1379.48	427.08	645.86	1385.71	1385.71	7.37	0.12
7	80.19	1061.31	560.54	78.49	1061.47	1061.47	5.32	0.82

(*continued*)

Table 3. (*continued*)

Pipe index	X value of ideal middle point/mm	Y value of ideal middle point/mm	Z value of ideal middle point/mm	X value of real middle point/mm	Y value of real middlepoint/mm	Z value of real middle point/mm	Position error/mm	Angle error/°
8	20.34	670.36	578.94	27.16	670.01	670.01	9.37	0.63
9	24.81	378.29	724.35	23.84	373.99	373.99	5.86	0.43
10	280.21	184.1	739.41	278.63	183.6	183.6	11.12	0.74
11	469.62	83.7	858.32	473.08	82.98	82.98	5.08	0.54
12	929.74	191.92	977.18	928.08	189.5	189.5	12.38	0.42
13	961.21	290.38	1024.56	965.25	290.57	290.57	17.43	0.36
14	1066.07	777.56	1168.04	1062.98	782.49	782.49	14.81	0.58
15	908.23	934.47	1296.95	909.7	934.29	934.29	12.79	0.13
16	511.77	1018.17	1351.69	515.81	1021.19	1021.19	11.02	0.53
17	444.34	866.86	1493.02	446.24	860.12	860.12	19.92	0.15
18	424.31	417.73	1695.77	430.85	413.58	413.58	12.45	0.11
19	544.85	442.4	1921.34	540.22	445.14	445.14	19.16	0.05
20	911.84	802.93	2348.31	913.27	798.43	798.43	21.08	0.67
Avg	/	/	/	/	/	/	10.31	0.49

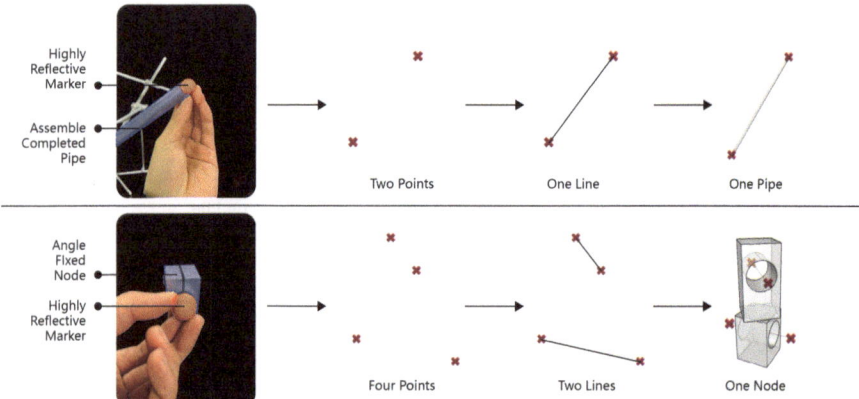

Fig. 10. Method used to get the 3D coordinate data of the constructed pavilion's pipes and nodes.

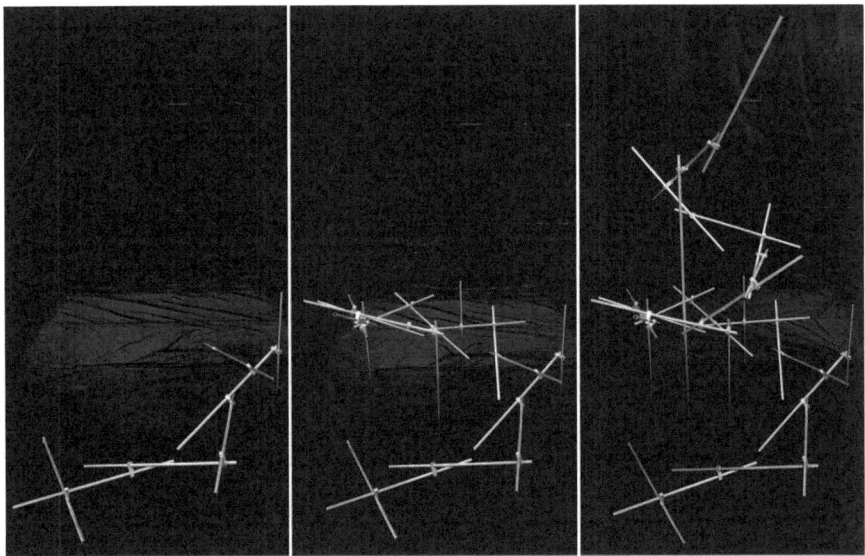

Fig. 11. Assembly progress 1/3 (Left), assembly progress 2/3 (Middle) and final assembly result (Right)

5 Conclusion and Discussion

This research pioneers the integration of a 6-DoF motion capture system into the workflow of AR-assisted assembly. It develops both a software framework and a custom-designed hardware device to implement image overlaying, demonstrating the workflow's high accuracy. Moreover, it provides precise pose tracking data within a larger spatial scope beyond the typical confines of fiducial markers, thus streamlining the assembly process and saving time overall. Optimization of structural design could mitigate deviations even more. Additionally, improvement of the custom-designed device is necessary to free up the hand, thereby paving the way for more precise and intuitive assembly workflows in the future (Fig. 11).

References

1. Yang, S., Richard, K., Shan, L.: Review and analysis of augmented reality (AR) literature for digital fabrication in architecture. Autom. Constr. **128**, 103762 (2021)
2. Côté, S., Myriam, B., Antoine, G., et al.: A live augmented reality tool for facilitating interpretation of 2D construction drawings. In: Augmented and Virtual Reality, 8853:421–27. Lecture Notes in Computer Science. Springer International Publishing, Cham (2014)
3. Gwyllim, J., Cameron, N., Nick, B.: Collaborative fabrication in mixed reality. In: Data, Matter, Design, 1st ed., pp. 239–47. Routledge (2020)
4. Gwyllim, J., Cameron, N., Matthew, B.: Making in mixed reality. In: Holographic Design, Fabrication, Assembly and Analysis of Woven Steel Structures, pp. 88–97. Mexico City, Mexico (2018)

5. Alexander, H.K., Arvin, H.X., Gwyllim, J., et al.: Augmented reality for high precision fabrication of glued laminated timber beams. Autom. Constr. **152**, 104912 (2023)
6. Sining, W., Dandan, L., Lujie, S.: Human-cyber-physical system for post-digital design and construction of lightweight timber structures. Autom. Constr. **154**, 105033 (2023)
7. Vincent, L., Francesc, M.N., Pascal, F.: EPnP: an accurate O(n) solution to the PnP problem. Int. J. Comput. Vision **81**(2), 155–166 (2009)
8. Xiao-Shan, G., Xiao-Rong, H., Jianliang, T., et al.: Complete solution classification for the perspective-three-point problem. IEEE Trans. Pattern Anal. Mach. Intell. **25**(8), 930–943 (2003)
9. Garvin, G., Kristof, C.: Augmented Reality-Based Collaboration - ARgan, a Bamboo Art Installation Case Study, pp. 313–322. Bangkok, Thailand (2020)
10. Eric, M., Hideaki, U., Fabien, S.: Pose estimation for augmented reality: a hands-on survey. IEEE Trans. Visualization Comp. Graphics **22**(12), 2633–2651 (2016)
11. Timothy, S., Jonas, B.: Object-based visual-inertial tracking for additive fabrication. IEEE Robot. Autom. Lett. **3**(3), 1370–1377 (2018)
12. Schall, G., Daniel, W., Gerhard, R., et al.: Global pose estimation using multi-sensor fusion for outdoor augmented reality. In: 2009 8th IEEE International Symposium on Mixed and Augmented Reality, pp. 153–162. IEEE, Orlando, FL, USA (2009)
13. Francisco, R., Rafael, M., Rafael, M.: Speeded up detection of squared fiducial markers. Image Vis. Comput. **76**, 38–47 (2018)
14. Maximilian, K., Acshi, H., Edwin, O.: Flexible layouts for fiducial tags. In: 2019 IEEE/RSJ International Conference on Intelligent Robots and Systems (IROS), 1898–1903. IEEE, Macau, China (2019)
15. Jack, C., Keyu, C., Weiwei, C.: Comparison of marker-based and markerless AR: a case study of an indoor decoration system. In: Lean and Computing in Construction Congress - Volume 1: Proceedings of the Joint Conference on Computing in Construction, pp. 483–490. Heriot-Watt University, Heraklion, Crete, Greece (2017)
16. Taihú, P., Thomas, F., Gastón, C., et al.: S-PTAM: stereo parallel tracking and mapping. Robot. Auton. Syst. **93**, 27–42 (2017)
17. Inês, S., Ricardo, S., Marcelo, P., et al.: Accuracy and repeatability tests on HoloLens 2 and HTC Vive. Multimodal Technol. Interact. **5**(8), 47 (2021)
18. Michail, K., Brennan, C., Sabrina, C., et al.: Fiducial markers for pose estimation: overview, applications and experimental comparison of the ARTag, AprilTag, ArUco and STag markers. J. Intell. Rob. Syst. **101**(4), 71 (2021)

Research on 3D-Printed Standardized Small-Scale Architectural Model Joints

Ren Tianye, Chia Hui Yen, Zhan Yucheng, Zhang Jie, and Zhu Ning[✉]

Department of Architecture, Tsinghua University, Beijing, China
zhuning@tsinghua.edu.cn

Abstract. This paper explores the design of standardized 3D-printed joints for small-scale architectural models, using the traditional mortise and tenon joint as a prototype. In comparison to traditional subtractive manufacturing for model production, 3D printing proves beneficial in saving time and materials. Therefore, many architectural students and design firms frequently use 3D printing for making models. Considering the limited size of 3D printers, convenience of transportation, and flexibility in demonstration, it is more advantageous to print model components with assembly joints rather than printing the entire model directly. This paper focuses on small-scale architectural models ranging from 1:50 to 1:200, conducting a typification, standardization and parameterization study of 3D-printed assembly joints. 13 standard joints are designed, forming a database and an interactive application.

Keywords: Architectural model · 3D printing · Assembly joints · Parameterization

1 Introduction

1.1 Background

In the field of architecture, 3D printing technology has brought a lot of convenience to the production process of architectural models. While traditional model making mainly adopts subtractive manufacturing, such as cutting foam board, milling and carving wood, etc. 3D printing, also named additive manufacturing, has significant advantages including high design flexibility, high material utilization, high production efficiency and high precision of details, especially for students' models.

3D printer with large printing area to make a whole model is very expensive. Usually, large objects are broken down into sub-components. The adoption of structural adhesives serves as a key method for joining 3D-printed components, but it brings problems of irreversibility of bonding, adhesive toxicity, limited weather resistance and structural strength (Morano et al. 2024). Finer models usually carry out special interface treatment like expanding the bonding surface (Knoll et al. 2003, pp. 37–38), which inspires us to add joints to interfaces. The advantages of joints include:

© The Author(s) 2025
H. Chai et al. (Eds.): CDRF 2024, *Symbiotic Intelligence*, pp. 91–105, 2025.
https://doi.org/10.1007/978-981-96-3433-0_9

1. Design change: When comparing multiple schemes throughout the design process, the detachable model can be partially modified without remaking an entire one.
2. Convenience of transportation: For large or complex architectural models, it is convenient to transport them as parts to avoid the risk of damage.
3. Flexible demonstration: in the demonstration of internal space, the detachable model can provide viewers with more design details.

Traditional Chinese wooden structures feature rigid joints: mortise and tenon. They are derived from the basic forms of "straight mortise" and "dovetail mortise" (Ye 2017, p. 15). Therefore, starting from these two basic forms, a series of standardized joints can be designed.

1.2 Research Objectives

As shown in Figs. 1, 2, 3, 4, 5 and 6,[1] 3D-printed assembly joints have been utilized in architectural models. Based on the Grasshopper platform, Zuo Wenkang establish a topology optimization framework for joints in spatial structures, which creates optimized performing joint with more aesthetic architectural appearance and lighter weight (2024). Enterprises of intelligent manufacturing and research institutions promote 3D-printed joints' design. Hubs, a digital manufacturing company, addressing material selection, fatigue avoidance, and advantages of 3D printing over injection molding (n.d.); MIT 's Digital Structures Lab study discretized and connected structures to overcome anisotropy limitation of 3D printing (2014). Their another project explores the effect of friction caused by the 3D-printed surface texture, and various wrapping accessories are designed to provide appropriate friction (2015). In addition, mortise and tenon joint has been developing towards intelligent processing. Tang Lin integrates numerical control technology and traditional joint sizing to automate processing (2021). Demi Fang studies the geometric potential and mechanical performance of joinery connections, and develop a joint design framework for optimal rotational stiffness (2018).

     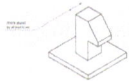   

Fig. 1. 3D-printed optimized joint in gridshell structure   **Fig. 2.** Experience of 3D printing snap-fit joints   **Fig. 3.** Discretized structure to overcome anisotropy

---

[1] Figures 1–6 source from the six references mentioned in this paragraph in sequence. The other figures are from the authors.

Fig. 4. Components wrapping the joint to provide friction

Fig. 5. CNC-machined model of joints

Fig. 6. Comparison of joinery connection geometry

3D-printed joints have various needs in the education, commerce, exhibition and other aspects of architectural design. However, existing parametric design of joints mainly aimed at large-scale models (such as furniture, ancient building construction, etc.). Neglecting models at scales of 1:50 or smaller, which are frequently used for education and exhibition. These small-scale models are composed of components 2–6 mm thick, requiring joint designs tailored to their accuracy and material strength, which cannot be directly adopted from large-scale models.

1.3 Research Contents

In this paper, the typification, standardization and parameterization of joints are studied for 1:50–1:200 architectural models. The three different types of joints, including face joints, line joints, and point joints, along with their dimensional parameters and printing configurations, are compiled into a database. The 1:50 model of villa savoy is taken as an example for 3D printing, in order to verify and demonstrate the research results above. The result shows that, compared with printing the whole model, printing components with joints reduces time, saves material and avoids non-detachable support. Besides, it also creates a structure that exhibits greater ductility because of the gaps between joints (Mueller et al. 2014) by combining traditional woodworking techniques with 3D printing technology, we aim to enhance the quality of architectural models and the simplicity of making models(Fig. 7).

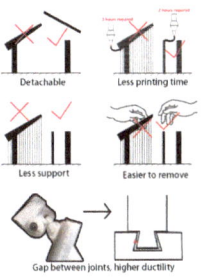

Fig. 7. Advantages of 3D printing for detachable models

2 Method

In the research process, we use *Rhinoceros 7* to do computer modeling and *Creality Slicer* to do model slicing. The printing material is *Ender-PLA* filament. *Creality3D Sermoon V1* is the 3D printer used for the entire process of the research. Considering the maximum printing size of 15 cm × 15 cm, precision of 0.2 mm, and a price range of approximately 2000–3000 RMB, this type of printer is commonly used by students. Our research process includes (Fig. 8):

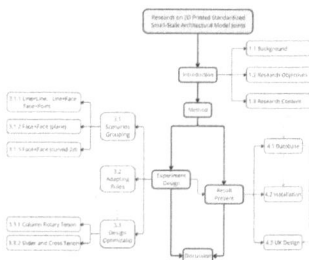

Fig. 8. Research framework

1. Experiment Design:

   - Scenarios Grouping: Categorizing the application scenarios of joints. Research for traditional mortise and tenon joints in corresponding scenarios. Preliminary designing the 3D-printed joints.
   - Adapting rules: Analyzing forms and size rules borrowable from traditional mortise and tenon joints. Summarizing rules for 3D-printed joint design
   - Optimization: Printing models. Conducting physical assembly experiments, and further refining joint design rules based on experimental results. Also optimizing joint design.

2. Result Present:

   - Database: Storing models of all joints as optional printing data. Summarizing dimension parameters, flexibility and printing direction to form a database.
   - Installation: Using the database to print samples—1:50 model of the Savoye Villa. Verifying the convenience and structural stability of joints.
   - UX Design: Compiling the joint generation algorithm to create a prototype design for a Grasshopper plugin.

3 Experiment Design

We have designed 13 joints suitable for various application scenarios, each with different levels of stability. Joint diagrams are presented in Table 1.

Table 1. 3D printing joint design result.

i. Dovetail End Buckle	ii. Large Column Rotary Tenon	iii. Small Column Rotary Tenon	iv. Concealed Magnet	v. Rotary Buckle
Line×Line	Line×Face	Line×Face	Face×Point	Face×Face (plane)
flexible	stable	stable	sup-flexible	stable
vi. Dovetail Tenon Slider	vii. Right-angle Tenon Slider	viii. U-shaped Tenon Slider	ix. Vertical Snap	x. External Corner Cross Tenon
Face×face (plane)	Face×Face (plane)	Face×Face (plane)	Face×Face (plane)	Face×face (plane)
sub-stable	sub-stable	sub-flexible	flexible	flexible
xi. Multi-face Tenon	xii. Curvy Surface Tenon	xiii. Snap-in Column Tenon		
face×face (multiple)	face×face (curved-2d)	Line×Face		
sub-stable	sub-stable	flexible		

3.1 Scenarios Grouping

3.1.1 Line × Line, Line × Face, Face × Point

The junctions in these three scenarios typically converge at a single point and can be applied as Fig. 9.

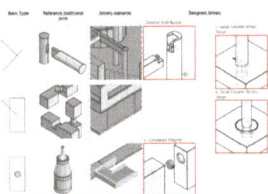

Fig. 9. Application scenarios: Line x Line, Line x Face, Face x Point

3.1.2 Face × Face (Plane)

This type of scenario involves the connection between wall slabs or floor slabs, making it widely applicable. As shown in Fig. 10, we have designed Face × Face (plane) joints at four levels of stability:

1. Stable: *v. Rotary Buckle*. Three steps to disassemble: rotation -- sliding -- pulling out.
2. Sub-stable: *vi. Dovetail Tenon Slider*, *vii. Right-angle Tenon Slider*. Two steps to disassemble: sliding -- pulling out.
3. Sub-Flexible: *viii. U-shaped Tenon Slider*. Two steps to disassemble: sliding -- pulling out, with the potential of a slight rotation during the sliding process.

4. Flexible: *ix. Vertical Snap.* One step to disassemble: pulling out.

Notes: When two walls are not at right angles or straight angle, the two walls should be divided into three parts: two rectangular blocks and a "joining medium" with special angles. The "joining medium" connects at a straight angle with the rectangular slabs so joints above can be used. When three or four walls intersect vertically, a combination of right-angle and straight-angle is required, leading to *xi. Multi-face Tenon.*

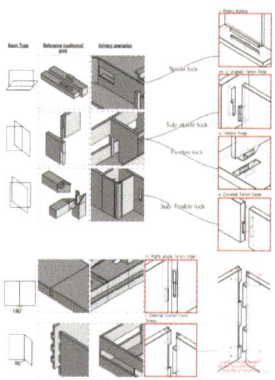

Fig. 10. Application scenario: Face ✕ Face (plane)

3.1.3   Face ✕ Face (Curved-2d)

To solve the problem of connecting curved walls, a program has been developed to automatically search for the joint with the minimum curvature and extract the curvature near the joint to create *xii. Curvy Surface Tenon* (Fig. 11).

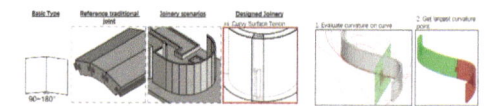

Fig. 11. Application scenario: Face × Face (curved-2d)

3.2   Adapting Rules

Traditional mortise and tenon joints have established rules regarding tenon scale, grain direction's impact on strength, and processing loss relative to tool thickness. For 3D printing, clearances, chamfers, and printing directions are constricted by the printer performance and printing object shape. Comparing tools and materials across two manufacturing methods, along with printing experiments, led to the formulation of 3D printing joint rules.

- Rule 1: For small scaled 3D-printed models, over large tenon may result in thin material on both sides of the mortise, and narrow mortise result in thin tenon, both potentially leading to printing failures and weak structure. Therefore, maintaining balanced thickness for tenon and mortise is important, which means the mortise width should be at around one-third of the component thickness.
- Rule 2: For traditional mortise and tenon joints, wood grain direction exhibits weak shear strength parallel to the grain, making it susceptible to failure, but it is hard to induce shear failure perpendicular to the wood grain. For 3D printing, layer-by-layer accumulation results in material textures similar with wood grain. If tenons and main components consist of different "material layers" and tenon root connecting to the main body is too small, breakage is likely to occur. If the tenon and main components are connected by the same "material layer", the resistance can be enhanced. It was found that the minimum width of the dovetail tenon root should be 0.9 mm when printing horizontally and as fine as 0.2 mm when printing vertically.
- Rule 3: The minimum tolerance in 3D printing is determined by the nozzle diameter of the printer. 3D-printed joints exhibit higher rigidity than wood, so larger tenons cannot be accommodated into smaller mortises through deformation. Therefore, a certain gap should be left. For joints fixed by friction, a side gap of 0.15mm ensures smooth interlocking and allows for mutual contact of surface textures, providing an ideal tolerance. Since the tenon head and mortise base do not contribute to the stability of the joint, the tolerance between them can be greater.
- Rule 4: Joints with angles greater than 45° or arcs reduce printing support for overhanging printing.
- Rule 5: 3D printers may occasionally generate rough surface due to uneven material flow. To reduce the risk of assembly failure, long tenons should be segmented into shorter ones, each measuring 10–20 mm, and arranged with intervals.
- Rule 6: When a single component requires multiple joints at different edges, every connecting edge should withdraw a certain margin to avoid conflicts among joints (Figs. 12, 13, 14, 15, 16 and 17)

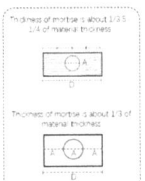

Fig. 12. Rule 1

3.3 Design Optimization

Taking the variants of the traditional dovetail tenon (ii. iii. vi. vii. viii. x. xiii.) as examples, the optimizations are as follows.

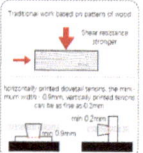

Fig. 13. Rule 2

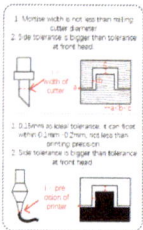

Fig. 14. Rule 3

Fig. 15. Rule 4

Fig. 16. Rule 5

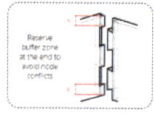

Fig. 17. Rule 6

3.3.1 Column Rotary Tenon

1. Joint Prototype: The tenon consists of two different-radius semicylinders, which, when vertically inserted into the mortise on the floor and rotate half a circle, can be locked by the larger semicylinder.
2. Shape Optimization: It is observed that the entire column, supported by two semi-cylinders with different radii, is easy to topple when printing. To address this, we draw inspiration from the balanced dovetail shape, so two symmetrical 1/4 cylinders will lock the column.

3. Support Optimization: Two overhanging 1/4 circular rings of the mortise require support during the printing process. However, removing support from the small groove is challenging. Following *Rule 4*, the two 1/4 cylinders on the tenon were modified to two 1/4 Frustums of cones.

4. Printing Settings: Following *Rule 2*, it is recommended to place the column flat on the printing bed.

5. Installation Optimization: Due to printing precision limitations, the tenon cannot be too small. Following *Rule 3*, if the tenon and mortise are printed separately, a gap of 0.15 mm (at least) should be left on each side. When the column diameter is smaller than 4 mm, the mortise on the floor is larger than the column, leaving a noticeable gap after installation. To address this, a "cover plate" was added to the column to conceal the gap (Fig. 18 and Table 2).

Table 2. Optimization result of column rotary tenon

Rotating installation, column diameter $\geq$ 4 mm	Rotating installation, column diameter < 4 mm	Vertical installation
ii. Large column rotary tenon	iii. Small column rotary tenon	xiii. Snap-in column tenon

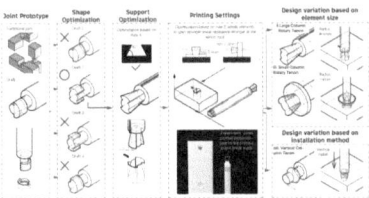

Fig. 18. Optimization details of column rotary tenon (ii. iii. xiii.)

3.3.2 Slider and Cross Tenon

1. Joint Prototype: Form directly extracted from the traditional dovetail joint.
2. Dimensional Optimization: The long end of the tenon should extend outward by at least 0.2mm on each side relative to the root of the tenon, corresponding to the diameter of the printer nozzle.
3. Scenarios Classification: When the dovetail joint is used to connect two walls at the external corner of a facade, following *Rule 5*, multiple dovetail tenons are evenly arranged. Based on the traditional *Cross Dovetail Tenon at a Right Angle*[2], thin plates are added to conceal the facade, resulting in *x. External Corner Cross Tenon*. If two walls are parallel, *vi. Dovetail Tenon Slider* can provide greater stability. Each tenon of the slider is 10mm in length adhering to *Rule 5*. Other varieties including *xi. Multi-face Tenon* and *xii. Curvy Surface Tenon*.

---

[2] 平板明榫角结合 in Chinese.

4. Thickness Classification: When two plane panels thinner than 6mm meet at a right angle, the mortise cannot be printed if the dovetail tenon symmetrically protrudes. Therefore, one acute angle is modified to a right angle, resulting in *vii. Right-angle Tenon Slider*.

5. Support Optimization: A wall slab or a floor slab may have multiple different joints which have to be printed in the same direction, but it is hard to ensure all joints without support. Following *Rule 4*, the tenon shape of the dovetail is transformed into a 1/4 arc to reduce or avoid support.

6. Application Expansion: By adjusting the tolerance dimensions between the tenon and mortise, the tenon can rotate slightly while sliding, increasing the flexibility of installation (Fig. 19 and Table 3).

Table 3. Optimization result of slider and cross tenon

Corner at periphery	Slab thickness ≥ 6 mm	Slab thickness < 6 mm	Slab printed flatwise
x. External corner cross tenon	vi. Dovetail tenon slider	vii. Right-angle tenon slider	viii. U-shaped tenon slider

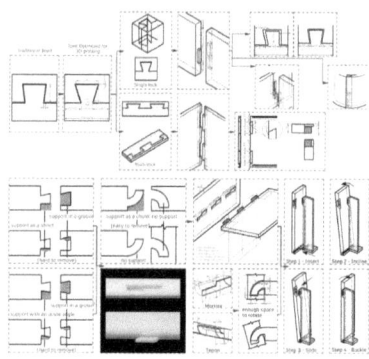

Fig. 19. Details of slider and cross tenon (x. vi. vii. viii.)

4   Result Present

4.1   Database

Refer to Table 4 for the joint allocation rules. Before users input their models into the database, they should initially decompose the components (shown in Fig. 20) and categorize them into '*Framework Zone*', '*Core Zone*', '*Transition Zone*', '*Periphery Zone*', '*Trivial Items*'.

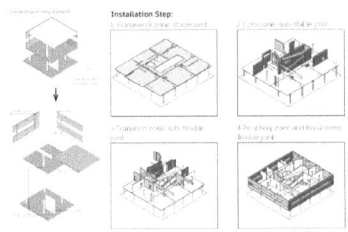

Fig. 20. Component categorization

1. The *Framework Zone* is composed of floor slabs and columns, requiring the highest stability.
2. The *Core Zone* consists of uniformly distributed walls as locators. Its stability is inferior to that of the *Framework Zone*.
3. The *Transition Zone* comprises other interior walls excluding the *Core Zone*, necessitating a certain degree of flexibility.
4. The *Periphery Zone* consists of all exterior walls and, for ease of assembly and disassembly, should possess the highest flexibility.
5. Trivial Items includes all unimportant decorative components.

Table 4. Joint allocation rules

Application scenarios	Framework zone	Core zone	Transition zone	Periphery zone	Trivial items
	Stable	Sub-stable	Sub-flexible	Flexible	Sup-flexible
Line × Line	ii	ix	ix	ix	iv
Line × Face Face × Point	iii			i	
Face × Face (plane)	v	vi	vi	viii	
			vii		
		vii	viii	x	
			xi		
Face × Face (curved-2d)		xii	xii	xii	

In the end, the database outputs components with joints, rotated to the printing direction and arranged in height-descending order. The database workflow is shown in Fig. 21.

4.2 Installation

We validated the design and adaptation algorithms of the above joints with the Savoy Villa 1:50 model, and the assembled results are shown in Fig. 26. After the component models

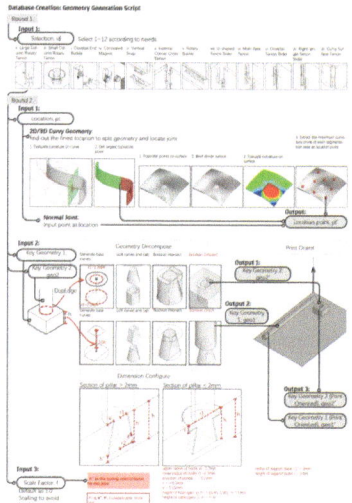

Fig. 21. Database workflow

are output from the database, they need to be arranged within the box corresponding to the printing bed size. Using Grasshopper plugins such as *Open Nest*, a relatively denser arrangement can be obtained.

As shown in Fig. 22, The assembly of the model is carried out in the order of '*from bottom to top*' and '*from Framework Zone to Trivial Items*'. Initially, all connections between components are stable. As the number of components increases, the space and direction available for manual operation decrease, but the increasing flexibility of joints helps to assemble smoothly.

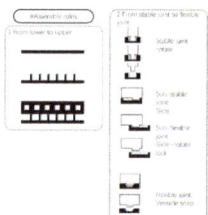

Fig. 22. Joint installation rules

Fig. 23. Printing results of universal ball

For *Line × Face (curved-3d)* unvalidated with the Savoy Villa model, it was verified using a model of *Universal Ball*, primarily using *ii. Large Column Rotary Tenon* and the assembly result is shown in Figs. 23 and 24.

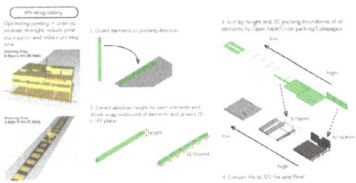

Fig. 24. Printing preprocessing

4.3 UX Design

UX design results are shown in Fig. 25. We have formed a design prototype for a Grasshopper plugin that includes four main operators. Users can directly obtain printable component models at the output end of the operator.

5 Discussion

The design and manufacturing of traditional wooden mortise and tenon were thoroughly explored in the past. The progress in material science provides more convenient ways to connect architectural model components. A prevailing approach is to incorporate simple mortise and tenon forms with adhesive materials. Nevertheless, as mentioned in Sect. 1.1, this method brings problems like irreversibility of bonding, adhesive toxicity, limited weather resistance and structural strength. It is more advisable to connect components with rigid joints.

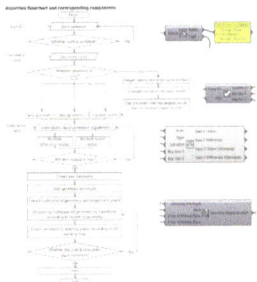

Fig. 25. Interaction design

For small-scale architectural models, directly shrinking traditional joints isn't feasible due to factors like material properties, printing accuracy, and manufacturing methods.

Additionally, as an additive manufacturing process, 3D printing has irreplaceable advantages, so adjustments need to be made in the design of connection joints to leverage its benefits.

Starting from the form of traditional wooden mortise and tenon, our optimization goals include adapting to the precision of 3D printers, accommodating the dimensions of small-scale models, and reducing unremovable supports. The iteratively optimized 3D-printed joints have advantages such as high success rates in printing, time and material saving, and convenient installation. Finally, we obtained joint solutions well-performing for the entire process from printing to assembly. These joints meet the scale requirements of architectural models, are suitable for Polymeric Laser Sintering 3D printing technology (PLS), and can be used for the connection of large architectural models (large in size but small in scale). Therefore, we break through the size limitations of 3D printers and improve printing efficiency (Fig. 26).

Fig. 26. Printing results of Savoy Villa

Apart from the advantages mentioned above, future research could explore optimization of the joints through multi-objective optimization algorithms to address these intricacies such as high aesthetic requirements and material performance demands. Additionally, this paper has not delved into a thorough exploration of complex 3d-curved surfaces, which are common features in current architectural design. If future research can organize the parameterization of seam search, joint adaptation, and printing settings for 3d-curved surfaces, it will further enhance the completeness of the database.

Acknowledgment. These two authors contributed equally: Ren Tianye, Chia Hui Yen.

References

Digital Structures MIT. Discretization and connections for 3D-printed trusses. http://digitalst ructures.mit.edu/page/research#discretization-and-connections-for-3d-printed-trusses (2014). Accessed 23 May 2024

Digital Structures MIT. Prototyping 3D-printed structural connections. http://digitalstructures. mit.edu/page/research#prototyping-3d-printed-structure-connection (2015). Accessed 23 May 2024

Fang, D., Mueller, C.: Joinery connections in timber frames: analytical and experimental explorations of structural behavior. The International Association for Shell and Spatial Structures (IASS), Cambridge, MA, USA. https://www.researchgate.net/publication/326674720_Joi nery_connections_in_timber_frames_analytical_and_experimental_explorations_of_struct ural_behavior (2018). Accessed 23 May 2024

HUBS. How do you design snap-fit joints for 3D printing? https://www.hubs.com/knowle dge-base/how-design-snap-fit-joints-3d-printing/#hubs-top-tips-tricks-for-designing-snap-fit-joints (n.d.). Accessed 29 Oct 2023

Knoll, W., Hechinger, M., Yuehua, L.: Jianzhu moxing zhizuo: moxing silu de jifa [Architectural Model Production, Inspiration for Model Concepts]. Dalian University of Technology Press, Dalian (2003) (Chinese)

Morano, C., Scagliola, M., Bruno, L., Alfano, M.: Crack propagation in adhesive bonded 3D printed polyamide: Surface versus bulk patterning of the adherends. Int. J. Adhes. Adhes. **131**(5) (2024). https://doi.org/10.1016/j.ijadhadh.2024.103660

Mueller, C., Irani, A., Jenett, B.: Additive Manufacturing of Structural Prototypes for Conceptual Design. The International Association for Shell and Spatial Structures (IASS), Brasilia, Brazil. http://digitalstructures.mit.edu/files/2015-07/pap234mueller.pdf (2014). Accessed 23 May 2024

Tang, L., Guan, H.: Jiyu canshuhua de sunmao chicun zhineng queding fangfa [Parametric-based intelligent determination method for joinery dimensions]. Beijing Linye Daxue Xuebao **43**(3), 145–154 (2021) (Chinese)

Ye, S.: In: Zhonghua sunmao: gudian jiaju sunmao gouzao zhi bashiyi fa [Chinese Traditional Wood Joinery, Eighty-One Methods of Traditional Furniture Joinery Construction], p. 15. China Forestry Publishing House, Beijing (2017) (Chinese)

Zuo, W., Chen, M., Cheng, B., Zhao, J.: Intelligent optimization and 3D printing of joints in spatial structures based on computer-aided technologie. J. Build. Struct. 04, 131–141 (2024) (Chinese)

Interactive Bricklaying: A Comparative Experiment on Human Involvement in Masonry Structure Design and Construction Through Augmented Reality

Weichen Zhang[✉] and Pierpaolo Ruttico

Politecnico di Milano, Piazza Leonardo da Vinci 32, 20133 Milano, Italy
weichen.zhang@mail.polimi.it

Abstract. Interactive bricklaying combines augmented reality technology with robot fabrication technology, utilizing both human intelligence and robot precision. Humans intuitively complete various differentiated bricklaying designs, while robot systems accurately execute bricklaying tasks. Traditionally, in such workflows, the design and fabrication processes are separated, with a lower level of human-robot interaction. By comparing with traditional interactive workflows through experimental studies, this research explores a higher degree of interactive workflow where design and fabrication alternate. It analyzes the impact, limitations, and potential of human involvement in linear robot fabrication processes. Using this new workflow, the study achieved the design and fabrication of a conical structure without prior design. During construction, the structure was gradually reduced in layers from the bottom up, with the number of bricks reduced and brick positions adjusted through manual observation, successfully controlling the tapering form and achieving closure of the circular brick wall. The new workflow utilizes human judgment to continuously determine the next design based on the current real construction situation, ensuring the rationality of the design and the stability of the construction process.

Keywords: Augmented Reality · Human Robot Interaction · Immersive Design · Robotic Bricklaying

1 Introduction

In recent years, the productivity growth in the construction industry has been slow compared to other manufacturing sectors (Statistical 2022). Simultaneously, the low sustainability of the construction industry is largely attributed to its low level of automation (Masson-Delmotte 2021). Digital fabrication methods have shown significant potential in integrating with architectural and engineering design practices, effectively enhancing productivity and reducing environmental impacts (Dörfler et al. 2016). Over the past decade, robotics manufacturing in the AEC (Architecture, Engineering, and Construction) sector has made significant advancements, presenting numerous new design

H. Chai et al. (Eds.): CDRF 2024, *Symbiotic Intelligence*, pp. 106–115, 2025.
https://doi.org/10.1007/978-981-96-3433-0_10

opportunities (Cai et al. 2019). However, relevant research still primarily occurs within universities and research institutions, remaining in the experimental prototype stage (Garcia de Soto and Skibniewski 2020), where the working environment, materials, and processes are clearly predictable, and the construction process for robots is linear (Mitterberger and Ercan Jenny 2022).

The complexity of the design and construction processes has always been a characteristic of the construction industry (Mitchell 2005), rendering linear and predictable robot construction processes inadequate. However, by integrating human tacit knowledge into precise linear cycles of robot construction, there is potential to address the complexity of the construction process with human intelligence (Mitterberger and Ercan Jenny 2022). This concept of human involvement in the robotic manufacturing process can be termed interactive fabrication (Willis et al. 2010), currently primarily achieved through augmented reality (AR) technology.

AR, proposed in 1997, is described as "real-time integration of 3D virtual objects into a 3D real environment" (Azuma 1997), with over 65% of human information acquisition stemming from visual sources, showcasing its evident potential (Chen et al. 2019). In recent years, the trend of AR technology in practical applications reveals its enormous potential (Chi et al. 2013).

AR finds extensive applications in architectural engineering (Wang 2009), such as generating more intuitive project management approaches by integrating Building Information Modeling (BIM) and AR (Meža et al. 2014). Furthermore, AR is employed directly in digital construction processes, categorized as 3D holographic teaching, AR data sharing, and AR for human-machine interaction (Song et al. 2021).

Numerous studies have demonstrated the feasibility of using AR to guide craftsmen in manual manufacturing processes (Mitterberger et al. 2020). Regarding using AR to assist human-robot interaction in interactive construction, research has been conducted on tasks such as wood milling and assembly (Kyjanek et al. 2019), plastering (Mitterberger and Ercan Jenny 2022), bricklaying (Song et al. 2023), proving the value of AR in robot construction processes. However, in the aforementioned studies, the design process and the robot construction process are separated, AR design precedes robot construction. Such an approach does not fully integrate human intelligence into the linear robot manufacturing process.

To explore how to more fully leverage human intelligence in linear robot tasks to better cope with the uncertainty in the construction process, this paper conducts a comparative experiment using bricklaying tasks as an example. The main contribution of this experiment is to compare the two interactive construction methods: "design before construction" and the more interactive "alternating design and construction," analysing changes in the type of construction goals, the experience of the construction process, and the accuracy of the construction results with increased interactivity. Subsequently, leveraging the characteristics of this new method, this study completes an example of constructing a conical structure, further demonstrating the significance of enhancing human-robot interaction in utilizing human intelligence to alleviate uncertainty in the construction process.

2 Methodology

This research project developed an enhanced interactive construction workflow. Firstly, a comparative experiment was conducted between conventional interactivity bricklaying and enhanced interactivity bricklaying. Two methods with different levels of interactivity were used to design and construct a parametric brick wall separately. Subsequently, to further understand the role of human intervention in the enhanced interactivity method, a design and construction experiment for a circular wall was added.

For experimental materials, small-scale wooden blocks (80 mm * 40 mm * 20 mm) were used, along with a homemade brick feeding ramp device.

In terms of hardware, this study primarily utilized an iPhone 8 Plus and HoloLens 2 as input and output devices for augmented reality; a pneumatic gripping fixture from Schunk was employed to pick up the bricks, along with a UR10e robot. Additionally, a laptop was used to adjust and connect these devices; all devices were connected to the same IP address on the same Wi-Fi router.

In terms of software, this study was mainly based on the Rhino and Grasshopper platforms, along with the Robots and Fologram plugins. Robots were used to control parameters such as the robot's motion path, motion speed, and motion type, as well as inputting script commands to control the opening and closing of the gripper. Fologram was used to project virtual 3D images onto hardware devices and to recognize input data generated by user gestures or screen taps.

In terms of task types, they can be broadly categorized into three categories: localization, interactive design/control, and fabrication. The software and hardware used for different tasks are illustrated in the following figure (Fig. 1).

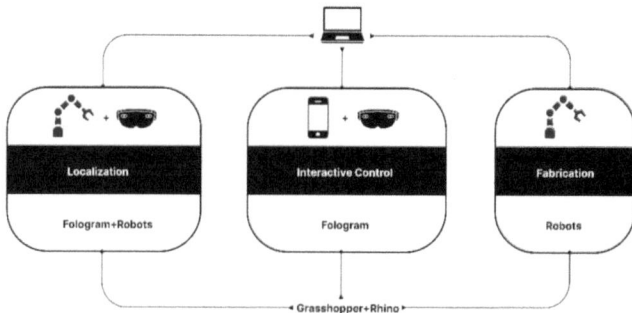

Fig. 1. Localization primarily involved the use of robots and HoloLens via the Fologram and Robots plugins. Interactive control/design mainly utilized HoloLens and iPhone via the Fologram plugin. Fabrication primarily employed robots via the Robots plugin

3 Comparison Experiment Between Traditional and Enhanced Levels of Interaction

3.1 Preparatory Operations

Before starting, it is necessary to align the AR environment with the real environment and establish a reference plane. The reference plane can be based on feedback from the robot's TCP (Tool Center Point) or on the HoloLens' recognition of QR codes. Aligning the virtual and real environments is generally achieved by placing QR codes in specific locations and using AR devices for recognition. Additionally, the Fologram settings provide an opportunity for manual fine-tuning of the position, facilitating more precise alignment results (Fig. 2).

When aligning the reference plane, firstly manually align the robot's TCP with the reference plane, then use a custom component to feed back the position data of the current TCP to the computer and save it. Repeat this process multiple times to obtain points on multiple reference planes, and then calculate the position of the reference plane. Additionally, place several QR codes on the reference plane and use AR devices to recognize their position information, registering them in the computer, which can also calculate the position of the reference plane. The first method mentioned above has extremely high accuracy and is preferred.

To avoid errors caused by excessive activity in AR device gesture recognition, the entire process is designed as several independent steps, with only the AR recognition of the currently running step being active. To switch between different steps, this study has created a series of interactive virtual buttons, each expressing its functionality in text. The buttons are color-coded: red indicates the inactive state, while green indicates the active state (Fig. 2).

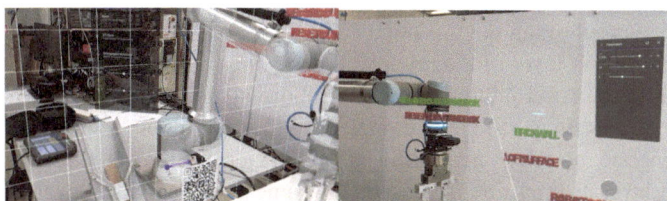

Fig. 2. Align virtual and real environments by scanning QR codes and manually align virtual and real environments (left), AR button controls switching between asynchronous steps (right)

3.2 Preliminary Comparison of Two Levels of Interaction

During the interactive control phase (Fig. 3): (1) Firstly, a basic curve needs to be drawn by gestures. HoloLens continuously recognizes the position of the user's hand and records a series of points. The basic curve is generated by passing through these points. After the curve is generated, (2) The user can further adjust the shape of the curve by moving control points on the curve. (3) Then, a similar curve is directly generated at a certain height. These two curves are used to create a surface through lofting. The newly generated curve also has several control points. Users can adjust the shape of the surface by modifying these control points. After obtaining the desired surface, (4) A parametric brick wall consisting of corresponding-sized bricks is directly generated based on the surface. While previewing the actual effect of the brick wall, users can continue to adjust the shape of the wall or parameters such as the distance between bricks by moving control points until the final state is achieved. (5) Once the design phase is completed, the "construction" button can be clicked to start the robot operation.

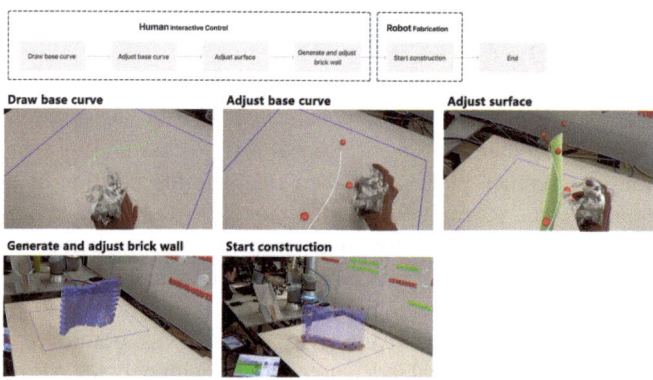

Fig. 3. The design and construction process of conventional interaction level

In enhanced interactive construction, the interactive control phase and the construction phase alternate layer by layer (Fig. 4): (1) In the interactive control phase, the first step remains to draw the basic curve using gestures. (2) Next, further adjust the shape of the curve through control points. (3) Based on the drawn curve, generate the bricks for the current layer of the brick wall and adjust parameters such as the spacing between bricks. (4) After confirmation, the robot will sequentially place the bricks in their respective positions. (5) Upon completion of construction, register the completed layer and generate the bricks needed for the next layer in augmented reality. The initial state of each layer of bricks defaults to be the same as the previous layer. (6) Based on observation of the completed part, further adjust the shape of the new layer by moving control points and shifting brick positions, ensuring changes in the brick wall's shape do not compromise structural stability. This fully demonstrates the role of human intelligence in intervening in the linear robot construction process. (7) After adjusting the new layer, proceed with construction, continually cycling until all layers are completed.

To provide a comparison of the characteristics between the different levels of interaction in construction methods, this study qualitatively analysed the features and distinctions of the two approaches based on the types of construction objectives, the experience of the construction process, and the accuracy of the construction results (Fig. 5).

In terms of the accuracy of construction results, the traditional interactive method, due to its uniform division of the complete surface into evenly spaced positions for each brick layer, exhibits very uniform variations, and the distances of brick shifts are also very tidy.

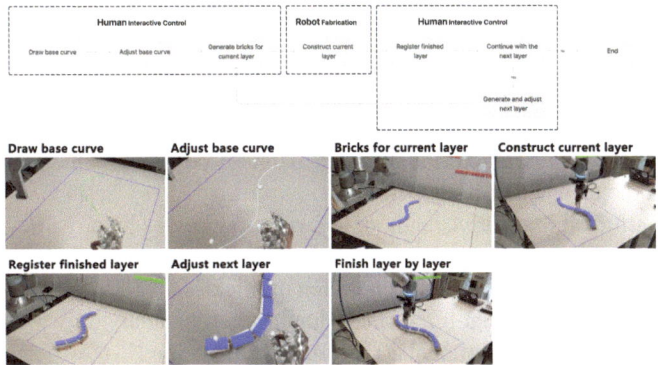

Fig. 4. The design and construction process of enhanced interaction level

Fig. 5. The construction results of conventional interaction level method (left) and enhanced interaction level method (right)

Conversely, the enhanced interactive method, as the shape variations between adjacent layers are entirely manually controlled, demonstrates highly unpredictable accuracy.

Regarding the experiential aspect of the construction process, the traditional interactive method can be decomposed into immersive design based on AR and fully automated robot construction. Sometimes, structural simulation is required before construction to ensure stability. In contrast, the experience of the enhanced interactive method is entirely different. As each layer's shape is manually determined separately, the efficiency is significantly lower. However, users have higher levels of involvement and control over the entire process.

Concerning the types of construction objectives, the traditional interactive method is highly suitable for linear parametric brick walls, primarily due to the need for high precision in their gradient forms, where the precision of robots can be fully utilized. Conversely, the enhanced interactive method is less suitable for such tasks. Excessive human intervention significantly reduces construction accuracy, resulting in poor gradient effects for linear brick walls.

To further explore the role of increased human intervention in the enhanced interactive method, this study has completed an additional construction example of a conical circular brick wall enclosure.

3.3 Further Comparison of Two Levels of Interaction

To emphasize the advantages of the enhanced interactive method, the new structure needs to incorporate constraints to reduce the freedom of manual control, thereby weakening the constraint of accuracy on design and construction. Therefore, this study chose a circular structure, with parameters that can be manually adjusted including the radius of the circle and the distance of brick shifts.

For the traditional interactive method, the workflow begins by providing several circles at various heights, manually adjusting the radius. Subsequently, the form is generated through lofting, followed by the generation of bricks. Once confirmed, construction begins. Other than the fact that the surface shape is not manually drawn, the workflow is similar to the previous experiments (Fig. 6).

When the degree of form variation is small and the number of bricks in each layer is consistent, the structure remains simple and stable. However, when the degree of surface variation is large and uneven, it is crucial to simulate the structure's stability before construction begins. Otherwise, due to the large number of bricks and the impact of overlapping, structural failures may occur due to some unreasonable factors that are inconvenient to observe.

Furthermore, the distance of brick shifts, as another important factor, can significantly affect the stability of the structure. This is especially true when the number of bricks in each layer is changing. Failures in construction may occur due to a brick being suspended and unable to be placed stably, particularly evident when bricks are placed along the direction of the shift on their short edges (Fig. 6).

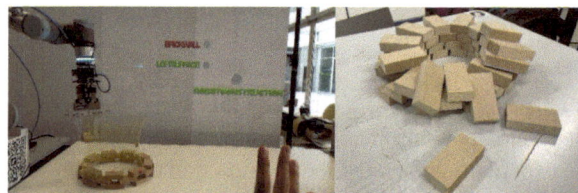

Fig. 6. The construction of circular structures using traditional interaction degree methods (left), Unpredicted structural instability factors leading to construction (right)

For the enhanced interactive method, the aforementioned issues can be effectively addressed. To verify this, a conical masonry structure was designed and constructed. Starting from the bottom, the circles are gradually reduced in size layer by layer, reducing the number of bricks and shifting the positions of the bricks. This is done to achieve a higher slope and complete the closure of the circular wall, forming a conical masonry structure. The workflow begins by generating the circle for the current layer, manually adjusting the radius of the circle. Based on observation of the completed part, the current layer's circle is manually reduced in size. After generating the bricks, the positions of the bricks are shifted. By examining the relationship between the current layer's AR image and the positions of the bricks from the previous layer, all bricks are ensured to be placed stably. Once confirmed, the robot operation begins. After completing the construction of the current layer, it is registered, and then transitioned to the next layer. The base circle for each layer defaults to be consistent with the previous layer, continually cycling until the circular wall is closed (Fig. 7).

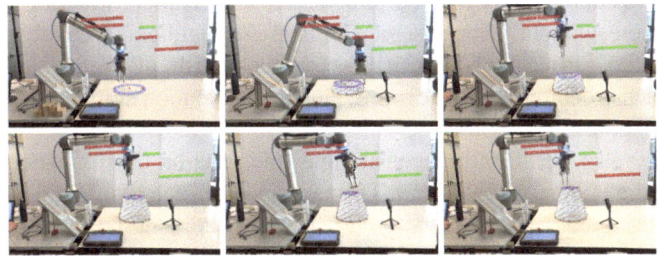

Fig. 7. Close the circular wall layer by layer through manual adjustment

The advantages of human intervention in the robot construction process are well demonstrated. From structural stability standpoint, human judgment replaces computer simulation, risks are identified through observation (Fig. 8). And the structure remains stable throughout the entire construction process. From an aesthetic standpoint, human intervention brings about a form of artificial aesthetics, which differs from the aesthetic value brought about by the precision of robots.

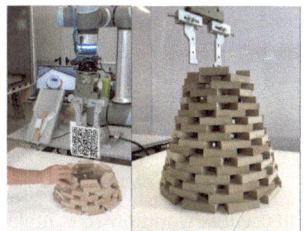

Fig. 8. The design and construction process of a circular structure with enhanced interaction methods (left), Construction achievements (right)

For the experiments on the design and construction of circular walls completed using two different methods, analysis is still conducted based on the types of construction objectives, the experiential aspect of the construction process, and the accuracy of construction results. Regarding the accuracy of construction results and the experiential aspect of the construction process, the results of the new experiment are similar to those of the previous one. The traditional interactive method exhibits higher accuracy, faster speed, and less human involvement. However, in terms of the types of construction objectives, the enhanced interactive method in the new experiment holds significant advantages. Firstly, after the freedom of manual operation is restricted, errors are effectively reduced. On the other hand, this method fully utilizes human intelligence, efficiently completing the conical masonry structure while considering aesthetic value, thus replacing computer simulation. In contrast, the traditional method may not guarantee structural stability in the absence of effective computer simulation.

4 Conclusion and Discussion

This study, using masonry structures as an example, proposed a workflow with enhanced interactivity and applied this method to complete a special construction instance of a conical structure without prior design. The analysis highlighted the advantages over traditional methods in terms of construction task types and the differences in accuracy and experience. It also demonstrated the role of human intervention in addressing uncertainty in linear robot construction processes.

This study also conducted demonstrations and tests on approximately 20 professionals from the Italian AEC industry. The participants successfully utilized the system to design bricks and carry out robot construction, validating the effectiveness of the system. Regarding their evaluation, participants generally believed that this AR-based human-machine interactive construction method has the potential to drive automation development in the industry. However, they also noted that it cannot be applied in actual construction scenarios in the short term. The reasons are multifaceted, such as insufficient accuracy of AR devices; the current experiments being limited to masonry structures only; absence of real construction environments and materials; the method being in the prototype stage; and the adequacy of human intervention in robot construction processes (Fig. 9).

Fig. 9. The subjects attempted to use this workflow to complete a construction of an instance

The future directions of this research could involve: exploring the application of AR-enhanced interactivity in other types of robot construction tasks and analyzing its effectiveness; attempting to complete 1:1 design and construction tasks using real-sized bricks to enhance system completeness; incorporating more guidance to facilitate autonomous usage; and seeking additional application scenarios where human intervention in linear robot construction processes can effectively leverage human intelligence.

References

Azuma, R.T.: A survey of augmented reality', presence: teleoperators and virtual environments. **6**(4), 355–385 (1997)

Cai, S., Ma, Z., Skibniewski, M.J., Bao, S.: Construction automation and robotics for high-rise buildings over the past decades: a comprehensive review. Adv. Eng. Inform. **42**, 100989 (2019)

Chen, Y., Wang, Q., Chen, H., Song, X., Tang, H., Tian, M.: An overview of augmented reality technology. J. Phys. Conf. Ser. **1237**(2), 022082 (2019)

Chi, H.-L., Kang, S.-C., Wang, X.: Research trends and opportunities of augmented reality applications in architecture, engineering, and construction. Autom. Constr. **33**, 116–122 (2013)

Dörfler, K, Sandy, T, Giftthaler, M., Gramazio, F., Kohler, M., Buchli, J.: Mobile robotic brickwork, robotic fabrication in architecture. Art Des. 204–217 (2016)

Entwicklung des Bruttoinlandsprodukts in Deutschland (Jahresdaten) n.d. DESTATIS

Garcia de Soto, B., Skibniewski, J.M.: Future of robotics and automation in construction, Construction 4.0 (2020)

Willis, K.D.D., Xu, C., Wu, K.-J., Levin, G., Gross, M.D.: Interactive Fabrication (2010)

Kyjanek, O., Bahar, B.A., Vasey, L., Wannemacher, B., Menges, A.: Implementation of an augmented reality AR workflow for human robot collaboration in timber prefabrication. ISARC Proceedings, pp. 1223–1230 (2019)

Masson-Delmotte, V., Zhai, P., Pirani, A.: 'The Physical Science Basis Working Group I Contribution to the Sixth Assessment Report of the Intergovernmental Panel on Climate Change Edited by', Climate Change 2021, vol. 2. The Physical Science Basis (2021)

Meža, S., Turk, Ž, Dolenc, M.: Component based engineering of a mobile BIM-based augmented reality system. Autom. Constr. **42**, 1–12 (2014)

Mitchell, W.J.: Constructing Complexity, pp. 41–50. Springer eBooks (2005)

Mitterberger, D., Dörfler, K., Sandy, T., Salveridou, F., Hutter, M., Gramazio, F., Kohler, M.: Augmented Bricklaying. Construction Robotics (2020)

Mitterberger, D., Ercan Jenny, S.: Interactive Robotic Plastering: Augmented Interactive Design and Fabrication for On-site Robotic Plastering. CHI Conference on Human Factors in Computing Systems (2022)

Song, Y., Agkathidis, A., Koeck, R.: Augmented Bricks an Onsite AR Immersive Design to Fabrication Framework for Masonry Structures, pp. 385–395 (2023)

Song, Y., Koeck, R., Luo, S.: Review and analysis of augmented reality (AR) literature for digital fabrication in architecture. Autom. Constr. **128**, 103762 (2021)

Wang, X.: Augmented reality in architecture and design: potentials and challenges for application. Int. J. Archit. Comput.Comput. **7**(2), 309–326 (2009)

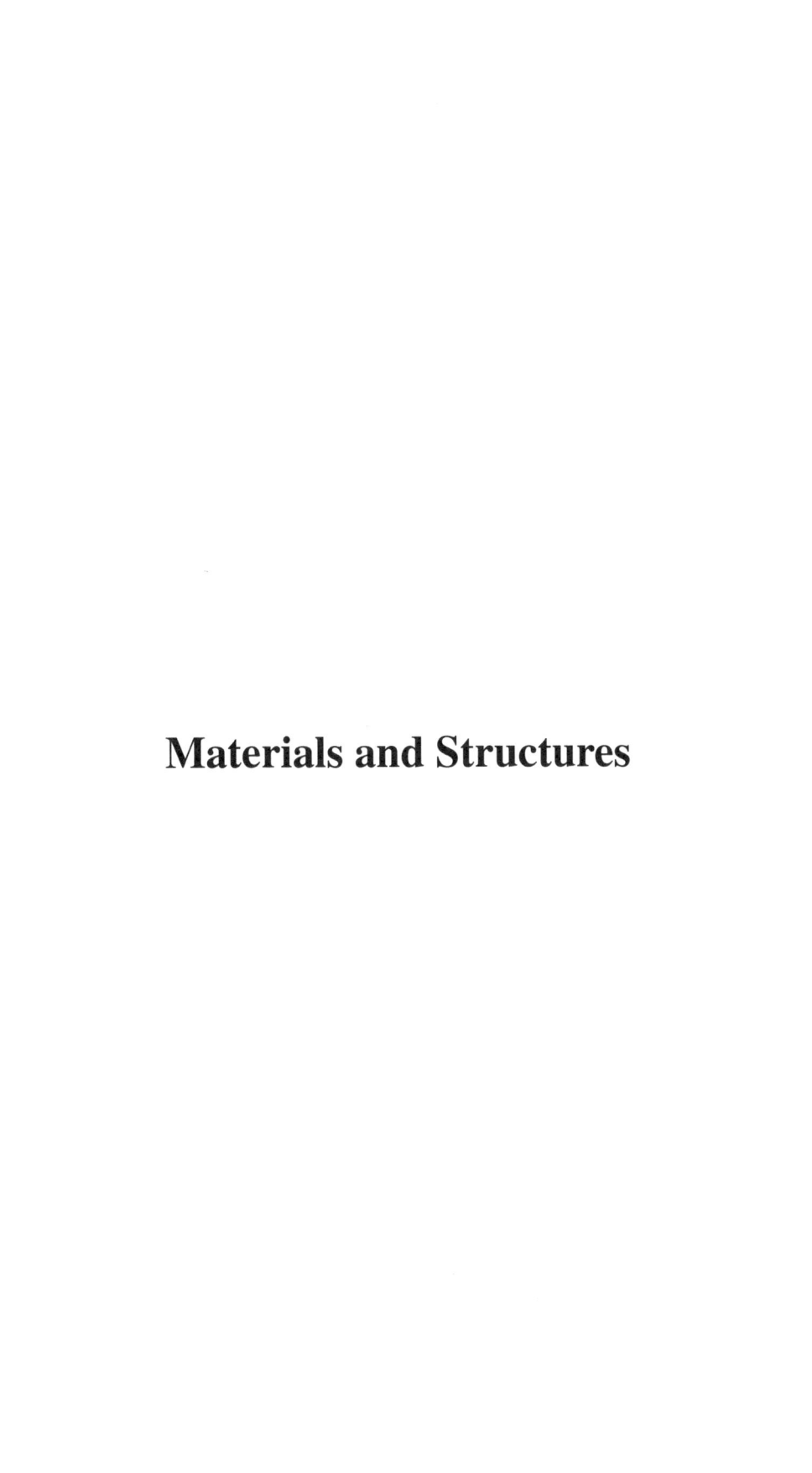

Materials and Structures

Moss Columns: Symbiosis Through 3D Printing Technologies

Yong Ju Lee[(✉)]

Seoul National University of Science and Technology, 232, Gongneung-ro, Nowon-gu,
Seoul 01811, Republic of Korea
yongjulee@seoultech.ac.kr

Abstract. Moss Columns are a series of prototypes to examine how to combine living organisms with architecture. A renewed perspective on the built environment has increased in the post-COVID-19 era. Even various methods have been employed for sustainable design, the construction material itself in contemporary architecture have remained largely unchanged. In these two different types of experiment, I am presenting a direct embedding approach of plants into the artificial materials. Mosses, chosen as the primary plant for these experiments, are suitable due to their non-vascular nature, which means they do not grow tall like other plants and use roots solely for anchoring rather than nourishment. To investigate high-res and complicated pattern for embedding, geometry is manipulated and generated through computational design tools. Advanced construction technologies are employed to realize the complex forms: such as two different types of large-scale 3D printers with an industrial robotic arm.

Keywords: Symbiotic System · 3D Printing · Robotic Fabrication

1 Introduction

Designers and architects have noticeably paid attention to the built environment after COVID-19. While various methods have been employed for sustainable design, the construction material itself in contemporary architecture has remained largely unchanged for a while. However, it has been convinced that the nature would increasingly become an integral part of our building environment. In the post-coronavirus era, architects need to re-design our world through the point of view of the viral ecosystems that co-exist with us (Pasquero and Poletto 2023). As digital technology expands its influence to the architectural field, related material research becomes more embraced, appearing as if it comes after geometric manipulation and production. Therefore, the statement about symbiosis by Kurokawa remains valid. Symbiosis in architecture means a new way of interpreting contemporary culture, considering humans as initial symbiotes within our society (Kurokawa 1994). In response of these perspectives, I aim to discover a new interconnected medium between humans and non-human or artificial and natural elements, which can facilitate stable coexistence and ultimately provide us with advantages in this time of environmental challenges.

H. Chai et al. (Eds.): CDRF 2024, *Symbiotic Intelligence*, pp. 119–128, 2025.
https://doi.org/10.1007/978-981-96-3433-0_11

This study focuses on active and radical solutions for integrating organic and inorganic matter through a single-body geometric system. This sustainable approach highlights the interaction and exchange of respiratory and photosynthetic by-products between people and nature. The utilization of eco-friendly architectural techniques in conjunction with embedded plants is anticipated to bring about significant transformations in the future construction industry. Here, I present the potential for sustainable implementation in the digital era, specifically with the use of moss as green organism, establishing new relationships among digital, physical, and, furthermore, natural elements.

2 Existing Studies

There are several prior studies that can be considered in relation to this concept, ranging from conceptual to realistic approaches. Kim developed a moss-based biomaterial for sound absorbing purpose in free-form structures. He produced a moss dough blended with ingredients such as beer and starch. Once applied to desired surfaces, the rhizoids of the moss adhere and assemble themselves, resulting a strong, cost-effective and environmentally friendly material for architecture (Kim et al. 2022). Ecologic Studio, a design firm based in London, presents experimental examples of incorporating living organisms into artificial and complicated environment in sculpture-like forms and scales. One noticeable example, H.O.R.T.U.S., a bio-architectural prototypes that demonstrate the feasibility of symbiosis. In this bio sculpture, photosynthetic cyanobacteria are injected into digitally fabricated parts, forming biological intelligence of the system powered by photosynthesis (Chen and Pasquero 2023).

Some attempts can offer practical advantages. An indoor vertical greenery system called Moss-iVGS, which utilizes moss, can serve as an alternative sustainable solution in specific conditions such as tropical environments. It helps in conserve water, control erosion, filter harmful chemicals and rainwater, and sequester carbon, withstanding up to 45 days without water supply or regular maintenance (Wang et al. 2018). Expanding this concept to an urban scale, green infrastructure including green walls in cites can support biodiversity by acting as a "corridor" or "stepping stone" to facilitate movement and dispersal of species (Angold et al. 2006). Furthermore, it can enhance the stability of urban biodiversity in the presence of disturbances and unpredictable events, providing resilience (Goddard et al. 2010).

One of the architectural possibilities for a sustainable future can be pursued by blending organic and inorganic elements. A key difference, compared to existing studies that bring natural elements into architecture for functional purposes, is that the approach described ahead considers the integration of architecture and nature as a cohesive entity, creating a symbiotic system. Now I am presenting a series of architectural prototypes utilizing moss for architectural applications to the intricate geometry fabricated using advanced technologies.

3   Why Using Moss?

Moss, a non-vascular plant belonging to the Bryophyte family, plays a beneficial role in our ecosystem. It not only thrives in natural environments such as rocks, cliffs, and bark, but also in human-made architectural environments (Jang and Viles 2022). Moss flourishes in the boundary layer above surfaces trapping heat and water vapor to create a favorable microclimate. This thin layer, including moss, absorbs water vapor from the atmosphere and maintains moisture to a significant capacity (Kimmerer 2003). This moisture regulation property plays a vital role in preventing flooding and drought on Earth (Anderson et al. 2010). Moss also exhibits the ability capture carbon through photosynthesis about 6.43 gigatons of carbon worldwide (Eldridge et al. 2023). This capability aids in reducing air pollution by storing particulate matter, heavy metals, and volatile organic compounds (Seo et al. 2023).

Fig. 1. The anatomy of moss shows the rhizoids and non-vascular leaces (left) and Smaller White-moss is a geo-local species in Korean Peninsula (right).

In this research, several characteristics of moss are beneficial when integrated with the inorganic structure. As a primitive plant, moss has thread-like rhizoids similar to true roots but no vascular tissue (Heim et al. 2014, shown as Fig. 1). Therefore, moss relies on atmospheric absorption, making it suitable for application on hard surfaces of artifacts. Designing the boundary layer is the essence of this symbiotic experiment of making suitable habitats for moss. These simple and hairy rhizoids also provide strength to act as a physical fixation on the substrate and to prevent erosion (Belnap 2006). Additionally, as a non-vascular plant, moss does not undergo significant growth or flowering (Kimmerer 2003). When integrated with human-intended designs, there is no concern of pollen dispersion and moss not only retains and showcases the initial design but also maintains its presence as a living organism.

4 Methodology

4.1 3D Printed Structure

Large-scaled physical prototyping allows architects to gain insight throughout the design and fabrication process. Pasquero and Zarucas introduced the concept of "weird media", referring to an unfamiliar-looking output that refracts the designer's input by realizing a new entity that acts as a strange medium (2016). When expanded on an architectural scale, this concept especially helps architects assert their designs as spaces to be perceived and experienced, rather than mere objects to be observed. In this study, a pillar, one of the most crucial architectural elements, is employed to demonstrate to function as weird media, exploring the desirable harmony of natural elements and artificial structure. Both presented prototypes draw inspiration from nature and are developed through algorithms to provide sophisticated habitats, maximizing the symbiotic effect. Moreover, to realize these intricate and unusual geometric forms, multiple advanced digital fabrication techniques are employed, including silica 3D printing, Fused Granulate Fabrication or robotics, offering high speed and accuracy. These technologies are essential for achieving practical symbiosis by providing suitable habitats for life and expanding human perception on an architectural scale.

4.2 Moss Preparation

Regarding the organic portion to integrate with the man-made structure, Smaller White-moss (the right image in Fig. 1), also known as bread moss (Leucobryum juniperoideum), is selected. This moss is geographically distributed throughout the entire Korean Peninsula, making it easy to find or purchase at a low price. It forms a dense cushion by tightly attaching one another, allowing it to store more moisture compared to other moss types and have a positive impact on its surroundings. It is also easy to cultivate due to its wide range of tolerances of temperature, moisture, and light control. Consequently, it is extensively utilized in artificial environments like terrariums and various landscape designs. The attachment and tight binding of the moss to hard surface such as rocks or man-made structures are referred to as adhesion (Jang and Viles 2022). For this process, the moss can be blended with water to create slurry-type mixture, or used as is. a small amount of plant adhesive, which is used in actual terrariums (George 2015) and primarily composed of an organic material called cyanoacrylate, is applied. I collected Smaller White-moss from my surroundings and also ordered some through the local farm. After washing and drying it with minimal moisture remaining, it was ready for adhesion. Finally, once the fabrication process is completed, symbiotic process begins as the moss adheres to minute geometric crevices in the man-made structure.

5 Prototyping

5.1 Moss Column I

Moss Column I is an intricate exoskeleton system with mosses embedded inside, showcasing how human and the living organism interact. The complicated outer structure wraps around human body, providing a spatial quality. A person inside exchange the breath with another life and perceive the symbiosis in eco-cycle. A single structure is designed by algorithm to simulate a living cell. To inject the nature into solid materiality, only a minimal amount of the triangulated pixel is subtracted to maintain structural stability. Physical sampling was conducted to study the stability of moss rooting and the visual effects within specific pixel types. Due to the free-formed structural single-body system, one of the key challenges of the project was contemplating how the moss would be attached to the structure. Therefore, the primary experiment was conducted using six different types of geometric embedding with pixelated green to determine the optimal shape of the structure. Each comparisos was differentiated by its axis, the frame thickness, form, length, and depth (Fig. 2). As a result, an equilateral triangle with a length of 15 mm, depth of 10 mm, and a horizontal axis frame thickness of 4 mm was chosen. This selection considered the number and area of threads for stable attachment, as well as the visual aesthetics resulting from the arrangement of polygons and gap thickness.

To realize this complex geometry, the only sand (silica) 3D printer in South Korea was employed. This particular printer is a binder jet type with the largest bed size of 1800 mm × 1000 mm × 700 mm and a resolution of 400 dpi with formaldehyde-free adhesive (Han 2017). The fine particles of sand provide high precision in dealing with the intricate form and enable the production of a monocoque structure layer by layer, with different concepts between the internal and external portions (Fig. 3). The material's grainy characteristic makes rhizoids anchor their threadlike hairs to the structure tightly. This printed output eventually decomposes into sand again when heated afterwards. It is printed in a half scale (Fig. 4).

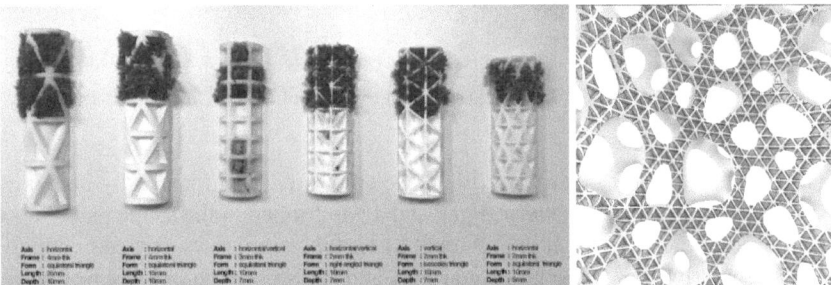

Fig. 2. Geometric embedding studies for pixelated green (left) and pixel Boolean-ed modeling for the final prototype (right)

5.2 Moss Column II

Moss Column II presents an updated perspective view on the blending methodology of artificial and natural elements, both theoretically and visually. Instead of using Boolean operations to attach mosses, as in the previous project, a subtler and more sophisticated shape and pattern are employed to create an inherent blend. This prototype provides an environment suitable for the life of moss similar to how it grows at the base of a tree or within gaps in bark, while also clarifying the structural role of the pillar. To achieve

Fig. 3. Sand 3D printing handles 400 dpi for the geometric complexity. The final result is excavated at the end.

Fig. 4. Moss embedded interior and structural exterior of Moss Column I in 1:2 scale

this, a natural shape was incorporated into the structure. Additionally, by twisting the base of the pillar to create a patterned gap and a gap formed in the shape itself, a better environment for moss to thrive is provided. Moss typically grows vertically and tends to grow irregularly in order to share nutrients and moisture with neighboring mosses. To accommodate this behavior, the prototype adopted a reaction-diffusion (RD) system. Originally developed by Turing, this is used to explain self-regulated pattern formation in the developing animal embryo. The RD model describes the interaction between two different substances, which autonomously and independently generate spatial patterns (Kondo and Miura 2010). By utilizing the algorithm of RD model in Rhino Grasshopper, three-dimensional pattern is generated on the base column-shape geometry.

Different from the previous prototype, it employs Fused Granulate Fabrication (FGF) extruder attached to a robotic arm. While using FGF printing with a 2 mm-diameter nozzle and PLA pellets results in a lower-resolution output, it allows for faster prototyping on a larger scale, with a height reaching 2.4 m. This printing method is implemented at the end of a 6-axis industrial robotic arm (ABB IRB-4600). This z-directional fused stacking process becomes a design constraint to establish a custom workflow (Fig. 5). The RD pattern creates a series of convex and concave linear geometry, enabling the exploration of appropriate furrow dimensions for moss by manipulating parameter values. The depth of the furrow, in which the moss would be stably active is set to 20 mm, while the width ranges from 10 mm to 35 mm (shown as ④ in Fig. 5). By applying this density to the twisted model C, living moss can be transplanted in a way that touches ever parts, creating a visual effect that is difficult for people to perceive the boundary between the foreground and the background. This approach strategically maximizes the unique attributes of the RD model through design, eventually achieving a smooth integration of artificial elements and plants. The low-resolution output resulting from the fused stacking also creates advantageous minute crevices for the rhizoids to adhere more strongly (Fig. 6). Therefore, this artificial living pattern establishes itself as a niche role, thickening the boundary layer and creating a microhabitat where moss can thrive and reaching a desirable symbiosis of nature and artifacts (Zhao et al. 2023). The final output is shown in Fig. 7.

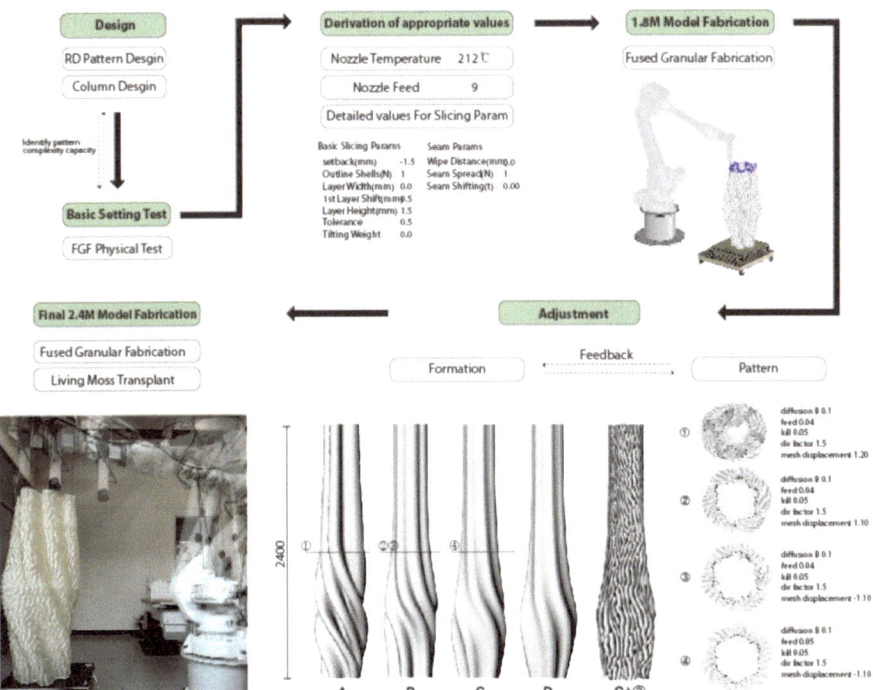

Fig. 5. A custom workflow is set to explore and fabricate appropriate relationship between the overall form and furrow dimension to blend natural and artificial elements through a nature-inspired pattern.

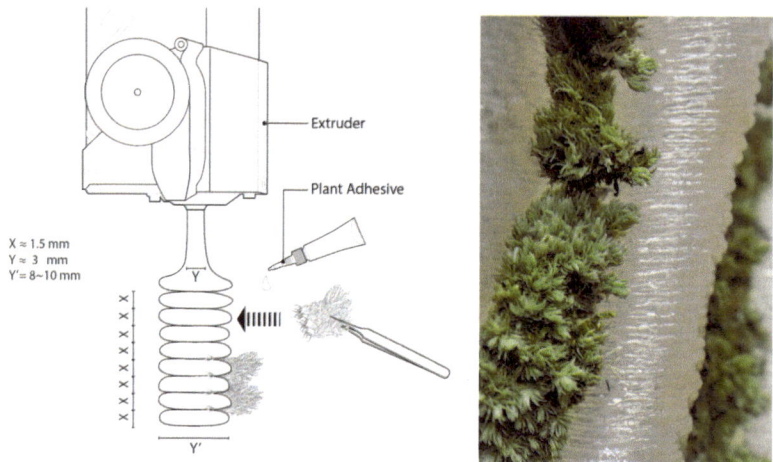

Fig. 6. The low-resolution output is considered advantageous for moss adhesion in the microscale of rhizoids.

Fig. 7. Moss Column II embraces nature within the man-made organic pattern.

6   Conclusion

This research envisions a novel approach that combines architectural elements and organic body system with contemporary advanced technologies. The proposal not only involves establishing a design protocol that incorporates moss into artificial structures but also celebrates the inherent aesthetic value of the designer's vision itself. Through two different prototypes, varied combination of form, material and construction technique are test to find the appropriate methodology about symbiosis in the architectural element. It is noted this research is still a theoretical and academic research project. Thus, further research and practice are needed to explore material and structural systems for the application in building construction. A promising research evolving from this work would be the reinforcement of architecture through moss's rhizoids. The enhancement of building performance through combination of artificial parts and living organisms can be considered as a solid concept of symbiosis in architecture.

References

Anderson, M., Lambrinos, J., Schroll, E.: The potential value of mosses for stormwater management in urban environments. Urban Ecosyst. **13**, 319–332 (2010)

Angold, P.G., Sadler, J.P., Hill, M.O., et al.: Biodiversity in urban habitat patches. Sci. Total. Environ. **360**, 196–204 (2006)

Belnap, J.: The potential roles of biological soil crusts in dryland hydrologic cycles. Hydrol. Process. **20**(15), 3159–3178 (2006)

Chen, H., Pasquero, C.: Making Matter: Small-Scale Biomorphogenic Prototype Based on Ulva-Algae-Biopolymer. Phygital Intelligence. CDRF 2023. Computational Design and Robotic Fabrication. Springer, Singapore (2024)

Eldridge, D.J., Guirado, E., Reich, et al.: The global contribution of soil mosses to ecosystem services. Nat. Geosci. **16**(5), 430–438 (2023)

Han, K.: Sandgraphy. Accessed from https://sandgraphy.com/sand3dprinter (2017)

George, M.: Modern Terrarium Studio: Design+ Build Custom Landscapes with Succulents, Air Plants+ More. Penguin (2015)

Goddard, M.A., Dougill, A.J., Benton, T.G.: Scaling up from gardens: biodiversity conservation in urban environments. Trends Ecol. Evol. **25**(2), 90–98 (2010)

Heim, A., Lundholm, J., Philip, L.: The impact of mosses on the growth of neighbouring vascular plants, substrate temperature and evapotranspiration on an extensive green roof. Urban Ecosyst. **17**, 1119–1133 (2014)

Jang, K., Viles, H.: Moisture interactions between mosses and their underlying stone substrates. Stud. Conserv. **67**(8), 532–544 (2022)

Kim, B.G., Yoon, M.H., Kim, J., et al.: All-natural moss-based microstructural composites in deformable form for use as graffiti and artificial-porous-material replacement. Materials **15**(24), 9503 (2022)

Kimmerer, R.W.: Gathering Moss: A Natural and Cultural History of Mosses. Oregon State University Press (2003)

Kondo, S., Miura, T.: Reaction-diffusion model as a framework for understanding biological pattern formation. Science **329**, 1616–1620 (2010)

Kurokawa, K.: The Philosophy of Symbiosis. University of Michigan. Academic Editions (1994)

Manso, M., Castro-Gomes, J.: Green wall systems: a review of their characteristics. Renew. Sustain. Energy Rev.v. **41**, 863–871 (2015)

Pasquero, C., Poletto, M.: Deep Green, no 112, pp. 24–30. Green Technologies (2023)

Pasquero, C., Zaroukas, E.: Design prototype. In: Aae 2016, vol. 1, pp. 96–108. Bartlett School of Architecture, University College London (2016)

Seo, Y.B., Dinh, T.V., Kim, S.J., et al.: CO_2 removal characteristics of a novel type of moss and its potential for urban green roof applications. Asian J. Atmos. Environ. **17**, 22 (2023)

Wang, C., Li, H., Neoh, S.: Moss-indoor vertical greenery system design protocol: using moss as an indoor vertical greenery system in the tropics. Indoor Built Environ. **28**(7), 887–904 (2018)

Zhao, X.Y., Zhao, M.Q., Dai, Y.Q., et al.: Influence of surface roughness on the development of moss-dominated biocrusts on mountainous rock-cut slopes in West Sichuan, China. J. Mt. Sci. **20**(8), 2181–2196 (2023)

Hygroflex: A Prototype of a Water-Responsive Facade System with an Innovative Integration of Timber and Biopolymer

Yifan Shi[✉] and Satyam Gyanchandani

The Bartlett School of Architecture, UCL, London, England
yifans1204@gmail.com

Abstract. This research presents a humidity-responsive facade fabricated using advanced techniques such as robotic 3D printing, CNC milled wood, and kerf application. Inspired by nature's adaptability, exemplified by how pinecones respond to humidity levels, the concept of 'animate materials' like wood, which can adapt and respond to their environment, was explored. A circular manufacturing approach was adopted, utilizing CNC milling and robotic 3D printing. Sawdust, a by-product, was transformed with Chitosan-based hydrogel to create a 3D printable material. This material was laid down in three distinct layers, each serving a specific function and forming an active gradient layer that allows for kinetic motion. A material library was established to customize designs, aiming to create spaces that interact with local ecosystems and respond to environmental conditions. This approach challenges the conventional timber industry by introducing innovative materials for construction.

Keywords: Animate Materials · Environmental Adaptability · Biomaterial · Robotic 3D printing · Circular Ecology Manufacturing · Biomimetic · Programmable material · Kinetic Facade

1 Introduction

Examples of Stimuli-Responsive Materials are ubiquitous in nature [1]. These materials systems have the capacity to alter their properties, such as size, shape, and colour in response to external stimuli such as temperature, light, humidity, ph [2]. Animate Materials, a subset of Stimuli-Responsive Materials made by humans designed to mimic the dynamics of responsiveness, have become a major area of research in fields such as bioengineering, mechanical engineering and the built environment [3]. This paper aim outlines a methodology to develop active animate material systems using a combination of material testing, digital fabrication, and computational design tools for applications in the built environment.

This paper delves deeper into Active Animate Materials, which can change their properties change their properties by taking a stimulus from the environment, such as energy from the sun, moisture, or other nutrients [1]. Hygroscopic material behaviour,

H. Chai et al. (Eds.): CDRF 2024, *Symbiotic Intelligence*, pp. 129–142, 2025.
https://doi.org/10.1007/978-981-96-3433-0_12

exhibited by wood and hydrogels, was identified as the stim-ulus and materials for exploration in this paper [4].

Hygroscopic actuation, a process that enables movement without the need for energy, essentially providing plants with inherent responsiveness. Volumetric changes of wood is a well-documented example hygroscopic actuation (commonly known as swelling and shrinkage) [5].

Hydrogel is widely used as a medium for drug delivery in the medical field, of-ten serving as a static container for microorganisms [6]. In medical 3D printing, there is interest in using it as a cell-laden material that can sustain microorganisms, as it can retain water, a vital resource for life. However, if it's left in a dry environment or without any moisture retention measures, it can dehydrate quickly. This presents a significant challenge for its use in medical 3D printing. Graeber et al. (2023) and Narayanan et al. (2018) document the Super Water Absorption Capacity (SWAC) of Hydrogels, with the later paper reporting a SWAC close to 100% [7, 8]. Chitosan was the hydrogel of focus in this paper.

Chitosan is an important composition for making hydrogel. It comes from the outer skeleton of shellfish, including crab, lobster, and shrimp. Its limited strength and high cost made it difficult to use as a construction material. A change in the stiffness of wood is observed when water is adsorbed and released. Hydrogel has the capacity to retain a specific amount of water, which it can transfer to wood when they are in contact. In relation to the steam bending principle, this paper explores the possibility of using hydrogels to bend wood. Therefore, a combination of wood and chitosan composite was selected as a material combi-nation to develop in this paper.

In the wider context of actuation in the built environment, the Institut du Monde Arabe, Jean Nouvel introduced 240 light-sensitive panels on the southern facade. These panels, equipped with sensors and 27,000 camera-like diaphragms, controlled the light entering the building [9]. This analogue yet technologically sophisticated structure echoes Arabic motifs and Mashrabiyas, creating a shading effect that is eye-catching and culturally significant [10]. However, this actuation system deployed was separate to the materials used for shading that would need to be controlled, maintained, and managed. Al Bahar Tower, located in UAE, uses shading screen which is computer-controlled and adjusts according to optimal solar and light conditions. Despite these investments, it did not result in a truly sustainable system for the building due to its high electric consump-tion and complex fabrication process. Therefore, by embedding the system of actuation within the material, leveraging the properties and behaviour to response would simplify the control, management, and maintenance of a façade.

Furthermore, advances in the fields of digital manufacturing and computation-al design enable an extensive design freedom to fabricate complex geometries, previously cost-prohibitive or unfeasible using conventional tools. The tools used in this paper include a combination of robotic material deposition and CNC milling.

Therefore, this paper aimed to develop a workflow that include the testing and hygro-scopic characterization of wood and chitosan, use these as parameters to generate a shading façade structure with an algorithm made in Houdini software and fabricate the design using digital manufacturing tools Grasshopper, Fusion 360 combine with CNC milling and Robotic 3D printing.

2   Methods

This section details the process of refining a concept from creating physical proto-types to gradual scaling. Initial experiments involved hydrogel and timber, hypothesizing that these materials could interact when combined. Small-scale experiments revealed a unique interaction between them. A simple test applied hydrogel to a 10 cm × 10 cm × 3 mm plywood, which bent after one day. This finding prompted further tests to explore the bending phenomenon's limits.

Tests were conducted with various combinations of Chitosan (extracted from seashells), Sodium Alginate (extracted from algae), and Methyl Cellulose (a compound derived from cellulose). The goal was to optimize the hydrogel composition to achieve desired properties for extrusion efficiency, printability, and functional behaviour [11].

2.1 Bi-layer Material Test

The test began by laser cutting plywood of 3 mm thickness into pieces measuring 10 cm by 5 cm (Fig. 1). The mixture of sawdust and hydrogel was composed of:

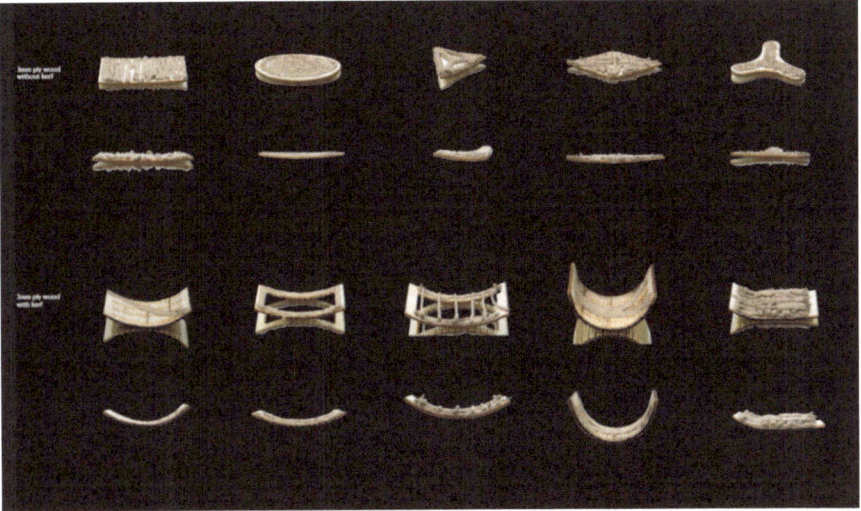

Fig. 1. Laser cut plywood of 3 mm thickness into pieces measuring 10 cm × 5 cm

- Swelling rate: 3% w/w Chitosan in 1% v/v acetic acid solution
- Rheology: 3% w/w Sodium Alginate in 0.1M NaOH solution
- Thickener: 5% w/w Methyl Cellulose in (CHI-3:1)
- Structural: 18% w/w Sawdust/hydrogel solution [11–14]

Shapes were constructed using triangles, circles, and rectangles, both with and with-out kerf. The results indicated that the pieces with kerf bent significantly more than those without, although all samples exhibited some degree of bending.

Same test was repeated on bigger sized timber with gradient in deposited mate-rial (Fig. 2) and gradient in thickness of material (Fig. 3), observations were recorded after the pieces dried out completely.

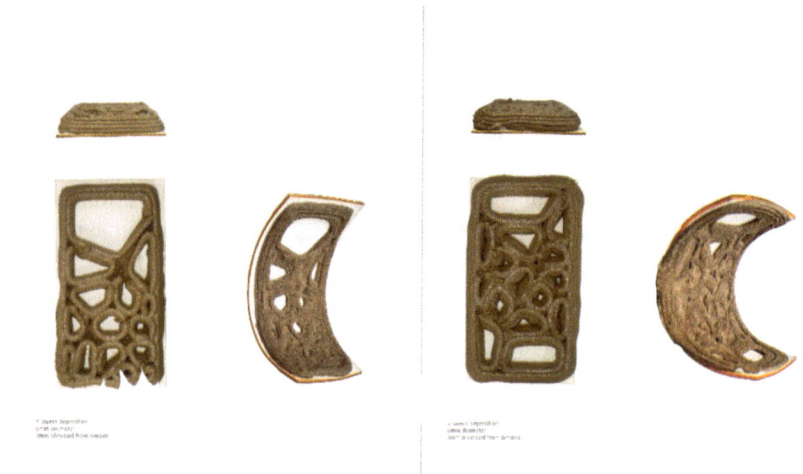

Fig. 2. Bending test with 3 mm plywood

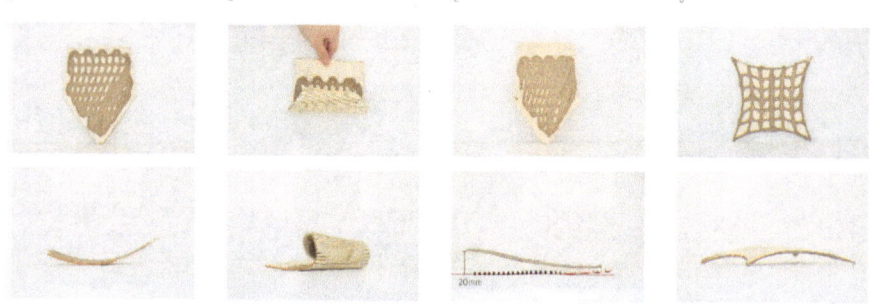

Fig. 3. Bending test with 3 mm plywood

2.2 Material Library Development

This research proposes a material library as a database to integrate computational design and digital fabrication, allowing for on-site customization. Through tailored experiments, a database of material behaviours was established. The relationship between the tool-path/printing density and patterns of hydrogel and the bending ratio and orientation was investigated, exploring as many possibilities as possible. Factors considered included graded printing density and graded kerf density. These material behaviours under various parameters were classified and archived in the material library, serving as a vital resource for customizing designs according to on-site requirements.

To achieve this, 10 cm × 5 cm cutouts were created on a 6 mm thick birch plywood sheet, maintaining one edge connected. This panel was subsequently printed with a hydrogel and timber composite. The following parameters were varied over a range of values:

- Kerf area
- Kerf density
- Printing area
- Printing density
- Printing layers

The setup was allowed to dry, and the deviation of each panel was recorded upon reaching the limit (Fig. 4).

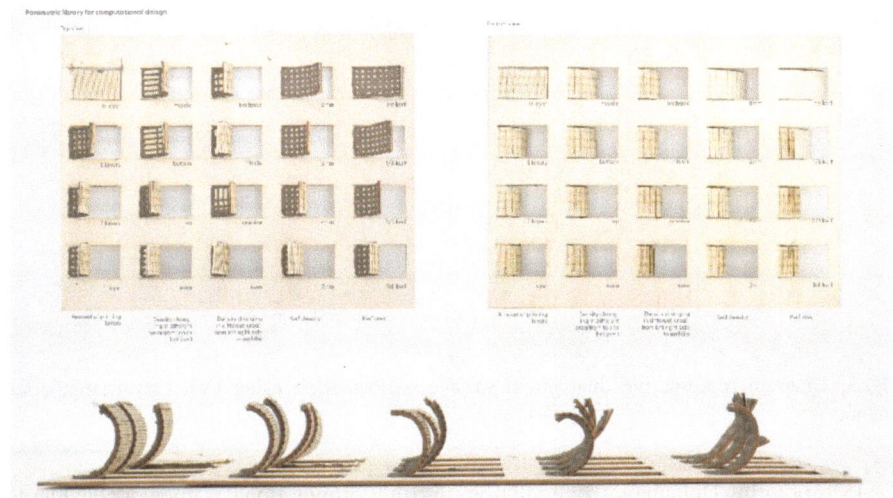

Fig. 4. Material library

2.3 Computational Aided Design

Conventional dynamic facade components are repetitive and modular and rely on specific mechanical joinery. These shading devices are limited to planar surfaces [15]. There is a pressing need to develop a new generation of adaptive building envelopes that can effectively adapt to the complex freeform geometries commonly found in contemporary architectural facades [16]. The parameters of Delaunay's math, driven by environmental analysis, allows for the creation of bespoke segments for optimal environmental response.

By taking inspiration from the peeling of a spherical surface and mapping onto a two-dimensional plane, we can design geometries by identifying optimal parameters within a limited space [17]. This process, when coupled with the utilization of dual graph math [18], enables the identification and optimization, yielding novel outcomes. An investigation of how surfaces can be peeled was done to create organic geometries for a component of a kinetic fa-cade. Existing computer programs and calculations were used to map geometry from a sphere (e.g., an orange) to a two-dimensional plane using optimization algorithms. This can be accomplished by mapping the skeletons and optimizing the peeling, which defines the parameters [17, 18].

The quest for optimal parameters necessitates the establishment of skeletal frameworks, which serve as the foundation for peeling optimization. Here we incorporate the skeletons done by the segmentation behaviour used by Delaunay's triangulations. Delaunay triangulation methodology segments surface into computationally efficient meshes composed of triangles or tetrahedra [19]. This segmentation behaviour not only furnishes us with the requisite skeletons for peeling optimization but also ensures the preservation of structural integrity and geometric fidelity.

DT = delaunay(x, y) creates a 2-D Delaunay triangulation from the points in vectors x and y. DT = delaunay(x, y, z) creates a 3-D Delaunay triangulation from the points in vectors x, y, and z (Fig. 5).

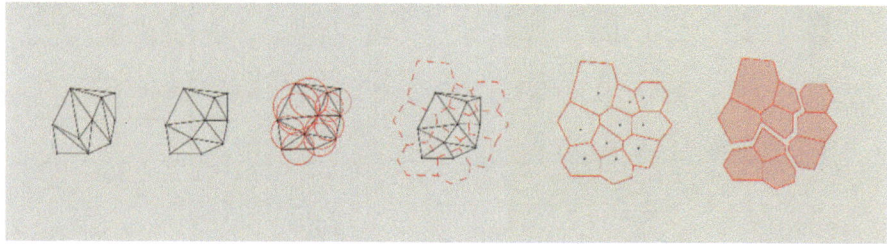

Fig. 5. Diagram representing dual graph surface segmentation using Delaunay triangulation method

Incorporating Delaunay's triangulations into the computational framework enables to navigate the intricate interplay between form and function in kinetic facade design. Thus, by fusing the principles of peeling, dual graph mathematics, and Delaunay triangulations, a course is charted towards the realization of adaptive building envelopes that seamlessly integrate with the complexities of con-temporary architectural landscapes.

Geometry for facade is created by splitting the surface based on Solar analysis data and noise which creates triangulated meshes. Solar analysis for the geometry is conducted using Ladybug plugin in grasshopper for specific location. This data is mapped over the surface in colour values. Since solar data is generates a concentrated colour map for the splitting operation, this map is mixed with a noise to give an even yet systematic spread. A bias is set between both set of data which governs the split based of higher solar radiance and noise concentration. The best results were found with anti-aliasing Simplex and Perlin noise, as the result of splitting created least number of islands from the split (Fig. 6).

These meshes are rearranged based on the Delaunay's triangulations logic. One key characteristic of the Delaunay is that it optimises vertex connections by adding to the triangulations one at a time. The splitting of primitive facets is ruled by noise and shadow

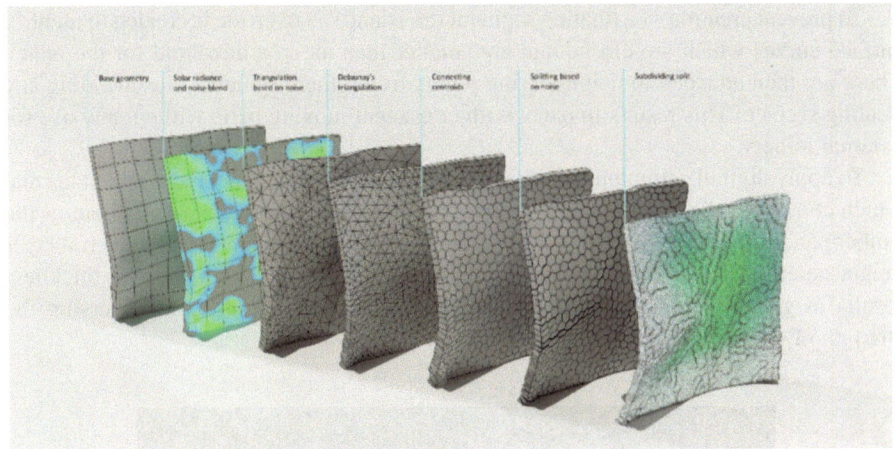

Fig. 6. Stages of splitting surface mesh

analysis. This parameter controls the spread of the split and higher values would incur greater number of moving segments and thus increased sunlight penetration.

To predict the motion, an algorithm is made in Houdini software to simulate the drying and rehydration cycle of the panel. To generate this simulation a spine for each of the peel segment is calculated based on the points on the boundary. Controls are based on a gradient normal from the ends of the spine, with higher values allocated closer to the ends of the segment. Graph and spread of normal on the neighbouring surface points guides the displacement characteristics. These parameters provides control to mimic the material behaviour and response to moisture fluctuation, to simulate the displacement (Fig. 7).

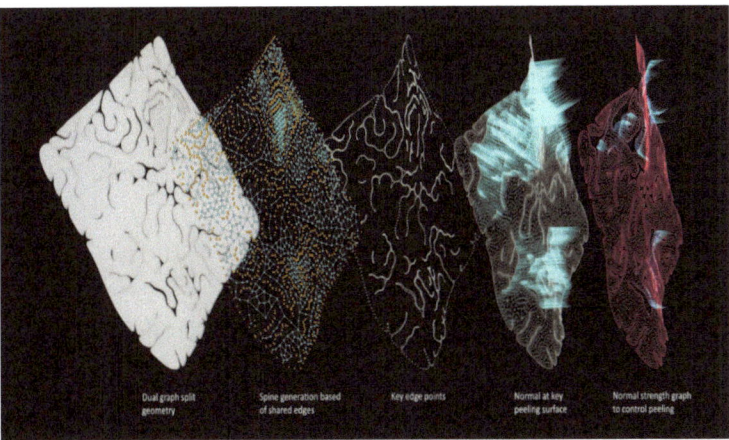

Fig. 7. Stages of surface evaluation to control and simulate surface peel

To prevent creation of a floating segment (an island) an override is created to identify surface curves which are closed and are smaller than an area threshold for the panel. These are then attached to neighbouring points from other segments thus avoiding any floating surface. This results in panel with consistent moving parts without any use for external joinery.

To apply digitally simulated motion to a fabricated timber panel the gradient normal which controlled the direction and motion is translated into a thickness gradient of the timber panel. As tested through material library the timber thick-ness and material layer height determined the deflection amount and direction of the fold. Smaller thickness results in greater fold. This is generated by extrusion gradience from reversing the strength of the normal from the end of the spine (Fig. 8).

Fig. 8. CAD data generation for CNC process

The sine curves are then generated for every distinct piece using its boundary for hydrogel 3d printing over the timber. The density of the curve can be defined according to the layer thickness of the print as they are directly proportional to the displacement. This workflow is used to generate geometries and toolpaths for fabrication using additive and subtractive manufacturing (Fig. 9).

2.4 Fabrication Method

This section explores the strategy of the circular economy using the subtractive-additive manufacturing loop method. Notable material waste was observed when a CNC milling machine processed a piece of wood stock in a workshop. The wood composite, abundant in fibre and large in volume with moisture-absorbing capabilities, was considered for recycling to be used as an aggregate in 3D printing with hydrogel. This method merges hydrogel and sawdust to create a 3D printing material, which is then printed onto the milled wood stock, extending the lifespan of the sawdust.

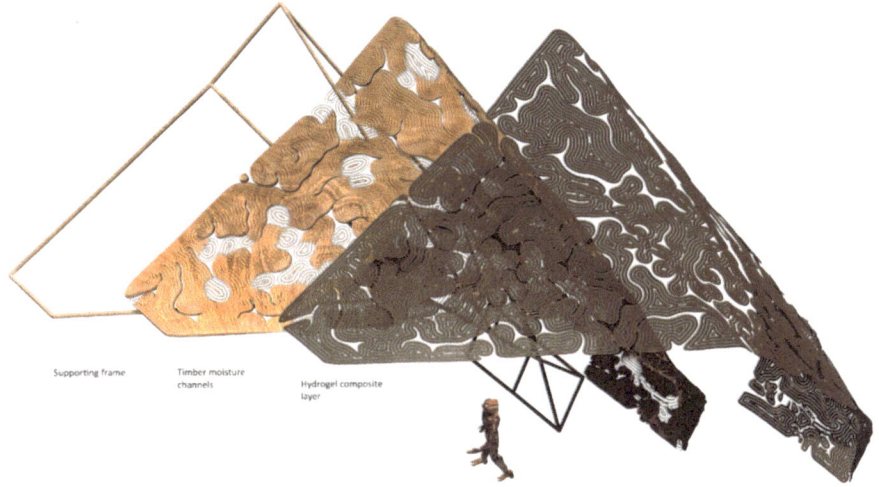

Supporting frame   Timber moisture
                        channels

Hydrogel composite
layer

Fig. 9. Exploded diagram of timber layer and printed layer assembly

Final Prototype

In the last section, the performance of hydrogel-timber bi-layer materials in a real-scale 1:2 application is examined utilizing Robotic 3D extrusion and CNC milling technology. The hydrogel is comprised of Chitosan, Sodium Alginate, and Methylcellulose. Lime tree wood, a deciduous broadleaf tree native to the UK and parts of Europe [20], serves as the support layer. The Felder CNC flatbed machine facilitates the milling process, while the ABB1600 handles non-planar 3D extrusion. The toolpath for CNC milling

Fig. 10. Functional prototype panel

originates from Fu-sion360, and the toolpath for Robotic 3D printing is derived from Houdini and Grasshopper programs. The two parts are joined using a wood joint and a bent metal component to stabilize the structure. After 48 h of complete dehydration, the assembly is finished (Fig. 10).

This protype was then left outside exposed to sunlight, rain and wind for 3 months, a photographic documentation was done during this period to observe the performance of the prototype.

3   Result

3.1   Bi-layer Material Test

The results from this section indicated that the pieces with kerf, bent significantly more than those without, although all samples exhibited some degree of bending (Fig. 2). A larger degree of bending was observed with more layers and varying printing density in different areas. It was observed that strong layer adhesion was achieved in the bilayer material, this synergy increased the bending of the sample.

In the graded material test, the kerfed sample displayed substantial bending compared to the untreated and thicker samples. Notably, the sample with a grad-ed thickness ranging from 6 mm to 20 mm showed no noticeable change (Fig. 3).

To prepare the chitosan mixture significant blending using magnetic stirrer was required to prevent the formation of lumps that would affect the quality of material deposition. This process was a time-consuming challenge and could be source of improvement. These tests helped us identify parameter that can affect the performance of the material and were used to develop a material library.

3.2   Material Library

The material library (Fig. 4) was developed based on classified material behaviours, revealing intricate relationships between tool-path/printing density patterns and the bending ratio and orientation. Key factors identified include graded printing density and kerf density. Utilizing one or two layers maximizes bending, while concentrating printing density near the pin area and along the middle line further increases bend curvature. Increasing kerf density and incorporating a larger kerf area significantly improve bending performance. This comprehensive material library supports customized designs for specific on-site requirements and provides a foundation for future research into additional parameters, optimizing structural applications requiring controlled bending.

The material tests were limited to a small sample of timber thickness and type and the resulting parameters were used to build the consequent prototype. The evaluation strategy for deviation of panel was limited to photographic documentation. These parameters were integrated into a computational workflow, aligning visual observations with digital simulations across a variable range. This integration created a system for producing designs based on these parameters (Fig. 7).

3.3   Computational Aided Design

Integrating Delaunay triangulations within the computational framework optimizes kinetic facade design by merging principles of peeling and dual graph mathematics. This approach creates adaptive building envelopes tailored to complex architectural forms, ensuring consistent movement by attaching closed sur-face curves and eliminating the need for external joinery. Digital simulations translate gradient normals into thickness gradients for timber panels, generating sine curves for hydrogel 3D printing. This work-flow efficiently produces geometries and toolpaths for both additive and subtractive manufacturing.

The form generated through this workflow doesn't account for flow of rainwater through the panel which might be useful to address to optimise efficacy of moisture adsorption. An optimal solution to segmentation and fixing of the panel could be inte-grated in the algorithm. Since the geometry is based on splitting the surface, facade does not provide complete environmental seal.

3.4   Prototype Fabrication

The panels of the prototype successfully opened as simulated digitally after 48 h of dry-ing. Prototype left exposed to environment over 3 months duration underwent approx-imately 25 cycles of rehydration and dehydration due to rain. A photographic docu-mentation of the panel was done to understand the performance. It was observed that motion and direction for the segment remained consistent. There was slight reduction in degree of deviation observed by overlap-ping the image from first and last day of the test (Fig. 11). Possible cause could be loss in adhesion between the Hydrogel and timber due to mechanical impact of the rain.

The fabrication process involved multiple steps of material processing and digital fabrication. Each milled timber panel required calibration for material deposition using a 6-axis robotic arm to enable 3D printing on non-planar surfaces. An offset from the material edge ensured that the material was deposited with-in tolerances.

Fig. 11. Transformation of prototype under environmental exposure

This test suggests the durability of the composite material for multiple rehydration cycles. A controlled test for material test on a surface level for chemical dissolution and mechanical impact deterioration needs to be conducted.

4 Conclusion

Drawing inspiration from nature's hygroscopic actuation mechanism and advancing fabrication techniques, this study endeavours to advance in paradigm shift towards "animate material," demonstrated by wood and chitosan. Through the fusion of advanced algorithms for digital manufacturing and CNC techniques milling and robotic 3D printing, the paper derives a circular manufacturing meth-od. This method repurposes wood sawdust to create optimized layers that work in synergy to achieve a controlled response by tailoring the design through environmental analysis of local ecosystems.

The research creates a material library by changing the features of material layers. This library is fed into a digital algorithm to create output for fabrication, to manufacture facade for a tailored response. The resulting prototype demonstrates the viability of the workflow and a closed manufacturing loop that can translate to create a functional hygroscopic module.

Further study can be conducted to assess and enhance durability of the material in external conditions to understand the number of rehydration cycles the panels can endure before it becomes dysfunctional. The material extruded through the cartridge of the 3D printer is required to be homogeneous and thus requires an efficient process of blending. The timber used for testing and prototyping was done with small samples of wood. A rigorous test on the type of timber for the panels and sawdust may further improve functionality and durability. A better evaluation strategy for example machine learning could be used to analyse the panel's efficacy over a period.

The segments of the panel are based on single-edge constraints, an exploration of different patterns such as auxetics can be done to create various opening systems for a facade. An optimisation of the design is required to maximise the moisture capture as well as create a seal from the external environment. The fabrication process involves several manual steps, such as material-making and calibration for 3D printing. These processes can be automated to achieve scalability. The responsive system achieved in this paper can be further explored as other architectural elements that can replace an active kinetic system.

The research provides a speculative glimpse into the future of kinetic facade systems and the potential for sustainable material using advancing manufacturing practices. By tailoring designs to respond dynamically to local ecosystems and environmental conditions, we aim to create spaces that actively engage with their surroundings, contributing to the evolution of resilient and adaptive built environments.

References

1. Rivera-Tarazona, L.K., Campbell, Z.T., Ware, T.H.: Stimuli-responsive engineered living materials. Soft Matter. **17** (2021)

2. Bril, M., Fredrich, S., Kurniawan, N.A.: Stimuli-responsive materials: a smart way to study dynamic cell responses. Smart Mater. Med. **3** (2022)
3. Khizar, S., Zine, N., Errachid, A., Elaissari, A.: Introduction to stimuli-responsive materials and their biomedical applications. In: ACS Symposium Series (2023)
4. Koch, S.M., Goldhahn, C., Müller, F.J., Yan, W., Pilz-Allen, C., Bidan, C.M., et al.: Anisotropic wood-hydrogel composites: extending mechanical properties of wood towards soft materials' applications. Mater. Today Bio. **22** (2023)
5. Peck, E.C.: How wood shrinks and swells. For. Prod. J. **7**(7) (1957)
6. Croisier, F., Jérôme, C.: Chitosan-based biomaterials for tissue engineering. Eur. Polym. J. **49** (2013)
7. Graeber, G., Díaz-Marín, C.D., Gaugler, L.C., Zhong, Y., El Fil, B., Liu, X., et al.: Extreme water uptake of hygroscopic hydrogels through maximized swelling-induced salt loading. Adv. Mater. **36**(12) (2024)
8. Narayanan, A., Kartik, R., Sangeetha, E., Dhamodharan, R.: Super water absorbing polymeric gel from chitosan, citric acid and urea: synthesis and mechanism of water absorption. Carbohydr. Polym. **1**(191), 152–160 (2018)
9. Decker, M., Zarzycki, A.: Designing resilient buildings with emergent materials [Internet] [cited 2024 May 31] (2014). Available from: https://www.researchgate.net/publication/271 329050_Designing_Resilient_Buildings_with_Emergent_Materials
10. Eduardo, S.: ArchDaily. Should Architecture Be Static? The Possibilities of Kinetic Buildings (2023)
11. Laia Mogas Soldevila by, Program in MedaA A. Water-based digital design and fabrication: material, product, and architectural explorations in printing chitosan and its composites. SMArchS Design Computation. (2015)
12. Correa, D., Papadopoulou, A., Guberan, C., Jhaveri, N., Reichert, S., Menges, A., et al.: 3D-printed wood: programming hygroscopic material transformations. 3D Print Addit. Manuf. **2**(3) (2015)
13. Klemmt, C., Meibodi, M.A., Beaucage, G., McGee, W.: Large-scale robotic 3D printing of plant fibre and bioplastic composites. In: Proceedings of the International Conference on Education and Research in Computer Aided Architectural Design in Europe (2022)
14. Panagiotidou, V., Koerner, A., Cruz, M., Parker, B., Beyer, B., Giannakopoulos, S.: 3D extrusion of multi-biomaterial lattices using an environmentally informed workflow. Front. Archit. Res. **11**(4) (2022)
15. Vazquez, E., Correa, D., Poppinga, S.: A review of and taxonomy for elastic kinetic building envelopes. J. Build. Eng. **1**(82), 108227 (2024)
16. Barozzi, M., Lienhard, J., Zanelli, A., Monticelli, C.: ScienceDirect the sustainability of adaptive envelopes: developments of kinetic architecture. Procedia Eng. **155**, 275–284 [Internet] [cited 2024 Jun 1] (2016). Available from: www.sciencedirect.com
17. Swart, D.: Orange Peel Optimization
18. Zhang, L., Li, X., Arnab, A., Yang, K., Tong, Y., Torr, P.H.S.: Dual graph convolutional network for semantic segmentation. In: 30th British Machine Vision Conference 2019, BMVC 2019 [Internet] [cited 2024 Jun 1] (2019). Available from: https://arxiv.org/abs/1909.06121v3
19. Weatherill, N.P.: Delaunay triangulation in computational fluid dynamics. Comput. Math. Appl. **24**(5–6), 129–150 (1992)
20. Logie Timber. Logie Timber. Lime (2024)

Optimizing Chitosan-Cellulose Composites for Sustainable 3D Printing

Junyao Hou, Hao Wu, Emily Liu, and Philip F. Yuan[(✉)]

College of Architecture and Urban Planning, Tongji University, Shanghai, China
philipyuan007@tongji.edu.cn

Abstract. This study aims to optimize the additive manufacturing of chitosan/cellulose materials, enhancing their applicability. According to the United Nations Environment Programme (UNEP) report, 40% of global greenhouse gas emissions come from the construction industry. Our goal is to reduce emissions in construction, improve material efficiency, and lower emission intensity. The traditional linear economic model in construction leads to significant CO_2 emissions and construction waste. Therefore, based on the principles of a circular economy, particularly in 3D printing technology, we investigated chitosan and cellulose as biodegradable materials, which are natural high-molecular-weight polymers. Past research has identified shrinkage issues in 3D printing using chitosan and cellulose. By optimizing the printing paths to increase the material's contact area with air, we aimed to reduce overall structural shrinkage, enhancing the application scenarios of this novel material in construction. The research covers three dimensions: materials, software, and hardware. Experimental results demonstrate that the addition of paper fibers not only reduces shrinkage and cracking but also improves strength, making the material more stable. We utilized Rhino-Grasshopper software for analysis and path generation and integrated robots and extrusion devices in 3D printing technology. Through the implementation of the material-design-build process, we aim to enhance the prospects of degradable materials in the construction industry, mitigate the negative impacts on the building lifecycle, and promote a more environmentally friendly, conscious, and energy-efficient construction sector.

Keywords: Bio-material · Additive manufacture · Sustainable Design · Optimize System

1 Introduction

The construction industry, characterized by its substantial consumption of materials such as concrete, steel, aluminium, glass, and bricks, is a significant contributor to carbon emissions, accounting for 9% of global carbon emissions as reported (Global-ABC 2022). In order to align with the United Nations Sustainable Development Goals (SDGs) and to effectively tackle the climate crisis, it is imperative for the building sector to devise proactive and actionable strategies aimed at curtailing carbon emissions, particularly by focusing on the reduction of emissions associated with construction

© The Author(s) 2025
H. Chai et al. (Eds.): CDRF 2024, *Symbiotic Intelligence*, pp. 143–153, 2025.
https://doi.org/10.1007/978-981-96-3433-0_13

materials. The exploration of low-carbon construction materials for achieving environmentally friendly construction is both feasible and effective, relying on breakthroughs in emerging construction materials.

Currently, in the field of 3D printing in construction, plastics (used for decorative panels) and concrete printing (used for structural elements) are commonly employed. However, both of these printing methods result in high carbon emissions. Plastic printing involves high energy consumption due to the heating process during thermal melting, while concrete printing requires a large amount of cement, contributing to high carbon emissions associated with cement production. In this context, biodegradable 3D printing technology has emerged as a powerful tool to address energy issues. 3D printing technology itself reflects high efficiency in resource utilization, reducing material waste through precise additive manufacturing. Furthermore, the use of biodegradable materials in 3D printing not only reduces dependence on traditional petroleum-based plastics but also allows the printed products to naturally decompose under degradation conditions, thereby alleviating environmental pressure.

The technology currently employed for 3D printing of chitosan composite materials at the architectural scale involves printing composite materials in a water-based state and subsequently drying them into shape. This process enables the completion of large-scale experiments with chitosan composite materials, achieving a printing process akin to plastic film production (Mogas-Soldevila and Oxman 2015). Moreover, the innovation of large-scale additive manufacturing printing has been introduced by utilizing chitosan solution combined with cellulose, marking the inception of a printing method for chitosan-cellulose composite materials in large-scale production (Burry et al. 2020; Sanandiya et al. 2018). Chitosan and cellulose materials are suitable for local manufacturing, thereby reducing carbon emissions during the transportation process. The unique additive manufacturing process allows materials to be used only where necessary. Compared to traditional subtractive manufacturing processes, this significantly reduces the use of raw materials and waste generation (Sanandiya et al. 2020).

Differences in material purity result in variations in formulations, indicating that there is still research space for 3D printing formulations. This study explores the performance optimization of chitosan/cellulose materials, testing formulations with wood chips, paper fibers, sodium alginate, and others. Through experimental conclusions, it is determined that the addition of paper fibers produces good compressive strength and reduces shrinkage. The study also involves experimentation with extrusion parameter tuning, angle rotations, weaving paths, etc., in the design of printing paths. Introducing porosity in the printing path design enhances interlayer drying areas. The research aims to summarize the impact of different printing paths on the shrinkage performance of natural materials.

2 Methods

This paper aims to explore the application of chitosan and cellulose materials in additive manufacturing through an integrated approach of material-design manufacturing. The study focuses on enhancing the printing methods for chitosan and cellulose in architectural applications. It involves the design and documentation of printing paths to observe changes in the shrinkage performance of biodegradable materials. Through this integrated design exploration, the research contributes to advancing the study of printing methods for chitosan and cellulose in architectural contexts (Fig. 1).

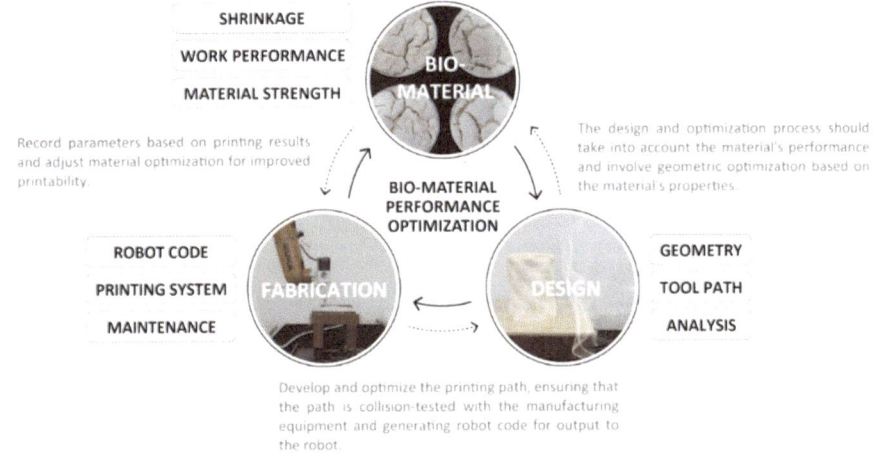

Fig. 1. "Materials-design-construction process" diagram

3 Preliminary Material Investigation

3.1 Characterization Images

SEM scanning electron microscopy was employed for characterization observations by gold-sputtering the surfaces. Figure 2A illustrates the characterization of films formed from a 3% chitosan solution after drying, observed at a magnification of 1 micron. Figure 2B displays the characterization of chitosan solution mixed with cellulose, while Fig. 2C shows the characterization of a mixture of chitosan solution, cellulose, and paper fibers.

Fig. 2. SEM (scanning electron microscope) characterization image

3.2   Material Preparation Method

See Fig. 3.

3.3   Material Formulation

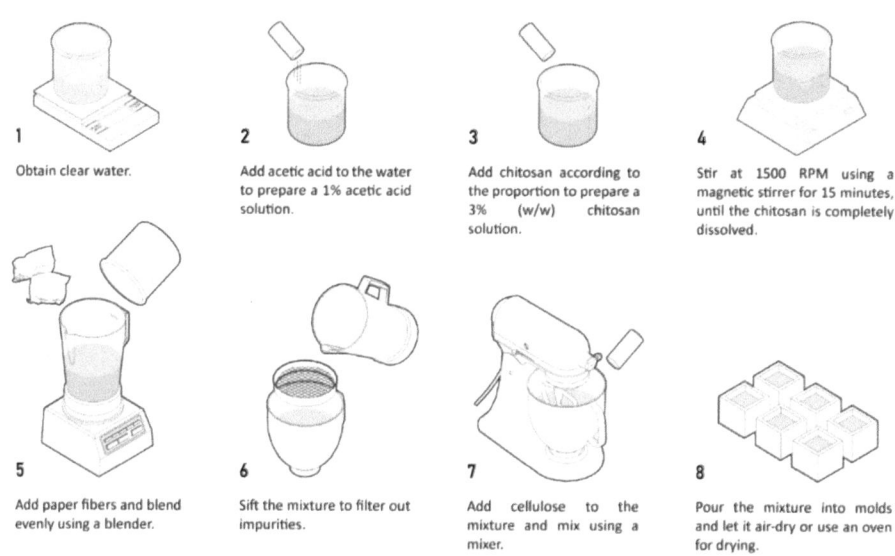

1 Obtain clear water.	**2** Add acetic acid to the water to prepare a 1% acetic acid solution.

Fig. 3.   Material preparation process

Phase One: Chitosan and Cellulose

A 3% chitosan solution is commonly acknowledged as the most stable and user-friendly composition in the literature (Giachini et al. 2020). In our experiments, we tested this chitosan solution with added cellulose to observe changes in various strengths and collected parameters.

Our findings indicate that raw material performances vary among different companies, making it challenging to obtain identical materials for formulation experiments. Therefore, this study conducted material formulation experiments using chitosan and cellulose readily available domestically, as depicted in Fig. 4. We also compared the variations in strength for each formulation with those reported by foreign teams (Dritsas 2022). In our experiments, the Chitosan:cellulose ratio of 1:17 exhibited optimal printing performance and strength.

Phase Two: Addition of Materials

The benchmark formulation for this study is based on a ratio of chitosan (dissolved in 1% acetic acid at a 3% w/w concentration) to cellulose at 1:17, determined through appropriate 3D printing experiments. Additional components, including xanthan gum, agar, sodium alginate, wood chips, corn flour, and paper fibers, were introduced into

Fig. 4. Chitosan and cellulose poured samples with varying ratios

the formulation as illustrated in Fig. 5 (Wei et al. 2022). Throughout the experimental process, xanthan gum, wood chips, cornstarch, and flour exhibited excellent flow characteristics. The components most effective in enhancing working performance were found to be wood chips and flour.

Upon drying, experimental results demonstrated excellent anti-shrinkage properties in wood chips, accompanied by a reduction in material weight. Xanthan gum contributed to improved working performance and increased material strength. Paper fibers were observed to reduce shrinkage and enhance overall structural strength.

Fig. 5. Poured samples with added composite materials

Phase Three: Incorporation of Interactive Additives

In this stage, the benchmark ratio is established with a combination of chitosan (dissolved in 1% acetic acid at a 3% w/w concentration) and cellulose in a 1:17 proportion. The key experimental materials for this group include xanthan gum, wood chips, and paper fibers. According to experimental observations as depicted in Fig. 6, xanthan gum exhibits advantages in enhancing structural performance, effectively increasing working performance, and improving strength. The combination with paper fibers addresses shrinkage issues and complements the working performance of paper fibers.

The mixture of wood chips and paper fibers proves effective in reducing material weight and minimizing shrinkage. Through a comparative analysis of two sets of experiments, it is observed that the combination of xanthan gum and paper fibers demonstrates superior compressive resistance and possesses good printing performance. On the other hand, the formulation involving wood chips and paper fibers shows better advantages in lightweight while maintaining excellent working performance.

Conclusion

Based on the results obtained from material experiments (see Table 1), this study found that a chitosan-to-cellulose ratio of 1:17 achieves the appropriate printing viscosity and strength for chitosan composite materials. Using chitosan as the base, wood chips were identified as an effective means to reduce weight and shrinkage. However, this led to

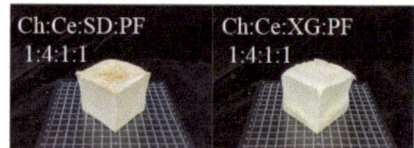

Fig. 6. Poured samples with a composite formula

a decrease in material strength. Xanthan gum strengthened the structural performance of the composite material but caused material dry shrinkage. Paper fibers were effective in reducing shrinkage and enhancing structural performance, but they resulted in poor flowability, making it challenging for easy printing. Other experimental materials exhibited excessive shrinkage and were deemed unsuitable for this study. In conclusion, paper fibers were identified as the primary additive material in this experiment, added to chitosan and cellulose in the ratio of 1:17:1.

4 Performance Strength

In this experiment, a universal testing machine (WDW-100KN) was used for compressive testing. The specimen surfaces were cut to the same dimensions (50 mm 50 mm 50 mm) and polished. The compressive tests were conducted at a constant speed of 4mm/min, under an ambient temperature of 20 °C.

Based on the material performance records and tests, as shown in Table 1, Sample03 exhibits the best compressive performance among the tested formulations, surpassing others significantly, with a maximum compressive strength of 58 Mpa, equivalent to the strength of reinforced gypsum (50 Mpa). Therefore, this study selects Sample04 as the final printing material and evaluates its printing performance.

Robot Printing System

While the experimentation, the KUKA company's robotic arm model KR6-R900 was employed. This six-axis robotic arm is characterized by a maximum payload capacity of 6 kg and an operational range extending up to 900 mm. The motor drive is facilitated by the YF-31, enabling a peak speed attainment of 999.9 RPM. In terms of the extrusion system, a pneumatic cylinder piston syringe was utilized, actuated by a compressed air system operating at a maximum pressure of 0.8 MPa.

The syringe, with a volumetric capacity of 2 L, features a discharge diameter of 10 mm, while the printing extruder incorporates an inner diameter of 5 mm see Fig. 7.

5 Print Performance Experiment

This study explores the potential improvement of the inherent shrinkage characteristics of natural materials by testing different printing paths. Through various control experiments, the material's design is optimized, ultimately leading to the derivation of an optimized performance method for printing paths.

Table 1. Performance Parameter Charts.

Ch = Chitosan
Ce = Cellulose
SD = Sawdust
PF = Paper fibers
XG= Xanthan gum
F   = Flour
Co = Cornstarch
SA = Sodium alginate

Sample	The ratio of materials		Decrease in weight (%)	Srinkage(%)	Compressive strength(Mpa)
01	Ch : Ce	1:17	61.28	61.28	16
02	Ch:Ce:SD	1:17:1	66.05	90.35	5
03	Ch:Ce:PF	1:17:1	66.10	86.41	58*
04	Ch:Ce:XG	1:17:1	65.75	90.61*	26
05	Ch:Ce:F	1:17:1	66.83	90.60	Damage caused by shrinkage
06	Ch:Ce:Co	1:17:1	66.26	89.87	Damage caused by shrinkage
07	Ch:Ce:SA	1:17:1	65.84	88.87	21
08	Ch:Ce:XG:PF	1:17:1:1	67.04	88.51	37
09	Ch:Ce:SD:PF	1:17:1:1	68.20 *	86.35	39

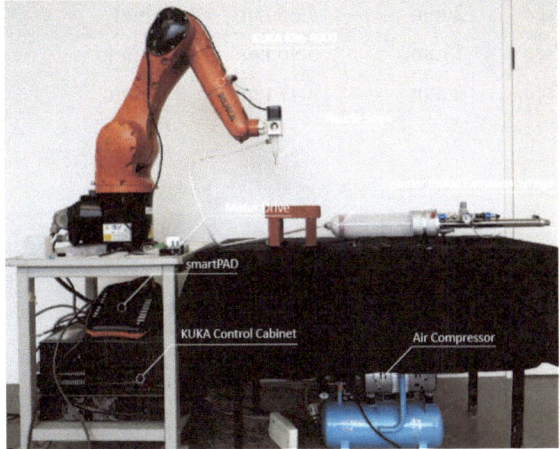

Fig. 7. Robotic printing system

5.1 Extrusion Basic Parameters

The focus of this set of experiments is to record the printing parameters of chitosan composite materials with a 5mm extrusion nozzle, pressurized using a 0.5Mpa air compressor. The stable extrusion of chitosan composite materials is recorded under controlled printing parameters. The observations and records, including print width, speed, and layer height, as shown in Fig. 8, are used to determine suitable printing parameters for the foundation of path design.

Table 2 presents the results of two groups of experiments. The first group maintains a fixed layer height of 2 mm, while the second group experiments with a fixed speed. In the first group, it is observed that slower speeds result in wider print widths and reduced shrinkage. However, practical observations reveal that excessively slow speeds lead to serious material stacking. At a speed of 30 mm/sec, the print width narrows, causing poor print quality, although shrinkage is minimized. In the second group, it is observed that increasing layer height optimally reduces shrinkage. While both groups of experiments indicate that higher speeds and layer heights can better minimize shrinkage, actual printing parameters should be chosen based on printing stability to achieve good printing performance.

Table 2. Printing basic parameters and test results

Sample	Printing speed	Layer height	Print width	Width after drying	Shrinkage rate
PD01	3 mm/sec	2 mm	15.74 mm	13.13 mm	83.42%
PD02	5 mm/sec	2 mm	12.32 mm	9.4 mm	76.30%
PD03	10 mm/sec	2 mm	7.64 mm	5.78 mm	75.65%
PD04	30 mm/sec	2 mm	4.73 mm	4.02 mm	84.99%
PD05	10 mm/sec	1 mm	9.37 mm	7.47 mm	79.72%
PD06	10 mm/sec	2 mm	7.65 mm	5.8 mm	75.82%
PD07	10 mm/sec	3 mm	5.96 mm	4.35 mm	72.99%
PD08	10 mm/sec	4 mm	4.47 mm	3.86 mm	86.35%

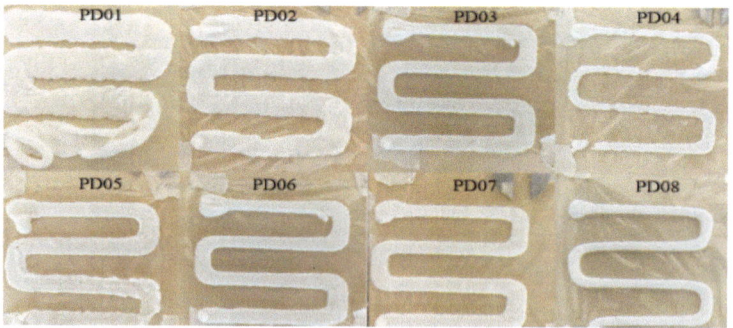

Fig. 8. Records of printing tests based on basic parameters

5.2 Angle Rotation

The experiment involved testing the effect of different angle rotations on the degree of printing shrinkage and cracking using a planar rotation method. The tests were conducted under the conditions of a speed of 10 mm/sec and extrusion pressure of 0.5Mpa. The experiment was grouped into rotations of 15 degrees, 30 degrees, 45 degrees, 60 degrees, 75 degrees, and 90 degrees. Based on the experimental results, it was observed that higher rotation angles led to more even drying and less deformation in the printed products see Fig. 11. Under angle rotation, the printing paths created different angles of stacking, compared to vertically stacked printing, allowing for increased air contact area. This contributed to a more stable drying process, reducing the likelihood of cracking and minimizing shrinkage (Fig. 9).

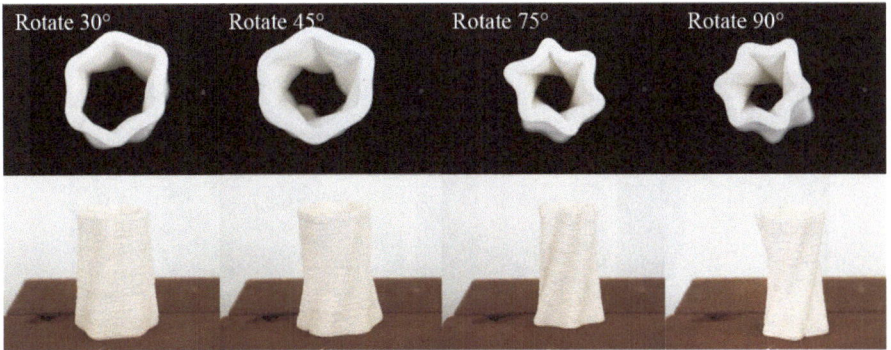

Fig. 9. Printing forming at a rotating angle

5.3 Overhang Printing

In the overhang angle experiment, we inclined geometric shapes horizontally at angles of 82.5 degrees, 75 degrees, and 67.5 degrees. During the testing process, we successfully achieved the overhang printing at 82.5 degrees, followed by the completion of the 75-degree overhang printing. The attempt at 67.5 degrees eventually collapsed. In this set of experiments, as illustrated in Fig. 10. The positioning of the center of gravity proved to be a crucial factor influencing the success of overhang printing. In addition to the horizontal direction, the consideration of the center of mass is also essential in the design.

Fig. 10. Printing forming at overhang test

5.4 Discussion

Degradable materials demonstrate significant sustainable development potential in the construction industry. The extrusion-based prototyping process enhances the customizability of materials in the architectural domain. Based on experimental results, control of shrinkage rates under various printing parameters is achieved and used as a benchmark to print cylindrical geometrical forms. Architectural application scenarios can be realized through unitized printing in this experiment. Additionally, a printed vase was placed in a pot for a month to test degradation, and the part buried in the soil has shown signs of structural loosening (Fig. 11).

Fig. 11. (Left) Print BIO-degradable cylinder based on experimental parameters, (middle and right) After a month of pot degradation testing.

References

GlobalABC.: 2022 Global Status Report for Buildings and Construction. Retrieved 9 Nov 2022, from https://globalabc.org/index.php/news/globalabc-releases-2022-global-status-report-buildings-and-construction (2022)

Bari, E., Morrell, J.J., Sistani, A.: Durability of natural/synthetic/biomass fiber–based polymeric composites. In: Durability and Life Prediction in Biocomposites, Fibre-Reinforced Composites and Hybrid Composites, pp. 15–26. Elsevier (2019). https://doi.org/10.1016/B978-0-08-102 290-0.00002-7

Lujan, L., Goñi, M.L., Martini, R.E.: Cellulose-chitosan biodegradable materials for insulating applications. ACS Sustain. Chem. Eng. **10**(36), 12000–12008 (2022). https://doi.org/10.1021/acssuschemeng.2c03538

Burry, J., Sabin, J., Sheil, B., Skavara, M.: Fabricate 2020: making resilient architecture. UCL Press (2020). https://doi.org/10.2307/j.ctv13xpsvw

Sanandiya, N.D., Vijay, Y., Dimopoulou, M., Dritsas, S., Fernandez, J.G.: Large-scale additive manufacturing with bioinspired cellulosic materials. Sci. Rep. **8**(1), 8642 (2018). https://doi.org/10.1038/s41598-018-26985-2

Wei, Y., Markopoulou, A., Zhu, Y., Martin, E.C., Kirova, N.: Additive manufacture of cellulose based bio-material on architectural scale. In Yuan, P.F., Chai, H., Yan, C., Leach, N. (eds.) Proceedings of the 2021 DigitalFUTURES, pp. 286–304. Springer Singapore (2022). https://doi.org/10.1007/978-981-16-5983-6_27

Dritsas, S., Vijay, Y., Dimopoulou, M., Sanadiya, N., Fernandez, J.G.: An additive and subtractive process for manufacturing with natural composites. Rob. Fabrication Archit. Art Des. **2018**, 181–191 (2019). https://doi.org/10.1007/978-3-319-92294-2_14

Dritsas, S., Halim, S.E.P., Vijay, Y., Sanandiya, N.G., Fernandez, J.G.: Digital fabrication with natural composites. Constr. Rob. **2**(1), 41–51 (2018). https://doi.org/10.1007/s41693-018-0011-0

Breseghello, L., Sanin, S., Naboni, R.: Toolpath Simulation, Design and Manipulation in Robotic 3D Concrete Printing (2021). https://doi.org/10.52842/conf.caadria.2021.1.623

Algae Reactor: A 3D-Printed Façade Module for Cultivating Chlorella with Indoor CO$_2$

Xinchang Chen[✉], Yue Zhou, Xini Chai, Muchun He, and Hao Hua

Southeast University, No.2 Sipailou, Nanjing, China
xinchang.chen@seu.edu.cn

Abstract. This paper introduces a 3D-printed façade module that cultivates Chlorella with CO$_2$ to purify indoor air for inhabitants and supply biomass. Algae produces oxygen and biomass and enhances carbon sequestration through photosynthetic processes. The idea of integrating algae into architecture as bioreactors has been developed in recent years. In this research, the façade module consists of a hybrid framework and an algae cultivation apparatus. The hybrid framework is composed of aluminum profiles, a 3D-printed skin, slim solar panels for the pump battery, and fasteners. The algae culture system regulates the photosynthesis inside the tubular liquid to interact with the indoor air. The module's freeform skin with grooves is 3D printed with large-scale Fused Granulate Fabrication (FGF). The convoluted tube along the grooves is always ascendant to make the air travel slowly but smoothly from bottom to top. The prototype installed on the building façade enables the algae organisms in the bioreactor to grow stably. This project introduces biochemistry processes into sustainable design toward metabolism in the built environment.

Keywords: Photosynthetic · Carbon Sequestration · Indoor Air Quality · Façade Module · 3D Printing · Algae

1 Introduction

The building and construction industries have been criticized for the large resource consumption and environmental pollution. Building operations also have a large amount of energy demand with a substantial carbon footprint. According to the UN Environment Programme 2022 Global Status Report for Building and Construction, operational energy demand in buildings (such as heating, cooling, lighting, and cooking) has grown to 135 EJ, which is an increase of around 4% from 2020 and exceeds the previous peak in 2019 by over 3% [20].

Hybrid bio-architecture system can reduce carbon emissions in the built environment and improve indoor comfort by incorporating biological elements with building structures and creating microbiological cycles in buildings. Using microalgae to treat waste gas, waste water and produce organic biomass has entered the research field. This work applies an algal biomass cultivation system to building façade through ancillary façade modules. The hybrid bio-architecture system as the medium builds a symbiotic relationship between humans and living organisms in the built environment.

© The Author(s) 2025
H. Chai et al. (Eds.): CDRF 2024, *Symbiotic Intelligence*, pp. 154–163, 2025.
https://doi.org/10.1007/978-981-96-3433-0_14

1.1 Algae Cultivation and Building Integration

Microalgae are aquatic photosynthetic microorganisms that form the largest group of primary producers and an essential source of oxygen on Earth, rendering them excellent carbon fixers. In a study of capturing CO_2 flue gas for cultivation of spirulina culture, the microalgae can fixate an average of 0.78 g CO_2 per gram of biomass and produce 39 g/m^2 of biomass per day [17]. Cultivation of the Spirulina platensis can reduce CO_2 in the atmosphere. Algae also produces biodegradable, renewable, non-toxic biofuels. Bioenergy has become the focus of research and development in various countries. Fast-growing microalgae can be a constant source of valuable natural compounds in food, animal feed, pigments, cosmetics, and pharmaceuticals [4]. Thus, our cities can benefit from the fast-growing algae which convert CO_2 into valuable nutrients.

There have also been pioneer projects that integrate algae cultivation into buildings. SolarLeaf is a façade system designed to cultivate microalgae in flat panel photobiore-actors to sequester carbon dioxide and harvest heat for residential heating [21]. The flat photobioreactors are used for algae culturing by compressing air from the bottom of the bioreactor into the water flow, creating large bubbles that stimulate the algae to absorb CO_2 and light. The generated biomass also renders the device a dynamic shading system. The Urban Algae Façade presented at the Milan Design Week in 2014 is a prototype made of two layers of plastic material for algae cultivation. Algae water is brought into each module through a primary inlet [2]. Once it fills up it will overflow into the main inlet that leads to the next module through a straw-like outlet tail. The proposed algae façade can generate energy and produce food. These experiments have shown the potential of combining algae culture with building façades.

1.2 Additive Manufacturing for Architectural Modules

Additive manufacturing, also known as 3D printing, is an emerging fabrication technology for the production of customizable components. Additive manufacturing can materialize complex forms efficiently, ideal for realizing multiple functionalities in a single element. Printing large-scale complex components often employs the robotic arm with six degrees of freedom and a large printing range. The toolpath for 3D printing is usually encoded in a specific robot's language, such as KUKA Robot Language (KRL). The toolpath specifies the trajectory of the nozzle, tool orientation (Euler angles), extrusion speed (kg/h), and feed rate (mm/min).

Thermoplastic polymers have shown advantages in printing transparent components. The application of thermoplastic 3D printing in functional facades has gained an increased interest. TU Delft developed 3D-printed facades with internal cellular structures for thermal insulation [14]. ETH Zurich developed a computational design tool for manipulating infill structures and assembly details of 3D-printed façades [19]. Researchers have also used translucent/transparent thermoplastic materials to create functional prototypes at an architectural scale. Tenpierik explored the combination of phase-changing materials (PCM) with thermoplastic 3D-printed elements to create a Trombe wall. Using smaller, stacked PCM modules could reduce convective heat transfer in the vertical direction [16].

2 Methodology

The biomimetic module, termed with algae reactor, realizes the long-term cultivation and regular recycling of algae organisms on building façade. Our approach includes energy recycling design, façade module development, and digital fabrication. Continuous bubbles within the algae culture tube are driven by an air pump to enhance the biochemistry processes. Solar panels are employed to provide electricity for the air pump. The cultivated algae can be collected and converted into biomass energy for multiple uses.

2.1 System Design of the Modular Reactor

The hybrid framework integrates the algae culture system and facilitates the installation of the module onto the interior side of the building façade (Fig. 1). Constructed from vertical and horizontal aluminum profiles, the framework consists of a 3D-printed skin with grooves, slim solar panels for powering the pump battery, and fasteners for assembly. The components are dimensioned for ease of installation. The 3D-printed skin with the appropriate notch size is fitted on the aluminum profile, which can be quickly disassembled and repaired; The air pump and battery pack are hidden in the bottom recesses on the back of the module.

The system features transparent corrugated tubes that contain water and Chlorella. These tubes are mounted onto the grooves of the 3D-printed skin, creating a visually appealing and functional element. The living Chlorella algae serves two purposes: purifying the indoor air and providing a source of biomass for potential use in biofuel production or other applications.

The geometry of the 3D-printed skin is crucial to the system's performance. By employing computational design techniques, the surface texture of the skin can be optimized to regulate the light intensity into the interior. This not only creates a comfortable

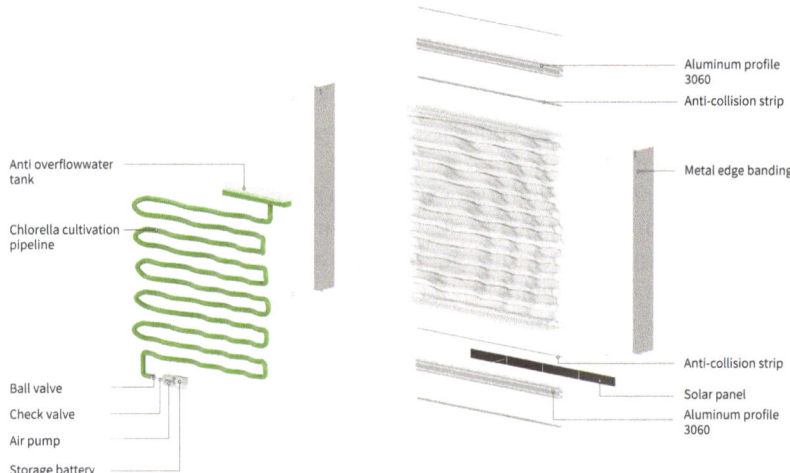

Fig. 1. System design of the modular algae reactor.

indoor environment but also helps algae cultivation by providing a better light source for photosynthesis.

Such an ancillary module for windows allows for scalability and flexibility in its application. Each modular unit can be seamlessly integrated into a large façade through node assembly. The frame is hidden behind the translucent skin and fixed on the structure of the building. A large matrix of modules installed behind the building façade can provide a significant surface area for the algae to thrive.

2.2 Algae Culture System

The algae cultivation system organizes biomass production and energy conversion (Fig. 2). During the first installation or maintenance, the mixed solution of Chlorella and nutrients is poured into the culture tube from the top notch of the module. The air pump provides a continuous air flow to the convoluted tube, allowing indoor gases to react with the Chlorella solution and enhance the biological reaction. After one cycle of cultivation, Chlorella can be collected by opening the control valve at the bottom of the module.

2.2.1 Algae Cultivation

Open ponds and photobioreactors (PBRs) are the two main methods for algae cultivation. There are both enclosed and open PBRs. Open PBRs with simple structure, convenient operation, and low operational cost, are commonly used in industrial cultivation. However, the algae species are susceptible to pollution and the culture conditions are not controlled, which is only suitable for limited species of algae. The PBR system usually consists of a feeding vessel that cycles the algae and a solar array that allows algal cells to be exposed to sunlight. As a façade element, the PBR system can filter air, regulate light, and provide acoustic and shading functions [9].

Algae grow under a variety of environmental conditions. They usually live in the temperature range of 29–35 °C and stop growing below 5 °C and above 35 °C [18]. Properly the light intensity is between 1000 and 10000 lx. Light drives photosynthesis, but strong light inhibits algae growth by causing photo-oxidation. In addition, algae cultivation requires a specific dose of nutrient solution consisting of nitrogen, phosphorus, carbon, and small amounts of potassium, magnesium, and calcium. Chlorella and Spirulina are the two most studied microalgae for biopolymer production from their high dry weight protein percentage [13]. Our experiment employed Chlorella as the algae for cultivation.

Fig. 2. Algae grow in the culture tube that embed in the grooves of the 3D-printed skin.

2.2.2 Algae Reactor

The algae culture system as an airlift bioreactor converts solar energy into biological energy storage. This system includes a convoluted tube for algae cultivation, an air pump (with battery) at the bottom of the tube, and an anti-overflow funnel at the top. The air pump draws indoor air into the tube steadily. The air flows from the bottom to the top of the convoluted, ascendant tube. This movement of air facilitates the interaction between the indoor air and the algae culture [9].

The solar panels of the system provide the necessary energy for the module's operation. The system is independent of other electricity supply. The system consumes carbon dioxide resulted from human and facility activities within the building. The indoor environment with stable and mild temperatures is ideal for Chlorella to convert carbon dioxide molecules into oxygen through photosynthesis.

The convoluted tube improves the efficiency of the system's biochemical process. The long and winding path of the tube increases the contact time between the air and the algal fluid so that the Chlorella has more time to absorb CO_2 and produce oxygen (Fig. 3). This feature enhances the overall effectiveness of the system in improving indoor air quality.

2.3 Robotic 3D Printing

The prototype of the algae reactor was mainly fabricated by a 3D printing system based on a KUKA robotic arm. Transparent recycled PETG (rPETG) granules were employed as printing materials for the freeform skin. The 3D printing toolpath was customized to integrate the culture tube and create visual attraction.

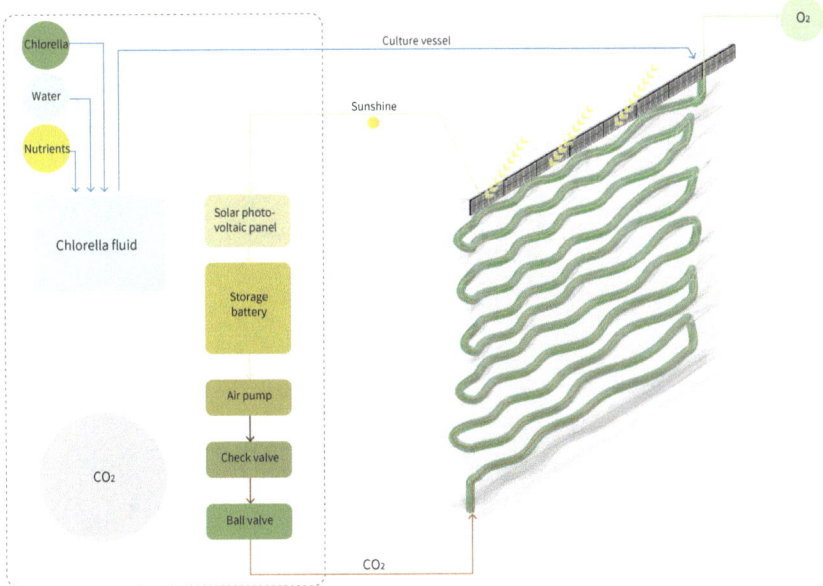

Fig. 3. The algae culture system involves a microbiological cycle in building.

2.3.1 Freeform Skin

The module's assembly and its installation rely on the design of the freeform skin. The surface texture imitates water ripples to regulate the light transmission of the 3D-printed PETG and contribute to a visually appealing façade. With six degrees of freedom of the robot arm, the ripple geometry can be efficiently constructed using robotic Fused Granulate Fabrication (FGF). The surface of the building skin reveals a sparkling texture in the sunlight. The 3D printing trajectory forms grooves whose geometry is coupled with the cultivation pipe. The geometric design and digital modeling were completed in Rhino and Grasshopper.

2.3.2 Robotic Fabrication

The complex structure of the module is fabricated using large-scale robotic FGF (Fig. 4). A pellet extruder is mounted onto a KUKA robot arm, resulting in a 3D printing system with six degrees of freedom. To optimize the module's performance, the toolpath of the robotic system is meticulously programmed to achieve structural integrity and other design features. The typical PETG for 3D printing has a density of around 1.3 g/cm^3; the melting temperature for extrusion is about 270 °C; the tensile strength is around 50 MPa.

The geometry of the module should accommodate the cultivation pipeline into a desirable shape and maintain its structural stability. The skin is printed with grooves that reinforce the skin structure, diffuse direct sunlight, and reshape the corrugated tube into a convoluted profile. By incorporating differentiated geometry and layered 3D-printed filaments, the module can convert direct sunlight into diffuse light, providing a better

Fig. 4. Left: robotic 3D printing (FGF) of the façade module; right: the differentiated skin of ripple pattern (front view) with customized grooves on the back side.

environment for algae growth. The light distribution within the module enhances the efficiency of photosynthesis and increases algae productivity.

The digital 3D model was sliced to create the printing trajectory which is further translated into the code (KRL) of the KUKA robot. For stability and precision, the toolpath is designed as a single continuous path connecting the curved surface and the internal cavities. The toolpath planning was implemented in Grasshopper (in Rhino). In the prototype printing, the cross-section of filament has a dimension of 6×3 mm (width $\times$ height). The skin component is 900 mm in height and 900 mm in width. 3D printing of one prototype took six hours.

3 Prototype

The prototype was installed on the interior side of a façade component of approximately 1 m by 1 m (Fig. 5). The corrugated tube has a diameter of 25 mm and a total length of 9 m, which can accommodate 44 L of algal liquid. Under a proper condition, a module absorbs 5 kg CO_2 and produces 2.7 kg biomass each year. The algae culture medium can be quickly filled and collected in the bioreactor. Solar energy and battery energy storage complement each other, providing the power for the air pump. The experimental prototype is integrated with the building system and the indoor environment. The algae reactor serves as an air purifier for the interior space and enhances the façade's visual effect for exterior viewers.

Fig. 5. Install the module on the interior of the building façade. Left: view from the interior; right: view from the exterior.

4 Discussion

In the process of algae culture system design, ensuring both the availability and simplicity of the system were the most critical issues. In order to facilitate the operation of algae substance delivery, cultivation, and collection, the system design included the addition of valves to adjust the air pressure and water pressure inside the pipeline. The final system achieved a simple way of infusion from top and col lection from bottom of the algae liquid.

The algae reactor project employed additive manufacturing and an algae culture system to create a microbiological cycle within buildings, which demonstrates the potential of implementing biological architecture in the built environment. However, there are still limitations in the current approach to scale the prototype for wider applications. First, the performance of our module relies on various conditions of the façade of the existing building. The application of the module as an independent facade system requires further structural research and experiments. The proposed tube culture mode provides flexibility of the façade photobioreactor, but it needs to control the internal environment of the biological cultivation liquid as the traditional plate photobioreactor. At present, this problem has been treated by simple means such as using air pumps, but more promising applications need to consider the economic, efficiency, and environmental impact of the system in a bigger picture.

5 Conclusion

Based on the Chlorella photosynthesis, the project demonstrates the energy conversion and carbon sequestration processes of a bio-architecture system that integrates building façades with biomass. This interdisciplinary design provides new insights for future urban planning strategies and building carbon sequestration. For larger-scale applications, the spatial and functional elements of the building also need to meet the conditions of the microbiological cycles within the building. Instead of relying solely on inorganic materials, architecture could evolve into a vibrant hybrid organism that harmonizes with its environment. This work suggests architects designing buildings with an ecological point of view, toward harmonious and living spaces. As a multidisciplinary development, the algae reactor explores the possibility of reshaping the long-term maintenance of buildings with systematic thinking. This module explores the potential for reshaping long-term building maintenance through systematic design thinking, offering inspiration for future interdisciplinary integration and development.

Acknowledgements. This work was supported by the National Natural Science Foundation of China [52278008].

References

1. Au, H.K.A.: Urban Algae: Redefining urban spaces and infrastructure through microalgae ecological and biophilic values. Doctoral Dissertation, University of Hawai'i at Manoa (2022)
2. Associati, C.R.: Algaetecture. Available at: carloratti.com/project/algaetecture/ (2014)
3. Cheibas, I., et al.: Light Distribution in 3D-Printed Thermoplastics. 3D Printing Addit. Manuf. **10**(6), 1164–1177
4. Gong, Y., Hu, H., Gao, Y., Xu, X., Gao, H.: Microalgae as platforms for production of recombinant proteins and valuable compounds: progress and prospects. J. Ind. Microbiol. Biotechnol. **38**(12), 1879–1890 (2011)
5. Strauß, H.: AM envelope: the potential of additive manufacturing for façade construction, thesis. TU Delft (2013). Accessed 26 April 2023 https://doi.org/10.7480/abe.2013.1
6. Hariyanto, L., Arifin, L.S., Damayanti, R.: Microalgae as a sustainable facade for occupants' health: a review. Adv. Civ. Eng. Sustain. Archit. **4**(1), 26–36 (2022)
7. Ievina, B., Romagnoli, F.: Potential of species as feedstock for bioenergy production: a review. Environ. Clim. Technol. **24**(2), 203–220 (2020)
8. Leschok, M., et al.: 3D printing facades: design, fabrication, and assessment methods. Autom. Constr. **152**, 104918 (2023)
9. Lei, L., Hongbing, Z., Wentao, L., Huixuan, L.: The research progress of the cultivation of microalgae reactor culture. Biotechnol. Prog. **10**(2), 117–123 (2020)
10. Malik, S.: Viscous boundaries: Algae-laden scaffolds for the built environment. Doctoral dissertation, UCL (University College London) (2022)
11. Malik, S., et al.: Robotic extrusion of algae-laden hydrogels for large-scale applications. Glob. Challenges **4**(1), 1900064 (2020)
12. Mungenast, M.B.: 3D-printed future façade. Doctoral Dissertation, Technische Universität München (2019)
13. Onen Cinar, S., Chong, Z.K., Kucuker, M.A., Wieczorek, N., Cengiz, U., Kuchta, K.: Bioplastic production from microalgae: a review. Int. J. Environ. Res. Public Health **17**(11), 3842 (2020)

14. Piccioni, V., Turrin, M., Tenpierik, M.J.: A performance-driven approach for the design of cellular geometries with low thermal conductivity for application in 3D-printed façade components. In: Proceedings of the Symposium on Simulation for Architecture and Urban Design (SimAUD 2020), pp. 327–334 (2020)
15. Proksch, G.: Growing sustainability—integrating algae cultivation into the built environment. Edinb. Archit. Res. J. **33**, 147–162 (2013)
16. Sarakinioti, M.V., Turrin, M., Konstantinou, T., Tenpierik, M., Knaack, U.: Developing an integrated 3D-printed façade with complex geometries for active temperature control. Mater. Today Commun. **15**, 275–279 (2018)
17. Setiawan, Y., Asthary, P.B., Saepulloh, S.: CO2 flue gas capture for cultivation of Spirulina platensis in paper mill effluent medium. In: AIP Conference Proceedings, vol. 2120, no. 1. AIP Publishing (2019)
18. Soni, R.A., Sudhakar, K., Rana, R.S.: Spirulina-from growth to nutritional product: a review. Trends Food Sci. Technol. **69**, 157–171 (2017)
19. Taseva, Y., Eftekhar, N.I.K., Kwon, H., Leschok, M., Dillenburger, B.: Large-scale 3D printing for functionally-graded facade. In: RE: Anthropocene, Design in the Age of Humans-Proceedings of the 25th CAADRIA Conference, vol. 1, pp. 183–192 (2020)
20. United Nations Environment Programme. 2022 Global status report for buildings and construction: towards a zero-emissions, efficient and resilient buildings and construction sector. Available from https://www.unep.org/resources/publication/2022global-status-report-buildings-and-construction (2022)
21. Wurm, J., Pauli, M.: SolarLeaf: the world's first bioreactive façade. Arq Archit. Res. Q. **20**(1), 73–79 (2016)

On the Application of Vector-Based Graphic Statics (VGS) for Structural Timber Optimisation – Pavilion Example

Denis Zastavni[✉], Sylvain Rasneur, and Jean-Philippe Jasienski

SST/LAB – Louvain Research Institute for Landscape, Architecture, Built Environment
UCLouvain – LOCI - Faculty of Architecture, Architectural Engineering and Urbanism Place du
Levant 1 (L5.05.02), 1348 Louvain-La-Neuve, Belgium
`denis.zastavni@uclouvain.be`

Abstract. Over the last two decades, various contributions have shown how the use of graphic statics made it possible to design remarkable engineering structures. Vector-based Graphic Statics (VGS) has been presented elsewhere extensively as a method and plug-in for Grasshopper. This contribution shows an application to the design of a timber pavilion based on Graphic statics and its benefits for the use of materials.

Keywords: Timber Construction · Structural Design · Digital fabrication · Parametric Design · 3D Graphic statics

1 Purposes of Graphic Statics

Abstract. Graphic Statics (VGS) is already known as a designer approach for structural design leading to outstanding results. Graphic statics was originally an analysis tool, and this first section recaps its main historical purpose and uses.

Historically, graphic statics was developed for the purpose of analysing structures. Its principle is to represent the full system of internal and external forces as a set of vectors characterised by directions and magnitudes of forces. Every operation in terms of combination or calculation is managed in a geometrical and graphical environment unified with the geometric definition of the structure itself (Fig. 1).

Its graphical approach enabled the mastery of geometrically complex problems and extremely efficient calculation methods compared with algebraic approaches, which were practically the only alternative available at that time. Graphic statics allowed extremely efficient analysis of isostatic trusses and beams. Because it is necessary to close the polygons of forces that translate the translation equilibrium, the graphical approach makes it possible to check that the graphical calculation has been carried out correctly. Although it is easy in this framework to calculate deformations in trusses, the analysis was most often limited to establishing the equilibrium of forces and their respective magnitudes. For these applications, the disadvantage was that the accuracy of the calculation depended on the scale of the technical drawing used for the analysis and

H. Chai et al. (Eds.): CDRF 2024, *Symbiotic Intelligence*, pp. 164–179, 2025.
https://doi.org/10.1007/978-981-96-3433-0_15

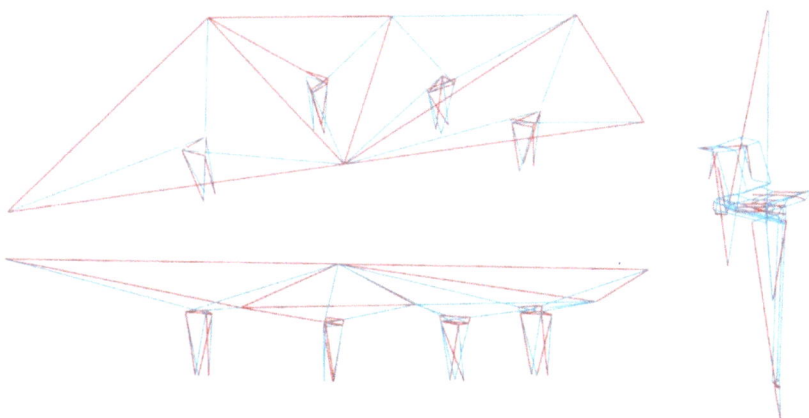

Fig. 1. 3D vector-based graphic statics: 3D form polygon F (on the left) and the magnitudes of the forces with 3D force polygon F* (on the right)

the accuracy of the drawing itself. In order to solve hyperstatic problems, it is necessary to consider the deformations of the various parts of the structure and their respective compatibilities. For these, with Graphic statics, the developments initiated by Mohr (1868) lead to a complete approach generalised by Suter (1932) and will be developed at least until the beginning of the 1940s (Han and Zastavni 2024), being progressively supplanted by the moment distributions method (Cross 1943) which proposes a more direct algebraic method. Nevertheless, the complexity of analysing statically indeterminate structures remained a challenge until the dissemination of numerical analysis tools in the 1970s, and Graphic statics was no exception. Nowadays the use of a CAD system removes the limitations due to the precision of the drawing and its scale. Combined with matrix analysis, the barrier of hyperstatic structures can be overcome for their analysis.

Combined with the lower bound theorem of the theory of plasticity (the "design theorem"), graphic statics can also be used to deal with the structural behaviour of continuous structures such as concrete walls or folded structures using materials that allow forces redistributions. The success of this approach depends on the use of strut-and-tie modelling (STM), taking advantage of the plastic redistribution capabilities of such materials. The same STM approach has been applicable, since its emergence (Mörsch 1908), to continuous type connections in concrete structures for which the method was engineered (Schlaich Jorg 1982). Recent work introduced how this same STM approach is also applicable to materials with limited redistribution capacities, such as timber, which has fragile tensile strength (Jasienski et al. 2024).

2   Graphic Statics for Structural Design

Abstract. This section presents historical and contemporary uses of graphic statics for design purposes in structural design.

A clear distinction must be made between structural analysis and structural design: in the first case, the geometry of the structure is given, in the second case, it must be defined. Because it offers a graphical representation of both form and forces, graphic statics prove a very powerful tool for structural design as well. Indeed, the history of structures provides us some examples where its use is encountered in the field of structural design, resulting in structure that are remarkable both for their efficiency and their aesthetic qualities. Examples by the likes of Swiss engineer Robert Maillart: Chiasso Shed (Zastavni 2008), Salginatobel Bridge (Fivet and Zastavni 2012), Vessy Bridge (Zastavni et al. 2014), but also in Maurice Koechlin's graphic design of the Eiffel iron tower (Fivet et al. 2015), Antoni Gaudi's work for the Catalan Park Güell (Gaudi-Groep Delft 1979) and Guastavino's Vaulting (Oschendorf 2010).

The last two decades have seen the emergence of significant structures designed with Graphic statics in a contemporary context and using modern tools. Some of remarkable examples are Conzett's Traversinersteg II (Mostafavi et al. 2006), SOM's Roof truss for a convention centre (Beghini et al. 2014), the Armadillo vault in Venice (Mele et al. 2016).

More recently, academic works showed how computer aided graphic statics can support structural design. Comprehensive research has been conducted to extend graphic statics to the third dimension, the two main approaches being the Polyhedron-based (Bolhassani et al. 2018; Konstantatou et al. 2018; Lee et al. 2018) and the Vector-based (D'Acunto et al. 2019; Jasienski et al. 2024).

The Vector-based 3D Graphic Statics (VGS) is a framework where the forces are represented by vectors which length corresponds to the force magnitude. It provides direct interactive feedback on forces' magnitudes and orientation and enables the design of reticular-type structures to be approached in a dynamic and interactive environment (D'Acunto et al. 2019; Jasienski et al. 2024; Rasneur et al. 2024) (Fig. 2).

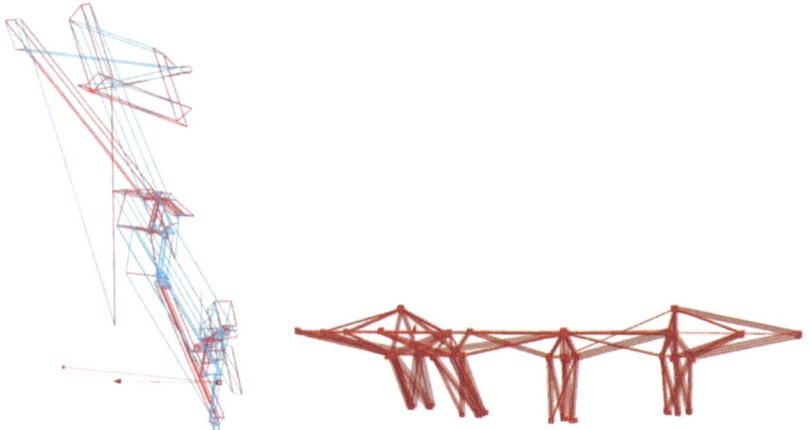

Fig. 2. 3d force diagram F* with wind and earthquake vectors (left) and the corresponding dimensioning including buckling forces (right): the consequences of longer forces segments in F* on parallel columns are particularly obvious

In most cases, architectural and structural design involves considering a vast array of requirements. These may be technical (strength, connections, available sections, constructability), spatial (useful heights, maximum overall dimensions, location of joints and interfaces) or architectural (volumetric and lighting qualities, aesthetic qualities). These are so numerous that only the designer's experience can integrate and order them properly. So, the real challenge is not the implementation of computational design protocols, but the development of environments where force-geometry interaction is efficient, visually understandable and practical for the designer. Beyond automatic structural optimization in the framework of computational design, this paper shows how the VGS plugin of Grasshopper (McNeel 2024) can be used for interactive structural design. The case study of the Guhua pavillion reveals strengths and weaknesses of this approach. It shows how this approach can support the optimization of structural behaviour and shows revision directions that architects or engineers can follow in a brain-driven optimised design process. This process showed significant structural improvements at each stage of its development.

3 The Structural Stakes of a Pavilion Design

Abstract. This section details the design objectives followed for the designs successively developed in Sect. 4 for a timber pavilion in the VGS framework.

Using the example of a 24 m × 14 m pavilion initially designed as a reciprocal-like system supported on a reduced number of 4 central supports, the article shows how VGS can be implemented to support the design of the structure and its main benefits. The case study is a pavilion submitted to our team in the framework of a National Key Research and Development Program of China (grant number:2022YFE0141400).

3.1 Efficiency Objectives for Structural Design

Structural design is a far more complex exercise than structural analysis and dimensioning. Design codes, for their part, are essentially geared towards the verification of structural elements. The verification stage alone is much simpler since it is straightforward: it avoids the phenomena of interdependence of the parameters that must be used for dimensioning. Dimensioning, even without questioning the geometry, involves these interdependencies of parameters. Structural corrections involve the geometry and reach an even higher degree of complexity. The design exercise itself combines the above with an even wider range of parameters.

Many parameters are involved in the design of structures, some of them contradictory in nature (Fig. 3).

For trusses, the following criteria could be considered for an optimization works on the geometry (typology) of trussed beams: the total amount of material used; the number of nodes; the number of bars per assembly node; the length of compressed bars (in relation with buckling issues); the magnitude of forces (which is influenced by the geometry); aim for constant force geometries (for rationalization); to rationalize topology, e.g., for constant and minimal sections; having same angles everywhere between bars to rationalize connections; meaning of combining several materials according to their intrinsic

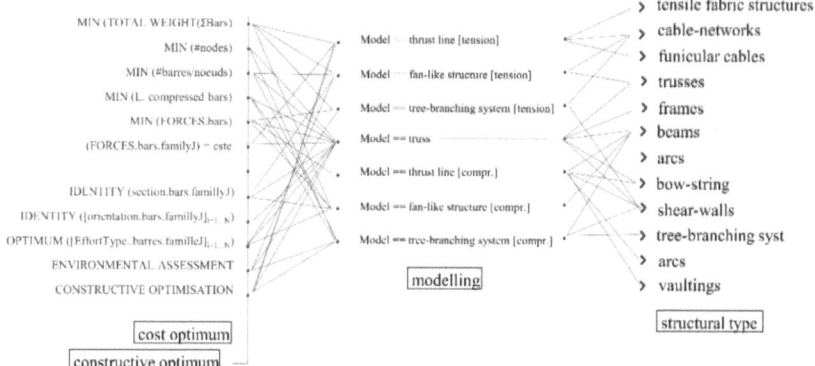

Fig. 3. A selection of design optimisation goals for bars or STM structural models

properties; rigid VS hinged connections; embedded or hinged supports; hyperstatic or isostatic systems; aim to foster specific structural working, etc. The designer is required to order and weight each of these parameters to obtain a "best" solution, that is always relative. Pareto Optimality is mentioned in this context as a criterion for the best option.

3.2   Pareto Optimality and the Design Parameters

The essence of the Pareto optimum is to aim for a state of "best" solution for multiple parameters. The term Pareto-optimality originates from his work on the distribution of resources with limited availability and conflicting goals (Pareto 1906; Pareto 1978). In the context of optimization, it can be formulated that a Pareto optimal solution cannot be improved for one goal without degrading another. Therefore, there is not one best solution, but many trade-offs between the extremes (Vierlinger 2022).

Pareto is today primarily attributed with the so-called 80–20-rule, after which 80% of the results can be achieved with 20% of the effort. This is the philosophy that had prevailed for this design exercise. Since the detailed geometry of the structure is the final goal, the forces acting on it are essential parameters along with geometrical parameters. Forces are at the heart of the challenge of the complete dimensioning of the structure. In this respect, the Graphic statics environment is perfectly suited to the task for revealing respective forces' magnitudes.

In the following structural implementation, the cross-sections will be considered square since the structural working in tension/compression only will be targeted and fostered for its efficiency. The main parameters are defined as follows: The material is glued laminated timber TCT28. Horizontal loading on the roof is 3 kN/m^2 including deadload and live load; wind force is 0,5 kN/m^2, earthquake forces are between 0.09 and 0.14 g: snow load will be considered as negligible with 0.3kN/m2 in Shanghai compared to vertical loading on the roof. This is an open structure composed of between 51 and 89 bars depending on the version of the design. A minimal dimension of structural members will be kept at 150 mm width, and their section will be considered constant along bars. For the most complex example, LVL and CLT will be considered as alternative to GLT.

4   The Assessment of the Guhua Pavilion in the VGS Framework

Abstract. This section details the elaboration of revised designs for a timber pavilion and presents the designs successively developed in the VGS framework with their main benefits for material saving on constructability.

4.1   Model 01 = Start Structure: "Full Reciprocal" Design

Assessment starts on a preliminary project defining the structure of the roof as such: 4 supports, each consisting of three bars, each carrying two bars which form the primary part of the roof structure itself (Fig. 4). These bars form the bisectors of slightly inclined triangular surfaces that are encountered four times, i.e. a triangle centered on each of the supports. Since the columns each carry two intersecting bars, the principle is like reciprocal structures, although not entirely so. To ensure that the structure behaves correctly overall, the bars must be rigidly connected to the supporting columns. The bars of the main structure are subjected to significant bending forces at the point where they are connected to the heads of the columns. If the triangles they support do not form closed frames capable of withstanding significant tensile forces, the main beams operate exclusively in bending. This also means that the connections must resist bending moments and must be designed accordingly.

In addition, due to the geometric complexity of the structure and its asymmetry, several joints and bars are warped (torsion, vertical and horizontal bending) and their behaviour leads to complex arrangements of joints between elements, and oversizing columns and bars.

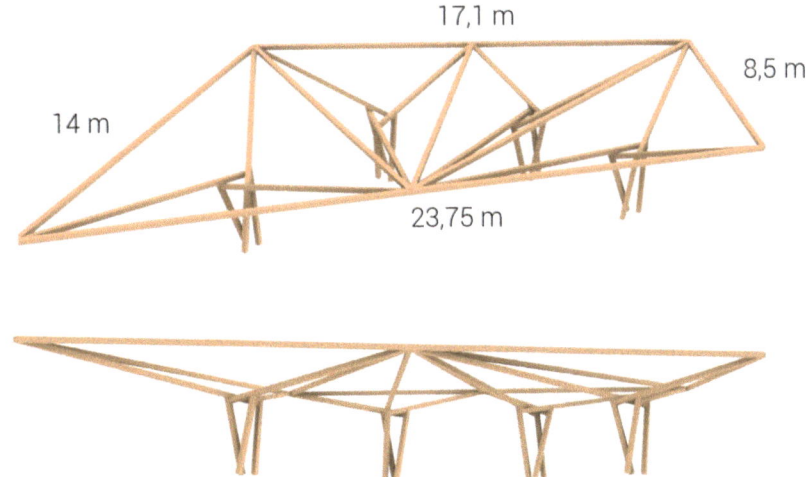

Fig. 4. Perspective view from top with general dimensions and perspective view from below of the pavilion (model 01)

Because of the importance of the bending forces, the width of the members including buckling and composed 2-axis bending, will be between 320 mm and 710 mm; the total volume of glulam timber is 18.2 m^3. The assessment was made using FEM analysis and custom dimensioning according to EC5 (Cen 2014), as summarised below:

Max.bending My/Mz/Mt (kN.m)	Max.traction (kN)	Min.compression (kN)	Min.sec. width (m)	Max.sec. width (m)	Max.sec. length (m)	Load Path (kN.m)	Volume (m3)
746,1/238/73,58	672,38	-713,11	0,15	0,63	14,73	103756,89	17,8

4.2   Model 02 = "Full Reciprocal" Design Including Upper Tension Ring

With the aim of favouring traction or compression forces for the structural working (so that the entire cross-section works at maximum stress, unlike bending), a peripherical tension ring is considered by connecting firmly the ends of all the upper beams. Enhancing the structural behaviour in that way, main roof elements develop compression forces alongside with bending forces although these are reduced. Maximum sections drop to between 150 and 370 mm and timber volume reduces by 9%. *Load path* magnitude (defined as adding the product of bar lengths and their forces magnitude) raises considerably since bending is not included in its calculation.

Max.bending My/Mz/Mt (kN.m)	Max.traction (kN)	Min.compression (kN)	Min.sec. width (m)	Max.sec. width (m)	Max.sec. length (m)	Load Path (kN.m)	Volume (m3)
103,3/113,9/39,2	359,49	-569,63	0,16	0,48	14,73	161671	16,43

Doubling the three bars making up each of the four support devices to give them a V-shaped design helps cancelling any negative effects linked to the asymmetry of the system. This strategy enables to cancel bending forces, which simplifies connections and enable contact joints. This has significant consequences in terms of magnitude of force, despite a limited impact on sections, working from now at 100% of their thickness. Next model will also go a step further in cancelling forces at the upper zone of support devices.

4.3   Model 03 = "Reciprocal System" with Stabilised Supports: Including Upper Tension Ring

Model 3 defines connexions as hinges to have traction and compression forces only while cancelling bending forces. Without additional modification, this destabilises the system. Therefore, additional members were added to the system (Fig. 5).

- two layers of structural rings at the head of tripod systems, respectively in compression and tension going downwards,
- maintaining a V-shape design of the element constituting the tripod systems of support
- bracing free nodes of bars and rings converging at the head of support systems.

At this stage of development, the analysis can be transferred to the Vector-based Graphic Statics [VGS] tool, which allows forces to be visualised in the form of 3D force polygons, enabling the absolute and relative magnitude of forces in the structural members to be assessed. For this, several points must be addressed. First, VGS is composed of a series of modules with various purposes. The "Evaluate Equilibrium" module relies on an algebraic calculation (equilibrium matrix) and so requires the structure to be stable to deliver results. In the process of hinging all connections, a careful attention must

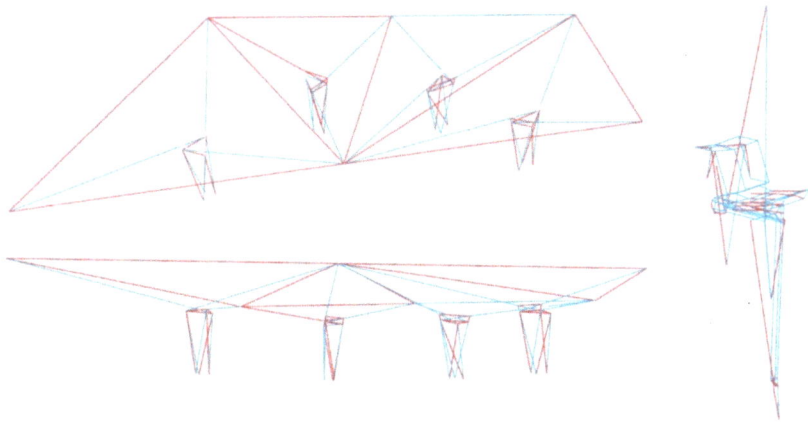

Fig. 5. Form diagram (left above and below) and forces diagram (right) of the pavilion (model 03) with members in compression (blue) and in tension (red)

be given to implement supplementary bars to guarantee equilibrium and the "Evaluate Equilibrium" module helps for this by providing direct feedback.

Second, the stable version of our hinged structure when symmetry is maintained in support tripods has an indeterminacy level of 3. This information is delivered again by the "Evaluate Equilibrium" module. For the indeterminacy to be solved, without deformation assessment as in classical FEM approaches, the user needs to implement *self-stresses*. A *self-stress* is tensioning (or compression) a bar, the effects of which will propagate throughout a structural (sub) system, like tensioning a tensegrity module for example. By extension, since the VGS environment is resolutely design-oriented, with connexions to the framework of plastic design, the approach is to impose the magnitude of several member forces which after allows to dimension the structure accordingly and overpass an indeterminacy for instance. To optimise the structural dimensioning, the minimal solution must be targeted. For this, an optimisation process has been carried out to define the magnitude of respective *self-stresses*. The volume of timber, with or without minimal dimension or buckling consideration, or using *load path*, or minimising the largest forces can be used as the main goal for the optimisation. *Load Path* is of interest as an indicator in the rest of the study and is frequently used in the framework of graphic statics. Another goal is particularly meaningful for the optimisation process: the compactness of the Force Polygon, provided as visual feedback of VGS. The force polygon quickly shows which forces are dominant and to which bars they correspond, which will guide the rest of the design process. In our case, the major forces are found in the support system.

Applied to model 3, some forces raise locally in the structure, but maximum sections drop to between 150 and 470 mm and timber volume reduces by 32%. *Load path* magnitude drops of 36%.

Max.bending (kN.m)	Max.traction (kN)	Min.compression (kN)	Min.sec. width (m)	Max.sec. width (m)	Max.sec. length (m)	Load Path (kN.m)	Volume (m3)
	2461,85	-2581,55	0,15	0,44	14,73	102630,88	11,11

Furthermore, the deployment of one *self-stress* as the only force in the model high-lights the level of propagation of a force in the structure. This turns out that propagation is particularly important in the case of a principle of structural reciprocity, that does not go in the direction of structural safety and robustness. Also, the weak point of that model is high forces to transmit between bars forming steep angles so that joints become problematic, what drive us to review the design.

4.4 Model 04 = Folded Model with V-Shape Folded Columns (Basic)

Model 4 implement larger supports arrangements and folding plate for designing structural elements. Support devices are braced and are supposed to represent two CLT plate connected as a V in plan. Currently, their orientation and inclination are random and could be analysed by an optimisation process. This stage also allows the use of CLT panels for the rest of the structure as an alternative to glued laminated timber. For the sake of consistency, each element will however continue to be modelled as a STM system consisting of bars. This mainly reduces max. forces (Fig. 6):

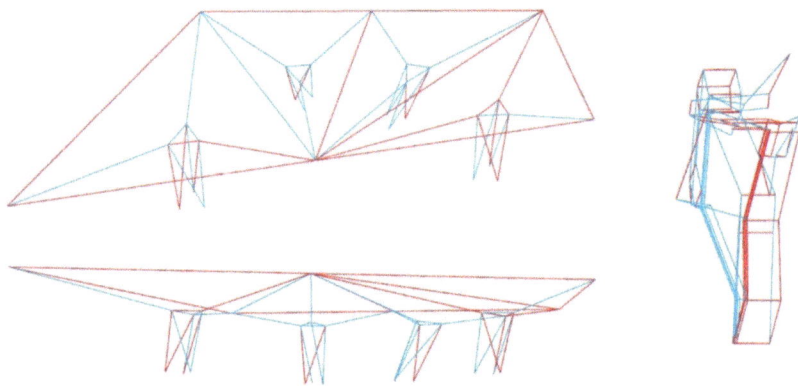

Fig. 6. Form diagram (left above and below) and forces diagram (right) of the pavilion (model 04) with members in compression (blue) and in tension (red)

Max.bending (kN.m)	Max.traction (kN)	Min.compression (kN)	Min.sec. width (m)	Max.sec. width (m)	Max.sec. length (m)	Load Path (kN.m)	Volume (m3)
	1481,83	-1292,48	0,15	0,34	14,73	100385,49	10,42

By examining the polygon of force, it is possible to try out a simple operation consisting of doubling the thickness of the roof, from 1 m to 2 m. This corrects the current weak point in the structure with a very low volume of material (−25.6%):

Max.bending My/ Mz/Mt (kN.m)	Max.traction (kN)	Min.compression (kN)	Min.sec. width (m)	Max.sec. width (m)	Max.sec. length (m)	Load Path (kN.m)	Volume (m3)
	876,92	-794,53	0,15	0,27	14,73	52997	7,75

4.5 Model 05 = Stared Folded Model with V-Shape Folded Columns (Basic + Flat Upper Side)

The thickening the roof system investigated above as a variant on model 4 radically changes the appearance of the structure and will not be retained for this reason, despite its effectiveness. However, a more secure approach involving tessellation of the areas with the greatest cantilevers has been devised in model 5. In this case, the main features of the structure are the following (Fig. 7):

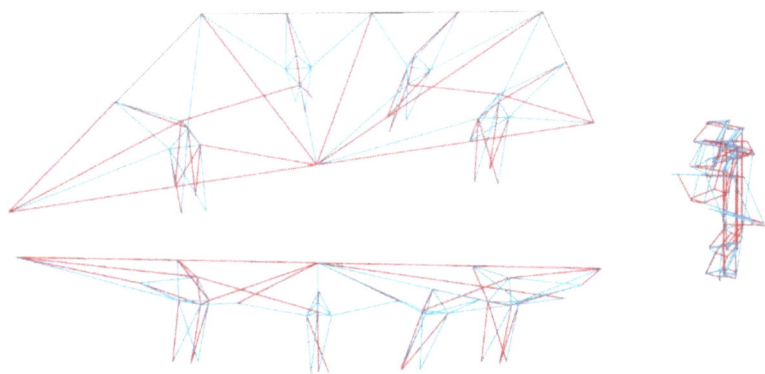

Fig. 7. Form diagram (left above and below) and forces diagram (right) of the pavilion (model 05) with members in compression (blue) and in tension (red)

Max.bending (kN.m)	Max.traction (kN)	Min.compression (kN)	Min.sec. width (m)	Max.sec. width (m)	Max.sec. length (m)	Load Path (kN.m)	Volume (m3)
	1333,32	-1778,16	0,15	0,33	14,73	102687,31	11,94

4.6 Stared Foldel Model with V-Shape Folded Columns (Basic+ Raised Top Support)

Finally, to address the problem of steep angles in joints, a last improvement is implemented by raising the converging connection points above columns. With this, a new improvement is reached with the following magnitudes (Fig. 8):

Max.bending My/ Mz/Mt (kN.m)	Max.traction (kN)	Min.compression (kN)	Min.sec. width (m)	Max.sec. width (m)	Max.sec. length (m)	Load Path (kN.m)	Volume (m3)
	1290,77	-1601,77	0,15	0,32	14,73	83739	10,77

Another optimisation goal can be used to minimise compression force involving buckling. This has a cost of +8% volume, but lowers this Min.force by two:

Max.bending My/ Mz/Mt (kN.m)	Max.traction (kN)	Min.compression (kN)	Min.sec. width (m)	Max.sec. width (m)	Max.sec. length (m)	Load Path (kN.m)	Volume (m3)
	816,08	-772,79	0,15	0,32	14,73	101175	11,71

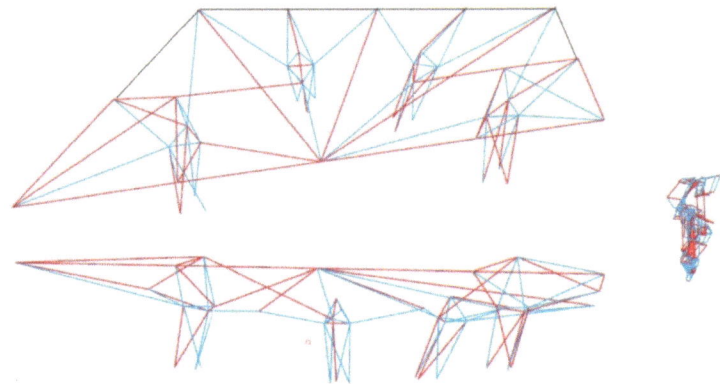

Fig. 8. Form diagram (left above and below) and forces diagram (right) of the pavilion (stared folded model) with members in compression (blue) and in tension (red)

The same structure opens to another way to redistribute forces on the roof, involving loading more intensively directly above the support tripods. Therefore, we see the magnitude of forces reduce again, as *Load path* and volume of timber. With this modelling, volume is lower of 23% and *Load path* divide by 3:

Max.bending (kN.m)	Max.traction (kN)	Min.compression (kN)	Min.sec. width (m)	Max.sec. width (m)	Max.sec. length (m)	Load Path (kN.m)	Volume (m3)
369,74	-438,10	0,15	0,23	14,73	28812,02	8,23

With this model, we have reached the limit of what can be achieved in terms of dimensioning gains. We are also faced with the limit of the minimum cross-sectional dimension for a proper joinery. Would we consider a minimum width reduced from 150 mm to 125 mm, that a saving on timber volume of 23% is again possible.

4.7 Wind Force, Earthquake, Snow Loads and Final Comparison

The analysis is based on a set of 7 models assessed through 20 simulations. The full analysis included several loading cases presented under §3.2. Since earthquake and wind forces are directionally applied, analyses showed that worse case is the occurrence of their simultaneous horizontal orientation. Nevertheless, wind force and earthquake are acting in opposite vertical directions. In the numbers, wind forces partially cancel the effects of gravity and the worst case for the structure remained vertical loads only, despite horizontal loading (Fig. 9).

Fig. 9. Axonometric front view of the pavilion (stared folded model) with faces

When comparing the versions of design model 6 (without redistributing forces), a consequent saving of 40.2% of timber volume is reached in comparison to model 1 (34.4% compared to model 2 having a ring effect around the roof and that should serve as a reference), even if the number of bars is practically doubled. The maximum section width can be divided by 3 with the great advantage of normalising sections for the benefits of joinery.

The evolution of the structural design under the guidance of VGS from first to latest models without bending showed that forces magnitudes can be reduced from enhancing the geometry of the structure. Final improvements on the structural efficiency could still include considering repositioning device supports, optimising their width, number, inclination and orientations. But these were not considered for this study.

5   Conclusion and Discussion

The initial analysis of the pavilion under study revealed the impact of asymmetry and bending mechanisms on its structural efficiency. Subsequently, we employed VGS as a suitable tool to revamp the pavilion as a sophisticated reticular system. This resulted in significant material savings ranging from 34 to 40%. Additionally, VGS has the advantage to reveal the structural parts that are most impacted by loadings and to guide the designer for his fundamental conceptual revisions. With the help of VGS, structural adjustments and load redistributions led to final reductions of up to 54,2% in volume and 72% in *load path*. A synthetic overview of these adjustment is proposed in a table in Appendix.

Throughout this design process, the VGS add-on for grasshopper demonstrated its full potential to drive structural improvements by providing outputs that highlight structural weak points in a user-controlled design process. Graphic statics' force polygon makes it possible to quickly identify and locate the highest force magnitudes, while also characterizing them from a geometrical point of view. Following the Pareto principle, VGS visual feedback based on the compactness of the forces diagram enables their distribution in a hyperstatic system to be optimized to within 15%. With feedback on volume, this brain-driven approach enables the optimal solution to be found to within 5% in a few minutes, as opposed to several hours with genetic optimizations.

VGS is user-oriented, fostering an interactive design process. Conversely algorithm-driven or computer-driven procedure carried at their conclusion present optimized structures such as very high Mitchell trusses, which are of very limited use in real-life situations.

For the pavilion, LVL (Laminated Veneer Lumber) and folded plate system (such as CLT, Cross-Laminated Timber) were considered. Alongside the environmental cost of LVL, this material did not lead to significant saving due to limitations linked to a minimal thickness of members for proper joinery. CLT addressed bucking issues (enabling a significant reduction of utile structural volume depending of the initial structural working) but necessitated fundamental revisions to connections. Nevertheless, the efficient timber structural core without buckling saw an additional 48% reduction compared to buckled members, even if volume comparisons become meaningless without considering the entire material volume, including the whole roof composition.

Moreover, beyond providing results of structural analysis and feedback, VGS also informed the designer of the state where an equilibrium stage was reached, in case of reviewing structural principles with systematic hinging. Implementing structural robustness through structural redundancy also impacts the number of *self-stresses* and too many of them really complexify the design with VGS.

Indeed, in the case of *self-stresses*, the first challenge is to be able to identify a number of independent *self-stresses* equal to the number of unknowns in the hyperstatic structural system. The task becomes more complex as the number of *self-stresses* increases. Then comes the question of evaluating their respective magnitudes, including in this the difficulty caused by their reciprocal interaction on the overall structural system represented by the polygon of forces. If the process is computer-driven, the optimisation times required can become hard to manage. At each stage, therefore, the designer's experience and common sense will be called upon and will guarantee the success of their definition.

In the design as presented above, computer-driven optimization could have been implemented also with the VGS-Transformation module, an interactive optimization process with several goals possible.

As a final conclusion, it should be noted that the final design has yet to reach its completion. Minimal dimensions (here 150 × 150 mm) and connection angles between bars were necessary to ensure proper force transmission, though these were considered in the designer's choices. The approach we followed eventually provided an interactively dimensioned structure, with different conclusions depending on whether the structure is thought of as a system of assembled bars, or as a folded system made up of CLT elements. Nevertheless, this contribution demonstrates the strong capabilities of the VGS-tool for designing, understanding and managing a structural system while the designer remains in control of the design process, and hence is able to develop enlightened design solutions.

Appendix

See Fig. 10.

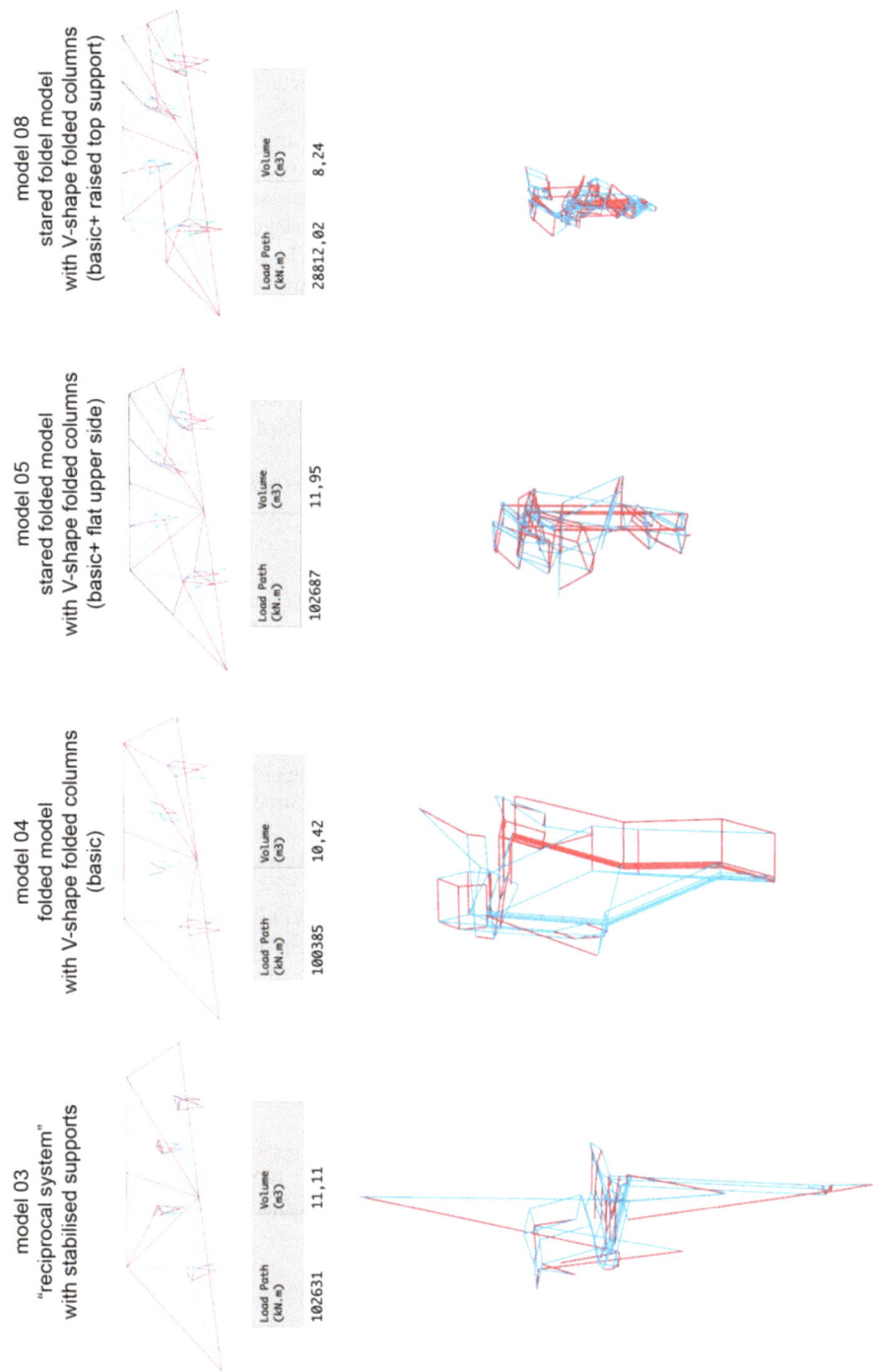

Fig. 10. Step by step view of the evolution of the Guhua Pavilion in the VGS framework.) Each model is identified by its structure, followed from top to bottom by the form diagram, the cumulative values of load path and volume and the force diagram.

References

Beghini, L., Carrion, J., Beghini, A., Mazurek, A., Baker, W.: Structural optimization using graphic statics. Struct. Multidiscip. Optim. **49**, 351–366 (2014)

Bolhassani, M., Akbarzadeh, M., Mahnia, M., Taherian, R.: On structural behavior of a funicular concrete polyhedral frame designed by 3D graphic statics. Structures **14**, 56–68 (2018)

CEN. Eurocode5: design of timber structures—part 1–1: general—common rules for buildings. In: Standardization, E.C.F. (ed.). European Committe for standardization (2014)

D'acunto, P., Jasienski, J.-P., Ohlbrock, P.O., Fivet, C., Schwartz, J., Zastavni, D.: Vector-based 3D graphic statics: a framework for the design of spatial structures based on the relation between form and forces. Int. J. Solids Struct. **167** (2019)

Fivet, C., Zastavni, D.: The Salginatobel bridge design process by Robert Maillart (1929). J. Int. Assoc. Shell Spat. Struct. **53**, 39–48 (2012)

Fivet, C., Zastavni, D., Ochsendorf, J.: The Papers of Maurice Koechlin (1856–1946). In: Brian Bowen, D.F., Leslie, T., Ochsendorf, J. (eds.) 5th International Congress on Construction History. Chicago (2015)

Gaudi-Groep Delft, B.P., Van Der Heide, R., Molema, J., Jos, T.: Gaudi, Rationalist Met Perfecte Materiaalbeheersing. Delft University Press, Delft (1979)

Han, S., Zastavni, D.: Graphic statics for continuous beams and frames: a review of the fixed-points method. Int. J. Archit. Heritage (On Press) (2024)

Jasienski, J.-P., Zastavni, D., Rasneur, S.: On the development of timber structures based on 3D interactive vector-based graphic statics (VGS). In: Yan, C., Chai, H., Sun, T., Yuan, P.F. (eds.) Phygital Intelligence, pp. 65–77. Springer Nature Singapore, Singapore (2024)

Lee, J., Mele, T., Block, P.: Disjointed force polyhedra. Comput. Aided Des. **99**, 11–28 (2018)

Mele, T., Mehrotra, A., Echenagucia, T., Frick, U., Augustynowicz, E., Ochsendorf, J., Dejong, M., Block, P.: Form Finding and Structural Analysis of a Freeform Stone Vault (2016)

Mörsch, E.: Der Eisenbetonbau—Seine Theorie und Anwendung (1908)

Oschendorf J., Freeman, M.: Guastavino Vaulting—The Art of Structural Tile. Princeton Archit. Press (2010)

Pareto, V.: Manuale di economia politica con una introduzione alla scienza sociale. Societa editrice libraria (1906)

Pareto, V.: Les Systèmes Socialistes. Œuvres complètes: Tome V. Librairie Droz, Genève (1978)

Rasneur, S.. Zastavni, D.; Jasienski, J-P.: On plastic development of timber structures based on 3D interactive vector-based graphic statics (VGS). Archit. Intell. (2024)

Schlaich Jorg, W.D.: Ein praktisches Verfahren Zum Methodischen Bemessen und Konstruieren im Stahlbetonbau. Comite Euro-International du Beton (CEB), Bulletin d'Inf.n N°150 (1982)

Vierlinger, R.: Octopus—Generative Design in Architectural Engineering (2022)

Zastavni, D.: The structural design of Maillart's Chiasso Shed (1924): a graphic procedure. Struct. Eng. Int. **18**(3), 247–252 (2008)

Zastavni, D., Cap, J.-F., Jasienski, J.-P., Fivet, C.: Load path and prestressing in conceptual design related to Maillart's Vessy Bridge. In: Pauletti, R.M. (ed.) IASS-SLTE 2014 Symposium "Shells, Membranes and Spatial Structures: Footprints". Brasilia, Brazil (2014)

Sand-Forming: Self-organization and Computational Optimization in the Creation of Flat Dune Sand Tilings

Marcus Farr[✉]

American University Sharjah, Sharjah, UAE
mfarr@aus.edu

Abstract. This paper undertakes an in-depth investigation into a hybrid digital and analogue design process employed in crafting sand-based panel systems conducive to computational optimization. Tailored for extreme desert environments abundant in dune sand, these systems leverage the material's inherent self-organizing properties. The methodology employs a multi-objective computational system strategically and culminates in an optimized, panelized architectural system. The approach emphasizes self-organization principles, initiated with physical experiments on natural dune sand piles. Advancing to controlled sand deposition on laser-cut planes facilitates precise configurations. The study systematically explores variables such as opening size and quantity, securing configurations with a binder and integrating them into diverse physical surface sequences. Computational analysis refines these sand patterns, identifying optimal configurations aligned with desert-specific contexts. This amalgamation of computational analysis and material processes enriches discussions on designing for extreme environments, aligning seamlessly with UN sustainability goals focused on sustainable communities, climate resilience, and responsible resource utilization.

Keywords: Material optimization · Dune sand · Self-organization · Panelized systems · Regional

1 Introduction

This study assesses the design of a dune sand material system and utilizes computational methodologies for optimization. The main goals include (1) the implementation of a novel material agenda centered on desert dune sand, (2) articulating an optimization approach for versatile panels adaptable to environmental requirements, (3) illustrating the significance of self-organization, and (4) presenting sustainable material alternatives within the global south. Sand is a critical global resource, with UN projections for 2060 identifying it as the most consumed construction material. Previous studies accurately forecasted the escalating demand for sand, indicating environmental challenges due to mining, habitat degradation, and widespread importation.

Throughout the global south, there is a substantial reliance on the importation of river sand to fulfil the requisites for many construction applications. This research explores

© The Author(s) 2025
H. Chai et al. (Eds.): CDRF 2024, *Symbiotic Intelligence*, pp. 180–188, 2025.
https://doi.org/10.1007/978-981-96-3433-0_16

the self-organization of dune sand with a water-based binder, revealing its manipulability under specific circumstances to create thin panels and tiles with unique geometries and openings. This work initiates a dialogue on the future of architectural materiality using dune sand and computation in the desert. The research underscores the viability of utilizing dune sand to fabricate prototypical elements, which can, for example, cater to the ventilation requirements in arid desert settings. The design workflow, utilizing both analogue and digital tools, aims to produce contextually relevant architectural outcomes with dune sand and predictive self-organization. The methodology reveals planar systems and discrete units, offering performance-based compositions calibrated using Grasshopper, Wallacei, and Ladybug. This approach provides adaptability based on solar and wind data, paving the way for innovative and sustainable architectural solutions.

1.1 Material Dynamics

Desert dune sand and river sand, both natural aggregates, possess distinct physical properties that influence their suitability for construction. Derived mainly from arid regions like the Arabian Desert, desert dune sand exhibits wind-blown characteristics, resulting in finer and well-sorted grains compared to river sand originating from water bodies [1]. The difference in grain size impacts packing density and stability, with dune sand's finer grains contributing to a denser arrangement. Despite its unique properties, desert dune sand faces limitations in construction due to inadequate cohesion and particle angularity resulting from aeolian processes [2]. River sand, with its angular particles, offers better stability and load-bearing capacity, making it preferred in construction for its mechanical properties [3].

Moreover, desert dune sand often contains a higher salt content due to its arid origin, posing significant durability challenges in construction, promoting corrosion and causing long-term structural issues [4]. River sand, with its diverse grain sizes and particle shapes, enhances interlocking, making it suitable for creating stable aggregates in concrete and mortar. Its proven use in construction applications, alongside availability and relatively consistent quality, contributes to its preference over desert dune sand. In summary, desert sand, characterized by unique attributes, differs significantly from its marine counterpart. River sand's angularity and compaction compatibility make it an efficacious constituent in concrete admixtures. In contrast, desert dune sand assumes a rounded morphology due to aerial abrasion, resembling a collection of marbles in configuration, with smooth surface textures and an exceedingly fine particle size. Its slightly alkaline nature and density akin to soil render it less amenable to contemporary construction methodologies.

1.2 Self-organization and Predictive Material Behaviour in Sand

The self-organizing behavior intrinsic to sand is exhibited during pouring leading to the formation of a characteristic funnel with a natural angle of repose (Fig. 1). The 34-degree angle of repose for sand, a critical aspect of granular material behaviour, is an important factor to understand regarding the self-organization phenomena [5]. It represents the maximum slope at which a pile of dry sand maintains stability before collapsing. As a self-organizing system, sand particles interact based on local rules, leading to emergent patterns and structures. In this research, that moment becomes the starting point for

Fig. 1. Angle of repose shown on poured sand on raised planes

predictive computational behaviour. As sand grains accumulate, their natural tendency to settle at the angle of repose contributes to the formation of piles and patterns. These piles reflect a balance between gravitational forces and friction, embodying the principles of self-organization, where simple interactions give rise to complex forms. For instance, Frei Otto's sand experiments explored self-organization through sand's inherent ability to find equilibrium shapes. The angle of repose is a tangible link between granular physics and the broader concept of self-organization in diverse natural and engineered systems [6, 7].

Fig. 2. Poured sand on raised planes demonstrating angle of repose as individual tiles.

2   Flat Plane Methodologies

The configuration of a flat plane that is either lifted from the surface or cut into specific shapes (Fig. 2) can influence the shape of a sand pile when dune sand is poured onto it. This is much different than making sand piles on an infinite plane, ground plane, or digitally on a computer. For example, in the physical world, a round, lifted plane creates a different pile than a rectangular, lifted plane, and so on. When sand is poured onto a lifted plane and accompanied by openings, the sand pile's organization is affected even more based on the opening and proximity to an edge. The organizational behaviour of sand allows it to be aggregated naturally based upon a specific set of criteria, mainly: resting plane, angle of repose, and gravity. This is a critical point in the project because it makes a controlled sand pile with specific depth and geometry, which is potentially a controllable architectural surface condition.

2.1   Simulating Digital Dune Sand Formations

A digital model illustrating the physical dynamics of sandpiles and their angle of repose was formulated using an array of 25 distinctive sand piles. To investigate the phenomenon of sand distribution, a series of digital simulations was systematically conducted, initiating the repositioning of two sand piles and progressively advancing through a range of pile adjustments until a total of 25 piles underwent modification (Fig. 3). From this model, it is discernible that any alteration to one pile induces consequential effects on neighbouring piles due to the resultant redistribution of sand. The digital simulations captured the sandpile's behaviour, depicting the top of the pile as a pattern density and the bottom as a plane shape. This approach provided valuable insights into the behavioural intricacies and interactions of a given set of sandpiles. A Grasshopper definition was introduced to the simulation, whereby the placement of points on a plane also corresponded with an angle of repose and a resultant cone replicating the material behaviour of sand. This establishes a generative condition for the sand that considers the point grid and the dynamics of piling conditions, including the 34-degree angle of repose. When arranged on a planar surface, these sand piles give rise to a Voronoi condition. Various simulations were conducted with diverse point quantities, revealing consistent behaviour in planar compositions created from 10 to 1000 points. In each iteration, the Voronoi condition exhibited a predictable evolution.

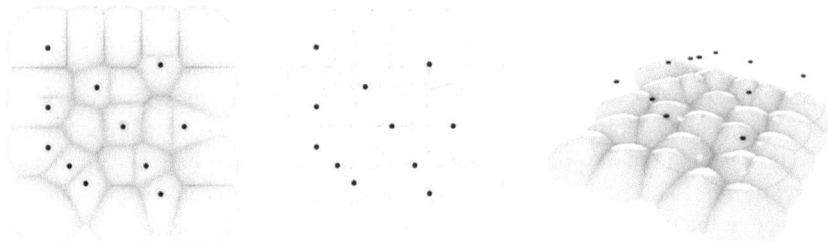

Fig. 3. Sand-pile computation model.

2.2   Physical Simulations of Sand Piles on Panels

A testing machine was built to compare physical piles with digital piles and attempt to control physical sand piles as three-dimensional surfaces with predictive intelligence.

The machine acts as a pouring mechanism that can assist in an ongoing series of sandpile creations with planes with different cut openings. The machine uses sliding trays and an auger system capable of accommodating laser-cut acrylic sheets with varying grids and geometries. The methodology continued (Fig. 4) with acrylic sheets subjected to laser cutting, guided by specific points derived from a series of tests generated through computational modeling. A reservoir for holding sand was positioned beneath each tray, complemented by an auger mechanism to regulate the sand's flow directly onto the acrylic sheet. This machine offers a comprehensive physical platform for investigating various aspects of sand piles, including topography, weight distribution, and geometric characteristics, and facilitates the manipulation of opening sizes on the plane, thereby offering different levels of performance and adaptability.

Fig. 4. Sand-pile generation with analog machine on flat planes.

3   Optimization of Openings and Data

The investigation into apertures of varying sizes within the planes was extended to clarify their role as a form-finding factor, in conjunction with the organizational characteristics of the sand. The specific placement of points, the introduction of apertures as potential openings for strategic ventilation, and variations in shadow densities are inherently interconnected. Additionally, an increased number of apertures correlates with a reduction in weight, a crucial consideration because sand is a heavy material. Also, a higher quantity of apertures results in more pronounced shadow patterns, which may be significant in desert architecture. This is particularly relevant as the primary objectives for an architectural surface in the desert involve safeguarding the building's facade while facilitating effective ventilation and heat dissipation.

3.1   Digitally Optimized Openings and Planes for Sand Tilings

Utilizing the diverse apertures and point placements for the sandpiles, an optimization strategy was implemented using the Wallacei plugin. Wallacei, is a multi-objective optimization tool, and executes evolutionary simulations by leveraging information related to the point placements for the Voronoi grids, the resultant sand piles, and the flat tile and

Fig. 5. Parakeet tiling sequences.

pattern sequences generated in Parakeet (Fig. 5). Such an approach enables informed decision-making at every stage of the sandpile system through predictive simulations. The optimization method enhances material efficiency and offers insights into how a future sandpile panels or tiling systems can be more effectively employed in a potential architectural project.

During the study, the algorithm applied fitness values to various tile layouts, starting with a limited set of solutions and progressing towards significantly higher values in the later stages of simulations. In this process, graphically, they start as an extensive range, and then become more concentrated, meaning an optimal solution exists, and the algorithm finds it. (Fig. 6). In the parallel coordinate plot, the more optimal solutions in the later generations indicate that they work well for both fitness objectives. This means that our fitness goals are not contradictory, and there can be a solution that meets more than one requirement.

Fig. 6. Digital optimizations of sand tilings with optimized openings from weather data.

In other words, we can make a sandpile composition from Voronoi points where we see the version best for sunlight and radiation and the best for an average condition. These were pulled from many different evaluations. The studies for this paper are based on a relatively small number of hours and generations, between 100 and 500, but this can be increased to much more.

Moving further, weather data from the UAE and sun trajectories were taken and entered into Wallacei with Grasshopper Ladybug. Around 500 h were selected from the year when the conditions were very hot and simultaneously involved high temperature, humidity, and low wind speed. Here, we can study specific latitude and longitude for a certain location in the desert along with dry bulb temperature, dew point temperature, relative humidity, direct radiation, horizontal radiation, wind speed, and wind direction.

These can all be used to make decisions about the performance of a sand pile composition while also comparing the aesthetic. Dry bulb temperature was a specific concern due to the high temperature of the ambient desert air. The expectation is that the physical dune sand compositions created by the original Voronoi point grid may act as shading devices, which can later be frozen with a binder to create a sand panel. The result is that the dune sand panels created for the study shaded up to 53% of the overall sunlight coming through in extremely hot times, creating enough evidence to continue the project and support the technology.

4   Conclusion

The research project explored the potential development of a regionally appropriate architectural system utilizing dune sand through an optimized approach. It investigates optimizing the process for future site-specific intelligence, centering exclusively on dune sand and the self-organization of sand piles as a generative mechanism. To facilitate this investigation, the research employed a combined analogue and digital methodology. The significance of this project arises from the infrequent utilization of desert dune sand in contemporary construction. Middle East and North African cities extensively import large quantities of river sand from Southeast Asia and Australia to meet the demand for concrete and masonry products.[8]

Fig. 7. Optimized tiling sequences made from self-organized dune sand.

The result is the integration of an analogue/digital workflow that incorporates this bio-material and is coupled with optimization to delineate distinct architectural surfaces responsive to local weather conditions (Fig. 7). The study highlights the feasibility of utilizing local dune sand material for crafting one-to-one typological prototypes reminiscent of mashrabiya and jalis found in arid climates. This presents an avenue for exploring viable, human-scale architectural prototypes as part of an ongoing series.

Additional research is warranted to delve into material responses to weather conditions, durability, and the structural constraints of the material.

The importance of sand as a resource is unparalleled, ranking as one of the most widely utilized materials globally, surpassed only by the elemental necessities of air and water [9]. This material pervades contemporary society, finding application in diverse products, from personal computers and glass to concrete and asphalt for road infrastructure. Despite its ubiquity, it is paradoxical that desert sand, the predominant source of this valuable resource, is largely overlooked for construction in desert regions. This underscores a critical imperative—to recognize the latent potential of desert sand and its relevance as a fundamental environmental concern, especially in regions closely connected to this invaluable resource.

In light of the profound impacts of both regional and global construction practices, it is fitting to consider this as an avenue for further scholarly investigation. Countries in the MENA region exemplify a reliance on predominantly imported materials or resources derived from foreign origins. Consequently, the necessity of leveraging indigenous resources and pioneering innovative architectural design paradigms using local sand emerges as a pathway toward transformative solutions aligned with the United Nations Sustainable Development Goal for terrestrial ecosystems.

Acknowledgements. Student Research Assistant: Zartaj Kamran Khan.
Wallacei Tutorial: Mesrop Andriasyan.

References

1. Ahmed, A.S., Al-Maadeed, M.H., Al-Kuwari, A.A.: Comparison of desert and river sands for construction purposes: a case study in Qatar. J. Mater. Civil Eng. **28**(4) (2016)
2. Ali, I.: Engineering properties of desert sands. Geotech. Geol. Eng. **19**(1), 65–83 (2001). https://doi.org/10.1023/A:1011168120592
3. Ahmed, M., Hossain, M.S., Rahman, M.A., Islam, M.M.: Influence of particle shape and gradation on the shear strength of river sand. Proc. Inst. Civ. Eng.—Ground Improv. **172**(4), 220–231 (2019). https://doi.org/10.1680/jgrim.18.00062
4. Al-Abdul Wahhab, H.I.: Performance of dune sand as a geotechnical material. Constr. Build. Mater. **36**, 640–645 (2012)
5. Otto, F., Rasch, B.: Finding Form: Towards an Architecture of the Minimal. Edition Axel Menges, Stuttgart (1995)
6. Hunt, M.L., Sweeney, R.E.: Effect of grain shape, roundness, and surface roughness on granular angles of repose. J. Res. Nat. Inst. Stand. Technol. **67A**(2), 155–158 (1963)
7. Llorens, P.: Self-organization and design: Frei Otto's legacy. Nexus Netw. J. **11**(2), 267–278 (2009)
8. Alghamdi, M.A., Siddiqui, S.K., Sayed, M.N.: River Sand importation and its implications on construction in Middle East and North African Cities. J. Constr. Eng. Manag. **142**(12), 04016077 (2016). https://doi.org/10.1061/(ASCE)CO.1943-7862.0001177
9. Dissanayake, C.B.: Sand: an indispensable resource. Science **283**(5408), 1155 (1999). https://doi.org/10.1126/science.283.5408.1155

Data-Driven and Algorithmic Design

A Novel AI-Driven Multi-objective Optimization Approach for Energy System Design in Industrial Zone

Jiesheng Yu, Yongming Zhang$^{(\boxtimes)}$, Zhe Yan, and Ziqi Li

Tongji University, Shanghai 200092, China
zhangyongming@tongji.edu.cn

Abstract. In the context of global energy conservation and emission reduction, it is imperative to improve energy efficiency and promote sustainable development. The industry sector has consistently been a pivotal industry in implementing low-carbon practices, particularly in large industrial zones characterized by high energy consumption and carbon emissions. A focal point of research in such contexts revolves around the configuration of energy systems in industrial zone. This paper proposes a novel AI-driven multi-objective optimization approach for energy system design in industrial zone. Economy, environmental sustainability, and safety are considered as optimization objectives, and the Pareto front solution set is calculated using the NSGA-II algorithm. The optimal system design is determined based on the composite optimization objective function under different priority scenarios. This method provides valuable insights that could serve as a blueprint for energy system design in industrial zones, offering guidance towards more efficient and environmentally conscious energy configurations.

Keywords: Building energy simulation · PV-battery system · Capacity configuration · Multi-objective optimization

1 Introduction

Reducing carbon emissions has become a global consensus, and various policies targeting carbon emissions have been successively developed. Against this backdrop, renewable energy, as an alternative to traditional fossil fuels, is being increasingly utilized. According to data released by the International Energy Agency (IEA), it is estimated that the investment in clean energy amounted to 872 billion dollars in 2023, marking an increase of 7.52% and 22.82% compared to the figures in 2022 and 2021, respectively [1]. Renewable energy sources, such as solar and wind power, largely depend on weather resources that are intermittent and volatile. To mitigate these fluctuations, battery energy storage systems (BESS) have been widely recognized as one of the potential solutions, offering advantages such as rapid response capabilities, continuous power supply, and geographical independence. Meanwhile, the construction industry, especially large industrial zones, has always been a high carbon-emitting sector. As an energy-intensive industry, it consumes about a quarter of the global total energy and accounts for 23% of

© The Author(s) 2025
H. Chai et al. (Eds.): CDRF 2024, *Symbiotic Intelligence*, pp. 191–201, 2025.
https://doi.org/10.1007/978-981-96-3433-0_17

the total greenhouse gas emissions [2]. Therefore, it is imperative to establish renewable energy systems in industrial zones to reduce energy consumption and carbon emissions.

Many researchers have studied renewable energy systems in industrial zones. Ximei Li et al. [3] investigated the techno-economic characteristics of energy system designs based on renewable sources, finding that in regions abundant in solar and wind energy, a combination of photovoltaic (PV)/wind power generation often has advantages over sole PV or wind power generation, with energy storage also playing a crucial role. Jiuping Xu et al. [4] developed a bi-level multi-objective model for the optimization of distributed energy resources in industrial park to respond to extreme events, ensuring 95.85% of demand for three consecutive hours after such events, maintaining over 50% of demand at all times, and minimizing financial losses due to power outages to as low as 0.23 thousand yuan. Muhammad Haseeb Rasool and his team [5] developed a genetic algorithm (GA) in MATLAB to analyze the impact of short-term and long-term energy storage systems (ESS) interactions on grid balance and used a comparative framework to assess multiple systems and ESS configurations to identify the best techno-economic configurations and strategies for grid balance.

From the perspective of artificial intelligence (AI) algorithms, GA is a powerful search algorithm inspired by natural selection, capable of finding optimal solutions for complex problems by exploring a large search space. And Non-dominated Sorting Genetic Algorithm II (NSGA-II) is an advanced version of GA specifically designed for multi-objective optimization, effectively handling multiple and often conflicting objectives while converging to the true Pareto front. Therefore, the NSGA-II algorithm is more suitable for solving complex system design problems in industrial zones.

Existing research primarily focuses on the economic and environmental aspects of renewable energy systems in industrial parks, as well as their capacity to respond to emergencies. It can't be ignored that during the operation of renewable energy systems, significant variations in grid transmission power and the substantial peak-to-valley differences in daily electricity consumption can increase grid volatility, thereby reducing the reliability of power supply and posing considerable safety risks. Therefore, this paper utilizes the NSGA-II algorithm to propose a novel AI-driven multi-objective optimization approach for energy system design in industrial zone, comprehensively considering economic benefits, environmental friendliness, and safety.

2 Methodology

The technical approach of this paper is illustrated in Fig. 1. Taking a certain industrial zone in Shanghai as the case study, renewable energy and energy storage devices are introduced to its existing energy system, constructing a Photovoltaic-battery (PVB) system. Initially, the weather and load data, along with the parameters of the PVB system, are inputted to establish a mathematical model of the energy system. With three objective functions as optimization goals, NSGA-II algorithm is employed to optimize the energy system based on the search ranges of the PVB system (PV and battery size). The optimization variables include five parameters, which consist of two system capacity allocation parameters and three operational parameters. After obtaining the Pareto front, three scenarios are set, focusing on the system's economic efficiency, environmental friendliness, and safety, respectively. Finally, the optimal configurations corresponding to the three scenarios are calculated, and the results are analyzed.

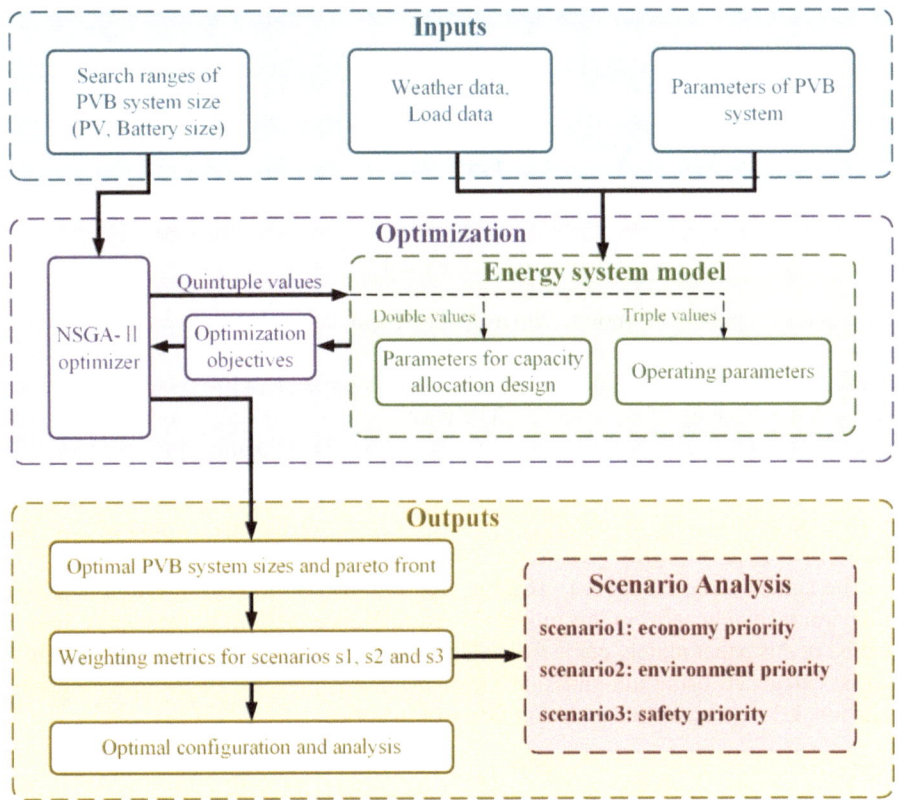

Fig. 1. Schematic diagram of the multi-objective optimization procedure

2.1 Optimization Objective

The annual total cost (*ATC*) is used in this paper as the economic optimization objective, aimed at evaluating the economic feasibility of the system's investment and operation [6]. The *ATC* is composed of the following four parts: annual investment expenditure (*INV*), operation cost (*OPE*), fixed maintenance cost (*FM*), and variable maintenance cost (*VM*), as shown in Eq. (1) to Eq. (5). For each design solution, the system configuration can be adjusted according to different operational scenarios. Therefore, the subscript k is used to differentiate between metrics in the calculation, such as *ATC*, *INV*, *OPE*, reflecting changes across different operational scenarios. The subscript i is used to distinguish parameters of various devices within the system, such as C and φ.

$$ATC_k = INV_k + OPE_k + FM_k + VM_k, \forall k \in K = \{s_1, s_2, s_3\} \tag{1}$$

$$INV_k = \sum_i (CRF_i \cdot C_i \cdot \varphi_i), \quad \forall k \in K, \quad \forall i \in I = \{pv, bat\} \tag{2}$$

$$OPE_k = \sum_{t=1}^{8760} \left(E_{grid,k,t}^{buy} \cdot c_{buy,t}^{ele} - E_{grid,k,t}^{sell} \cdot c_{sell,t}^{ele} \right), \quad \forall k \in K \tag{3}$$

$$FM_k = \sum_i \left(C_i \cdot \varphi_{fix,i} \right), \quad \forall i \in I \tag{4}$$

$$VM_k = \sum_{t=1}^{24} \sum_i (E_{i,k,t} \cdot \varphi_{var,i}), \quad \forall k \in K, \forall i \in I \tag{5}$$

$$CRF_i = \frac{m(1+m)^{n_i}}{(1+m)^{n_i} - 1}, \quad \forall i \in I \tag{6}$$

s_1, s_2 and s_3 represent economic, environmental and safe operation scenarios, respectively. CRF denotes the capital recovery factor, a ratio to calculate the present value of an annuity. m is the annual real interest rate, n is the system life period in years. C_i indicates the capacity of equipment i. The terms φ_i, $\varphi_{fix,i}$, and $\varphi_{var,i}$ represent the unit price, the fixed and variable maintenance cost coefficients of equipment i, respectively. $E_{grid,k,t}^{buy}$ and $c_{buy,t}^{ele}$ indicate the amount and price of the electricity purchased from the utility grid, respectively. While $E_{grid,k,t}^{sell}$ and $c_{sell,t}^{ele}$ indicate the amount and price of the electricity sold to the utility grid, respectively.

The annual carbon emission (ACE) is used to represent the optimization objective of environmental benefits in this model [7]. The carbon emissions of the energy system studied in this paper mainly come from electricity purchased from the grid, so the ACE can be calculated using the emission factor method [8], as shown in Eq. (7). In this equation, ξ^{ele} represents the emission factor of the grid.

$$ACE_k = \sum_{t=1}^{8760} \left(\xi^{ele} \cdot E_{grid,k,t}^{buy} \right), \quad \forall k \in K \tag{7}$$

The annual maximum power purchase (AMP) is used as the safety optimization objective in this study. The smaller its value, the smaller the difference between the peak and valley of the power exchange between the park and the grid, the smaller the fluctuation of the power, and the higher the reliability of the system.

The Pareto front is obtained by optimizing the above three objectives using the NSGA-II algorithm. Therefore, for different operating scenarios, a composite optimization objective function F can be formed by weighting the system's optimization objectives, as shown in Eq. (8).

$$F = \omega_{ATC} \cdot \underbrace{\frac{ATC_k - ATC_{min}}{ATC_{max} - ATC_{min}} \times 100\%}_{ATC_k^{nor}}$$

$$+ \omega_{ACE} \cdot \underbrace{\frac{ACE_k - ACE_{min}}{ACE_{max} - ACE_{min}} \times 100\%}_{ACE_k^{nor}}$$

$$+ \omega_{AMP} \cdot \underbrace{\frac{AMP_k - AMP_{min}}{AMP_{max} - AMP_{min}} \times 100\%}_{AMP_k^{nor}}, \quad \forall k \in K \qquad (8)$$

In Eq. (8), ω_{ATC}, ω_{ACE} and ω_{AMP} represent the decision weights for economic, environmental and safe objectives, respectively. ATC_{min}, ACE_{min} and AMP_{min} are the minimum values for the economic, environmental and safe objectives, respectively, and represent the ideal values of objectives. On the contrary, ATC_{max}, ACE_{max} and AMP_{max} are the maximum values for the economic and environmental objectives, respectively, and represent the worst-case expected values of objectives. Correspondingly, ATC_k^{nor}, ACE_k^{nor} and AMP_k^{nor} denote the ATC_k, ACE_k and AMP_k after normalization, respectively.

2.2 System Model

The energy system framework of the industry zone studied in this paper is shown in Fig. 2. The AC loads are directly connected to the AC bus, which is connected to the utility grid through a transformer to obtain electricity. Meanwhile, DC loads such as PVs, BESS, and electric vehicles (EV) are connected to the DC bus, which is linked to the AC bus through a DC/AC converter. The entire system can be controlled and operated through an energy management system (EMS).

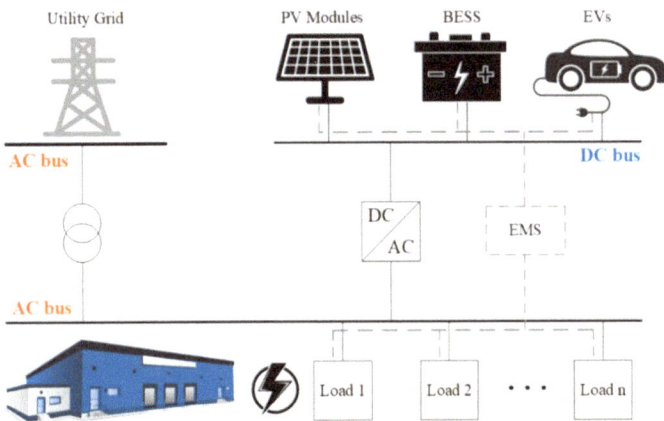

Fig. 2. System architecture diagram

Based on this system architecture, the mathematical model of the system and energy constraints can be constructed, as shown in Eq. (9) and Eq. (10) respectively.

$$\begin{cases} \begin{cases} P_{pv} = \eta_{pv} A_{pv} G_\beta \\ \eta_{pv} = \eta_\gamma \eta_{pc} \left[1 - \beta \left(T_c - T_{cref} \right) \right] \\ T_c = T_a + \left[(NOCT - 20)/800 \right] G_\beta \end{cases} \\ \begin{cases} SOC = \frac{E_{bat}}{C_{bat}} \\ P_{bat} = \alpha P_{bat,ch} + (1 - \alpha) P_{bat,dis} \\ SOC(t+dt) = SOC(t) + \alpha \frac{\eta_{ch} P_{bat,ch} dt}{C_{bat}} + (1 - \alpha) \frac{P_{bat,dis} dt}{\eta_{dis} C_{bat}} \end{cases} \end{cases} \quad (9)$$

$$\begin{cases} P_{pv} + P_{grid}^{buy} = P_{load} + P_{bat} + P_{grid}^{sell} \\ SOC_{min} \le SOC(t) \le SOC_{max} \\ P_{bat,dis}^{max} \le P_{bat} \le P_{bat,ch}^{max} \end{cases} \quad (10)$$

where P_{pv} represents the actual power generation of the PV modules, η_{pv} denotes the efficiency of PV generators, and A_{pv} is the area of the PV panels (m^2). G_β indicates the solar radiation on tilted module plane (W/m^2). η_γ represents the reference module efficiency, and η_{pc} denotes the power conditioning efficiency, which equals 1 under maximum power point tracking (MPPT). β corresponds to the temperature coefficient of the generator's efficiency, which ranges from 0.004 to 0.006 (°C-1) for silicon cells. T_{cref} indicates the reference cell temperature (°C), while T_c is the cell temperature (°C) which can be estimated based on the ambient temperature T_a (°C) and the solar radiation G_β. NOCT represents the nominal operating cell temperature (45 °C).

In Eq. (9), The state of charge (SOC) is commonly used to describe the charging and discharging state of a battery, defined as the ratio of the battery's energy content (E_{bat}t) to its rated capacity (C_{bat}). P_{bat} represents the actual operating power of the battery, with $P_{bat,ch}$ and $P_{bat,dis}$ corresponding to the battery's charging and discharging power, respectively. α is a binary number, where 1 indicates the battery charging state, and 0 indicates the battery discharging state. η_{ch} and η_{dis} respectively represent the charging efficiency and discharging efficiency of the battery.

2.3 System Operation Strategy

Considering the need to reduce the AMP to enhance the system's stability, a system operation strategy for orderly battery discharge is utilized in this paper. As illustrated in Fig. 3, (a) represents the conventional baery operation mode, where the battery discharges from t1 to t2 until it is depleted. In this operating mode, the P_{grid}^{buy} during the peak usage time is not reduced. The highest grid power purchase (P_{grid}^{max}) is equal to the maximum load power (P_{load}^{max}). (b) depicts the orderly battery discharge mode used in this paper, which is set to initiate discharge process only when the P_{load} exceeds a pre-set value, which is represented by P_{bat}^{pre}. The battery discharges between t3 and t4, effectively reducing the P_{grid}^{max}, significantly lower than the P_{load}^{max}. To express clearly in the subsequent optimization process, R_{pre} is defined, representing the ratio of P_{bat}^{pre} to P_{load}^{max}, as shown in Eq. (11).

$$R_{pre} = \frac{P_{bat}^{pre}}{P_{load}^{max}} \quad (11)$$

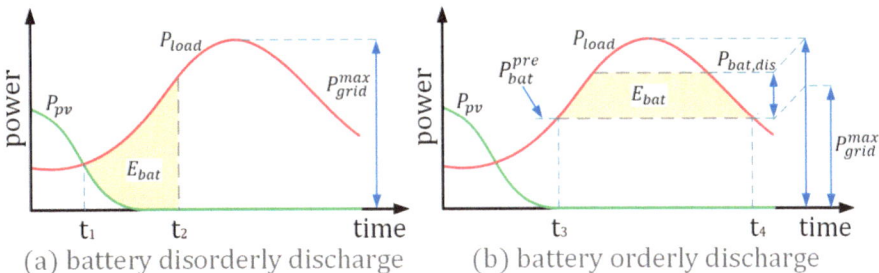

(a) battery disorderly discharge   (b) battery orderly discharge

Fig. 3. Schematic diagram of orderly discharge of batteries

3 Simulation Results and Discussion

An industrial park in Shanghai is selected as the case study for this paper, with its annual electricity load and local solar irradiance data illustrated in Fig. 4.

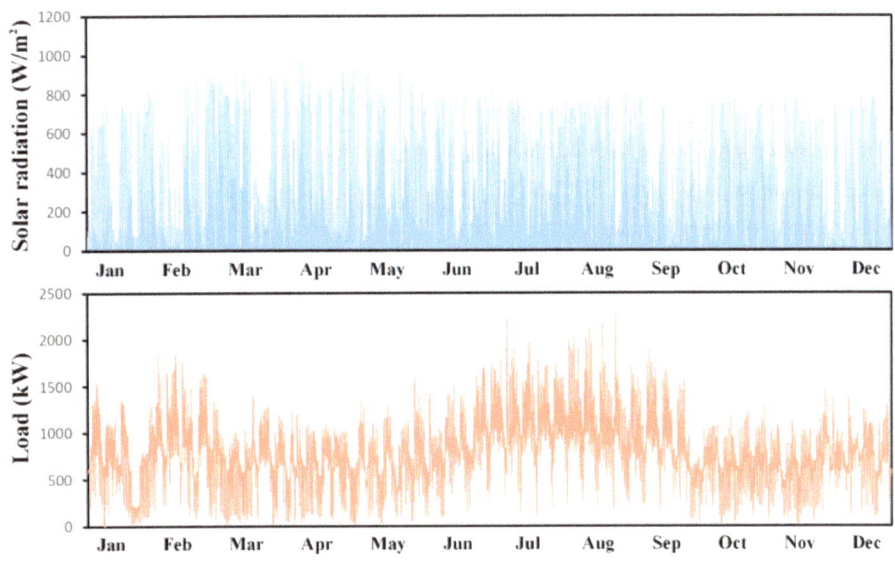

Fig. 4. Load and solar irradiance data

Considering the significant seasonal variations, R_{pre} can be classified as R_{pre}^{sum}, R_{pre}^{spr}, and R_{pre}^{win}, corresponding to summer, spring/autumn, and winter, respectively. Similarly, the allocation of the system's capacity can also be represented using dimensionless variables, as shown in Eq. (12). The capacity of PV can be expressed by the PV penetration rate (R_{pv}), defined as the ratio of PV power generation (E_{pv}) to the industrial park's electricity consumption (E_{load}). The storage capacity can also be represented by the relative storage rate (R_{bat}), defined as the ratio of storage capacity (C_{bat}) to the diurnal average electricity consumption (E_{load}^{diu}).

$$\begin{cases} R_{pv} = \frac{E_{pv}}{E_{load}} \\ R_{bat} = \frac{C_{bat}}{E_{load}^{diu}} \end{cases} \tag{12}$$

The population size of the NSGA-II algorithm in this study is set to 100 individuals, with a total of 200 generations specified. The crossover rate and mutation rate are set at 0.8 and 0.2, respectively. The other relevant parameters used for simulation and optimization are shown in Table 1. Taking the parameters as the input for the PVB system model and NSGA-II algorithm, with R_{pv}, R_{bat}, R_{pre}^{sum}, R_{pre}^{spr} and R_{pre}^{win} as the optimized objects and ATC, ACE, and AMP as the optimization objectives, the calculated Pareto front solution set and its distribution are shown in Fig. 5.

Table 1. The parameter settings

Parameters	Value	Unit
The search range of R_{pv}	0~2.0	/
The search range of R_{bat}	0~2.0	/
The search range of R_{pre}^{sum}, R_{pre}^{spr}, and R_{pre}^{win}	0~0.5	/
The maximum value of battery SOC (SOC_{max})	0.95	/
The minimum value of battery SOC (SOC_{min})	0.10	/
Max charging/discharging rate	0.5	C
Charging and discharging efficiencies (η_{ch}/η_{dis})	95	%
Battry cycle times	6000	Times
The unit price of PV (φ_{pv})	8000	CNY /kW
The unit price of BESS (φ_{bat})	2500	CNY /kWh
The fixed maintenance cost coefficient of PV ($\varphi_{fix,pv}$)	0.01	CNY /kWh
The fixed maintenance cost coefficient of BESS ($\varphi_{fix,bat}$)	0.01	CNY /kWh
The variable maintenance cost coefficient of PV ($\varphi_{var,pv}$)	0.01	CNY/kWh
The variable maintenance cost coefficient of BESS ($\varphi_{var,bat}$)	0.02	CNY /kWh
Annual real interest rate (m)	8	%
PV system life period (n_{pv})	20	Years
Battery energy storage system life period (n_{bat})	15	Years
Emission factor of the grid (ξ^{ele})	0.96	kg/kWh

As illustrated in Fig. 5, each pink dot in the three-dimensional coordinates represents an optimal solution. The distribution of the optimal solution set is summarized in the graph on the right, with the optimal configuration range for R_{pv} and R_{bat} being approximately 1~2 and 0.25~1.25, respectively. Although the maximum values of parameters

R_{pre}^{sum}, R_{pre}^{spr} and R_{pre}^{win} could reach 0.5 when set, their optimal ranges are all below 0.35 after the optimization process.

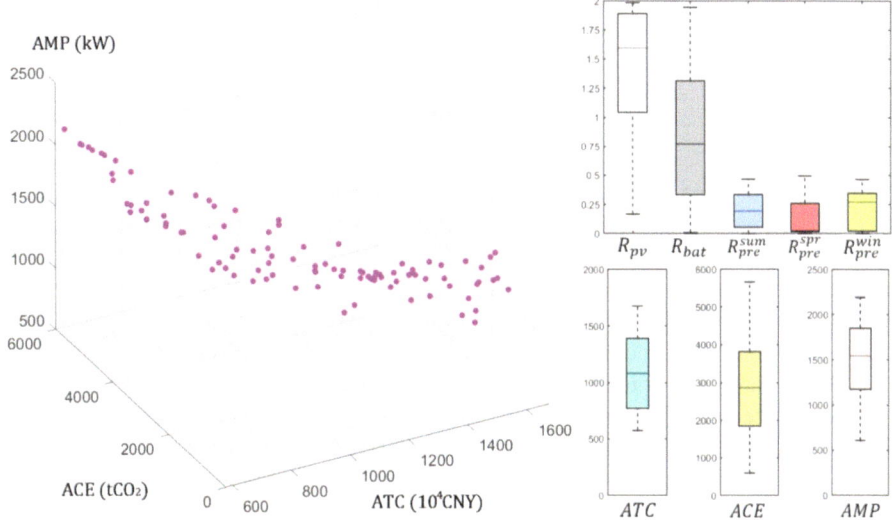

Fig. 5. Distribution of pareto optimal solution set

After obtaining the set of optimal solutions, it is necessary to determine the decision weights for each objective based on the three scenarios set in this paper. Table 2 is constructed from the pairwise comparison matrix based on scores given by experts for each objective. The scoring range is from 1 to 9, where a score of 1 indicates two equally important objectives, and a score of 9 signifies one objective being significantly more important than the other [9]. Considering the varying levels of importance different stakeholders place on each objective, this study accounts for decision weights of three groups of stakeholders with different parties.

Table 2. Decision matrix of different scenarios

s_1: Economic scenario				s_2: Environmental scenario				s_3: Safe scenario			
	ATC	ACE	AMP		ATC	ACE	AMP		ATC	ACE	AMP
ATC	1.00	3.00	2.00	ATC	1.00	0.25	0.50	ATC	1.00	0.50	1.00
ACE	0.33	1.00	4.00	ACE	4.00	1.00	4.00	ACE	2.00	1.00	0.50
AMP	0.50	0.25	1.00	AMP	2.00	0.25	1.00	AMP	1.00	2.00	1.00
ω	0.513	0.330	0.158	ω	0.133	0.655	0.211	ω	0.264	0.329	0.407

Based on the weights in Table 2, the composite optimization objective function F for three scenarios can be derived, which enables the calculation of the optimal system configuration and related parameters as shown in Table 3.

Table 3. The optimal configurations for scenarios s_1, s_2 and s_3

	R_{pv}	R_{bat}	R_{pre}^{sum}	R_{pre}^{spr}	R_{pre}^{win}	$ATC(10^4CNY)$	$ACE(tCO_2)$	$AMP(kW)$
s_1	1.062	0.5096	0.1073	0.0085	0.0168	801	2381	1968
s_2	1.933	0.9747	0.0467	0.0271	0.0124	1192	887	1635
s_3	1.899	0.9303	0.3356	0.2682	0.3066	1299	3310	763

The data in Table 3 reveals that under the economic priority scenario (s_1), the allocation capacities for PV and battery are smaller than those in the other two scenarios. A lower renewable energy capacity effectively enhances the system's economy (ATC) but comes with higher carbon emissions (ACE) and the highest peak electricity load (AMP). Both environmental and safety priority scenarios (s_2 and s_3) have higher allocation of PV and battery capacity, as well as higher ATC. However, the R_{pre} values in s_3 are significantly larger than those in s_2, implying that the BESS is almost exclusively used to balance peak electricity consumption, thereby substantially reducing AMP and significantly enhancing system stability. In s_3, during low power loads, the BESS will not discharge to balance the load and will remain fully charged until the next charging cycle, leading to lower renewable energy utilization and increased ACE. In contrast, the lower R_{pre} values in s_2 maximize battery power utilization, enhancing renewable energy utilization and significantly reducing ACE. Comparing parameters R_{pre}^{sum}, R_{pre}^{spr} and R_{pre}^{win} across the three scenarios shows that R_{pre}^{sum} is consistently higher than R_{pre}^{spr} and R_{pre}^{win}, indicating the abundance of solar resources in summer and higher electricity consumption, which is more effective in reducing peak loads compared to the other seasons.

4 Conclusion

In this paper, taking a certain industrial zone in Shanghai as a case study, a novel AI-driven multi-objective optimization approach for energy system design in industrial zone is proposed. This method optimizes five parameters of the energy system from three aspects: economy, environmental protection, and safety. For three scenarios, composite optimization objective function F is constructed based on the weights of optimization objectives to calculate the corresponding optimal configurations. The results show that under the economic priority scenario, the system's allocation capacities for PV and batteries are lower. The environmental priority scenario has the highest utilization of renewable energy, significantly reducing carbon emissions. In contrast, the safety priority scenario reduces electricity purchasing loads substantially at the cost of sacrificing renewable energy utilization, thereby enhancing the system's stability. The method proposed in this paper provides valuable insights for the design of industrial parks, offering guidance for more efficient, environmentally friendly, and safe energy configuration design.

Acknowledgments. The study was supported by the Shanghai Science and Technology Plan Project (Grant no. 22DZ1201707), Suzhou Science and Technology Plan Project - Carbon Peak and Carbon Neutrality Special Project (Grant no. ST202313) and Open Research Fund Program of Anhui Province Key Laboratory (Grant no. IBES2022KF12).

References

1. IEA: Clean Energy Investment by Region, 2018–2023. IEA, Paris https://www.iea.org/data-and-statistics/charts/clean-energy-investment-by-region-2018-2023. IEA. Licence: CC BY 4.0
2. IEA: The Role of CCUS in Low-Carbon Power Systems. IEA, Paris (2020). https://www.iea.org/reports/the-role-of-ccus-in-low-carbon-power-systems. Licence: CC BY 4.0
3. Ximei, L., Jianmin, G., Shi, Y., et al.: Optimal design and techno-economic analysis of renewable-based multi-carrier energy systems for industries: a case study of a food factory in China. Energy **244**(PB) (2022)
4. Jiuping, X., Yalou, T., Fengjuan, W., et al.: Resilience-economy-environment equilibrium based configuration interaction approach towards distributed energy system in energy intensive industry parks. Renew. Sustain. Energy Rev. 191114139 (2024)
5. Haseeb, M.R., Onur, T., Usama, P., et al.: Comparative assessment of multi-objective optimization of hybrid energy storage system considering grid balancing. Renewable Energy **216** (2023)
6. Zhe, T., Xiaoyuan, L., Jide, N., et al.: Enhancing operation flexibility of distributed energy systems: a flexible multi-objective optimization planning method considering long-term and temporary objectives. Energy **288** (2024)
7. Chen, H., Liu, S., Kuang, Y., et al.: Decomposition analysis of regional electricity consumption drivers considering carbon emission constraints: a comparison of Guangdong and Yunnan Provinces in China. Energies **16**(24) (2023)
8. Roy, A., McCabe, Y.B., Saxe, S, et al.: Review of factors affecting earthworks greenhouse gas emissions and fuel use. Renew. Sustain. Energy Rev. 194114290 (2024)
9. Das, R., Wang, Y., Putrus, G., et al.: Multi-objective techno-economic-environmental optimisation of electric vehicle for energy services. Appl. Energy **257**(C), 113965–113965 (2020)

The Application of Machine Learning Methods in the Identification of Rural Landscape and Rural Planning of Shanghai

Ni Xie, Yiru Huang[(⊠)], and Yuanxiao Kuang

Tongji University, Siping Rd. 1239, Shanghai, China
huangyr@tongji.edu.cn

Abstract. At present, studies of rural texture lack practical guiding significance for rural planning, resulting in a single form of new rural residential areas, causing serious damage to the original rural texture, moreover, there are many research methods for spatial planning, of which the method of machine learning has been very common in the design field, especially the ability to learn spatial features and the quantitative study of features. The method of using machine learning to influence decision-making in planning and design has not been deeply studied and applied in rural areas. The GANs (Generative Adversarial Networks) model is established in this paper by creating a four-level scale model of the boundary definition and factor-labeling paradigm of the rural space, clarifying the boundary, factor, and calculation form that affect the scale at all levels, and annotating the samples using the standard as a reference. Through the case study of a rural area in Shanghai, the labeling method was adjusted, various sample augmentation methods were continuously used to increase the sample size, the sample quality was adjusted by screening samples, and multiple pieces of training were finally generated to meet the needs of the new residential areas planning scheme in the given area. Through the establishment of area indicator and spacing indicator, the rationality of the scheme is quantitatively evaluated. The findings of this study show that machine learning approaches have a lot of potential for new settlement planning in rural areas; the GANs model is very effective in creating planning schemes, and this research method generates fresh ideas for rural area planning.

Keywords: Machine learning · GANs model · Rural landscape · Rural planning

1 Introduction

Rural residential land in China constitutes a substantial proportion of the nation's total land, yet its utilization efficiency in rural areas remains low. As an example, rural residential land in Shanghai occupied 45% of the city's total residential land, while the agricultural household registration population constituted merely 10% of the city's overall registered population during the same period, indicating an incongruity between the scale of rural residential land and the agricultural population (SMPG 2017). Consequently, rural residential planning has emerged as a critical juncture for tackling issues of rural land efficiency and housing demands.

© The Author(s) 2025
H. Chai et al. (Eds.): CDRF 2024, *Symbiotic Intelligence*, pp. 202–216, 2025.
https://doi.org/10.1007/978-981-96-3433-0_18

Rural residential planning fundamentally requires a comprehensive understanding of rural area textures, as the texture of rural residential areas directly influences the rural landscape's form, thereby making it an essential consideration in such planning processes. Current research on rural textures, particularly in regions like the southern part of the lower Yangtze River basin where Shanghai is situated, has provided qualitative descriptions of rural attributes (SMBPNR 2019; Zhou and Wang 2021), alongside quantitative analyses of villages' overall form, scale, and distribution patterns through parametric technology and other technical measures (Tong 2016). However, prevailing research primarily centers on the macroscopic morphological traits of villages, neglecting quantitative investigations into their internal spatial characteristics and spatial texture, and thus limiting their applicability in planning practices.

Over recent years, the feasibility of machine learning methods in the realm of architecture has been extensively validated. Typically, machine learning methodologies rely heavily on manual feature engineering. Given the complex, often indistinct objectives inherent in architectural design and urban planning tasks, machine learning can statistically summarize large amounts of empirical data. Notably, Generative Adversarial Networks (GANs) excel in generating analogous graphical outcomes from data (Huang and Zheng 2018). Prior studies have shown that training GANs with image labels enables computers to understand the relationships between spatial elements and the logical connections between spatial elements and boundaries (Liu, Fang et al. 2021). Some recent studies have successfully applied GANs to traditional layout generation tasks, proving its efficacy. In these studies, they marked functional areas with solid colors for image labeling purposes, a technique that significantly enhances the quality of house layout generation and underscores GAN's potential for capturing relations between functional areas (Peters 2018). Despite this, this methodology is restricted to target spaces with well-defined boundaries and similar scales. Solid boundaries and uniform scales provide a relatively fixed foundation for output generation, rendering it challenging to extend this machine learning method to cases without clear boundaries.

Given that rural spaces typically exhibit vague boundaries and pose difficulties in spatial feature extraction, GANs model-based research in rural areas is relatively scarce. The objectives of this paper include transforming traditional planning's descriptive architectural language into measurable indices by extracting spatial features from the study subjects and refining the topological relationships within the spatial texture. Another objective is to leverage machine learning to enhance the efficiency of quantitative research by training models capable of generating village group layouts. This would support planners and designers in enriching the database of rural landscape design, and potentially reveal overlooked layout principles in conventional planning and design practices. Through this approach, we aim to bridge the gap between the limitations of current methods and the complexities inherent in rural landscapes.

2 Background

Whether dealing with a concrete depiction of spatial texture or an abstract representation of cultural texture, the act of quantifying descriptive vocabulary is pivotal for genuinely guiding planning and design endeavors. Urban space consists of both tangible and intangible components: the material aspect pertains to functional space, whereas the immaterial aspect concerns the "theoretical" investigation of spatial texture. This "theoretical" texture of urban space is often overshadowed by the more evident functional and structural dimensions.

In 1784, Giambattista Nolli created the Nolli Map of Rome, wherein the white areas represent the "figure" formed by open and public spaces, while the black areas denote the "ground" composed of private spaces. The map graphically expresses the boundaries and continuity of these spaces. The figure-ground theory embodies the dichotomies of interior-exterior, public-private, and closed-open in urban environments. Investigating the constitutive elements of spatial fabrics serves as the foundational step in urban space research. Conzen posited that urban landscape constituents comprised town plans, building forms, and land use; his concept of the plan unit paved the way for spatial pattern studies (Conzen 1966; Conzen 1975). Kar Kropf further argued that the urban fabric was an organic amalgamation of all constituent structures, including the elements of structure, room or space, plot, street, and block or plot series (Kropf 1996). However, spatial composition transcends the mere analysis of individual elements; instead, it involves a holistic examination of the relationships between them, giving rise to quantitative studies. Such studies might calculate ratios like street width to building height (Yoshinobu 1983), analyze the correlation between street and square scales, and building heights (Sitte 1889), or introduce street-related metrics such as interface density and interface continuity (Zhou, Zhao et al. 2012; Fang, Zhang et al. 2017). Furthermore, various regulations ensure the continuity of streets and the integrity of street outlines within urban spaces, providing a foundation for the quantitative assessment of spatial scales in both urban and rural contexts.

Drawing upon this theoretical body of knowledge, this paper proposes to establish an extraction logic rooted in feature units rather than single administrative or geographical boundaries. It aims to identify the texture characteristics impacting the formation of rural residential layout paradigms, subsequently establishing the categorization concepts and calculation formulas pertinent to rural spatial scales. This groundwork lays the foundation for data analysis and model construction in rural planning and design.

2.1 Boundary Determination

The objective of defining boundaries lies in identifying a consistent unit of a comparable scale that is advantageous for residential planning research. While city administrative boundaries are typically stable, well-defined, and amenable to analytical approaches, those in rural areas tend to be less distinct. Traditional village boundaries emerge from a combination of natural and anthropogenic influences and can evolve over time due to changes in both environmental and social conditions. A significant number of rural perimeters encompass substantial expanses of farmland or water bodies, necessitating that the demarcation of rural boundaries integrate administrative borders, natural boundaries, and additional elements to create a coherent spatial unit.

Moreover, the delineation of the unit boundary should be driven by the specific research objectives and the inherent characteristics of the research subject. When the aim is to actualize new residential community planning within village zones, the uppermost scale boundary is generally anchored to the extant administrative village territory. The method employed for boundary definition in this context centers around the characteristic features of residential clusters within the studied cases. Specifically, the unit is derived by referencing the naturally enclosed areas defined by road networks and rivers, thereby forming the basis for the residential group's spatial organization.

2.2 Feature Extraction

Whether it is human learning or machine learning, the feature extraction method serves as a means of image processing, with the ultimate outcome being the analysis of a paradigm designed to tackle a specific problem. The primary objective of feature extraction is to scrutinize the correlation between rural texture and the underlying paradigms guiding rural group layouts. Within well-defined boundaries, the abstract paradigm transitions into a tangible, generative one. Figure 1 demonstrates this analytical process using AutoCAD, incorporating aspects of factor marking and image processing.

Acquiring a factor labeling paradigm is crucial to address the deficiency of spatial characteristics in rural areas. By examining the current status of residential groups, one can devise a paradigm suitable for generating new residential layouts. Implementing machine learning methodologies enhances the efficiency and precision of feature extraction and labeling, thereby enabling a more accurate analysis of the interactions between individual factors.

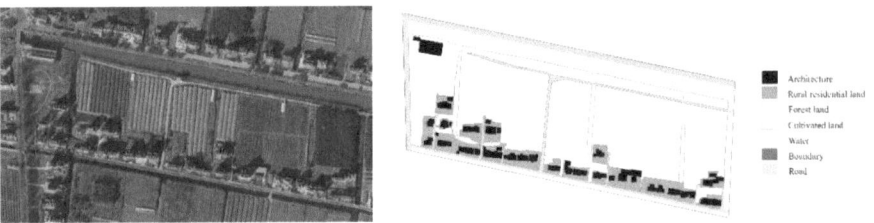

Fig. 1. Satellite (left), map depicted by AutoCAD (right)

2.3 Classification of Rural Space Scale

Table 1 illustrates that the spatial scale of rural areas is partitioned into four hierarchical levels, with each level having its respective elements and measurement criteria defined. At the fourth and most granular scale, the homestead scale, the basic unit is the homestead block serving as the boundary, representing the living quarters of individual households. The homestead density is utilized to compute the building coverage of each unit. Key elements within this scale include the main building area (MBA), auxiliary building area (ABA), homestead area (HA), building orientation, building length (BL), and building width (BW). The homestead building density ($De_{(H)}$) is calculated as the ratio of the building's footprint area to the total homestead area.

Advancing to the third scale, known as the street scale, which is delineated by the aggregation of homestead plots, the elements involved are street length (SL), street width

(*SW*), and building interface length (measured as the building length (*BLi*)). The richness of the street interface is a vital texture evaluation metric, exemplified by the building density of the main street interface (($De_{(MSI)}$)). Within a village-scale context, the street hierarchy commonly comprises the main street, secondary street, and walkways. Main streets serve as the primary thoroughfares between villages, often acting as the boundary of the first scale. Secondary streets cater to intra-village traffic and typically define the second scale boundary, while walkways are central to villagers' daily activities. The term "main street interface" here refers specifically to that of the main roads. The building density of the main street interface is denoted as ($De_{(MSI)}$).

Figure 2 elucidates the variables: *SL* signifies the street's length, corresponding to the length of each block along the street frontage. *BLi* represents the length of the building facade, also interpreted as the building interface length. θ depicts the angle between the building facade and the street. α indicates the angle between each building facade and the horizontal plane, which is equivalent to 90 degrees minus the azimuth angle. Lastly, β defines the angle between the street facade and the horizontal line.

Fig. 2. Graphical representation of the density of the main street interface

The second scale considered is the block scale, which arises from the natural enclosures created by roads or rivers. As depicted in Fig. 3, the key elements encapsulated at this scale consist of the block area (*BA*), homestead area within the block (*HAB*), cultivated land area within the block (*CLAB*), water area within the block (*WAB*), and permanent basic farmland area within the block (*PBFAB*). The block's building density ($De_{(B)}$) and block interface building density ($De_{(BI)}$) serve as descriptors of the block's textural qualities.

In this context, *Li* represents the projection of the homestead along the interface edge onto the main street's midline. *SL* symbolizes the street length, which corresponds to the length of each block segment facing the main street. *Wi* denotes the projection of the homestead along the interface edge onto the midline of the secondary street. Meanwhile, *SW* represents the street width and is associated with the length of each block segment bordering the secondary street.

The first scale under consideration is the village scale, with the administrative village as its defining perimeter. Illustrated in Fig. 4, the components at this scale include the village area (*VA*), homestead area within the village (*HAV*), cultivated land area within the

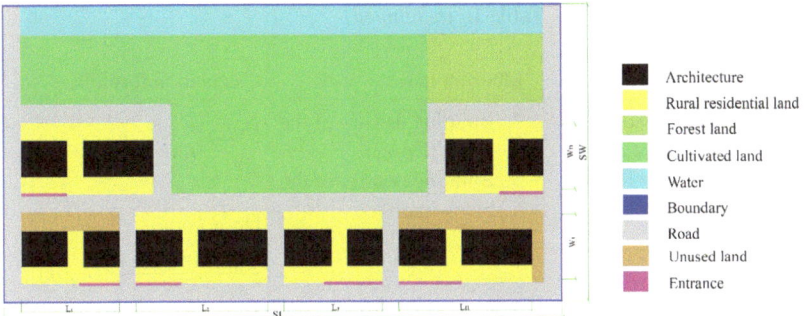

Fig. 3. The factor-labeling paradigm of the block scale

village (*CLAV*), water area within the village (*WAV*), and permanent basic farmland area within the village (*PBFAV*). The village's building density (*BDV*) serves as an indicator, representing the ratio of the combined area of all homesteads within the village to the village's total area, exclusive of permanent farmland and roadway areas.

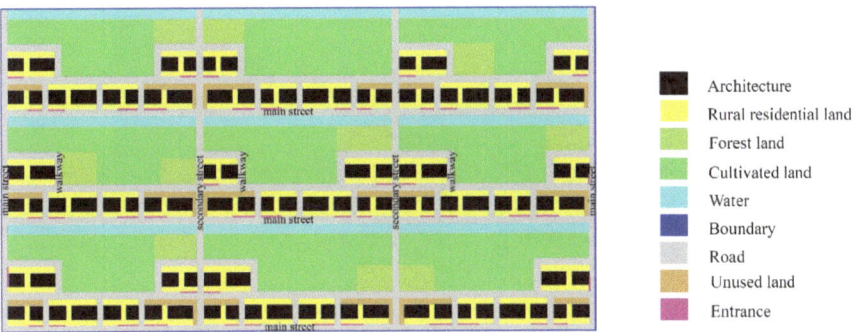

Fig. 4. The factor-labeling paradigm of the village scale

Table 1. Classification of rural space scale

Scale level	Scale type	Measure index	Calculation formula
The fourth scale	Homestead scale	1. Building density of the homestead ($De_{(H)}$)	$De_{(H)} = \frac{MBA+ABA}{HA} \times 100\%$
The third scale	Street scale	2. Building density of the main street interface ($De_{(MSI)}$)	$De_{(MSI)} = \sum_{i=1}^{n} \frac{d_i \times \cos(\alpha_i + \beta)}{SL} \times 100\%$

(*continued*)

Table 1. (*continued*)

Scale level	Scale type	Measure index	Calculation formula
The second scale	Block scale	3. Building density of the block ($De_{(B)}$) 4. Building density of the block interface ($De_{(BI)}$)	$De_{(B)} = \dfrac{HAB}{BA-CLAB-PBFAB} \times 100\%$ $De_{(BI)} = \dfrac{\sum_{i=1}^{n} Li + \sum_{i=1}^{m} Wi}{SL+SW} \times 100\%$
The first scale	Village scale	5. Building density of the village ($De_{(V)}$) 6. Water surface ratio (WSR) 7. Permanent basic farmland ratio (PBAR)	$De_{(V)} = \dfrac{HAV}{VA-CLAV-PBFAV} \times 100\%$ $WSR = \dfrac{WAV}{VA} \times 100\%$ $PBAR = \dfrac{PBFAV}{VA} \times 100\%$

3 Training Method

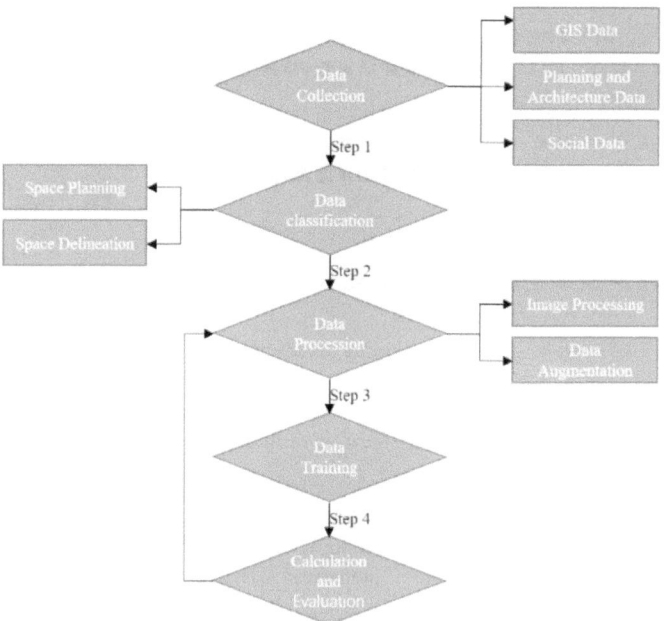

Fig. 5. Training process

As illustrated in Fig. 5, the training procedure is divided into four sequential steps. In this study, Unmanned Aerial Vehicles (UAVs) are employed to gather fundamental image data and produce 3D models, which serve as the groundwork for our research endeavor.

Utilizing the mask R-CNN model, the contours of each element are extracted from the satellite images. This data is then integrated with GIS and land planning data furnished by the planning department, culminating in a more exhaustive and comprehensive dataset. In Step 1, the gathered data is classified appropriately. Step 2 involves processing this data through a combination of image processing techniques and data augmentation. This phase includes distilling and summarizing the key rules that concern architects and planners, and identifying those rules that can be effectively digitized. Upon identification, digitizable rules are programmed and executed through scripts. Subsequently, sample images are segmented and labeled to create a training dataset. According to these rules, abstract data is extracted from the complex environmental dataset. Moving on to Step 3, the model undergoes training, followed by a check on its input-output functionality. Then, in Step 4, the quality of both the input and output is rigorously evaluated. Based on the findings from this evaluation, the data processing and training processes are continually refined and optimized.

3.1 Model Architecture

In this research project, a software known as Pix2Pix, which has been empirically proven effective by the scientific community for "image-to-image" generation tasks (Isola, Zhu et al., 2017). This software integrates a Generative Adversarial Network (GAN) into a supervised "image-to-image" translation framework. The input to the system is an abstract representation of a village group image. Both the output generated by the generative network and the ground truth image are subjected to scrutiny by the discriminative network, whose role is to differentiate between the synthetic images produced by the generator and the genuine data distribution.

Throughout multiple iterative cycles, the generator consistently generates a multitude of images. Ultimately, the final image must successfully pass the discriminative network's assessment. A satisfactory image will exhibit indistinguishable element relationships when compared to the ground truth. Moreover, once fully trained, the generator becomes adept at producing additional images that meet the same stringent criteria regarding element relations.

3.2 Datasets

3.2.1 Sample Collection

As a representative example of a village enveloped by a water network in Eastern China, akin to many dwindling rural communities, the selected case study—Waizao Village located in Pudong New Area of Shanghai—does not possess any historic structures or profound cultural heritage requiring preservation, hence it is not constrained by the conservation stipulations of historical villages.

On one hand, the traditional water network texture in Waizao Village remains intact, offering a valuable resource for groups that necessitate a comparable scale for model training, thereby enhancing computational efficiency. On the other hand, Waizao Village is among the typical locales currently undergoing active development for new residential settlement construction. Consequently, the research outcomes can directly furnish a guiding foundation for planning and design endeavors.

Employing the previously discussed boundary definition method and applying the factor-labeling paradigm, the village was systematically divided into 42 groups in accordance with its water network pattern and road divisions. Figure 6 showcases 42 sample images representative of these groups.

Fig. 6. Sample image of 42 village group

3.2.2 Augmentation

Currently, the machine learning process relies solely on data from the sampled villages, hence the training dataset is relatively constrained. To expand the scope of the training set, several typical data augmentation strategies are implemented. Per the outlined guidelines, a village can theoretically be split into approximately 20 groups; however, this quantity proves insufficient for robust training purposes. As a solution, the images are horizontally flipped and subsequently added to the database, thereby increasing the variety and volume of the training data. This aspect of the data augmentation process will be elaborated upon further in the subsequent section dedicated to training.

3.3 Sample Labelling

For machine learning to effectively process images, they must be rendered machine-readable. As depicted in Fig. 7, through the recreation of the village group's image, an abstract representation distills essential layout information about the village. Adhering to four core principles is imperative:

1. Solid colors are utilized to fill in rivers, streets, buildings, and agricultural zones.

2. The dimensions and positioning of each individual feature within the representation must remain consistent.
3. Feature shapes are simplified in order to accurately convey relational aspects.
4. During the selection process of features, insignificant elements like pathways within agricultural fields, minor water bodies, and tiny plots within residential areas are intentionally omitted.

This ensures that the resulting images provide a clear and standardized format conducive to machine learning algorithms without compromising critical structural details.

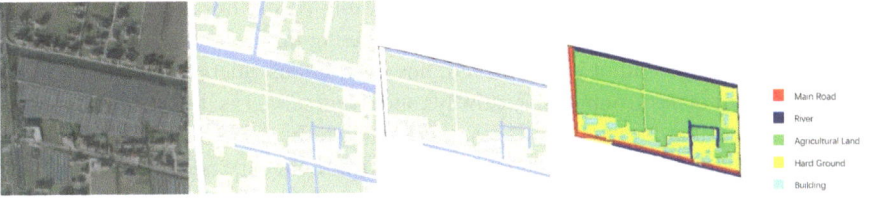

Fig. 7. Image labeling process

4 Training and Analysis

In the process of three trainings, the element and boundaries of elements were constantly adjusted, and the parameters of the model were modified through the investigation of the actual planning of the case village, the layout of the houses, and the living habits of the villagers.

4.1 Training Process

As depicted in Fig. 8, during the initial training phase, the input images were filled with solid colors, thus leading to output images that were also constituted of abstract hues. For data augmentation, a set of 20 villages was selected as the exemplar dataset, and each building group within these villages was horizontally and vertically flipped. Consequently, the test dataset comprised 10 such augmented samples. Throughout the learning process, the model learned to position buildings alongside roads and waterways. However, the guideline employed to demarcate agricultural land seemed inadequate, as patches of solid ground intended for construction were sometimes allocated within the agricultural land zone. This could potentially be attributed to an insufficient sample size.

During this preliminary training, a notable issue arose concerning the placement of buildings, as they were at times positioned on both sides of streets or rivers. In reality, most buildings tend to be constructed on one side of rivers. Given that traditional Chinese architectural practices generally follow a north-south alignment, with the building entrance typically facing southward, flipping the samples vertically for augmentation inadvertently led to the creation of several inaccurate configurations.

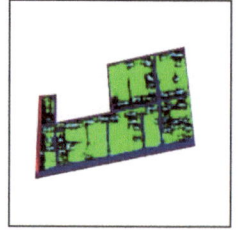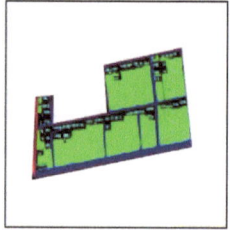

Fig. 8. First training image output (left) and the ground truth (right)

As displayed in Fig. 9, during the second training phase, horizontal flipping was implemented to augment the training dataset. Following this second round of training, a more rational image output was achieved; however, issues emerged pertaining to the method of obtaining the groupings. Roadways and rivers served as dividers to demarcate the groups, given that in natural villages, there tends to be no distinct natural contour separating adjacent groups. Nonetheless, the practical boundary adopted was the centerline of rivers and roads, which inadvertently bisected these features, rendering them visually akin to smaller watercourses. It is worth noting that locals conventionally avoid constructing buildings close to major rivers and roads. Consequently, the manner in which the groups were partitioned inadvertently conveyed misleading information.

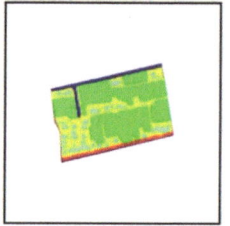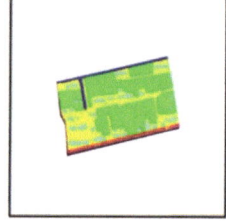

Fig. 9. Second training image output (left) and the ground truth (right)

In the third training iteration, as depicted in Fig. 10, groups were divided more precisely, and the outlines of rivers and roads were offset to encapsulate the entirety of these features, thereby providing the model with a comprehensive understanding of the environmental context around the boundaries. The resultant final images closely mirrored the actual conditions (ground truth). Observations from the test dataset reveal that each selected element is correctly positioned in the layout, and the model effectively learns the proportional relationships between area and location.

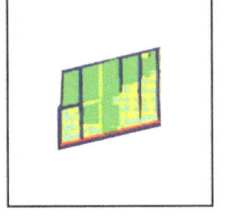 

Fig. 10. Third training image output (left) and the ground truth (right)

5 Quantitative Indicators and Reasonableness Evaluation

When assessing the suitability and innovation of generated rural planning and design schemes, relying solely on similarity comparisons is insufficient to provide a thorough evaluation of how well they integrate into the rural spatial fabric and embody fresh qualities. To guarantee that proposed plans effectively conserve and bolster the original spatial attributes of the village while aligning with prospective development requirements, a detailed quantitative research analysis is indispensable.

For example, by employing area indices, distance ratios, along with other pertinent metrics such as green space coverage ratio, road network density, and the percentage of public open spaces, a comprehensive evaluation framework can be established. This system enables a rigorous examination to ascertain whether the design proposition not only preserves but also improves the quality of the rural living environment, concurrently promoting sustainable development and the conservation of cultural heritage within the village.

5.1 Objective Evaluation Analysis of Rationality Assessment

In this investigation, it is posited that an evaluation of the generated scenarios necessitates a systematic approach utilizing objective indicators to scrutinize both the schemes developed by the GAN model and those derived through conventional methods on a uniform plane, thereby ensuring a holistic appraisal of the rationality of rural spatial textures and stylistic traits.

The Area Indicator serves to gauge the reasonableness of design schemes in relation to land use efficiency and adherence to regulatory frameworks. By calculating pivotal metrics such as the building footprint area, residential land allocation, floor area ratio, and the proportion of residential land usage, it juxtaposes these figures against the national and local planning control benchmarks. The execution of this assessment yields quantifiable results, where if all the generated buildings conform to the national and local planning control standards, the maximum rating is assigned. Conversely, the lower the degree of standard conformity, the lower the score awarded.

The Spacing Indicator, on the other hand, predominantly concentrates on the inter-building distance dynamics within rural landscapes, particularly verifying the adequacy of sunlight exposure through sunlight analysis to ensure optimal lighting conditions and the overall quality of the living environment. Specialized sunlight simulation software is employed to conduct calculation simulations, acquiring sunlight duration data for every building unit. This data is then utilized to evaluate the rationality of the design scheme concerning the minimum required sunlight hours in winter, adequate summer shading, and other related aspects.

5.2 Evaluation and 0ptimization

The sunlight conditions of the schemes produced by the GAN model were assessed at particular locations. Due to the lack of information on building heights and floor plan layouts within the generated designs, architects in this study arranged them according to the specifics of each scheme. For consistency, a reference model typical of densely populated residential areas in southern regions was adopted—a common six-story building configuration with approximate 100 square meter floor plans, housing two households per unit. In this study, prevalent building standards in southern regions were utilized to compute the sunlight exposure of buildings on the winter solstice day.

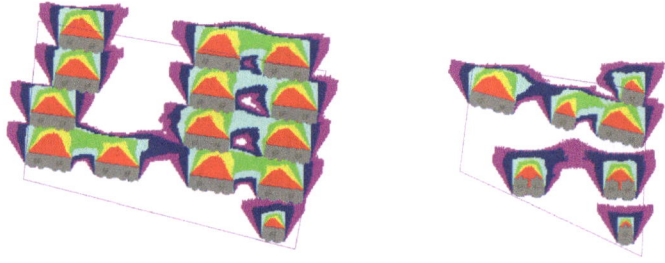

Fig. 11. Evaluation of sunlight obstruction in case 1 (left) and case 2 (right)

As Fig. 11 illustrates, it is evident that during the preliminary generation phase, the GAN model has effectively learned how to position south-facing entrances and handle the site's surrounding water systems, as well as incorporating certain green spaces within the layout (a transformation based on cultivated land found in the training set). Consequently, the objective indicators for the generated schemes, including floor area ratio and building density, mostly comply with the prescribed requirements. Thanks to the employment of buildings that are not excessively high and a dot-like building layout, even under stricter sunlight conditions, there appears to be no significant issue with sunlight exposure on a larger scale.

6 Conclusion

This paper introduces a methodological framework for boundary definition and the factor-labeling paradigm applicable to rural contexts, which can be effectively utilized in analyzing the spatial morphology of villages across different regions. Furthermore, the successful application of GANs models in identifying, extracting, and generating spatial textures within rural landscapes has been demonstrated.

The preliminary planning schemes generated through machine learning methodologies contribute significantly to the provision of theoretical grounding and stimulate innovative thinking. Leveraging these algorithmically enhanced and rationally validated planning approaches, planners can gain fresh perspectives on the underlying structure and evolution of rural environments. The optimized blueprints generated by such means not only unveil a range of potential spatial layout configurations but also uphold theoretical relevance and practical feasibility.

Regardless of whether in rural planning or architectural design, the crux of the matter consistently revolves around preserving the unique local culture and addressing the genuine needs of residents. Thus, any preliminary scheme produced by machine learning must be tightly coupled with local realities in subsequent phases. This involves comprehensive communication and empirical research by professional teams and local communities, leading to continuous refinement and improvement. By doing so, the finalized planning strategies and architectural designs will not only manifest the benefits of advanced technology but also genuinely encapsulate local characteristics and humanistic concerns, thereby driving sustainable development and enhancing the quality of life in rural communities.

Acknowledgment. This work was supported by the National Key Research and Development Program of China "Technologies of spatial planning and decision for rural communities", Grant number 2019YFD1100802.

References

Conzen, M.R.G.: Geography and townscape conservation. In: Uhlig, H., Lienau, C. (eds.) Anglo-German Symposium in Applied Geography, Giessen-Würzburg-München, 1973, pp. 95–102. Giessener Geographische Schriften, Giessen (1975)

Conzen, M.R.G.: Historical townscapes in Britain: a problem in applied geography. In: Jw, H. (ed.) Northern Geographical Essays in Honour of Ghj Daysh, pp. 56–78. Department of Geography, University of Newcastle upon Tyne, Newcastle upon Tyne (1966)

Fang, Z., Zhang, D., Xiong, C.: Quantitative analysis about indexes of street interface's functional continuity based on the human scale. New Archit. **05**, 116–121 (2017)

Huang, W., Hao, Z.: Architectural drawings recognition and generation through machine learning. In: Annual Conference of the Association for Computer Aided Design in Architecture (ACADIA) (2018)

Isola, P., Zhu, J.Y., Zhou, T., Efros, A.A.: Image-to-Image Translation with Conditional Adversarial Networks. IEEE (2017)

Kropf, K.: Urban tissue and the character of towns. Urban Des. Int. **1**, 247–263 (09/01 1996). https://doi.org/10.1057/udi.1996.32

Liu, Y., Fang, C., Yang, Z., Wang, X., Zhou, Z., Deng, Q., Liang, L.: Exploration on machine learning layout generation of Chinese private garden in Southern Yangtze. In: Proceedings of the 2021 DigitalFUTURES, pp. 35–44 (2021)

Peters, N.: Enabling Alternative Architectures: Collaborative Frameworks for Participatory Design. Harvard University Graduate School of Design (2018)

Sitte, C.: City Planning According to Artistic Principles. The Birth of Modern City Planning, Camillo Sitte (1889)

SMBPNR, Shanghai Municipal Bureau of Planning and Natural Resources: The Character-Defining Element of Shanghai Vernacular Architecture in Rural Areas. Shanghai University Press (2019)

SMPG, Shanghai Municipal People's Government: The Interpretation of Shanghai Master Plan 2017–2035 (2017)

Tong, L.: Parametric analysis and reconstruction of villages' spatial texture and its planning application research. Doctor, Zhejiang University (2016)

Yoshinobu, A.: The Aesthetic Townscape, Translate By: E. Riggs L. The MIT Press, Cambridge; London (1983)

Zhou, Y., Wang, H.: The characteristics of housing and living in the countryside of Shanghai. [In Chinese]. Housing Sci. 4(07), 18–25(2021). https://doi.org/10.13626/j.cnki.hs.2021.07.004

Zhou, Y., Zhao, J., Zhang, Y.: Street interface density and planning control of urban form. City Planning Rev. 36(06), 28–32 (2012)

Weighing Elements Affecting Indoor Openness with Machine Learning

Chuang Lyu, Weisun Xu[✉], Fuyi Lai, and Shuang Yu

College of Civil Engineering and Architecture, Zhejiang University, Hangzhou 310058, China
xuweishun@zju.edu.cn

Abstract. Indoor openness offers a critical reference to how users perceive an interior space. Past studies have accumulated a diverse range of impacting factors based on visual evidence and empirical data through simulated simplistic environments, yet have failed to associate such evidence with complex real-world spaces to provide evaluations. This study introduces a framework that combines computer vision and machine learning to quantitatively assess perceptions of openness in office spaces based on digitally generated visual data, feature extraction, and perception rating data. Machine learning models are trained to predict openness ratings based on the proportions of different elements. Our results demonstrate a method to integrate multiple elements for evaluating the perception of openness, and highlight that collective understanding of the concept of openness influences users' perception.

Keywords: Indoor Openness Evaluation · Machine Learning · Computer Vision · Office Space

1 Introduction

In architectural practice, the widely used concept of "openness" seems to hold universal implications in both interior design and environmental psychology (Xu and Xia 2020). Currently, there have been numerous approaches proposed for quantitatively assessing the openness of indoor spaces. However, most existing methods only effectively reflect the impact of one or a few elements on the perception of openness, or fail to relate quantification of spatial qualities with human perception. In this paper, we propose a quantitative approach that combines machine learning to simultaneously evaluate multiple visual factors influencing the openness of indoor spaces. Additionally, we have observed that such perception is influenced by how people understand the concept of openness, which was not given enough stress in most previous researches.

1.1 Development of Quantitative Openness Evaluation

Researches on quantification of indoor openness started in the 1950s and were generally executed with simulated environments with controlled parameter variations and limited user perception data. As digital modelling tools evolved, a later line of research was also

H. Chai et al. (Eds.): CDRF 2024, *Symbiotic Intelligence*, pp. 217–226, 2025.
https://doi.org/10.1007/978-981-96-3433-0_19

developed by extracting objective spatial features. So far, there has been a lack of studies in bridging the gap between user empirical data and complex space quantification based on real-world scenarios.

Early studies primarily explored quantification based on user's feedback to various spatial elements such as window area and space size. In the 1950s, Hediger proposed that the openness originates from animals' instinctual need to avoid threats and ensure survival. He introduced the concept of "flight distance" as a characteristic value to gauge the impact of room width on the perception of openness (Hediger 1955) Subsequently, other scholars adopted similar approaches in comparing specific spatial elements with empirical data to investigate the correlation space and perception, and included factors such as light intensity, space size, aspect ratio, and the size of the space surrounding the opening (Inui and Miyata 1973; Stamps III 2011).

Instead of relying on user feedbacks, later studies focus on establishing objective measurements of openness from features extracted from spaces with algorithmic models. For example, the isovist approach transformed human perception of spatial quality into visual representations of visible areas from certain viewpoints (Benedikt 1979). Later studies developed methods to deal with higher dimension conditions. 2D planar views were extended to the spatial cone view, which is used to measure the visible space of a viewpoint in 3D space to evaluate the openness sense (Fisher-Gewirtzman and Wagner 2006).

Existing literature based on user feedbacks links the perceptions of openness to a wide range of factors, including room size, open area, floor height, and boundary complexity, mostly relying on linear regression to correlate one factor at a time (Stamps III 2010). Therefore, as more factors are added to the list, it becomes increasingly difficult to assess the overall spatial performance with such fragmented results. Meanwhile, methods of algorithmic metrics often neglect the crucial feedback from humans, which leads to the lack of empirical support for real-world applications.

1.2 Opportunities of Quantifying Space Attribute Using Machine Learning and Machine Vision

The advancement of machine vision technology has significantly improved the ability to process images of complex environments for extraction of spatial elements (Badrinarayanan et al. 2017). Relevant studies starts utilizing segmentation to analyze spatial quality in urban environments (Tang et al. 2016). Emerging techniques in perception quantification and space metrics extraction by combining computer vision and machine learning in areas such as urban studies have exhibited high promise in linking complex visual information to human feedback at large scale, covering spatial perception such as cover dimensions such as safety, liveliness, and depressive feelings (Dubey et al. 2016). Ye Yu, Zhang Lingzhu, and others have combined Baidu Street View images with machine learning to measure and analyze street space quality and accessibility in Shanghai on a large scale (Ye et al. 2019; Shi et al. 2020). Zhang Fan and others have used deep learning methods to explore the objective and subjective relationship between street quality and visual elements (Zhang and Liu 2021).

Similar methods have not been applied to the quantification and analysis of indoor space quality or attributes. One critical reason is that they rely heavily on image segmentation methods to process visual data. However, general-purpose computer vision models like SegNet still face challenges in recognizing and classifying all indoor elements with enough accuracy. In addition, indoor spaces have fewer standardized data sources compared to outdoor street view data, making datasets harder to obtain.

1.3 Assumptions About the Definition of Openness

Despite discussions on indoor openness across varied contexts, a consensus regarding its definition and theoretical underpinnings within the field of architecture has yet to be reached. Researchers' cultural cognitive backgrounds have an impact on their understanding of spatial openness (Stamps III and Krishnan 2006). Such difference often gets neglected in questionnaire-based empirical data collection experiments targeting at controlled variables in simplistic environments. Combining such fragmented empirical evidence from such disciplinary variances can result in incoherence in creating complex evaluation models. Therefore, we hypothesize that the definition of hypothetical openness has the potential to influence evaluation.

2 Method

Our literature review highlights two main issues. Firstly, the lack of unified understanding regarding the definition of spatial openness results in metric fragmentation for complex real-world problems. Secondly, early algorithm-based metrics neglect human feedbacks. The combination of machine learning and computer vision offers a promising and potential framework for connecting perception with the quantification of complex spaces.

Therefore, we propose a methodological framework to weigh elements affecting indoor openness with user evaluation through digital modeling, feature extraction, and machine learning. Our method includes generating visual data samples, extracting relevant features, and scoring these attributes based on participants' evaluations, as well as data mining through machine learning. This structured approach allowed us to systematically analyze the impact of various spatial elements on the perception of openness and explore the effectiveness of machine learning algorithms in predicting such perception. Additionally, control group method with and without openness definition hints are applied during perception data collection phase to examine whether collective understanding of the concept of openness affects model performance.

2.1 Perception Data Collection

Image Sample Generation: We prepared adjustable digital models using 22 existing office interior spaces. These models allow for easy changes in interior layout and spatial boundary conditions, such as removing furniture or adjusting the height of the space. We used Enscape renderer to capture a series of scenes in the models with consistent view height and perspective angle width. A total of 1,000 image samples were obtained by rendering. After calibration for brightness and considering factors such as overly dark lighting in the rendered images and unreasonable space within the field of view, the samples were manually filtered, resulting in a final collection of 861 visual samples.

Feature Extraction - Delphi Method for Space Element Categorization: In collaboration with interior design experts (10 interior designers from XXX institution, and the name is removed for blind review) through Delphi method, indoor space design elements affecting the attribute of openness, such as furniture, windows, floors, or ceilings, were classified into a total of ten categories. Elements in the model are assigned to different layers according to the category results, with each corresponding layer matched to a fixed RGB color, and channel maps that correspond exactly to the position in the rendered images are output (Fig. 1). Finally, the feature maps were standardized in size to produce images with a resolution of 427 × 360 pixels. The proportion of each color in the image (called pixel percentage in the following text) in the channel map is extracted through a customized script. The channel map itself and the proportions of different categories of elements serve as two features of the corresponding space in the machine learning process.

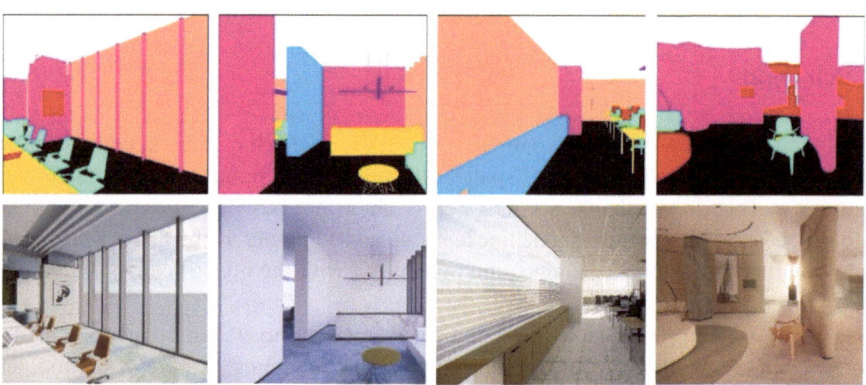

Fig. 1. Selected rendering images and respective channal maps

2.2 Data Filtering and Processing

Evaluation Participant Selection: A total of 8 architectural design students from third year to recent graduates are recruited in the evaluation stage of the study. Architecture students are selected because we assume that they are more familiar with the concept of openness despite not being explicitly informed of its definition.

Hypnotizing that the definition of openness could influence the results, we divide the participants into two groups. One randomly chosen group of students (referred to as the informed group in the following text) are told that the concept of openness is defined as the freedom of movement or escape in an interior space, a concept similar to that of Stamps (Stamps III 2011). The other group (referred to as the uninformed group) is not given any definition.

Quantification: Using a custom-designed Java program, both groups are told to use their own judgments to compare a pair of randomly selected rendering images shown on the screen and select the image representing an office interior with the higher level of openness (Fig. 2). On average, each participant makes 1,000 comparisons based on rules in similar research from the MIT Place Pulse 1.0 dataset, with each image being rated 15 times on average (Salesses et al. 2013). Images that are matched fewer than 10 times are redrawn and matched against each other until each image has a minimum comparison count of 10. Then, all match results are processed by the ELO (Elo rating system) algorithm, converting the comparison results into scores. Each sample photo is initially assigned a score of 1500, and then iterated based on comparisons to obtain the final scores. The ELO scores are then tiered and normalized into four categories: 1, 2, 3, and 4.

Fig. 2. Graphical users interface of the comparison program

2.3 Machine Learning

Two machine learning strategies are adopted based on the dataset. One approach uses the channel maps as features, while the other uses the pixel percentages corresponding to the channel maps as features. In both cases, the tiered openness scores serve as labels for training.

For the first strategy, we utilize a Convolutional Neural Network (CNN) algorithm to process the channel maps directly, adjusting parameters to optimize results. This approach takes advantage of the spatial information inherent in the channel maps, allowing the model to learn patterns related to the distribution of elements within the space.

In the second strategy, we employ various algorithms, including Random Forest, Support Vector Machine (SVM), Logistic Regression, K-Nearest Neighbors (KNN), Decision Tree, and Artificial Neural Networks (ANN), to analyze the pixel percentage data extracted from the channel maps. This approach focuses on the relative proportions of different elements within the space, providing a different perspective on how these elements contribute to the perception of openness.

Model evaluation for both strategies are conducted using k-fold cross validation, with repeated training and validation using randomly generated subsamples. The prediction accuracy (correct prediction percentage) of the generated evaluation models allows us to compare the effectiveness of the two machine learning approaches in predicting the perceptions of openness in office spaces.

3   Analysis

3.1   Effectiveness and Algorithmic Superiority

To verify the effectiveness of our framework, we compare our test results random guessing prediction correctness, which is 21% for both groups. In both groups, the best models achieve an accuracy of over 50%, with data trained from the informed group achieving an accuracy of 57% when trained with Artificial Neural Network (Table 1). The results demonstrate that our proposed system can assist in quantitatively evaluating the perception of openness.

Table 1. Training result

Informed group			Uninformed group		
Use percentage and label			**Use percentage and label**		
Classification method	Precision	Recall	Classification method	Precision	Recall
Random forest	0.501	0.4942	Random forest	0.4815	0.4903
Support vector machine	0.4995	0.4749	Support vector machine	0.3771	0.3833
Logistic regression	0.5623	0.5598	Logistic regression	0.4499	0.4517
K nearest neighbor	0.42	0.4208	K nearest neighbor	0.3736	0.3629
Decision tree	0.4497	0.4517	Decision tree	0.3981	0.3900
Artificial neural network.	0.5767	0.5598	Artificial neural network	0.3928	0.3977
Random guess	0.21	0.0526	Random guess	0.21	0.0526
Use image and label			**Use image and label**		
Classification method	Precision	Recall	Classification method	Precision	Loss
Convolutional neural network	0.4419	4.007	Convolutional neural network	0.3895	4.2293

3.2   Comparison of Results in Informed and Uninformed Groups

Compared with results from the uninformed group, the model trained with data from the informed group exhibits superior performance in all categories regardless of the training method used. This suggests that the informed group exhibits higher level of perceptual consistency of indoor openness, even among architecture students who have been previously exposed to the concept.

Furthermore, heat maps (Fig. 3) provide a visual representation of the correlations between different spatial elements in an office environment and the perceived openness of the space, as determined by participants in the study. The elements along the horizontal axis include desks, chairs, windows, lamps, and other components of the office space. The vertical axis at the bottom row is marked "label," which denotes the overall openness rating assigned by par-tic pants. The color intensity on the map indicates the strength of the correlation: darker shades represent higher correlation, and lighter shades indicate a lower correlation. According to the heat map, with the uninformed group, no significant single element is found except for wall area, which suggests a highly complex nature of the perception of openness under natural conditions. In contrast, with the informed group, the proportions of ground and windows show a strong correlation with the tiered ELO ratings. The result indicates that after being informed that the level of openness is related to the freedom of movement, participants' perception of the openness is more related to floor and ceiling visibility. The result supports our hypothesis, which underscores the importance of clear communication and definition when conducting research in this area. It is crucial to consider how the definition of openness is framed, as it can significantly impact the results and interpretations of the study.

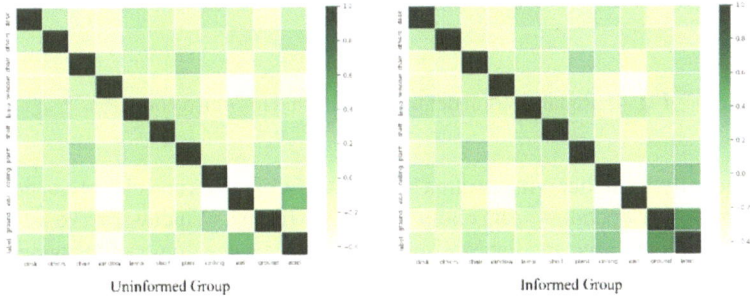

Uninformed Group   Informed Group

Fig. 3. Heat map under uninformed and informed condition

The radar charts (Fig. 4) depicted below illustrate the average proportions of various spatial elements associated with different levels of perceived openness in office spaces. These elements are plotted on the axes of the radar chart, with each axis representing a different element such as walls, ceilings, chairs, windows, and others and the length of the spoke corresponds to the average proportion of that element in the space. The area enclosed by the lines connecting these points offers a graphic representation of the spatial composition that characterizes each level of perceived openness. For both the informed and uninformed groups, these diagrams allow for an at-a-glance comparison of how various spatial elements contribute to the perception of openness across different ratings. By analyzing the radar chart in Fig. 4 and heat map in Fig. 3, we can examine the variation in the proportion of elements in indoor spaces with varying degrees of openness (higher values indicating greater openness). It becomes evident that walls, ceilings, and floors are crucial indicators for assessing the perceptions of openness. The proportion of walls exhibits a negative correlation with the perceived openness of the interior space, while ceilings and floors demonstrate a positive correlation. Our conjecture is that walls restrict the field of view within a room, obstructing people's line of sight and impeding the visual perception of the space's extension and continuity. The proportion of ceilings and ground, on the other hand, correlates with the perceived size of the space, with larger spaces evoking a stronger sense of indoor openness.

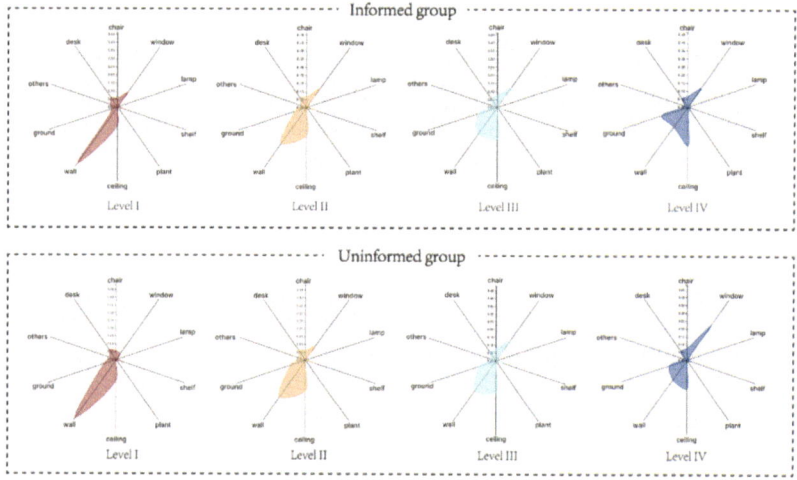

Fig. 4. Radar chart under uninformed and informed condition

Additionally, a smaller window area and an excess of tables and chairs can negatively the perceptions of openness, especially in smaller rooms. Windows serve as conduits between the indoor space and the external environment, providing visual connections to natural light and the surrounding landscape. Tables and chairs, as functional elements within the interior, also impact the perceptions of openness. An excess of tables and chairs encroach upon the indoor space, curtailing freedom of movement and visual perception, leading to a crowded and constricted ambiance that diminishes the experience of openness. Conversely, a judicious arrangement and quantity of tables and chairs can furnish adequate functional support while preserving the openness and fluidity of the space, thereby augmenting the perceived perceptions of openness.

Other elements such as lamps, plants, and shelves do not exhibit a clear association with the perceptions of openness.

4   Results and Discussion

Our research indicates that while it is feasible to quantitatively predict the perception of openness using complex visual inputs, the understanding individuals have of the concept significantly influences the outcomes. As previous researches have not established a universally accepted definition of openness, this assumed concept demands critical examination and possible revision. Future studies should also consider conducting comparative analyses with previous researches especially with regards to consistency of participants' evaluation.

As an initial phase study, our framework has room for further development and validation. Specifically, the Delphi method employed to categorize spatial elements into layers, which subsequently influences the grouping of elements in the channel maps, was not adapted to accommodate varying definitions of openness. This may have resulted in critical elements being categorized into the same class as unimportant objects. Future research should explore the impact of element categorization on model performance and investigate methods to align the segmentation process with the intended definition of openness.

Our perception data acquisition method is also subject to further improvement. Since each participant had to perform 1,000 comparisons, some participants exhibited fatigue and reflected that their level of performance dropped as the experiment went on. We did find inconsistency in certain image pairs through manually checking the comparison results. This may have negatively impacted on our model performance.

Acknowledgement. This research is supported by the National Natural Science Foundation of China under Grant 52208036, China's Research and Development Project of Ministry of Housing and Urban-Rural Development under Grant 2022-K-004, and Center for Balance Architecture at Zhejiang University.

References

Badrinarayanan, V., Kendall, A., Cipolla, R.: Segnet: a deep convolutional encoder-decoder architecture for image segmentation. IEEE Trans. Pattern Anal. Mach. Intell. **39**(12), 2481–2495 (2017)

Benedikt, M.L.: To take hold of space: isovists and isovist fields. Environ. Plann. B. Plann. Des. **6**, 47–65 (1979)

Dubey, A., Naik, N., Parikh, D., Raskar, R., Hidalgo, C.A.: Deep learning the city: quantifying urban perception at a global scale. In: Proceedings of the 14th European Conference on Computer Vision—ECCV 2016. Springer, Amsterdam, The Netherlands 196–212 (2016)

Fisher-Gewirtzman, D., Wagner, I.A.: The spatial openness index: an automated model for three-dimensional visual analysis of urban environments. J. Archit. Planning Res. 77–89 (2006)

Hediger, H.: Studies of the psychology and behavior of captive animals in zoos and circuses (1955)

Inui, M., Miyata, T.: Spaciousness in interiors. Light. Res. Technol. **5**(2), 103–111 (1973)

Salesses, P., Schechtner, K., Hidalgo, C.A.: The collaborative image of the city: mapping the inequality of urban perception. PLoS ONE **8**(7), e68400 (2013)

Stamps, A.E., III.: Use of static and dynamic media to simulate environments: a meta-analysis. Percept. Mot. Skills **111**, 355–364 (2010)

Stamps, A.E., III., Krishnan, V.V.: Spaciousness and boundary roughness. Environ. Behav. **38**, 841–872 (2006)

Stamps, A.E., III.: Effects of area, height, elongation, and color on perceived spaciousness. Environ. Behav. **43**, 252–273 (2011)

Shi, C., Yuan, Q., Pan, H., Ye, Y.: Measurement of pedestrian suitability and design guidance for street space—a case study of Jing'an Temple area in Shanghai. Urban Planning Shanghai **05**, 71–79 (2020)

Tang, J.X., Long, Y., Zhai, W., et al.: (2016) 'Measurement, evaluation of changes, and identification of influencing factors of street space quality: based on the analysis of large-scale multi-temporal street view images.' New Archit. **05**, 110–115 (2016)

Xu, W.S., Xia, B.: A review of quantitative methods for the openness of architectural indoor space. Archit. Cult. **01**, 161–163 (2020)

Ye, Y., Zhang, Z.X., Zhang, X.H., Zeng, W.: Measurement of street space quality at human scale—a large-scale, high-precision evaluation framework combining street view data and new analysis techniques. Int. Urban Planning **34**(01), 18–27 (2019)

Zhang, F., Liu, Y.: Street view images—methods and applications based on artificial intelligence. J. Remote Sens. **25**(05), 1043–1054 (2021)

Flexible Plot-Scale Urban Design Using Quadratic Programming

Qian Hu[1,2], Yujiao Wang[1,2], and Peng Tang[1,2(✉)]

[1] School of Architecture, Southeast University, Nanjing 210096, China
{huqian,tangpeng}@seu.edu.cn
[2] Key Laboratory of Urban and Architecture Heritage Conservation (Southeast University),
Ministry of Education, Nanjing 210096, China

Abstract. This research aims to tackle the problem of the generative urban design of residential areas using a general-solving machine of mathematical programming. Residential areas on university campuses are taken as examples. As a type of urban design problem, the layout of residential areas on campuses is subject to multiple indicators and various boundary shapes. Quadratic Programming (QP) offers a representation of this problem, and with the assistance of cutting-edge mathematical programming solvers, the urban design problem with quadratic constraints can be automatically tackled. However, the difficulties lie in formulating complex boundaries, flexible building templates, and directional variability. To overcome these challenges, this research combines inside-model techniques of representation and outside-model modules utilizing geometric methods to enhance the main model of QP. A pipeline is provided to apply the approach in real urban design projects. The generated results validate the effectiveness of the enhanced model and the pipeline.

Keywords: Generative Urban Design · Universal Solving Machine · Quadratic Programming · Residential Areas on Campus

1 Introduction

Generative urban design refers to the use of generative methods in urban design scenarios and has attracted significant attention from academic communities, professionals, and government agencies in recent years (Jiang et al., 2023). Generative urban design can benefit from the recent development of universal solvers, which can automatically tackle linear and other related mathematical programming problems (Meindl and Templ 2012). The approach has been approved to be useful in dealing with design problems (Keatruangkamala and Sinapiromsaran 2005). This frees designers from having to Fig. Out how to solve problems and allows them to focus on what to solve. In other words, designers are only required to formulate urban design problems with mathematical expressions, without calculating. However, the cost is that the problem description must be highly formalized under specific rules of the universal solver for efficient problem-solving.

Quadratic Programming (QP) is a type of mathematical programming with quadratic constraints that can be solved using a cutting-edge universal solver (Gurobi Optimization

© The Author(s) 2025
H. Chai et al. (Eds.): CDRF 2024, *Symbiotic Intelligence*, pp. 227–235, 2025.
https://doi.org/10.1007/978-981-96-3433-0_20

Inc., 2012). Compared to other methods such as evolutionary algorithms, QP shows better performance in adhering to explicit constraints. It has been applied to urban design scenarios with a data structure of a special ordered set (Hua et al. 2019), where constraints of irregular boundaries are bypassed, and fixed templates are used to represent buildings. Other generative design using mathematical programming focused on floor plans and thus did not have to face complex shapes (Wu et al. 2018). A breakthrough regarding complex boundaries has been reached with highly modularized data structure, but it is consequently not flexible for real-world size (Peng et al. 2014). To extend the flexibility of using QP in generative design may be the next challenge to tackle.

Designing residential areas on Chinese university campuses, including placing buildings, is a typical example of urban design with multiple constraints, such as boundaries, floor area ratios, building density, distances between buildings etc. Generative design methods can improve the effectiveness and efficiency of this design process, and a QP solver can be used for decisions under urban planning indicators not directly related to operational knowledge of urban design. However, it is challenging to deal with complex residential area shapes and the flexibility of dormitory buildings with a QP model and to generate results similar to manual design schemes.

This research aims to enhance the flexibility when using QP to solve design problems, through a solution for generative design of residential areas on Chinese university campuses. By formulating main considerations inside the QP model and providing extra modules outside, the problem is decomposed and reorganized, generating valuable results for real design practice. The programs are written in Python with Gurobi used as the mathematical programming solver, and the results are visualized using Rhino and AutoCAD.

2   The Main Model

Designing residential areas on Chinese university campuses poses integrated problems where the boundaries' shapes, floor area ratios, building densities, and distances between buildings impose strong restrictions on the design solutions. In the meantime, the locations, heights, and lengths of dormitory buildings can be flexible within permitted domains. According to experienced designers, manual design is an iterative process, where the solution moves from one state to another, until feasible solutions are found. To supersede this process, Quadratic Programming (QP) is introduced.

QP is a type of mathematical programming that formulates all constraints with equations and inequalities with degrees not exceeding 2. With the aid of cutting-edge solvers, problems with quadratic constraints can be tackled automatically. While the requirement makes problem formulation more challenging, it also promotes the efficiency of solution-finding.

2.1   Building Formulation

In the residential areas, the majority of buildings are dormitories (Fig. 1(a)) characterized by a corridor-dominated structure. The depth of a building is determined by the presence of rooms on both sides of the corridors, while the length is dependent on the number of room units in each row. The building height is controlled by the number of above-ground floors, and the distance between buildings in the main direction is closely related to the height to ensure adequate natural lighting and direct sunlight for most of the rooms.

In addition to the strip-shaped buildings with a corridor-dominated structure, there are also box-shaped buildings with relatively stable sizes, such as the canteen building (Fig. 1(b)). These buildings typically have no more than three floors, with similar depths and lengths.

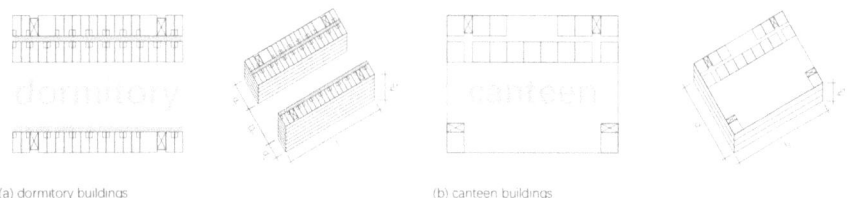

(a) dormitory buildings   (b) canteen buildings

Fig. 1. Dormitory building and canteen building

To define a dormitory building, we need variables for its location and its size. To further represent the variability and fixation of the building, (d, l, h) are replaced through the below formulas.

$$d = d_{cor} + (1 + \xi) \times d_{room} \quad (\xi \in \{0, 1\}) \tag{2.1}$$

$$l = n_l \times l_{unit} \quad (n_h \in N^*) \tag{2.2}$$

$$h = n_h \times h_{unit} \quad (n_h \in N^*) \tag{2.3}$$

Here n_h and n_l are the main variables. In the meantime, the range of the main variable is limited according to real design needs. For instance, $5 \le n_h \le 6$ and $20 \le n_l \le 27$.

For the sake of convenience, the definition of a canteen building is extended from a dormitory building, while some constants should be changed and the variables n_h and n_l are usually fixed or fluctuating in narrow ranges.

2.2 Building Relationships

Once the buildings are defined, the primary task is to describe the relationships between them, with the main one being to maintain distance. Considerations include fireproofing distance and sunlight distance based on Chinese regulations. The former requires the distance in any direction (D_x) should be more than 6, 9 or 13 m, according to specific features of the buildings. The latter requires the distance between buildings in the main direction (D_m), usually south-north, to be more than the minimum distance for adequate direct sunlight for the building in the north. The formula is below.

$$D_m \ge h \times k \tag{2.4}$$

The value of constant k is determined by the location of the city, and it is mainly 1.35 in this study. However, if the orientation of the buildings does not face south, local regulations may give slight variations in this value.

To express these requirements in a QP model, we introduce the concept of a control area (Fig. 2) which utilizes variables $(D_{iN}, D_{iS}, D_{iW}, D_{iE})$ to expand a building's shape to include its actual area of influence. The central principle is that each building must not encroach on another building's control area, while allowing for overlap between control areas. For example, when considering building-1 and building-2:

$$\sum_{i=1}^{4} \xi_i = 1 \quad (\xi_i \in \{0, 1\}) \tag{2.5}$$

$$\xi_1 \times (x_i - x_j - l_j - D_{jE}) + \xi_2 \times (-x_i + x_j - l_j - D_{jW}) + $$
$$\xi_3 \times (y_i - y_j - d_j - D_{jN}) + \xi_4 \times (-y_i + y_j - d_j - D_{jS}) \geq 0 \tag{2.6}$$

Here, binary variables ξ_i are used to guarantee building-i would keep away from the control area of building-j from either N/S/W/E side. Also note that building subscripts (i and j) are exchangeable, which means two groups of constraints are set for every building pair.

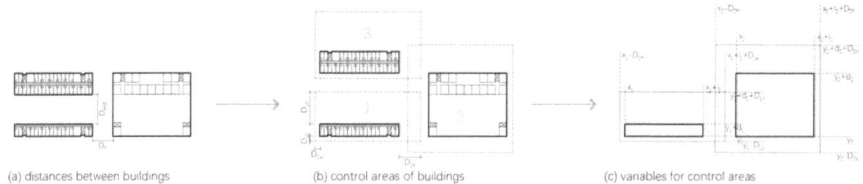

(a) distances between buildings   (b) control areas of buildings   (c) variables for control areas

Fig. 2. Control areas of buildings

2.3 Boundary Formulation

This section aims to illustrate how QP models can represent various shapes of boundaries (polygons) and their relationships with building rectangles. Different boundary shapes for residential areas are demonstrated in Fig. 3, and we do not consider holes in polygons in this section.

To confine a building to a specific area, we can approximately constrain its key sample points, which include the four corner points. In relatively complex situations, we may need to extend these points to 12 or more to avoid errors (Fig. 3(d)). This is achieved by formulating inequalities for each sample point $P(x, y)$ and a polygon $\{Q_1, Q_2, Q_3, \ldots, Q_n\}$. Note that these expressions are valid through this section.

The simplest shape to consider is a rectangle, and the corresponding inequalities are easy to write. More generally, for an irregular convex and a sample point, the formula is 2.7 (representing the equation of line Q_iQ_{i+1} by $a_ix + b_iy + c_i = 0$). Take the opposite numbers for (a_i, b_i, c_i) in the expression of Q_iQ_{i+1} when $a_ix_0 + b_iy_0 + c_i \geq 0$ does

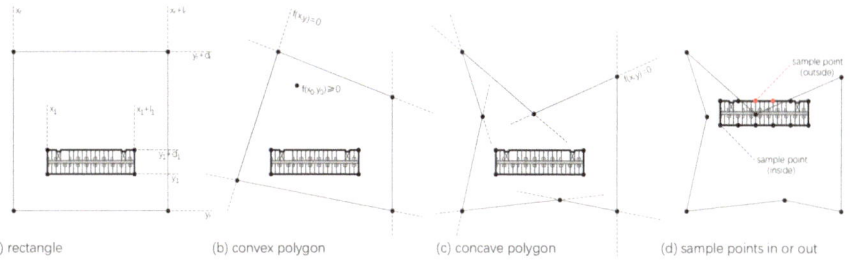

Fig. 3. Different shapes of residential areas

not work for the center point $Q_0(x_0, y_0)$ of the convex polygon. The inequalities for not-in-convex, namely to represent that a sample point $P(x, y)$ is not in a convex polygon $\{Q_1, Q_2, Q_3, \ldots, Q_n\}$ is 2.8. The rules for taking opposite numbers are the same as in-convex.

$$\forall i \in \{1, 2, 3, \ldots, n\}, a_i x + b_i y + c_i \geq 0 \tag{2.7}$$

$$\sum_{i=1}^{n} \xi_i = 1(\xi_i \in \{0, 1\}) \wedge \sum_{i=1}^{n} \xi_i(a_i x + b_i y + c_i) \leq 0 \tag{2.8}$$

The problem is more complex for concave polygons. The equations would likely exceed the limitation of polynomial degrees (no more than 2). To overcome this difficulty, we use a concavity processing method. As shown in the below codes and Fig. 4, a concave polygon is transformed into IN convex polygon and NOT-IN convex polygon:

WHILE $\exists Q_i \in C_0, < \overline{Q_{i-1}Q_i}, \overline{Q_iQ_{i+1}} > < 0$

    IF $\exists a, b \in \{1, 2, 3, \ldots, n\}, a \leq b, \forall i \in \{a, a+1, \ldots, b\}, < \overline{Q_{i-1}Q_i}, \overline{Q_iQ_{i+1}} > < 0$

    $\wedge < \overline{Q_{a-1}Q_a}, \overline{Q_aQ_{a+1}} > \geq 0 \wedge < \overline{Q_{b-1}Q_b}, \overline{Q_bQ_{b+1}} > \geq 0$

        $C_0 = C_0 - \{Q_a, Q_{a+1}, \ldots, Q_b\}, T = T \cup \{\{Q_{b+1}, Q_b, \ldots, Q_{a-1}\}\}$

    END

END

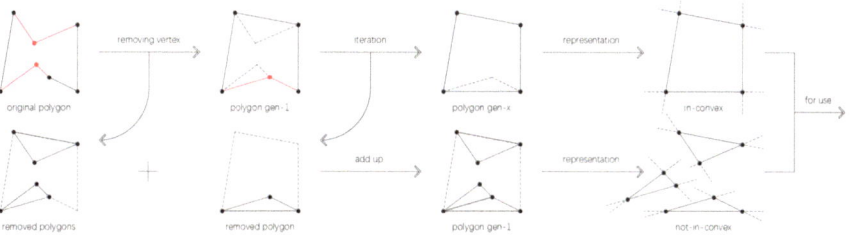

Fig. 4. Concavity processing method

This process is iterated until the polygon becomes convex. The resulting IN convex polygon and NOT-IN convex polygons can be represented in a mathematical program respectively, allowing for accurate descriptions that do not exceed the degree limitation.

2.4 Overall Constraints and Solving

The cutting-edge solver has the capability to provide optimal solutions, such as maximizing the floor area, if an objective is defined. Alternatively, it can generate possible solutions if the objective is set to be nonsense. In Sect. 2.1, we have already considered the variables and constraints related to individual buildings and the relationships between buildings. However, there are still some overall indicators that need to be inputted before the solver is executed.

These overall indicators include the floor area ratio (F), building density (δ), and height limitation. The consideration for overall height is the same as the height requirements for individual buildings. For {builing-1, builing-2, ..., building-n}, the other two can be formulated as below.

$$\delta_{\min} \leq \sum_{i=1}^{n} d_i \times l_i \leq \delta_{\max} \tag{2.9}$$

$$F_{\min} \leq \sum_{i=1}^{n} d_i \times l_i \times n_{hi} \leq F_{\max} \tag{2.10}$$

As shown in Fig. 5, We have generated some schemes, using real-world conditions in Nanjing, China. These buildings are designed to have rows of rooms on both sides of the corridor, which can be called double-depth.

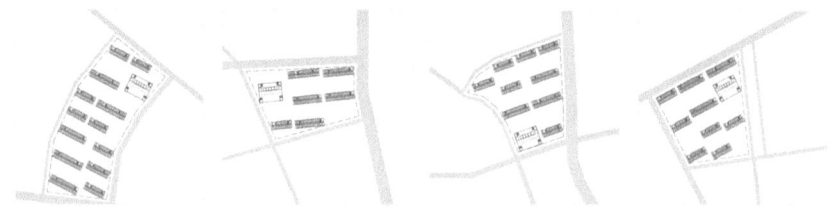

Fig. 5. Possible solutions for residential areas in Nanjing City

3   Extra Modules and a Pipeline

3.1   Building Alignment

Buildings tend to be grouped or aligned in urban design, which can be easy to describe in a grammar-based generative system. Here, QP is proved to be similarly helpful in keeping specific types of alignment.

In Figs. 5 and 6, it can be observed that the patterns of these buildings are relatively casual. This is because they do not adhere to any specific grammar, but instead respond reasonably to the complexity of the boundaries.

We conducted an additional experiment to illustrate that the QP solver could also control building alignment in a similar way to rule-based approaches. As is shown in Fig. 6, a specific alignment rule is extracted from an example and applied to another using QP. Figure 6(a), (b) shows a typical example of how building alignment works in an irregular area. The buildings are grouped in pairs and aligned on at least one side, allowing the other side to respond freely to the irregular boundaries. The equations and inequalities for building-i and building-j are given below, which are grouped in pairs and aligned to each other.

$$\xi_a + \xi_b = 1 \quad (\xi_a, \xi_b \in \{0, 1\}) \tag{3.1}$$

$$\xi_a \times (x_i - x_j) + \xi_b \times (x_i - x_j + l_i - l_j) = 0 \tag{3.2}$$

$$y_i - y_j \leq 0 \tag{3.3}$$

$$2 \times k \times h_{unit} \times (n_h)_{min} + y_i - y_j \geq 0 \tag{3.4}$$

The result generated for another area can be seen in Fig. 6(c), (d), where the buildings are designed to have a row of rooms on the south side of the corridor, which can be called single-depth. Figure 8 compares the result without and with alignment requirements. It can be seen that the grouping method works both for single-depth buildings and double-depth buildings.

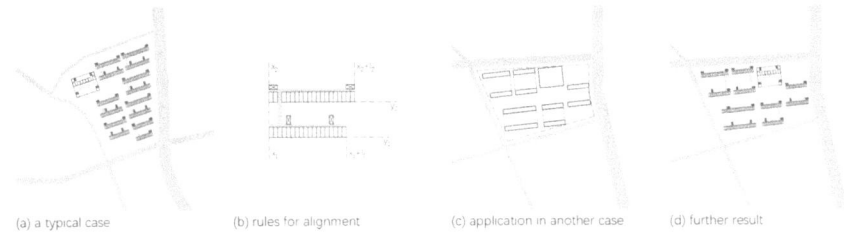

(a) a typical case   (b) rules for alignment   (c) application in another case   (d) further result

Fig. 6. The process of extracting and applying alignment rules from one case to another

3.2 Other Modules and the Pipeline

Some steps in this pipeline have not been introduced in this paper because of the limitation of 10 pages, including vector field generation and directional coordination of buildings. These steps are added because QP models could not provide multi-directional schemes directly.

Additionally, we provide a pipeline to illustrate how these modules can be combined to produce schemes. Figure 7 presents an overview of a modelling pipeline, which demonstrates a possible way to integrate the QP model and geometric methods discussed in Sects. 3.1 to 3.3, for real design scenarios. The pipeline allows users to automatically generate buildings by defining key parameters and area boundaries. Constraints are formulated in a QP model and then solved by a mathematical programming solver. Geometric methods are applied both before and after the main model to enhance the framework's flexibility and applicability.

This pipeline has been successfully tested in several campus design projects, enabling efficient building generation in residential areas based on preliminary land layout schemes. Information about ongoing campus design projects is not disclosed in this paper to preserve confidentiality.

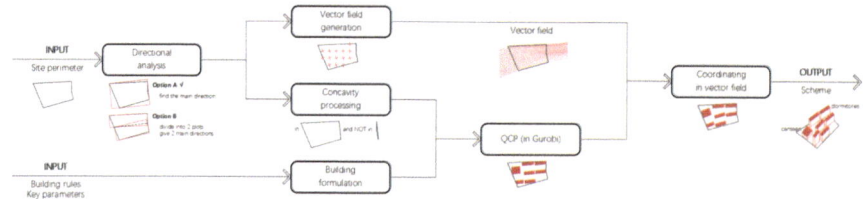

Fig. 7. A Modelling pipeline

4 Conclusion

This paper demonstrates the effectiveness of utilizing QP and cutting-edge solvers to address the urban design issues prevalent in residential areas within Chinese universities. By incorporating inside-model techniques and outside-model modules that employ geometric methods, the model can effectively handle complex boundaries, flexible building templates, and directional variability. The flexibility of QP models and general solvers is apparently enhanced. This approach, which is founded on a comprehensive understanding of urban design problems, has proven to be beneficial in practical urban design scenarios.

However, it is worth noting that this study does not fully capture the flexibility of buildings that are primarily dominated by corridors, as rule-based approaches can do. Moreover, the program's speed may be compromised when dealing with large-scale problems.

References

Hua, H., Dillenburger, B.: Packing problems on generalised regular grid: levels of abstraction using integer linear programming. Graph. Models, [online] **130**, 101205 (2023). https://doi.org/10.1016/j.gmod.2023.101205

Hua, H., Hovestadt, L., Tang, P., Li, B.: Integer programming for urban design. Eur. J. Oper. Res. **274**(3), 1125–1137 (2019). https://doi.org/10.1016/j.ejor.2018.10.055

Jiang, F., Ma, J., Webster, C.J., Chiaradia, A.J.F., Zhou, Y., Zhao, Z., Zhang, X.: Generative urban design: a systematic review on problem formulation, design generation, and decision-making. Prog. Planning, [online]. 100795 (2023). https://doi.org/10.1016/j.progress.2023.100795

Keatruangkamala, K., Sinapiromsaran, K.: Optimizing architectural layout design via mixed integer programming. In: Computer Aided Architectural Design Futures, pp. 175–184 (2005). https://doi.org/10.1007/1-4020-3698-1_16

Meindl, B., Templ, M.: Analysis of commercial and free and open source solvers for linear optimization problems. repositum.tuwien.at (2012). [online] Available at: http://hdl.handle.net/20.500.12708/37465. Accessed 18 Dec 2023

Peng, C.-H., Yang, Y., Wonka, P.: Computing layouts with deformable templates. ACM Trans. Graphics 33(4), 1–11 (2014). https://doi.org/10.1145/2601097.2601164

Sun, Y., Dogan, T.: Generative methods for Urban design and rapid solution space exploration. Environ. Planning B: Urban Analytics City Sci. 239980832211421 (2022). https://doi.org/10.1177/23998083221142191

Wu, W., Fan, L., Liu, L., Wonka, P.: MIQP-based layout design for building interiors. Comput. Graphics Forum 37(2), 511–521 (2018). https://doi.org/10.1111/cgf.13380

Zhang, Q., Li, B., Li, H., Tang, P.: Towards Integration and Hybridization in Urban Generation: An Extendable Urban Generative System for Better Natural Ventilation (2023). https://doi.org/10.52842/conf.ecaade.2023.2.379

Environmental Data-Driven Optimization of Building Skin Design by Coupling Genetic Algorithm and Neural Network Algorithm -Taking Shaanxi Xi'an Garment Office Building as Example

Mingchunjian Shi[1], Liming Kong[1], Yongzhong Chen[2(✉)], and Xiang Li[1]

[1] Xi'an University of Architecture and Technology, State Key Laboratory of Green Building, Xi'an, China

[2] School of Architecture and Urban Planning, Huazhong University of Science and Technology, Hongshan 430074, P.C, China

m15797876958@163.com

Abstract. It is complex to design facade skin for different building, in view of the high operational energy consumption that accompanies buildings with excessively large window-to-wall ratios. For public buildings with, the energy load cost relatively large. There are a characteristic of facade which have high wall window rate to consume a lot of energy or increase Insulation cost due to the influence of interfaces. For the treatment of shading in summer, excessive overhang of the eaves often increases the load and structural cost of the roof or increases the external sunshade structure to increase structural cost. Integration of surface energy at the interface level, combined with the shading construction of light materials. In the current process of shortening the carbon cycle of buildings, study are exploring innovative exploration ways, focusing on building integrated sunshade and insulation for building with high window-wall-ratios. study consider the changes between winter and summer differentiated needs to propose optimized design solutions for reducing cooling and heating energy consumption under photo-thermal comfort conditions.set hina xian project as the case. Study show by new toughness skin design improving Performance and efficient cut of Energy cost. Solving photo-ermal objective analysis and selection problems based on ANN neural network learning prediction feedback, bench marking of environmental parameters, and parameter definition of evolutionary solvers. The results show that the solver can reach convergence at an early stage, and the validation of the chosen solution proves the effectiveness of the strategy-guided morphology. Based on learning prediction can be more accurately coupled with existing simulation trends. The total annual energy consumption of design scenario for this case skin, which is 8.4% more energy efficient than the conventional scenario and more than 5.3°C.

Keywords: Multi-objective optimization · Skin morphology · Photo-thermal conversion · Building energy efficiency

H. Chai et al. (Eds.): CDRF 2024, *Symbiotic Intelligence*, pp. 236–248, 2025.
https://doi.org/10.1007/978-981-96-3433-0_21

1   Introduction

Intervention from the perspective of architectural, Design improvements have been made in optimization methods, ranging from the synergy of morphology and parameters to the utilization of genetic algorithms for optimizing and screening effective information [1, 2]. Form optimization of built form typologies integrated energy-emerge approach [3, 4]. The generation of skin isclosely related to the environment. It is necessary to establish a building skin design process to cope with environmental changes from the perspective of the whole process of design [5, 6]. On the basis of optimal control of photo-thermal targets, it is essential to tap the dual energy- saving potential of sur-face form and material properties, The aim of to explore the from design methods and technology integration of shape and materials into energy consuming elements. That is optimization of shape and facade of an office building using the genetic algorithm. Optimization of building form to reduce incident solar radiation and the design at the facade renovation design plays a crucial role in the overall architectural scheme [7]. This study is aimed at design skin of excessive window area considering the light performance.

2   Study on Prototype and Adaptation Type of Out- Layer Skin

2.1   Research Background

Although good coordination between architectural form design and mechanical equipment can change efficiency, the role of the surface layer cannot be ignored. Different skin as the inner and outer boundary layer have different temperature distribution and heat transfer process. Specifically, for buildings with larger body shape coefficients which transparent spatial design for public Mid-rise buildings attributes for urban exhibition spaces with public openness such as exhibition halls, foyers and waiting areas. The facade of the whole building to form a open pubic boundary. Due to the high cost of maintaining interface materials have advantages in display and pubic transparency, design decisions are usually aimed at meeting the standards even insufficient. This facade should give full consideration to the problems. Facade strategies should be used as a means to optimize overall comprehensive benefits, defining feasible boundary layer design strategies by predictive optimization. For regions with hot summers and cold winters, the response research methods include transparent interfaces and cavities, as well as internal design shading systems such as variable shading blinds. The building skin system's response to the environment and user needs, is an effective way to realize the balance between low energy consumption and high comfort in buildings [8]. Revealing the potential of building adaptive skins for trade-off improvements in building performance [9]. Some scholars have proposed the application of "Dynamic Climate Analysis" to develop the adaptive skin design of buildings, aiming to improve the accuracy of the environmental response of the adaptive skin of buildings [10]. Research on user-centered adaptive skin design for buildings has received increasing attention [11–13]. A study has developed a system for evaluating the operation of adaptive skins, which is centered on user satisfaction [14]. Several studies have shown that building adaptive skins have a good potential to reduce building energy consumption and carbon emissions, and to create a healthy and comfortable indoor environment [15]. The adaptive skin morphology of buildings has different impacts on the green performance of buildings, such as energy consumption and photo-thermal performance, which need to be weighed and considered, and there is an urgent demand for multi-objective oriented design of adaptive skin morphology facing with the requirements of various aspects of buildings [16, 17].

2.2 Research Technical Route

We use the advantages of building boundary layer waste heat recovery, shading, insulation, and heat insulation as response methods. As a solution, we have comprehensively designed the surface morphology and materials of the transparent interface. Although the design intervention of the outer transparent skin will increase the initial investment cost of the project, By using new skin the problem of high energy consumption, poor comfort, and difficulty in improvement in existing built spaces can be effectively solved, the subsequent benefits can effectively shorten the payback period. Establishing platform workflow (Fig. 1) to introduce the research process and application tools. The type of the skin of prototype research and application research are shown in defining research subjects. Introduced the considered conditions and operating conditions. The description of the characteristics of the skin panel in photo-thermal feature optimization introduced the changes in research objects and objectives.

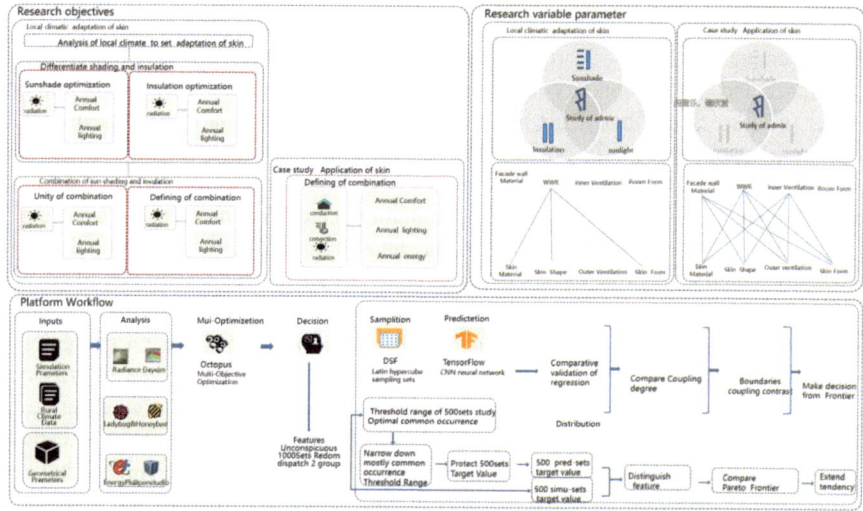

Fig. 1. Analyzing process

2.3 Skin Prototype Design by Passive Strategy

This article takes Xi'an, the cold region (Zone II) of China, as an example to observe passive strategies. As shown in Fig. 2. Focus on reducing heating energy consumption shading and heat insulation on the outer can really improve indoor conditions. It can be concluded that it is necessary to pay attention to both shading in summer and solar heat gain in winter, as well as ventilation, to achieve the best improvement effect.

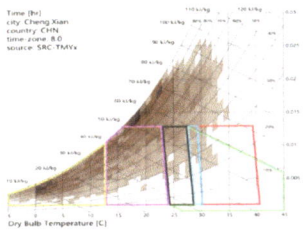

Fig. 2. Analyzing Psychrometric

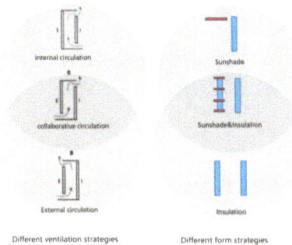

Fig. 3. Establishment prototype

Using the numerical information on the interface of skin as an indicator to divide different characteristics of the skin. According to the balance of heat radiation losses and gains in winter and summer, the area with sufficient sunshine is chosen to directly convert solar heat energy. Followed design conditions by the selection strategy for adjusting climate passive strategy as shown in Fig. 3.

2.4 Establish Parametric Skin of Adaptation Type

The study conducted research on transparent materials, first analyzing the differences in the influence of different transparent materials as skin materials, and comparing them, it was found that their interval distribution (Figs. 5 and 6). Based on local adaptability conditions of materials and forms, Fig. 4 establishes all possible situations to cope with diverse differences in changes.

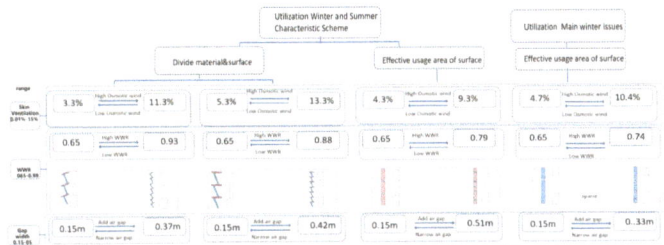

Fig. 4. Typical adaptation to local epidermal experiments

3 Case Study of Application Skin Type

As Fig. 8, by compare Type1–4 Based on this parameter model to solve the outstanding problem of heat loss of Excessive WWR. Photo-thermal performance of the local skin is solved by select Type- 1 to application case shading in summer and solar heat gain in winter Because sundials with different angles.

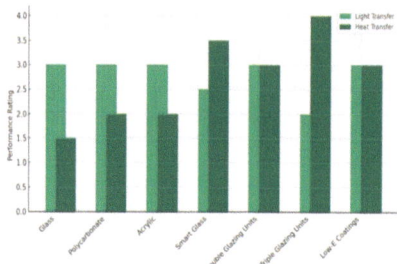

Fig. 5. Typical transparent materials by compare transparent

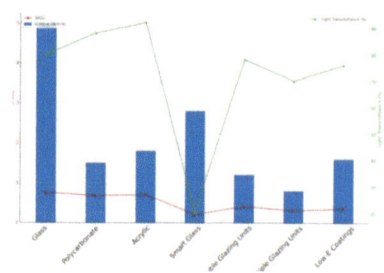

Fig. 6. Typical transparent materials by compare U &shgc &transmittance

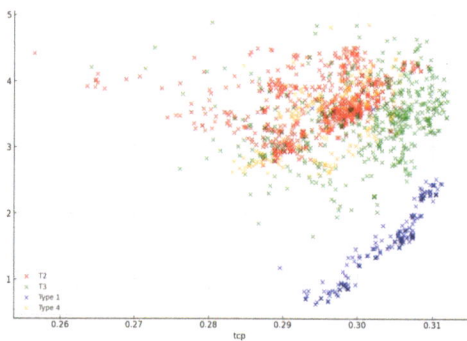

Fig. 7. Simulation of Typical adaptation

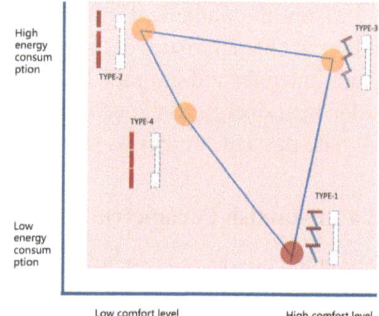

Fig. 8. Range of designers choices

3.1 Narrowed Parameter Interval Optimize Parameter Design

The units of design skin are 6.6*6.6*3.6 m, defining the material and forma characteristics of the transparent and non-transparent areas (shown in the Fig. 6). Based on the parameter characteristics of the material set, select the conventional range of heat transfer and transparency performance according to the median interval distribution, and use the interval range as the initial condition for finding the target (Figs. 9 and 10). The multi-objective optimization design for photo-thermal &comfort based on Type- 1 ranges of changes efficiency synergism witch have lower energy consumption and more comfort level.

Tables 1, 2, 3, 4, 5 and 6 shows the definition of the variation range of the skin material and the design range of the skin form and material. The purpose is to design parameters for the optimization results. For day-lighting, which can reduce excessive intensity of light for office buildings. The second is for heat gain, which remains unchanged. The advantage of shading in summer is that it can greatly promote the reduction of cooling energy consumption in summer Fig. 11.

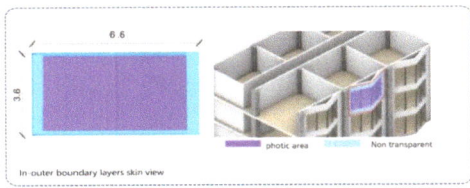

Fig. 9. Design inner window unit

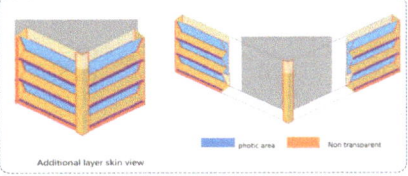

Fig. 10. Design outer skin unit

Table 1. Threshold distribution of Sampling parameters developed based on previous search

Type	WWR	U1	SHGC 1	t_vis1	U2	SHGC 2	t_vis2	WWR2	ang.
Thre-lower	2	2	0.079	0.31	0.216	0.4	14.8	2	5
Thre-upper	3.2	3.2	0.383	0.6	0.345	0.67	21.6	3.2	17

Table 2. Simulation and learning sampling parameter Threshold distribution

Thre-Lower	2	2	0.079	0.31	0.216	0.4	14.8	0.2	7
Thre-Upper	3.2	3.2	0.383	0.6	0.345	0.67	21.6	0.37	12

Table 3. Finally select the parameter Threshold distribution with high return ratio of tcp

Resulte	2	2	0.079	0.31	0.216	0.4	14.8	0.25	10

Table 4. Threshold distribution of Photo-thermal targets before simulation

Type	Energy	Udi2000+	Udi300+	Tcp	Energy2	Udi2000+2	Udi300 +	Tcp2
Thre-Lower	20	0.05	0.08	0.11	25	0.05	0.08	0.17
Thre−Upper	32	0.3	0.383	0.06	3.2	0.3	0.38	0.31

3.2   Sets Organization to Designing Neural Networks Models

The optimization of building photo-thermal performance is a multi-objective optimization problem, which strives to achieve the optimal state of both building energy consumption and comfort. It requires the use of multi-objective genetic algorithms to analyze the problem, using the Octopus software Hype-reduction algorithm for the optimization

Table 5. After simulation and threshold distribution of Photo-thermal targets

Thre-Lower	20	0.05	0.08	0.11	25	0.05	0.08	0.17
Thre−Upper	32	0.3	0.383	0.06	3.2	0.3	0.38	0.31

Table 6. Finally threshold distribution for the Photo-thermal target with high return ratio of tcp

Resulte	21	0.212	0.279	0.31	0.21	0.421	0.14	0.307

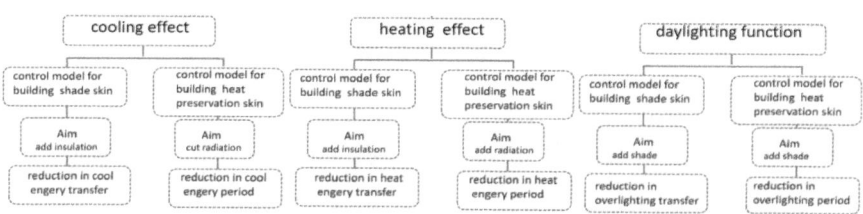

Fig. 11. Classification of performance optimization

parameters. The elite degree is set to 0.5; The probability is set to 0.5; The mutation rate is set to 0.2; The crossover rate is set to 0.6, the population size is 100, and the iterations is set to 50. A total of 1000 solutions were analyzed data Threshold distribution as Tables 2 and 4 showhigh precision and fast prediction model for commonly used supervised machine learning prediction as SVM, artificial neural network and neural network. Latin hypercube sampling (LHS) is used to extract feasible solutions in order to obtain the operational results that make the sampled data have global coverage before training the prediction model and optimizing parameters. In this study, 1000 groups of data were sampled and simulated to obtain the training samples needed by the model, of which 500 groups were used for training and testing. It is necessary to improve the accuracy of SVM prediction model. There are two important parameters for SVM support vector machines: regularization parameter c (also known as penalty coefficient) and kernel function parameter Gamma. In the experiment, the range of Gamma is set to [0, 100] and the range of penalty coefficient is set to [1, 100] according to the sample number. The objective of optimization is the linear correlation coefficient and mean square error (MSE). The Established the mechanical learning and prediction ways In Fig. 12.

The linear coefficient points reflects the effectiveness of prediction data. The linear correlation coefficient is set to prevent the model from being over- fitting. Ultimately, minimizing mean squared error and maximizing linear correlation coefficient, mainly analyzing its dispersion, which shows a good normalized trend overall.

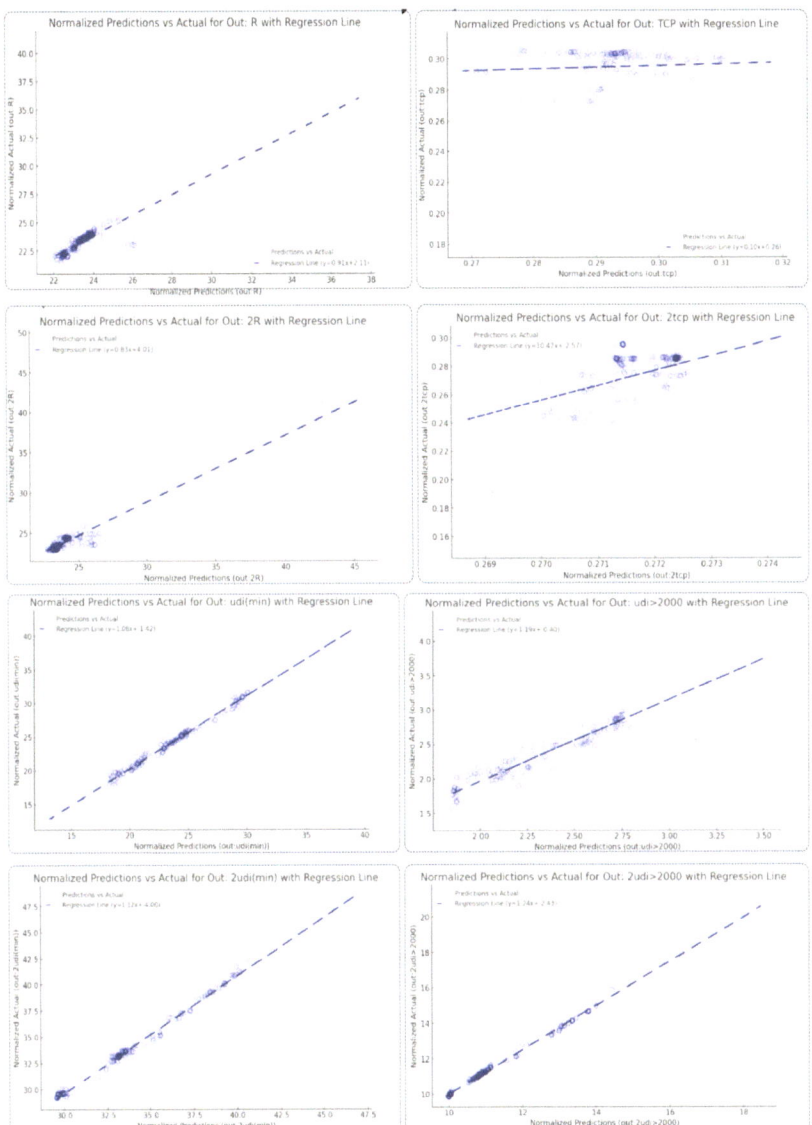

Fig. 12. Related mean square error regression of machine learning

3.3 Validation of the Efectiveness of Predictions

The prediction effect is not good enough when LUX is out range of 2000 + and TCP. Its accuracy needs to be further corrected for features. But other indicates are fit the distribution of trends Blue dashed line , which mean point concentrate on the trend online rather than dispersed. This study mainly combines the features of lighting with LUX300-,300–2000 and energy consumption for comprehensive decision-making. In other words that the prediction some indicates is relatively Deviant, but the overall lighting and energy consumption features are more scientific to do further research for compare the difference, as shown in Figs. 13 and 14. For interval distributions, it can be concluded that there is a strong similarity between them and simulated data.

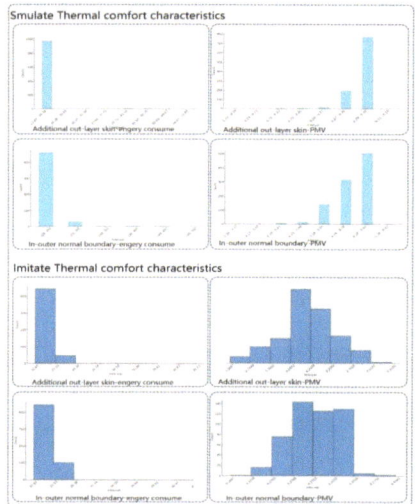 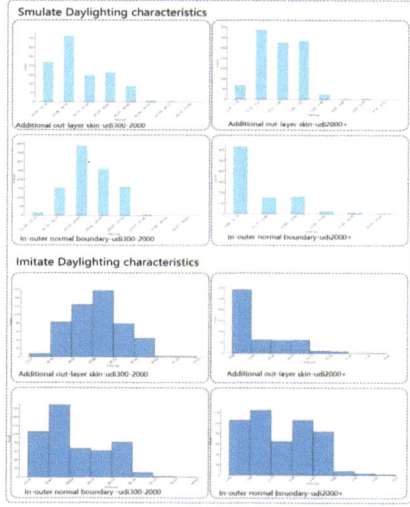

Fig. 13. Distribution of simulation　　**Fig. 14.** Distribution of network learning

3.4 Decision-Making by Compare Pareto Frontier of Predictions

When the frontier data cannot establish a wide range of interval relations for the target range. The comfort range of the frontier solution is 0.294–0.33% with an energy consumption of 20.75–22.25 kwh/m^2 which is the optimal distribution under the comfort range. Forecast expanding to the front-line, the curve rises significantly, and the upward trend of the curve becomes slower, which reflects the importance of the influence level of comfort improvement. This means that improving comfort consumes more efficiency than improving energy (Figs. 15 and 16).

It is necessary to provide appropriate targeted solutions to improve comfort, in summary when it is difficult to observe the trend of leading edge by using multi- objective optimization simulation. Through machine learning, establishing data coupled with frontier characteristics can quickly fill in the broader frontier range of frontier characteristics, forming a trend feature, which is meaningful for rapid decision-making in practical projects. This case To balance energy consumption and comfort indicators, choosing Type 1 focuses on Choose the one with higher profit ratio, as the frontier curve becomes slower that meaning prioritizing trend make decision due to the energy consumption to increase the same comfort level time radio(%) is less cost energy consume than before, which is better Photo- thermal performance. Due to increasing convection and radiation, the indoor air temperature is affected. It is characterized in that the temperature

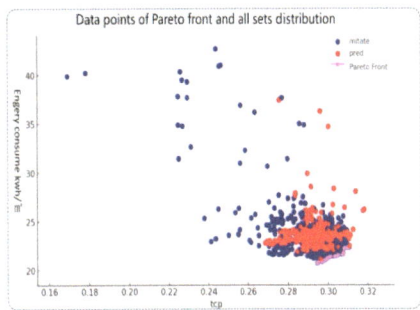

Fig. 15. Feature value distribution

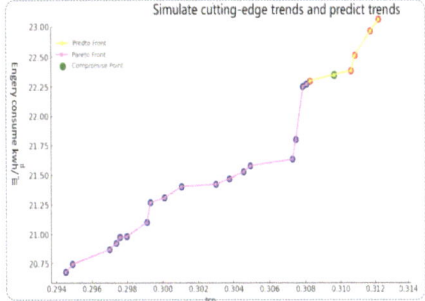

Fig. 16. Pareto Frontier distribution

drops by 0. 12 °C. In summer, it drops more significantly by 0.89 °C, indicating that the design strategy to block direct sunlight in summer has a positive impact on improving the indoor photo-thermal environment, especially in the dominant time period of summer. The parabolic curve of the temperature field on that day will tend to smoother. to narrow the comfort range of temperature change between day.

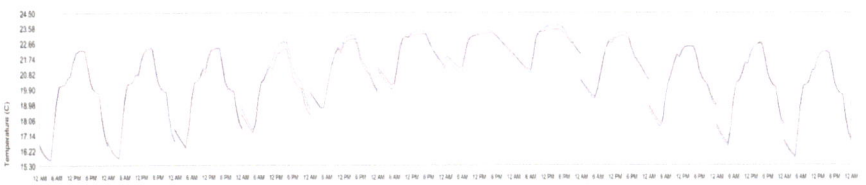

Fig. 17. Select target of energy feature value compare without

Due to the calculation of overall energy consumption different condition, the peak temperature decreased. The change in air temperature affects the perceived temperature, temperature fluctuations throughout the day are more smooth. Compared to the design of the new skin and the original glass scheme, Ideal comfort is improved by 5.3%. And due to the narrowing of the thermal peak and cold peak thresholds, it is more suitable for human comfort to rise (Figs. 17 and 18).

Fig. 18. Target of add time of acceptable zoom compare without T-1skin

In this case, the energy consumption can be saved by new skin. Increase lighting energy consumption 37% as the cost of sacrifice, which bring the total energy consumption can be saved8.4%and the indoor temperature can be increased by more than 5.3% (Fig. 19). In this case, the annual total energy consumption of the design is established by application Type- 1. Its saw louver skin ∘This method will locally increase the additional cost of materials, which can improve the overall photo-thermal performance. The design provides guidance according to the energy-saving characteristics of 32.3 kWh/m^2 on the south side of the building of 1,052 square meters and the data as Tables 3 and 6 shows. From local needs, Air gap barrier skin acts as buffer (Fig. 20) Summer shading and winter insulation have achieved ideal compromise benefits. Creating advantages in cold and hot heat exchange, thermal convection, and heat conduction during winter and summer.

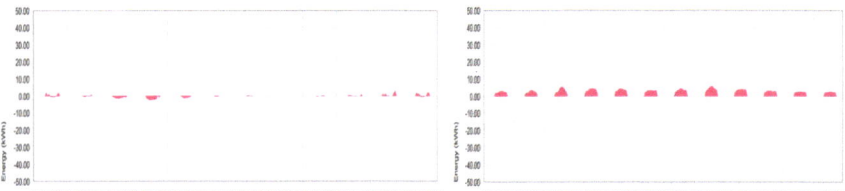

Fig. 19. Energy changes compare without skin

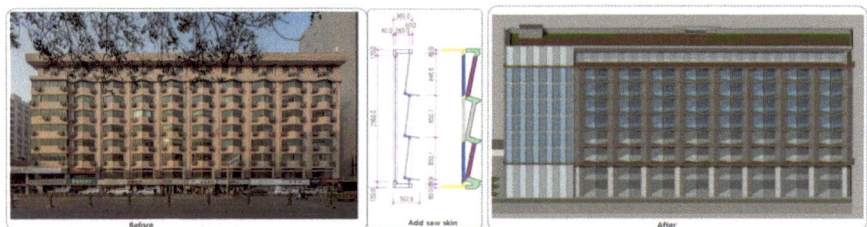

Fig. 20. Compare the transformation results of add skin

4 Conclusion

Through multi-objective optimization first and then optimizing the application form through machine learning in new features. By the integrated thermal design of skin thermal insulation and heat storage, the adaptive potential of the skin unit is explored. Without the definition of thermal inertia effect, heat transfer efficiency is used as a static index to further track the skin morphology of light-heat coupling through surface energy adjustment under the influence of material changes and lighting and heating performance. Through predictive coupling research, early design decisions can be guided through three steps. The first step is to simulate typical features and highlight cutting-edge features through prediction. The second step is to observe the validity of the prediction, and combine the effective prediction results with the simulation values. The third step is to observe the cost of comfort improvement for the same energy consumption, select the target value with the largest return ratio, and go back to its parameters for experimental verification. Through the integrated thermal design of skin thermal insulation and heat storage, the adaptive potential of the skin unit is explored. Without defining thermal inertia effect to calculate heat transfer efficiency, energy & TCP chat is used as a static indicator to further track the skin morphology of light-heat coupling through surface energy adjustment. It is necessary to conduct post-occupancy evaluations of public buildings that have been built during the rapid development of cities aim to deal with the lack of green performance orientation. Such design can generate positive public cognitive feedback and add cognitive Significance of light and heat energy conservation.

References

1. Adamski, M.: Optimization of the form of a building on an oval base. Build. Environ. **42**(4), 1632–1643 (2007)
2. Jin, J.T., Jeong, J.W.: Optimization of a free-form building shape to minimize external thermal load using genetic algorithm. Energy and Buildings **85**, 473–482 (2014)
3. Jalali, Z., Noorzai, E., Heidari, S.: Design and optimization of form and facade of an office building using the genetic algorithm. Sci. Technol. Built Environ. **26**(2), 128–140 (2020)
4. Taleb, S., Yeretzian, A., Jabr, R.A., et al.: Optimization of building form to reduce incident solar radiation. Journal of Building Engineering **28**, 101025 (2020)
5. Konis, K., Gamas, A., Kensek, K.: Passive performance and building form: an optimization framework for early-stage design support. Sol. Energy **125**, 161–179 (2016)
6. Dino, I.G., Üçoluk, G.: Multiobjective design optimization of building space layout, energy, and daylighting performance. J. Comput. Civ. Eng. **31**(5), 04017025 (2017)
7. Taleb, S., Yeretzian, A., et al.: T. Optimization of building form to reduce incident solar radiation. Journal of Building Engineering **28**, 101025 (2020)
8. Juaristi, M., Loonen, R., Isaia, F.: Dynamic Climate Analysis for Early Design Stages. Sustainable Cities and Society (60), NO102232 (2020)
9. Hosseini, S.M., Mohammadi, M., et al.: T: Improving Visual Comfort Based on Dynamic Daylights. Building and Environment (165), NO106396 (2019)
10. Abdollahi, R.R.: A user detective adaptive facade towards improving visual and thermal comfort. Journal of Building Engineering (33), No101554 (2021)
11. Morteza, H.S., Masi, M., Torsten, S.: Integrating interactive kinetic façade design with colored glass to improve daylight performance based on occupants, position. J. Build. Eng. (31), NO101404 (2020)
12. Environmental Building; New Findings from University of Liege Describe Advances in Environmental Building. Ecology Environment & Conservation (157), 268–276 (2019)
13. Ascione, F., et al.: The evolution of building energy retrofit via double-skin and responsive façades: a review. Solar Energy **224**, 703–711 (2021)
14. Ding, F., Ahsan, K.: Tall buildings with dynamic facade under winds. Engineering **6**(12), 1443–1453 (2020)
15. Mohammad, J., Alice, A.: Review of approaches, opportunities, and future directions for improving aerodynamics of tall buildings with smart facades. Sustainable Cities and Society (2021). https://doi.org/10.1016/J.SCS.2021.102979
16. Tabadkani, A., Isaeed, T.M.: Integrated parametric design of adaptive facades for user's visual comfort. Automation in Construction (106) (2019)

17. Rana, R. A., et al.: A user detective adaptive facade towards improving visual and thermal comfort. Journal of Building Engineering (33), NO101554 (2021)

Artificial Intelligence in Design and Simulation

Constructing a Knowledge Graph for Extreme Climate Architecture Based on Large Language Models (LLMs)

Shaotsu Tu[1], Weimin Zhuang[1(✉)], and Fei Ren[2]

[1] School of Architecture, Tsinghua University, Beijing, China
zhuangwm@tsinghua.edu.cn
[2] Architectural Design and Research Institute of Tsinghua University Co., Ltd., Beijing, China

Abstract. With the escalation of climate change, extreme weather events have become increasingly common, posing significant challenges to the architectural domain. Only focusing on the design methods and characteristics of buildings in typical climates is no longer enough to cope with future weather challenges. Therefore, the study of architectural design under extreme climates has become an emerging and important topic. However, China's research and engagement on extreme climate architecture lag behind many other countries, resulting in related architectural knowledge in this domain being scattered and fragmented, and even in the gray area of information retrieval, making it difficult for architects to use. This paper explores how to systematically categorize, organize, and present information on extreme climate architecture in a way that is easily accessible and beneficial to architects, thereby supporting well-informed design decisions. It proposes a top-down approach to constructing a knowledge graph for extreme climate architecture. By leveraging architectural programming theory, the study constructs an ontology model and employs ChatGPT for the extraction of knowledge from unstructured data. Additionally, it uses web crawlers to gather relevant information from general encyclopedias, integrating these into a triplet form. Compiled Data is then stored in Neo4j, facilitating efficient domain knowledge querying and visualization. Furthermore, this research presents a Q&A application demo named Extreme-Architecture-Graph based on the constructed knowledge graph. This application aims to transform extreme climate architectural knowledge into actionable insights, enabling architects to acquire a comprehensive understanding of design objectives under extreme climate conditions in the early stages of planning and design. In summary, this study constructs a knowledge graph for architecture in extreme environments to enhance preparedness for the unforeseen risks posed by extreme weather conditions, and explores the usability of multi-source heterogeneous data in the field of architecture in the era of artificial intelligence.

Keywords: extreme climate architecture · polar architecture · knowledge graph · Large Language Models (LLMs)

H. Chai et al. (Eds.): CDRF 2024, *Symbiotic Intelligence*, pp. 251–261, 2025.
https://doi.org/10.1007/978-981-96-3433-0_22

1 Introduction

In recent years, the escalating effects of climate change have manifested in more frequent extreme hot and cold events(Cohen et al., 2021), with increasingly significant negative impacts on the urban built environment. A growing body of research shows that only focusing on buildings under typical climates is no longer sufficient to cope with future scenarios of weather change. As a result, studying the design and construction methods of buildings under extreme climate has become as a critical issue.

International research on extreme climate architecture dates back to the International Geophysical Year 1957–1958. During the IGY, several countries, including the United States, the United Kingdom, Australia, and Germany, undertook studies on a series of planning, design, and construction building methods in Antarctica. This effort led to the accumulation of rich experience and knowledge, as well as the formation of databases for further study. However, China's involvement in this field began later around 1985 (Ren, 2005). Despite the prior knowledge, accessing relevant information remains a challenge due to fragmented data, inconsistent structures, and retrieval complexities within the domain of extreme climate architecture, making it very difficult for architects to integrate, use and understand.

The concept of a knowledge graph, proposed by Google in 2012, offers a promising approach to address these challenges. By structuring complex information into "entities," "relationships," and "attributes," a knowledge graph creates a network of structured data, provides a solution to eliminate data silos and facilitates efficient data integration and extraction of key insights (Qi et al., 2017). This study aims to leverage the knowledge graph framework to construct a comprehensive understanding of extreme climate architecture, focusing on the "Polar Architecture" as a representative type. Through the integration of data, information, and knowledge, architects can gain valuable insights early in the design process, aiding in informed decision-making and improved design outcomes under extreme climate conditions.

By exploring the complete chain of "data-information-knowledge-application," this research contributes to bridging the gaps in understanding and utilizing data for effective design solutions in extreme climate architecture.

2 Related Work

Knowledge graphs have found widespread applications in various fields such as medicine, finance, healthcare, news, and education(Zou, 2020). However, their use in architecture remains limited, primarily due to the challenges associated with storing large amounts of raw data uniformly and structurally within the architecture domain. To address this and improve the accessibility of architectural data in the era of artificial intelligence, knowledge graphs offer a viable solution.

Constructing a knowledge graph typically involves four steps: ontology construction, knowledge extraction, data fusion, and data storage(Yang et al., 2018). Among these, knowledge extraction plays a pivotal role and involves the identification and extraction of three types of objects: entities, relationships, and attributes.

For instance, Zhang et al. constructed a knowledge graph of Huizhou architecture using a semi-automatic knowledge graph construction method that combines BiLSTM-CRF model with a Huizhou architectural style dictionary(Zhang et al., 2021).Similarly, Li et al. proposed a model for joint extraction of entities and relationships using deep learning algorithms, specifically focusing on extracting complex and multiple semantic relationships from Chinese building codes(Li et al., 2020). While these methods perform well, they often require extensive data labeling or existing domain corpora to improve the accuracy of knowledge extraction models, which can be time-consuming and resource-intensive.

Recent advancements in large-scale pre-trained Language Models (LLMs) like Chat-GPT offer new avenues for improving the efficiency of knowledge graph construction, particularly in extracting knowledge from unstructured data based on natural language. Wei et al. achieved zero-shot knowledge extraction using multiple rounds of question and answer training with prompts in ChatGPT(Wei et al., 2023). Additionally, Yang et al. utilized ChatGPT to extract knowledge about Yue Opera and developed an intelligent question answering system prototype using the constructed knowledge graph, showcasing the potential of this approach for further development(Yang et al., 2024).

These related studies demonstrate the ongoing efforts and advancements in knowledge graph construction methodologies, which inform and contribute to the study of constructing a knowledge graph for extreme climate architecture based on LLMs.

3 Methodology

This paper proposes a top-down approach for constructing a knowledge graph in the domain of extreme climate architecture. The process is divided into two steps (Fig. 1):

(1) Data Layer Construction: The foundation of constructing knowledge graphs lies in the quality and quantity of data. The data sources for this study include COMNAP, ATCM and other open-source polar building databases, relevant design standards, global extreme climate databases, and a polar building database developed by the Polar Architecture Research Center of Architectural Design & Research Institute of Tsinghua University Co., Ltd after a long-term research and collation. Given the current inadequacies of semantic databases in the architecture domain, this study explores a knowledge extraction method based on ChatGPT, which is capable of obtaining a triplet database of polar building entities and their relationships with minimal or even zero training samples.

(2) Semantic Layer Construction: In order to build the ontology model, several core concepts of polar architecture are clarified, drawing upon the main content outlined in the five steps of architectural programming(Zhuang, 2016). The core concept includes various types of "entities", each described in detail with "attributes". The abstract relationships among these entities are expressed through "relationships", thus establishing the semantic layer of the knowledge graph for extreme climate architecture.

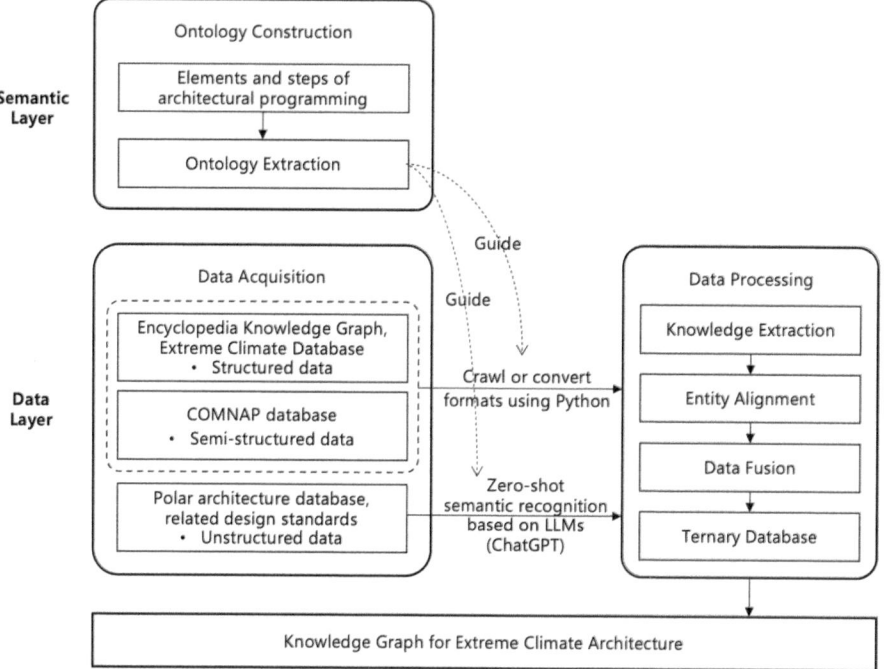

Fig. 1. Methodology for constructing an extreme climate architectural knowledge graph

3.1 Data Sources

The data used in this study are as follows:

1. Unstructured Data: The Polar Architecture Case Database, organized internally by the Polar Architecture Research Center of Architectural Design & Research Institute of Tsinghua University Co., Ltd (Ren, 2005). This database includes a comprehensive collection of case studies and examples of polar architecture, offering a rich source of unstructured data.
2. Semi-Structured Data: Polar architecture database released by COMNAP[1] in 2017(Council of Managers of National Antarctic Programs, 2017). This database provides a valuable compilation of semi-structured data related to the construction of Antarctic scientific research stations from multiple countries, making it easier access and analysis.
3. Structured Data: Including general knowledge graph databases such as OwnThink and CN-DBPedia[2](Xu et al., 2017), The Database of Antarctic Conservation Biographic Regions (ACBRs) (Terauds and Lee, 2016) published by the Australian Antarctic Data Center, and The Environmental Domains of Antarctica Database (Morgan et al., 2007)

---

[1] Council of Managers of National Antarctic Program.

[2] OwnThink and CN-DBPedia both contain millions of entities and billions of relationships, making them currently the two most important Chinese general knowledge graphs, providing APIs for free use.

published by the New Zealand government. These sources offer a range of structured data, from general knowledge graphs to specific environmental and biogeographic information, supporting detailed analysis and research in polar architecture.

3.2 Ontology Construction

Based on the principles of architectural planning and the foundational knowledge of architecture(Zhuang, 2016), this paper identifies five core concepts essential to the construction of a knowledge graph in the field: internal conditions, external conditions, spatial conception, technical conception, and economic planning. Each concept encompasses a series of entities, such as "research station", "countries", "structural materials", each entity will further characterized by attributes or relationships.

For example:
[Entity: Kunlun Station] - [Relationship: Country] - [Entity: China],
[Entity: Kunlun Station] - [Relationship: Building Foundation Type] - [Entity: Elevated],
[Entity: Kunlun Station] - [Relationship: Structural Materials] - [Entity: Steel],
[Entity: Kunlun Station] - [Attribute: Capacity] - [26],
[Entity: Kunlun Station] - [Attribute: Area Under Roof] - [558 m^2],
[Entity: Kunlun Station] - [Attribute: Average Temperature in February] - [−41.2 °C],
[Entity: Kunlun Station] - [Attribute: Annual Average Wind Speed] - [3.9 m/s]…
In this way, the ontology model which is also the semantic layer of extreme climate architecture knowledge graph was built (Fig. 2).

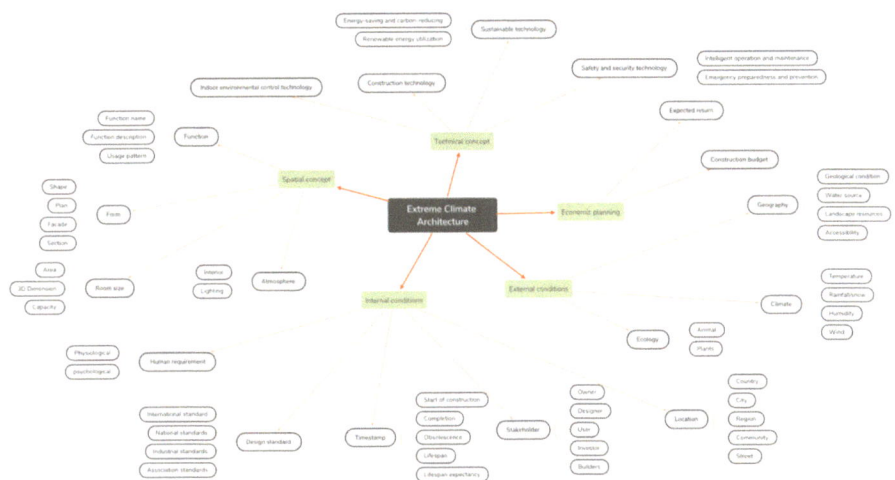

Fig. 2. Semantic layer of the extreme climate architecture knowledge graph

3.3 Knowledge Extraction

The polar architecture domain exhibits a pronounced phenomenon of data silos, characterized by the absence of a standardized method for storing and describing information related to extreme climate conditions, geo-environmental factors, and building design (including dimensions, forms, construction methods, etc.). Furthermore, the existing data pool is limited, particularly lacking in crucial data that could assist architects in making preliminary design decisions. More importantly, the field suffers from a scarcity of existing corpora to serve as a priori knowledge for labeling and training models. Consequently, traditional methods for constructing knowledge graphs tend to be inefficient, whereas approaches that based on Large Language Models (LLMs) such as ChatGPT, demonstrate significant advantages.

The knowledge extraction method combined with LLMs (ChatGPT) employed in this study unfolds in two steps (Fig. 3):

Step1 Describe the task context to ChatGPT. This entails explaining polar architecture terminology to facilitate the model's comprehension of domain-specific knowledge and a clearer understanding of the task objectives.

Step2 Clarify the requirements for knowledge extraction using prompts and access GPT-4 through the OpenAI API to undertake the knowledge extraction task. At this stage, the design of the prompts critically influences the outcomes of the knowledge extraction process. The Prompt designed in this paper consists of three parts, the first is the input task text, the second is the list of entity, relationship and attribute defined in the ontology model, and the third is the desired output format of the knowledge extraction results.

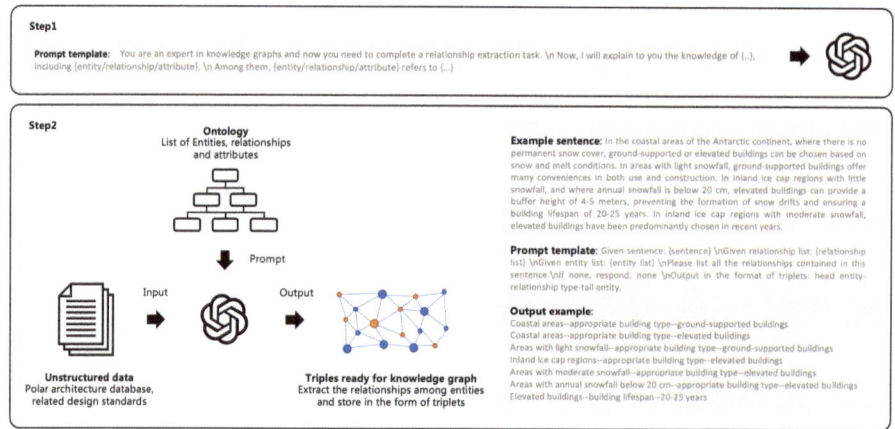

Fig. 3. Knowledge Extraction Process Based on LLMs (ChatGPT)

3.4 Data Fusion

For structured generic knowledge graph data, we directly retrieve data using the API provided by the respective websites, obtaining triplets in CSV format. The polar building database provided by COMNAP, presented in PDF format, constitutes semi-structured data, encompassing images, text, and tables. Initially, we convert the tables from the PDFs into Excel format, and subsequently, we employ the Pandas module in Python to transform the semi-structured data into structured data, which is also stored in CSV format.

So far, a total of 432 entities and 1491 relationships have been collected from 3.1 data sources. Given the manageable diversity of entities, manual alignment was employed to eliminate near-synonyms and synonyms, ensuring consistency and clarity in the dataset.

3.5 Data Storage and Visualization

This study utilizes Neo4j to store knowledge graph of extreme climate architecture. Neo4j is a graph database that represents knowledge in property graph database model and uses Cypher, also known as CQL, as its graph query language. The property graph database model is a directed graph consists of nodes, edges, labels, relationship types, and attributes.

Figure 4 presents an example of Neo4j's property graph model, which shows that the building units' organizational mode of the Kunlun Station is relatively independent, and the type of building foundation is elevated. Furthermore, "id" serves as the unique identifier of a node, "name" represents the name of the entity, and "label" indicates the category or subcategory of the core concept. The visualization of the knowledge graph enables users to understand the association rules between entities more intuitively and clearly.

The fused extreme climate building data was organized into two csv files: the first one stored the entity names and entity attributes, named after the "label" name, which was then used to create the nodes of knowledge graph; the second one stored the entity-to-entity relationships, which was used to create the relationships of knowledge graph. Finally, all csv files were imported into Neo4j using Python and the Py2neo module, completing the data storage and visualization process for the Extreme Climate Architecture Knowledge Graph (Fig. 5).

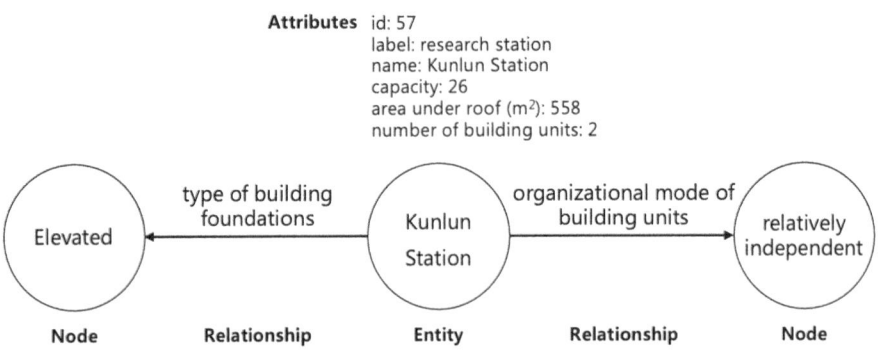

Fig. 4. An example of Neo4j's property graph model

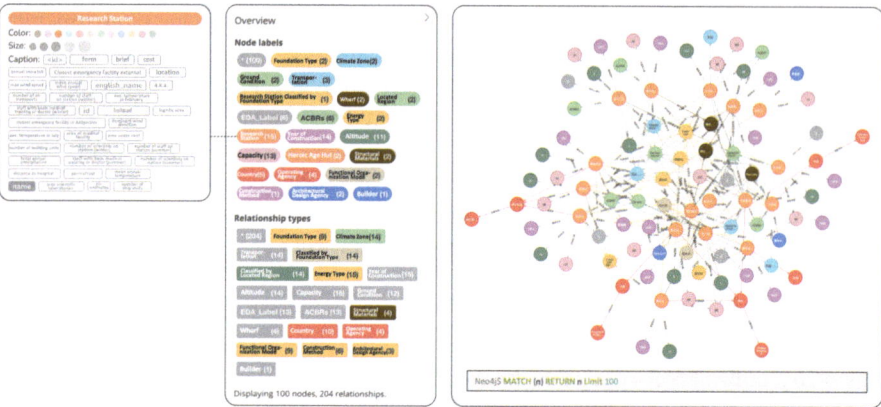

Fig. 5. Extreme climate architectural knowledge graph stored in Neo4j (partial)

4 Extreme-Architecture-Graph: A Knowledge Graph Question Answering System Based on ChatGPT

The Knowledge Graph Question Answering (KGQA) System offers users natural language dialogue interfaces, based on a deep understanding of semantic relationships, it could help people quickly understand a certain field and achieve the conversion from data to knowledge (Omar et al., 2023). Based on the constructed knowledge graph

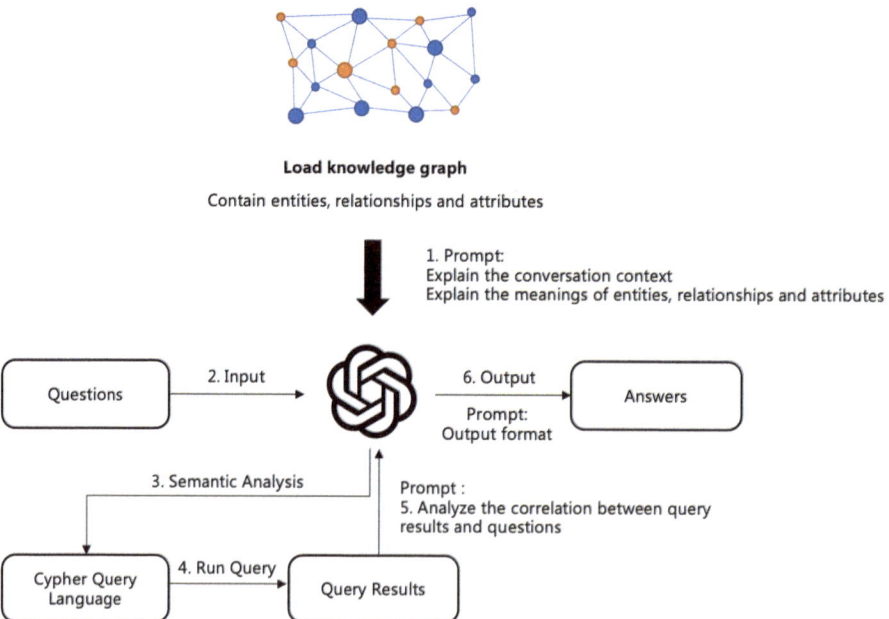

Fig. 6. Method of constructing a knowledge graph Q&A system based on ChatGPT

of extreme climate architecture, this study developed a prototype Q&A system demo named Extreme-Architecture-Graph, and explained the development method. This approach uses ChatGPT as the "hub" to convert natural language query statements into CQL, and obtain query results from Neo4j. The correlation between the question and the query result underpins the generation of the final answer. Additionally, prompts can be utilized to further standardize the format of the output answers, and even provide analysis reports to enhance the logic and comprehensiveness of the answers (Fig. 6).

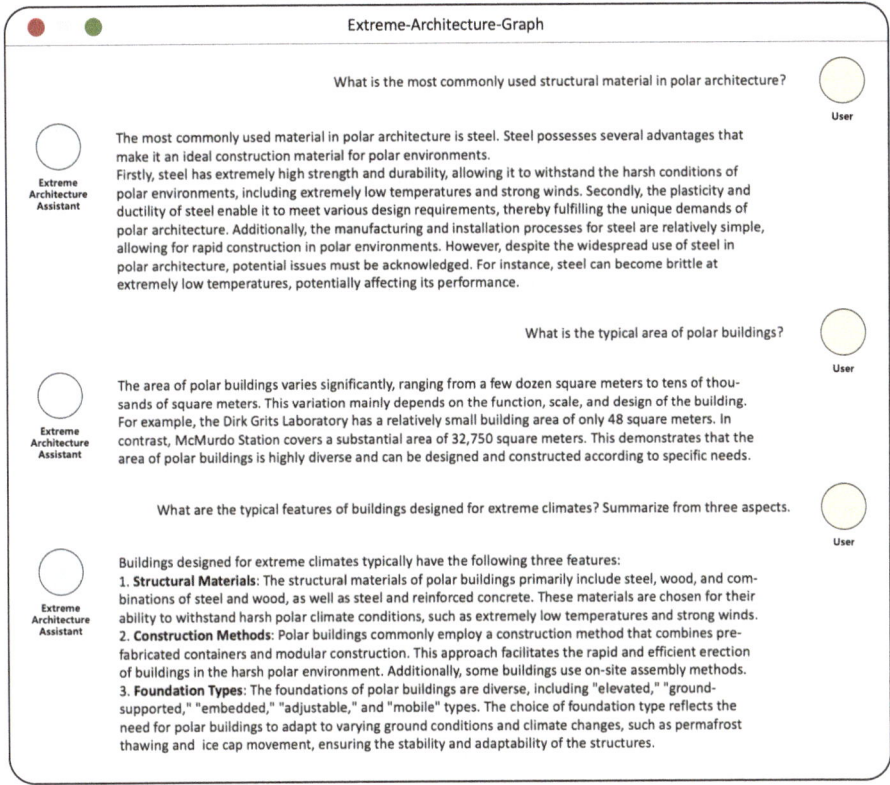

Fig. 7. Usage scenarios of the Extreme-Architecture-Graph demo

Figure 7 illustrates several potential usage scenarios of the Extreme-Architecture-Graph demo, showcasing the user interface as a conceptual design. In the first dialogue, the user inputs a question " *What are the most commonly used structural materials for polar buildings?*". After execution, the system translates this inquiry into a Cypher Query Language (CQL) command: *"MATCH (p:research stations)-[:structural materials]- > (m:structural materials) RETURN m.name, COUNT(m) AS usage ORDER BY usage DESC LIMIT 1"*. The query was then put into Neo4j and the result *"The most common material used in polar construction is steel. "* generated. By prompting *"Please output a report of not less than 100 words based on the query results"* as output format, the system will further generate the final Q&A response that fully articulates the logic behind the answer.

5 Conclusion and Future Work

In response to the increasing frequency of extreme climate and the current gaps of knowledge in architectural domain regarding coping strategies for extreme weather, this study proposes a methodology based on Large Language Models (ChatGPT) to construct a knowledge graph of extreme climate architecture. Then, the application value of this research is demonstrated with the example of constructing a knowledge graph in the domain of polar architecture. The contribution of this study includes:

- The integration of databases related to extreme climate buildings and further mining of knowledge in existing information with the assistance of ChatGPT;
- The construction of a domain ontology for the knowledge graph based on architectural programming theories, forming a semantic network that aligns with architects' cognitive processes;
- The development of a Q&A system demo leveraging ChatGPT and the constructed knowledge graph;
- The expansion of the architecture domain's knowledge graph to enhance preparedness for the unforeseen risks posed by extreme weather conditions.

Employing the methodology proposed in this study enables the rapid transformation of heterogeneous data from multiple sources into actionable knowledge. Moreover, the knowledge graph can be open-source and updated, providing a path for the intelligent development of the architecture industry in the era of artificial intelligence. Nonetheless, this study acknowledges certain limitations:

- There is a need to unearth more data related to extreme climate architecture. Extreme climate architecture encompasses not only polar buildings but also structures in high latitude and extremely hot areas, etc.
- The scale of the ontology model is limited, there is scope for extracting more entities and relationships.
- The fuzzy information retrieval and intelligent reasoning capabilities of the Q&A system need to be improved.

Acknowledgement. The authors gratefully acknowledge the financial supports by the National Natural Science Foundation of China (Major Program) under Grant No. 52394225 as well as the Polar Architecture Research Center of Architectural Design & Research Institute of Tsinghua University Co., Ltd.

References

Cohen, J., Agel, L., Barlow, M., Garfinkel, C.I., White, I.: Linking Arctic variability and change with extreme winter weather in the United States. Science **373**, 1116–1121 (2021). https://doi.org/10.1126/science.abi9167

Council of Managers of National Antarctic Programs. Antarctic Station Catalogue (2017)

Li, F., Song, Y., Shan, Y.: Joint extraction of multiple relations and entities from building code clauses. Appl. Sci. **10**, 7103 (2020). https://doi.org/10.3390/app10207103

Morgan, F., Barker, G., Briggs, C., Price, R.: Environmental Domains of Antarctica Version 2.0 Final Report. Manaaki Whenua Landcare Research New Zealand Ltd. (2007)

Omar, R., Mangukiya, O., Kalnis, P., Mansour, E.: ChatGPT versus Traditional Question Answering for Knowledge Graphs: Current Status and Future Directions Towards Knowledge Graph Chatbots (2023). https://doi.org/10.48550/arXiv.2302.06466

Qi, G., Gao, H., Wu, T.: The research advances of knowledge graph. Technology Intelligence Engineering **3**(1), 4–25 (2017). https://doi.org/10.3772/j.issn.2095-915x.2017.01.002

Ren, F.: Research on Ecological Strategy of Architecture Design for Buildings in Antarctica: Extension and Rebuilding Design for Zhongshan Station of China in Antarctica. Tsinghua University, Beijing (2005)

Terauds, A., Lee, J.R.: An update of the Antarctic Conservation Biogeographic Regions (ACBRs). Australian Antarctic Data Centre (2016). https://doi.org/10.1111/ddi.12453

Wei, X., et al.: Zero-Shot Information Extraction via Chatting with ChatGPT (2023)

Xu, B., et al.: CN-DBpedia: A Never-Ending Chinese Knowledge Extraction System. Presented at the International Conference on Industrial, Engineering and Other Applications of Applied Intelligent Systems, pp. 428–438. Springer, Cham (2017). https://doi.org/10.1007/978-3-319-60045-1_44

Yang, S., et al.: YueGraph: A Prototype for Yue Opera Lineage Review Based on Knowledge Graph. In: Fang, L., Pei, J., Zhai, G., Wang, R. (eds.) Artificial Intelligence, pp. 435–441. Springer Nature, Singapore (2024). https://doi.org/10.1007/978-981-99-9119-8_39

Yang, Y., et al.: Accurate and Efficient Method for Constructing Domain Knowledge Graph. Journal of Software (2018)

Zhang, R., Yang, C., Yin, L., Zhang, Y.: Semi-automated build of Huizhou architectural knowledge graph. Journal of Anhui Jianzhu University **29**, 13–19 (2021)

Zhuang, W.: Architectural Programming and Design. China Architecture & Building Press, Beijing (2016)

Zou, X.: A survey on application of knowledge graph. J. Phys.: Conf. Ser. **1487**, 012016 (2020). https://doi.org/10.1088/1742-6596/1487/1/012016

Exploring Optimized Generation Methods for Post-War Cityscapes Restoration Based on Stable Diffusion Model

Jiqian Huang[1], Shuo Yu[2], Hehan Zhou[1], Guoguang Wang[1], and Hao Zheng[3(✉)]

[1] School of Architecture, South China University of Technology, Guangzhou, China
202220104605@mail.scut.edu.cn
[2] School of Architecture, Harbin Institute of Technology, Harbin 15001, China
22S134162@stu.hit.edu.cn
[3] Architectural Intelligence Group, Department of Architecture and Civil Engineering, City University of Hong Kong, Hong Kong SAR, China
hazheng@cityu.edu.hk

Abstract. Nowadays, frequent local wars have inflicted severe damage on urban built environments, presenting substantial challenges for post-war restoration. Moreover, the scarcity of architectural imagery further exacerbates these challenges. In this context, virtual restoration techniques have shown significant advantages in speed and accuracy over traditional experience-based methods. This paper aims to explore the potential of artificial intelligence in the restoration of architectural ruins and the generation of visual predictions. Specifically, we compared the performance of pix2pix GAN and Stable Diffusion Models in architectural restoration, then further applied Stable Diffusion Models based on a modern style to the entire post-war restoration process spanning time. Notably, the optimization of its U-NET module through rule-enhanced learning and the precise mapping of image features through ControlNet improved the accuracy and coherence of restoration. Experimental findings indicate that Stable Diffusion Model surpasses traditional machine learning approaches in preserving architectural characteristics and styles, effectively addressing the issues of paired training data scarcity and minor facade feature dissipation, while astutely retaining selective elements indicative of war-induced architectural damage and aging.

Keywords: Deep learning · Stable Diffusion Model · Post-war debris restoration · Visual prediction

1 Introduction

The restoration of urban environments has garnered significant attention in architecture, particularly against the backdrop of contemporary socio-political turmoil marked by frequent local conflicts. These conflicts have inflicted severe damage on cityscapes. According to the latest Ukraine-Rapid Damage and Needs Assessment [12], jointly published by the World Bank and others, the conflict resulted in 14.6 million people

© The Author(s) 2025
H. Chai et al. (Eds.): CDRF 2024, *Symbiotic Intelligence*, pp. 262–273, 2025.
https://doi.org/10.1007/978-981-96-3433-0_23

displaced, 150,000 homes destroyed, and extensive damage to infrastructure. The total cost of dismantling and removing the debris is estimated to amount to $5.6 billion. Significantly, housing emerges as the most critical area of need, representing 17% of the total estimated costs.

However, for most war-induced building damage, previous restoration methods relied heavily on empirical knowledge, requiring architects to design analogically and speculatively from objects and memories [3]. This approach is case-by-case, difficult to extrapolate, and not scalable to other reconstruction cases. Besides, tools like Geographic Information Systems (GIS) [1], remotely sensed data, and graphic damage detection technologies [4] primarily serve to assess the current state of post-war urban areas and digitally preserve cityscape elements, rather than guiding architectural design or style decisions. Although these methods provide insights for large-scale architectural restoration, they rely on rich image data. Less research has been done on generation and restoration where image data is scarce or where textual descriptions are available.

Meanwhile, recent advances in Artificial Neural Network (ANN) technology have ushered in new possibilities for envisioning and reconstructing cityscapes across diverse domains and scales. Notable among these are Variational Autoencoders (VAEs), Generative Adversarial Networks (GANs), Autoregressive Models (ARMs), and Diffusion Models (DMs). While each model boasts unique architectural features and data requirements, they share a common challenge in training: the number of instances in the dataset [15]. Presently, Diffusion Models stand out for their ability to generate high-quality image samples within multimodal and context-free settings, showcasing their superiority in crafting detailed and realistic future cityscapes.

In the realm of architecture and urban planning, DMs have paved the way for innovative research approaches. For instance, Immanuel Koh has ingeniously applied DMs to create architectural visuals by feeding the model a food-related lexicon [8]. Siyuan Zhang has taken this innovation a step further by using the DM for the restoration of Traditional Chinese Garden designs. This process entailed creating a dataset of text and images, training the model to generate visual designs from the text, and converting these designs into 3D models for detailed restoration projects [14]. Similarly, Yuki Mugita has utilized semantic segmentation to identify individual buildings, generate mask images for these structures, and create visuals depicting the buildings' appearance post-removal and reconstruction based on DMs [10]. These pioneering studies highlight the capability of DMs to translate textual representations into detailed building designs, suggesting a rich field for further exploration and experimentation in restoration methodologies.

Therefore, this paper aims to explore the potential application of the stable diffusion (SD) model in the restoration of post-war urban built environments. Given the limited availability of historical images of post-war cityscapes, we investigate the unique advantages of the SD model in terms of data requirements and its accuracy in capturing architectural details, features, and styles through a comparative analysis with the existing generative model, pix2pixGAN. The SD model is further optimized using a rule-based reinforcement learning method to enhance the accuracy and visual coherence of the model-generated images. This approach offers a novel perspective and technical methodology for the digital, end-to-end restoration of post-war cityscapes, suggesting a promising direction for the digital whole-process restoration.

2   Methodology

GANs and DMs represent the main image generation techniques in deep learning [5]. In this study, we first conduct a comparative analysis between the pix2pix GAN and SD model, assessing their performance in post-war building restoration. As shown in Fig. 1, we compiled a dataset of pre-war and post-war cityscape images to train both models under identical conditions. Notably, given that the Stable Diffusion Model was originally a text-to-image model, we adapted it for our image-based application by integrating Low-Rank Adaptation (LoRA) for style specialization and ControlNet for maintaining input image consistency throughout the conversion process. Finally, we systematically compared the output images and loss curves from both models.

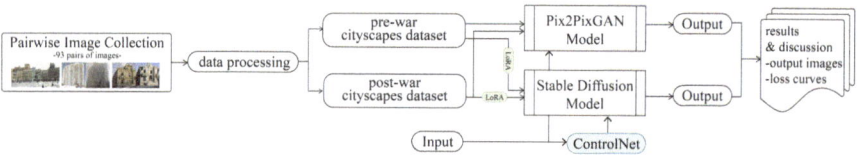

Fig. 1. Comparative analysis methodology between pix2pix GAN and SD model

2.1   Data Processing

The initial phase of our study involved curating datasets of paired pre- and post-war cityscapes. We selected images from regions significantly impacted by recent conflicts or major explosions, including Damascus, Aleppo (Syria), and Korsky (Ukraine). Our dataset encompasses a wide range of urban elements, such as building façades, streetscapes, and interior views, to offer a detailed representation of the urban devastation. The collection comprises 83 pairs of pre-war and post-war images (Fig. 2), which underwent cropping and perspective adjustment to achieve visual consistency and accurate correspondence between each pair.

Fig. 2. Part of the pre-and post-war cityscape dataset images

2.2   Training the Pix2pix Model

Pix2pixGAN, embodying a conditional Generative Adversarial Network (GAN) framework, integrates a U-Net-based generator with a discriminator characterized by a convolutional Patch GAN classifier [7]. It relies on supervised learning with pairings of input and output images. During training, these networks engage in adversarial learning, mutually enhancing to continuously optimize model performance.

For the practical implementation of pix2pix training, we opted for Google Colab, capitalizing on its high-performance GPU capabilities. Employing the TensorFlow framework, our training parameters were set as shown in Table 1. The generator's objective is to replicate images that mirror the pre-war environment accurately, while the discriminator strives to discern between the generated images and authentic pre-war photographs. Notably, the losses for both components displayed a converging pattern over the course of training, with both L1 and total losses demonstrating a decline, indicative of model refinement [2].

Table 1.   Parameter settings during pix2pix training

Parameter	Value	Parameter	Value
Batch Size	1	Learning Rate Policy	Linear Decay
λ (LAMBDA)	100	Beta_1	0.5
Learning Rate	2e-4	Total Training Steps	50000

Upon evaluating the trained pix2pix model against our test set, we encountered limitations concerning the visual clarity of the generated images. Moreover, when processing images not included in the training dataset, the outputs tended to blur, thereby obscuring distinct architectural details (Fig. 3).

Fig. 3.   Images generated in Pix2pix training with different steps

2.3   Training the Stable Diffusion Model

This phase will detail the transformation of post-war images to their restored states using the SD model, emphasizing the pivotal role of ControlNet in enhancing image generation, the tailored training of the LoRA model for various war ruin styles, and the analysis of corresponding output images and losses.

2.3.1   Precise Mapping of Image Features via ControlNet

Since the use of prompt control alone cannot satisfy the need for the details of the generated images to be consistent with the input images, we integrated ControlNet, a novel neural network concept developed by Lumin Zhang at Stanford University in 2023. The principles are explained in Fig. 4. ControlNet enhances pre-trained models by incorporating additional elemental inputs for end-to-end training, overcoming the constraints of traditional methodologies [13]. By extracting and controlling architectural details such as contours, door and window openings, and structural components, ControlNet ensures that the output closely mirrors the architectural features of the input images. For complex scenes such as those with obvious front-to-back relationships in the frame, the input image is processed using the LeReS depth information estimation method to generate a spatially more accurate image of the building.

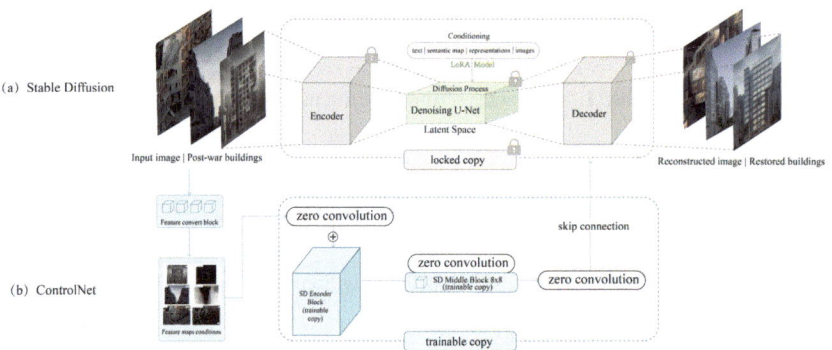

Fig. 4. Schematic diagram of the principles of Stable Diffusion and ControlNet

2.3.2   Pre-experimental Testing

Initially, our objective was to empower the SD model with the capability to generate images stylized as post-war ruins. Preliminary tests utilizing the general stylization Model 1.5 checkpoint revealed limitations in architectural detail and realism. Therefore, it is necessary to use a large number of real photos with building elements to train the Huang 1.0 checkpoint model, so that the SD model can achieve more accurate specialization of the building facade style image generation, as shown in Fig. 5.

2.3.3   Training Stylized the LoRA Model and Generating

In architectural restoration, preserving the original characteristics of the buildings is paramount. The traditional data enhancement methods tend to change the attributes such as color and geometry of architectural images to increase the number of images. We utilized the same pre-war image dataset as for the pix2pix model to train the LoRA model A. We employed the BLIP image annotation model for initial annotations, followed by manual fine-tuning, and then introduced a style control mechanism within the Spatial

Fig. 5. Images generated using different checkpoint models

Transformer of the U-Net framework [6]. The training parameters are shown in Table 2, carried out on kohya_ss. This approach safeguards the authenticity of the restoration work, ensuring originality.

Table 2. Parameter settings during SD training

Parameter	Value	Parameter	Value
Batch Size	1	Text Encoder learning rate	5e-5
Learning Rate	0.0001	U-net learning rate	0.5
Network Rank	128	total optimization steps	4000

The above concludes the comprehensive description of the final stage in the Stable Diffusion workflow for post-war building restoration.

2.4 Compare the Two Models

In our comparative analysis of the pix2pix and Stable Diffusion (SD) model, we found that SD outperforms pix2pix in generating images with superior architectural feature quality and clarity (Fig. 6). Conversely, the pix2pix model frequently produced images where different façade styles were indistinct, leading to challenges in discerning specific building features. This issue may stem from the dataset's image pairing and the variability in the number of instances. If the dataset could be more accurately paired with similar photo angles, viewpoints, and aspect ratios of the buildings, a more accurate and clearer output may ultimately be obtained.

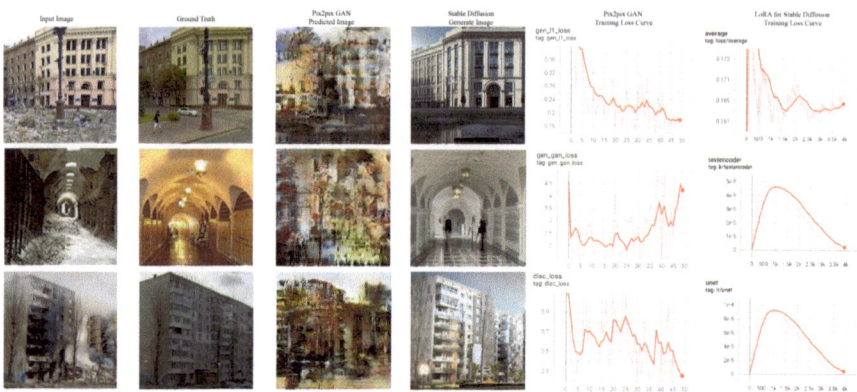

Fig. 6. Comparative illustration of restoration effects using pix2pixGAN and Stable Diffusion

3   Prediction of Post-War Restoration of Modern Urban Architectural Styles

Post-war restoration embodies a complex, long-term, and contradictory process. It demands the removal of architectural debris and the swift erection of shelters using contemporary materials, all while deeply considering the traumatic experiences of affected populations alongside local architectural and cultural legacies. Therefore, building upon the SD post-war building restoration workflow we developed, we endeavored to apply this model across the broad spectrum of post-war restoration activities, spanning various time frames. To assess the model's versatility and precision, we selected three case studies showcasing different urban perspectives—elevation, street view, and axonometric. These were utilized as test samples to evaluate the trained network's applicability and accuracy across diverse situations.

3.1   Theoretical Background

In response to a series of very complex and ever-changing problems such as the economic situation and living needs in the post-war period, it is very important to construct a sustainable reconstruction model. E. L Quarantelli has proposed the stages of post-war reconstruction including emergency shelter, temporary shelter, temporary housing, and permanent housing [11], which provides a theoretical framework for post-war reconstruction. Based on this, we use SD to visualize and simulate post-war restoration according to these stages, thus trying to predict the process of restoration possibilities of damaged urban buildings and their surroundings after the war (Fig. 7).

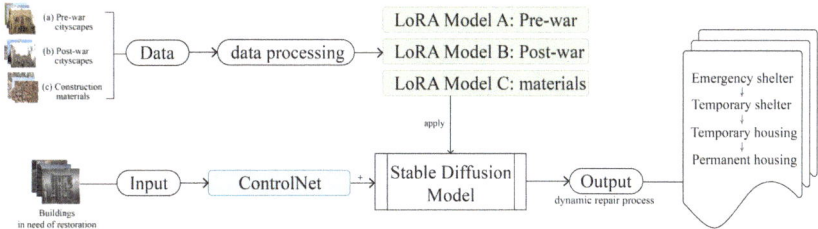

Fig. 7. Using the SD workflow post-war cityscapes restoration process

3.2 Generate 4 Stages of Post-War Restoration

In refining our SD workflow to preserve specific elements of war-induced building damage and aging, we select a subset of images depicting post-war destruction for training Lora Model B. Additionally, images depicting building materials used for temporary and rapid reconstruction are used to train Lora Model C. Together, Lora Model A, B, and C embody diverse stylistic elements, which are integrated into a pyramid structure [9]. The influence coefficients for each model are adjusted between 0 and 1, based on extensive image generation experiments aimed at enhancing the restoration of buildings. This methodology's strength lies in its ability to use a limited number of images to precisely control the generated images' parameters. Consequently, this allows for a minimized yet highly selective modulation of architectural features impacted by the war, facilitating a more nuanced restoration approach.

Specifically, according to the experience in 2–2, the basic prompts for describing the image are true, cinematic, cinematic lighting hyper-detailed, 8k, architectural photography, with a step count of 20, a DDIM sampler, and a CFG scale of 7. The ControlNet preprocessor is selected Canny. By adding different prompts, adjusting ControlNet parameters, and mixing Lora model A (pre-war), Lora model B (post-war), and Lora model C (construction material), we simulate different stages of post-war building restoration and generate different effects. The characteristics of each phase are as follows (Fig. 8):

1. **Emergency shelter:** In the early days after the war, the city was sparsely populated and a large number of buildings were abandoned and unmaintained, often covered by vegetation such as weeds. In this phase, natural elements such as "overgrown" and "deserted" were added as cues in the image generation experiments. By adjusting the influence coefficients of the Lora model A = 0.6, B = 0.4, C = 0.7, and the control mode is Balanced, Figures A-02, B-02, and C-02 are generated.
2. **Temporary shelter:** As residents return and social order gradually restores, temporary shelters are erected amidst the remnants of damaged buildings. The buildings in this period need to meet the basic living needs of the residents, such as airtightness and basic public facilities. What is changed in this stage is the style control factor in the pyramid structure, increasing the influence coefficient of LoRA models A and B, and decreasing the weight of C, generating Figures A-03, B-03, and C-03.
3. **Temporary Housing:** Meeting the elevated demands for personal privacy and quality of life, this stage sees the reinforcement of temporary structures with more durable materials like steel and concrete. Adjusting the pyramid's influence coefficients to

Fig. 8. Experimental results of all phases of post-war restoration

favor model B depicts the transition towards more permanent living solutions (Figures A-04, B-04, and C-04).

4. **Permanent Housing:** Focusing on the revival of pre-war cultural and architectural elements, this final stage utilizes a higher coefficient for LoRA model A (set to 0.8) and a control mode emphasizing the importance of our prompts. Through these adjustments, we simulate the restoration of urban spaces to their pre-war state, capturing traditional cultural motifs and modern architectural cues (Figures A-05, B-05, and C-05).

3.3 Text to Video Using Deforum

In the culmination of our study, we ventured beyond static image generation to produce a dynamic representation of the post-war building restoration process over the identified five-phase period. Leveraging the Deforum plugin for Stable Diffusion, we embarked on creating videos from textual descriptions (Fig. 9). Employing the four-stage prompt settings previously outlined, each stage's generated still image via Stable Diffusion served as a guided frame. The Deforum plugin facilitated the creation of additional, sequential frames, ensuring a seamless transition and consistency in visual detail across the video sequence. Furthermore, the introduction of rapid modularization in subsequent prompts enabled the AI to generate an expanded visualization, encompassing a bird's-eye view of the post-war cityscape in transition. See the link below for specific video effects: https://www.bilibili.com/video/BV1xk4y1T7c7/?spm_id_from=333.999. 0.0&vd_source=6a74023fb4084e0154eec90346c5703d.

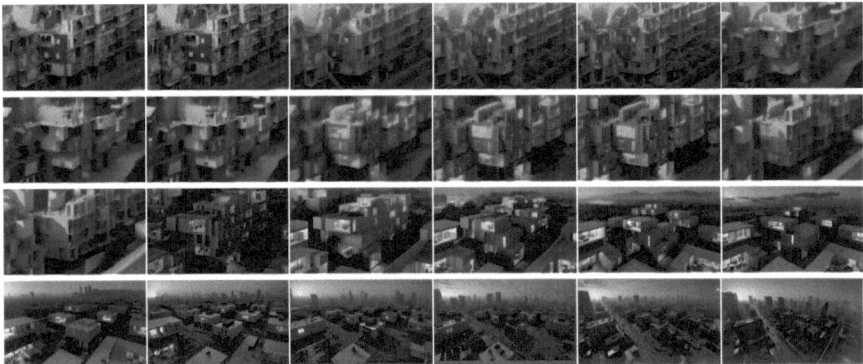

Fig. 9. Each frame of the restoration video generated using Deforum

4 Conclusion and Future Work

One of the primary objectives of this research is to explore artificial intelligence methodologies for post-war cityscape optimization generation. Based on existing machine learning methods and the restoration approach of pix2pixGAN, a workflow for post-war architectural restoration using Stable Diffusion has been established and expanded for application across all stages of the post-war restoration process, with a particular emphasis on adhering to the authenticity principle in building restoration. Our comparative analysis reveals that Stable Diffusion surpasses pix2pixGAN in generating images with superior detail, material features, and architectural style readability and interpretability. Notably, Stable Diffusion achieves this with a smaller dataset and without the need for data enhancement methods, thus preserving the original images' color and geometric attributes. This confers an inherent advantage in terms of restoration realism.

Through the lens of post-war urban landscapes in Syria, Ukraine, and other cities, our study explores the potential of AI to visually predict architectural styles in the early stages of architectural design. By enabling AI to assimilate elements of pre-war architecture and post-war debris, we facilitate the analysis and synthesis of these elements to predict urban architectural styles at various post-war stages. This methodology underpins the development of a sustainable reconstruction model over time.

Currently, our study concentrates on enhancing the clarity and recognizability of images generated by the initial machine-learning restoration method, and the scope of the study is limited to the modernist style of buildings. Human intervention and modulation of the style control weights are still required at several stages of the restoration process, which improves the visual effect of the images but limits the prediction range. Future research will explore the use of image similarity analysis tools and algorithms for recognizing transformed images and assessing their cultural appropriateness relative to pre-war buildings. Based on this, the post-war building restoration methodology will be improved and adapted to be applied to a wider range of research subjects.

References

1. Belal, A., Shcherbina, E.: Heritage in post-war period challenges and solutions. IFAC-Papers on Line **52**(25), 252–257 (2019)
2. Çiçek, S., Turhan, G.D., Taşer, A.: Deterioration of pre-war and rehabilitation of post-war urbanscapes using generative adversarial networks. Int. J. Archit. Comput. **21**(4), 695–711 (2023)
3. Chizzoniti, D.G., Lolli, T.: Urban Morphology, Identity, Heritage, and Reconstruction Processes in Middle East Post-War Scenarios: The Case of Mosul Old City. Land **12**(12), 2140 (2023)
4. Gorsevski, V., Kasischke, E., Dempewolf, J., Loboda, T., Grossmann, F.: Analysis of the impacts of armed conflict on the eastern afromontane forest region on the south sudan—uganda border using multitemporal landsat imagery. Remote Sens. Environ. **118**, 10–20 (2012)
5. Ho, J., Jain, A., Abbeel, P.: Denoising diffusion probabilistic models. Adv. Neural. Inf. Process. Syst. **33**, 6840–6851 (2020)
6. Hu, E.J., et al.: Lora: Low-rank adaptation of large language models (2021). arXiv preprint arXiv:2106.09685
7. Isola, P., Zhu, J.Y., Zhou, T., Efros, A.A.: Image-to-image translation with conditional adversarial networks. In: Proceedings of the IEEE conference on computer vision and pattern recognition, pp. 1125–1134 (2017)
8. Koh, I.: AI-Bewitched architecture of Hansel and Gretel. In: Proceedings of the 28th International Conference of the Association for Computer-Aided Architectural Design Research in Asia (CAADRIA), Vol. 1, pp. 9–18 (2023)
9. Liang, L.: Exploration and improvement of the stable diffusion model in the field of image generation. Adv. Comp. Comm. **4**(3), 163–166 (2023)
10. Mugita, Y., Fukuda, T., Yabuki, N.: Future landscape visualization by generating images using a diffusion model and instance segmentation. In: Dokonal, W., Hirschberg, U., Wurzer, G. (eds.), Digital Design Reconsidered - Proceedings of the 41st Conference on Education and Research in Computer Aided Architectural Design in Europe (eCAADe 2023), Volume 2, pp. 549–558 (2023)
11. Quarantelli, E.: Sheltering and housing after major community disasters: Case studies and general observations (No. 29). Federal Emergency Management Agency (FEMA) and Disaster Research Centre at the Ohio State University (1982)
12. World Bank: Ukraine - Third Rapid Damage and Needs Assessment (RDNA3): February 2022 - December 2023. Washington, D.C.: World Bank. Report No.: 187700 (2023)
13. Zhang, L., Rao, A., Agrawala, M.: Adding conditional control to text-to-image diffusion models. In: Proceedings of the IEEE/CVF International Conference on Computer Vision, pp. 3836–3847 (2023)
14. Zhang, S., Li, Y., Zhang, S., He, X., Tian, R.: Text-to-garden: generating traditional Chinese garden design from text-descriptions at scale with multimodal machine learning. In: Koh, I., Reinhardt, D., Makki, M., Khakhar, M., Bao, N. (eds.), Human-Centric - Proceedings of the 28th CAADRIA Conference, pp. 79–88 (2023)
15. Zhou, L., Pan, S., Wang, J., Vasilakos, A.V.: Machine learning on big data: opportunities and challenges. Neurocomputing **237**, 350–361 (2017)

Evaluating AI-Generated Design Schemes from Professional and Non-Professional Perspectives

Shuyang Li[1,2], Qian Cao[2], Junyi Wen[3], Hongxiu Liu[4], and Rudi Stouffs[1(✉)]

[1] National University of Singapore, Singapore 119077, Singapore
stouffs@nus.edu.sg
[2] Singapore-ETH Centre, Singapore 138602, Singapore
[3] China Southwest Architectural Design and Research Institute Corp, Ltd., Chengdu 610041, China
[4] Southampton University, Southampton SO17 1BJ, UK

Abstract. Artificial Intelligence (AI) technology has been widely used in architectural design. However, many architects remain skeptical about AI-generated design schemes. This study aims to explore the reasons behind architects' dissatisfaction with the AI-generated design. We invited professional and non-professional participants to evaluate three hospital design schemes: two generated by AI application EvoMass and one by experienced architects. Visual behavior data and oral description of participants during the evaluation were collected and analyzed. The results show significant differences in focal points and observation patterns. Professional participants paid more attention to the building footprint and the junction of buildings and city roads. These findings indicate that understanding of architectural design principles is crucial to enhancing Generative AI's capabilities.

Keywords: Artificial Intelligence · Generative Design · Design Evaluation · Eye-Tracking · Cognition

1 Introduction

Architectural design is a process of solving complex problems (Rowe, 1991), which demands design intelligence. AI technology can facilitate the design intelligent by tackling various challenges in the design process and generating substantial innovative solutions, rather than relying on limited rule-based approaches (Tang et al., 2019). AI-supported applications such as Delve, Giraffe, HyperAI, PlanFinder, XKool, and EvoMass can produce design schemes as references for architects. However, a significant gap remains in directly applying these generated design schemes to practical projects. While AI-generated design schemes can deceive the public and even gaining praise, architects tend to doubt about the generated design schemes, whether these design schemes meet the requirements of multi-objective optimization or based on extensive practical projects as training data (He and Yang, 2020). This skepticism may stem from

© The Author(s) 2025
H. Chai et al. (Eds.): CDRF 2024, *Symbiotic Intelligence*, pp. 274–284, 2025.
https://doi.org/10.1007/978-981-96-3433-0_24

the current limitations of AI technology in discovering the underlying design principles, especially these design principles are vague and difficult to quantify or encode (Leach, 2022).

Hidden design principles can be revealed by investigating human cognition of design schemes (Li et al., 2021; Sun et al., 2022; Zhu et al., 2022), which could enhance AI's ability to generate design schemes. This study employs a verbal questionnaire within an eye-tracking experiment to gather sufficient quantitative and qualitative data on human cognition when they evaluate design schemes. Taking a hospital design project as a case study, we invited both professionals and non-professionals to evaluate two AI-generated design schemes and one scheme contributed by senior architects. Twenty-nine participants took part in this evaluation, half of whom had academic backgrounds in urban planning or architecture, including several senior architects. Participants were asked to assess the three design schemes based on aerial view renderings, general layout renderings, and human-eye view renderings. For data analysis, gaze maps, heat maps, areas of interest derived from the eye-tracker device and the verbal reports will be integrated to investigate these participants' assessment process.

2   Experiment Design

2.1   AI-Generated Design Schemes

In this study, we chose a hospital design project to examine the quality of design schemes. Hospital design is challenging due to its complex functional requirements, diverse medical processes, and stringent design specifications. These aspects necessitate problem-solving skills based on professional knowledge and unquantifiable design principles, posing challenges for AI's generative capabilities.

The hospital design scheme created by senior architects was constructed in September 2023 (as shown in Fig. 1). The hospital includes emergency, outpatient, inpatient, medical technology, and research office departments. Apart from organizing of various departments, the architects focused on ensuring sufficient sunlight and reducing noise interference in the inpatient department.

For comparison, two design schemes were generated by EvoMass, an AI-supported Grasshopper plugin capable of generating and optimizing architectural volumes based on various performance indicators (Wang, 2022). Sunlight and noise were set as the primary optimization objectives. Numerous design schemes were produced using both additive and subtractive generative approaches. Then, we selected one design scheme from each generation approach respectively for further research (as shown in Fig. 2).

2.2   Representation of Design Schemes

Considering the differences in cognitive abilities between professionals and non-professionals (Colaço and Acartürk, 2019; Jam et al., 2022), we use the rendered images instead of line drawings to eliminate bias in evaluation. Firstly, we simplified the 3D model of the man-made design scheme, to ensure the appearance of 3D model of three

Fig. 1. Hospital design schemes contributed by senior architects

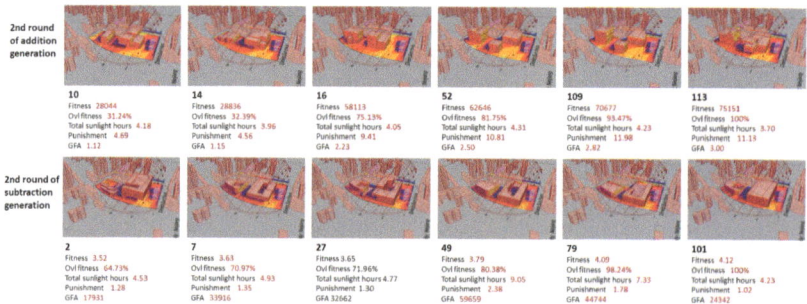

Fig. 2. Additive and subtractive generation process of EvoMass

design schemes at the same Level of Detail (LOD), with same facade elements to maintain consistency. Then, line drawings, including aerial views, layouts, and two outdoor perspectives, were exported from 3D design software to present the design schemes.

The rendered images were generated based on the line drawings using the same prompt words and parameters in a stable diffusion model. One crucial step in the image generation is the selection of sampler, which directly affects the interpretation of input images and the quality of output images. We evaluated different samplers (Euler a, DPMS Karras, DDIM, UniPC, LMS, etc.) and determined that 'Euler a' best met our requirements. We then selected the highest-quality rendered images with minimal modifications compared to the line drawings. These 12 rendered images will be presented on the monitor in the eye-tracking experiment (as shown in Fig. 3).

2.3 Participants

Individuals with or without an academic background in architecture may assess a design scheme differently. We hypothesize that those with professional knowledge can make more rational judgments compared to non-professionals. By investigating the judgment process, we aim to understand the human condition in evaluating design schemes and find out the direction to improve the generative capabilities of AI.

We recruited 29 participants, dividing them into a professional group with 15 people and a non-professional group with 14 people. Both groups had a roughly equal number

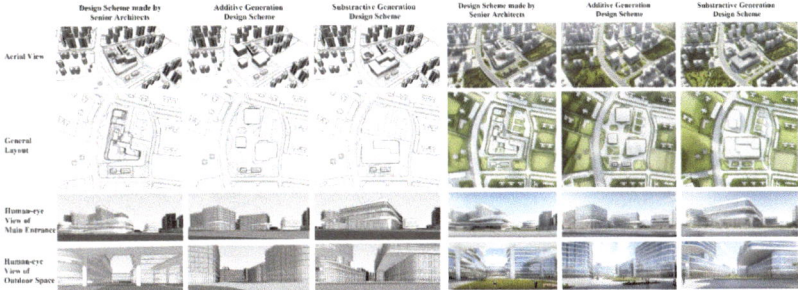

Fig. 3. Line drawings and renderings of three design schemes

of males and females, ensuring no gender bias in the experiment. The participants were well-educated and had normal or corrected-to-normal vision.

2.4 Experiment Process

A pilot experiment was conducted to test all the steps of the eye-tracking experiment. We optimized the experimental procedure according to participant feedback. Each image was displayed for 40 s, making the entire experiment process takes 8 min. In this experiment, the three design schemes were anonymized and labelled as Design Scheme A (contributed by senior architects), Design Scheme B (additive generation by EvoMass), and Design Scheme C (subtractive generation by EvoMass), ensuring the participants did not know which design scheme is generated by the AI-supported application.

To better understand human cognition, we set questions to guide participants in observing the design schemes. After hearing the questions, participants will quickly observe some areas in the images, the elements and features in these areas could affect their assessment (Keskin et al., 2023; Kim and Lee, 2020). For each rendered image, participants were required to answer 2 to 4 questions and provide brief descriptions for evaluation, such as whether the buildings looked like hospitals, whether these hospital buildings are suitable for the site, and whether the main entrance of the hospital is appropriate.

Quantitative and qualitative data were collected in this experiment for analysis. Quantitative data included gaze trajectories, heat maps of observation, and observation duration were recorded by eye-tracker device. Qualitative data were derived from oral reports, including yes or no responses, good or bad evaluations, and some keywords of evaluation. Additionally, participants were asked to rank these three design schemes (Fig. 4).

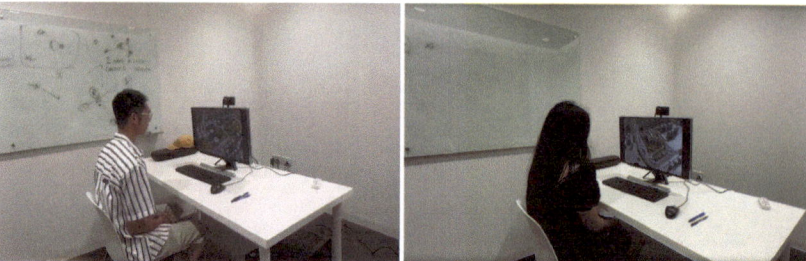

Fig. 4. Eye-tracking experiment

3   Data Analysis and Explanation of Human Cognition

3.1   Gaze Map

The gaze map illustrates the sequence of participants' observations, indicating that the areas prioritized could be crucial for exploring cognitive processes. Significant differences in observation sequences were found between professional and non-professional participants (as shown in Fig. 5). For the aerial view of Design Scheme A, when evaluating the compatibility of the hospital buildings with the site, professional participants first observed the outdoor space on the left side of the hospital and its connection with the main road. In contrast, non-professionals initially focused on the façades and roofs of the hospital buildings.

The observing trajectories also showed notable discrepancies between professionals and non-professionals (as shown in Fig. 5). In terms of the general layout of Design Scheme A, when asked whether the buildings appeared to be a hospital, professionals primarily focused on the footprint of the hospital buildings, whereas the observing trajectories of non-professionals are dispersed and random, extending even to the surrounding green spaces.

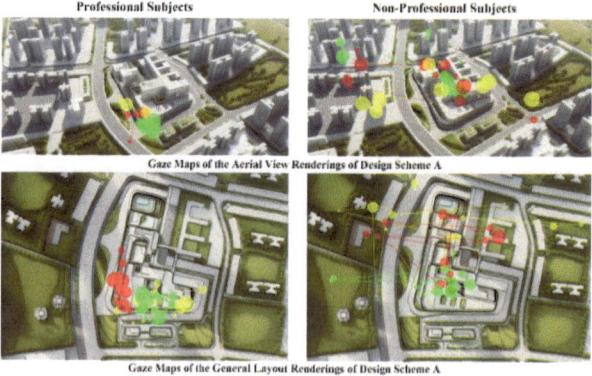

Fig. 5. Gaze maps

3.2 Heat Map

The distribution of participants' attention can be visualized using heat maps, where red indicates areas of high attention concentration, and green represents areas of lower attention. By examining these heat maps, we can analyse the different focal points of observations between professional and non-professional participants.

The heat maps of the human-eye views renderings are similar between professionals and non-professionals. However, the heat maps of the general layout renderings show significant differences (as shown in Fig. 6). For the general layout renderings of Design Scheme C, professional participants focused mainly on the largest building at the bottom of the image, whereas non-professional participants concentrated more on the buildings at the top of the image. According to the heat maps of the human-eye view renderings of main entrance of Design Scheme B, professional participants preferred to observe the building in the middle of the image, whereas non-professional participants tended to focus on the building at the left of the image. Regarding the heat maps of the human-eye view renderings of outdoor space of Design Scheme C, both professional and non-professional participants concentrated on the outdoor space between the two buildings.

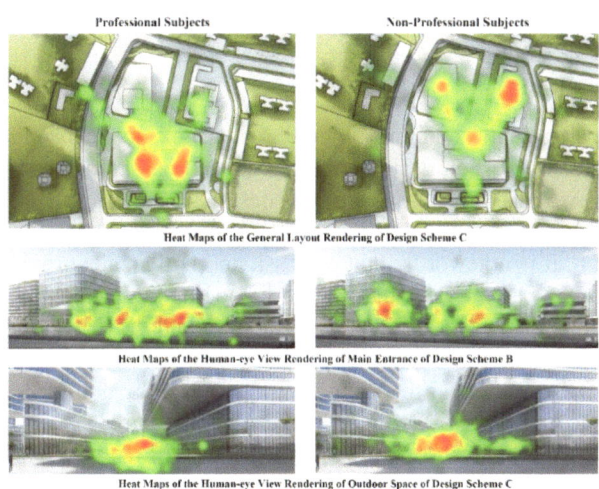

Fig. 6. Heat maps

3.3 Area of Interests

Area of Interest (AOI) analysis is an effective method to evaluates the attractiveness of various elements (such as buildings, green spaces, and roads) by accumulating the observation durations of multiple participants. Considering potential biases in understanding the general layout renderings between professional and non-professional participants, we chose aerial view renderings for AOI analysis.

Taking the aerial view rendering of Design Scheme C as an example, we compared the aggregated observation durations across different areas between professionals and non-professionals. Non-professional participants spent more time observing the "green space" (green areas in Fig. 7) and "hospital building" (red areas in Fig. 7) than professional participants. However, the observation duration on "entrance and outdoor space" (blue areas in Fig. 7) of professional participants reached approximately 16,500 ms, nearly double the observation duration of non-professional participants, indicating that this area contains important or appealing information for them.

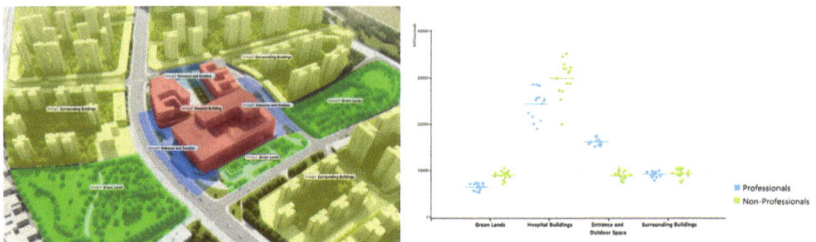

Fig. 7. Observation duration analysis for Design Scheme C based on the AOIs

3.4 Verbal Reports

To explore the participants' opinions on the three design schemes and determine their ability to distinguish between design schemes created by senior architects and those generated by AI, we asked them to score each design scheme from 1 to 3. The average scores given by the professional and non-professional participants were calculated, as depicted in Fig. 8. It is noteworthy that professional participants provided consistence evaluations across the aerial view, the general layout, and the human-eye view renderings. They regarded the Design Scheme A as the best and the Design Scheme B as the worst, indicating that Design Scheme A likely made by senior architects. Conversely, we discovered a discrepancy in the averages scores given by non-professional participants for different renderings of Design Scheme A and Design Scheme C. Nevertheless, non-professional participants deemed Design Scheme B as the worst, aligning with the evaluation of professional participants.

The verbal reports provided valuable information beyond our expectations in this experiment. Most professional participants rated Design Scheme B the lowest according to the aerial view rendering, stating the building mass were too independent of each other and lacking the necessary internal transport connections. However, some non-professional participants gave positive evaluations for Design Scheme B according to the aerial view rendering, appreciating the scattered distribution of buildings. Besides, both professional and non-professional participants evaluated the outdoor spaces from thermal comfort and healing perspectives, but some professional participants noting the necessity of treating outdoor spaces as assembly areas for emergency evacuation.

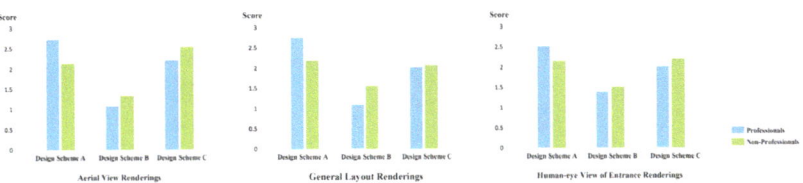

Fig. 8. Average scores for three design schemes

4   Discussion

4.1   Cognitive Deviation and Judgement Process

Numerous disparities exist in the evaluation of design schemes between professional and non-professional participants, which may stem from cognitive deviation between the two groups. Compared with non-professional participants, professional participants generally offer reasonable assessments through a rational analysis process. This suggests that AI-generated design schemes are less likely to deceive professionals, particularly experienced architects.

Based on the analysis of verbal reports, it becomes evident that professional participants grasp the overall layout renderings, whereas many non-professional participants express confusion when observing such renderings. Consequently, professional participants can extract critical information such as the building footprint and approximate location of the main hospital entrance from the general layouts. Although the cognitive deviation could be reduced when the professionals and non-professionals observe the aerial view rendering of design schemes, the professional participants could speculate the position of various department of the hospital and the organization of these departments, which is almost impossible for non-professionals participants to achieve it.

Furthermore, the observation duration data revealed that professional participants allocate over 60% of their time to some key areas such as the entrance of the hospital, the tower buildings, and the podium building, whereas non-professional participants' attention tends to wander towards green spaces and surrounding buildings (as shown in Fig. 9). Moreover, we found the reciprocal observation behaviour predominantly occurs among professional participants rather than non-professional participants (as shown in Fig. 5). Professionals repeatedly observed the podium buildings of hospital, the adjacent outdoor spaces, and the city roads. Their verbal reports also demonstrate a heightened concern for the connectivity between buildings and city roads.

4.2   The Deficiencies of AI-Generated Design Scheme

Through the analysis of participants' visual behaviour and verbal reports, we investigated how professional participants distinguish the origins of design schemes. Professional participants identified significant deficiencies in AI-generated design scheme that were absent in those crafted by experienced architects, leading them to determine the design schemes were generated by AI-supported application. These deficiencies, while

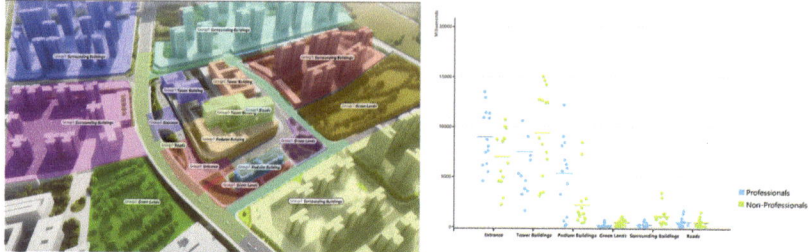

Fig. 9. Observation duration analysis for Design Scheme A based on the AOIs

overlooked in AI-supported generation methods, constitute crucial aspects of hospital design.

Firstly, the inappropriate connection between buildings and city roads represents a critical flaw in AI-generated design schemes. Professional participants emphasize their importance of the outdoor spaces between the buildings and the city roads in both aerial view renderings and general layout renderings. Given the concentration of people and vehicles at hospital entrances, these outdoor spaces serve not only as buffer zones but also as crucial access points. However, quantifying and defining the accessibility and buffer capacity pose challenges.

In addition, insufficient internal transportation system within hospital buildings as another notable deficiency for professional participants. For instance, in the Design Scheme B, the four buildings are standalone and dispersed across the site, a feature that drew attention from professionals. In fact, hospital buildings typically feature multiple connections, such as overhead bridges or corridors, to accommodate patient visitation routes and optimize the efficiency of medical staff.

5   Conclusion

This study investigated professionals' and non-professionals' evaluation of hospital design schemes through an eye-tracking experiment, utilizing a combination of quantitative and qualitative methodologies. Twenty-nine participants were recruited to assess one design scheme contributed by senior architects and two design schemes generated by AI-supported application. The research elucidates the visual behaviour of both professional and non-professional participants during their assessment of the design schemes, and discuss the influencing factors and judgement process of their evaluation. The finding underscores that individuals with professional knowledge are adept at identifying deficiencies in the AI-generated design schemes and providing reasonable evaluations. A notable revelation is that shortcomings related to site planning and internal transportation within the hospital significantly impact the quality of design schemes, which hitherto have not been encoded in generative AI. These results imply the necessity for deeper comprehension of professional evaluations of design schemes and exploration of qualitative aspects to enhance AI's capacity for generating architectural design schemes.

Acknowledgement. The research was conducted at the Future Cities Lab Global at the Singapore-ETH Centre, which was established collaboratively between ETH Zurich and the National

Research Foundation Singapore. This research is supported by the National Research Foundation Singapore (NRF) under its Campus for Research Excellence and Technological Enterprise (CREATE) programme.

References

Colaço, C.A., Acartürk, C.: Visual behaviour during perception of architectural drawings: differences between architects and non-architects. In: Design Computing and Cognition'18, pp. 381–397. Springer International Publishing (2019)

Jam, F., Azemati, H.R., Ghanbaran, A., Esmaily, J., Ebrahimpour, R.: The role of expertise in visual exploration and aesthetic judgment of residential building façades: an eye-tracking study. Psychol. Aesthet. Creat. Arts **16**(1), 148 (2022)

He, W., Yang, X.: Artificial intelligence design, from research to practice. In: Proceedings of the 2019 DigitalFUTURES: The 1st International Conference on Computational Design and Robotic Fabrication (CDRF 2019) 1, pp. 189–198. Springer Singapore (2020)

Keskin, M., Krassanakis, V., Çöltekin, A.: Visual attention and recognition differences based on expertise in a map reading and memorability study. ISPRS Int. J. Geo Inf. **12**(1), 21 (2023)

Kim, N., Lee, H.: Evaluating visual perception by tracking eye movement in architectural space during virtual reality experiences. In: Human Interaction, Emerging Technologies and Future Applications II: Proceedings of the 2nd International Conference on Human Interaction and Emerging Technologies: Future Applications (IHIET–AI 2020), April 23–25, 2020, Lausanne, Switzerland, pp. 302–308. Springer International Publishing (2020)

Leach, N.: AI and the limits of human creativity in urban planning and design. In: Artificial Intelligence in Urban Planning and Design, pp. 21–37. Elsevier (2022)

Li, J., Wu, W., Jin, Y., Zhao, R., Bian, W.: Research on environmental comfort and cognitive performance based on EEG+ VR+ LEC evaluation method in underground space. Build. Environ. **198**, 107886 (2021)

Rowe, P.G.: Design thinking. MIT Press (1991)

Sun, C., Li, S., Lin, Y., Hu, W.: From visual behavior to signage design: a wayfinding experiment with eye-tracking in satellite terminal of PVG airport. In: Proceedings of the 2021 DigitalFUTURES: The 3rd International Conference on Computational Design and Robotic Fabrication (CDRF 2021) 3, pp. 252–262. Springer Singapore (2022)

Tang, Y.C., et al.: A review of design intelligence: progress, problems, and challenges. Frontiers of Information Technology & Electronic Engineering **20**(12), 1595–1617 (2019)

Wang, L.: Workflow for applying optimization-based design exploration to early-stage architectural design–Case study based on EvoMass. Int. J. Archit. Comput. **20**(1), 41–60 (2022)

Zhu, S., Qi, J., Hu, J., Hao, S.: A new approach for product evaluation based on integration of EEG and eye-tracking. Adv. Eng. Inform. **52**, 101601 (2022)

A Symbiotic Database Framework for Chinese Ancient City Spatio-Temporal Information Modelling

Xin Yan[1], Keyang Tang[1(✉)], Mengyao Li[2], and Zheng Zhang[3]

[1] Future Laboratory, Tsinghua University, Beijing 100084, China
tangkeyang@mail.tsinghua.edu.cn
[2] Academy of Arts and Design, Tsinghua University, Beijing 100084, China
mengyao-21@mails.tsinghua.edu.cn
[3] School of Architecture, Tsinghua University, Beijing 100084, China
zhangzhe23@mails.tsinghua.edu.cn

Abstract. Researchers of ancient cities often rely on textual descriptions, limited archaeological excavations, and scattered map materials to infer urban forms due to the absence of intuitive materials. However, urban descriptions in historical texts are often composed of interconnected information. The intricate and complex relationships between these pieces of information form a three-dimensional and multidimensional spatiotemporal model of the ancient city. In the conventional research context, researchers need to manually combine these historical materials to make reasonable conjectures, but such conjectures are often limited by the researcher's personal data and view, making it difficult to form a rapid and precise generation mechanism. This research introduces big data processing and large language model technologies to construct an effective symbolic database framework for conveniently analyzing, demonstrating, and utilizing Chinese ancient city information in historical texts. This framework allows for the rapid generation of different "representations" of ancient city at specific historical moments, facilitating scholarly assessments of historical scenarios, which promotes the integration of computational technology, historical urban studies, and architecture, and offering new tools and perspectives for related academic work.

Keywords: Spatio-temporal Information Modelling · Historical Textual Data · Urban Morphology · Symbiotic Database · Chinese Ancient City

1 Introduction

In contemporary times, morphological derivation of ancient cities is typically executed through the cross-verification of various urban information or descriptions from different sources, such as historical records, biographies, stories, poetries, maps, archaeological excavations, etc. A lot of significant research about Chinese urban morphology have been developed based on the laborious artificial management and deductions with diverse data. In theory, these pieces of information with varying attributes collectively constitute a

H. Chai et al. (Eds.): CDRF 2024, *Symbiotic Intelligence*, pp. 285–296, 2025.
https://doi.org/10.1007/978-981-96-3433-0_25

multi-dimensional spatio-temporal model of ancient cities. The accuracy of this model may never truly be verified. What researchers typically obtain in the study of ancient cities are not specific conclusions but an intersection of numerous possibilities within the intricate web of spatio-temporal relationships.

However, there are always many information gaps among the fragmentary historical texts, which requires a lot of human labours and times to search the relevant spatial-temporal information coincidence (Fig. 1) [1]. And as a result, constrained by the limited data and manual computational capacity of individual researchers, such time-space deduction work often falls to a slow and tedious job, and fail to generate a sufficiently accurate ancient city model, which involves the dynamics of changing people, objects, and events. Taking *Chang'an* and *Luoyang* in *Sui and Tang* Dynasties as examples, there are many scholars from the fields of archaeology, history, and architecture conducting textual excavations, like collecting information about ancient cities [2–5], deducting history events [6–11], or constructing historical city sceneries [12–20].

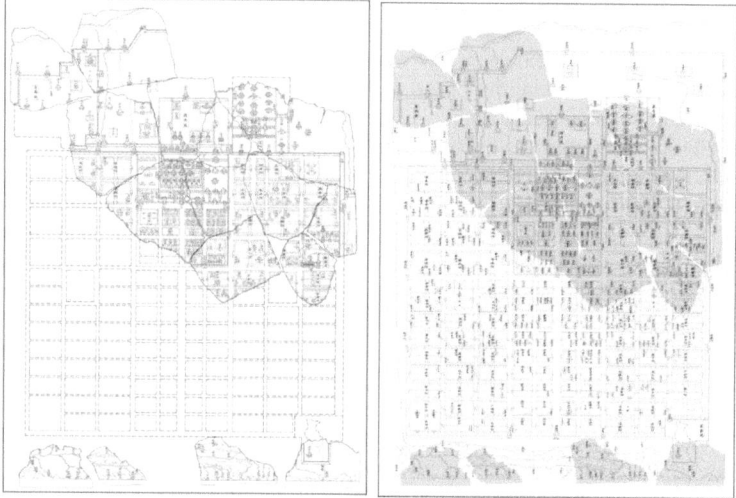

Fig. 1. Remaining fragments (left) and supplementary drawingm (right) of LV Dafang's Chang'an City Map [1].

In consideration of the above architectural research situations, this paper introduces an innovative symbiotic database framework based on interactive database technology to easily save, update and analyze the Chinese ancient city information. With this framework, it is convenient for architectural or ancient urban researchers to establish a multi-dimensional, symbiotically updated, cross-indexable digital database for Chinese ancient cities. Specifically, this study chooses the typical and famous Chinese traditional city of *Chang'an* in the *Sui and Tang* dynasties as the sample to detailly illustrate the database framework.

2   Historical Text Analysis

2.1   Recognition and Organisation of Textual Information

The information processing of various forms of textual data is a primary issue addressed in this study. This portion of data should not be confined to historical texts but should also encompass diverse data formats such as historical maps, tablet inscriptions, book artworks, etc. In response to current issues in historical texts, such as abstraction in content, difficulty in digital reading, and noise in excessive information, this study has developed three methods for constructing historical text data (Fig. 2).

The first method sets up an online text input platform, which allows users to taxonomically input textual information such as time, place, people, events, etc.

Secondly, the method involves using Computational Vision (CV) algorithms to clean and mine existing historical text data. Since most existing historical texts are in the form of scanned images, this study introduces digital technologies such as Digital Image Processing Technique (DIPT) and Optical Character Recognition (OCR) to identify text messages.

Additionally, for data related to historical maps, the third method is implemented with techniques such as coordinate linear interpolation to achieve the mapping and alignment of ancient maps with geographic information.

After collecting the digital information, an algorithm then forms internal associative logic within the data, creating an interwoven social network of people and events in spatio-temporal relationships.

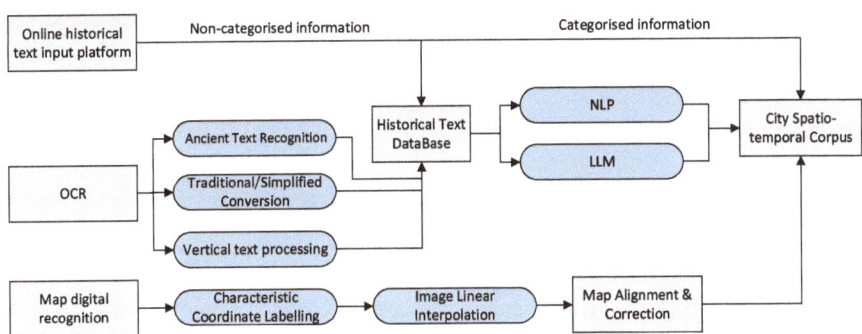

Fig. 2. Basic framework of textual information recognition and organization.

2.2   Text Semantic Analysis

Historical texts passed down from various dynasties in China are replete with materials that describe cities. Among these, some focus on recording changes in urban form or the evolution of built-up areas, such as *"Zhou Li Kao Gong Ji"*, *"Tang Liang Jing Cheng Fang Kao"*, *"Liang Jing Xin Ji"*, *"Chang'an Zhi"*, and *"Chang'an Zhi Tu"*; others emphasize biographical historical books or fictional notes that narrate people, events, and objects within urban spaces, such as "*Dong Jing Meng Hua Lu*", "*Luo Yang Qie Lan Ji*", "*Tai Ping Guang Ji*", "*Jiu Tang Shu*", "*Xin Tang Shu*", "*Quan Tang Wen*", etc. The former typically record temporal or spatial information about urban elements such as states, prefectures, counties, cities, towns, gathering places, pavilions, streets, neighborhoods, markets, courtyards, gardens, paths, and roads. The keywords related to time and space in these texts are relatively clear and fixed, and the structural similarity of these texts facilitates recognition and processing by computer programs. In contrast, the latter are often originally intended to narrate stories occurring within cities, not purely for recording urban temporal and spatial information. As a result, they contain a significant amount of textual "noise" and their sentence structures are complex and varied, making it difficult to perform simple procedural processing for information topology.

Combining classical Natural Language Processing (NLP) with Large Language Models (LLM), this research has developed a semantic analysis algorithm framework for historical text data, which is applied to disassemble and analyze the meaning groups in Classical Chinese texts. It accomplishes functions such as word segmentation, named entity recognition, and dependency parsing for sentences in Classical Chinese. Currently, toolkits for semantic analysis specifically targeting Ancient Chinese are not common. After comparative evaluations, this study primarily adopted two tools: JIAYAN and HanLP 2.0 to conduct a secondary calibration with the historical literature corpus related to cities that the team had organized. JIAYAN is an NLP toolkit focused on Ancient Chinese, supporting the construction of a Classical Chinese lexicon, word segmentation, part-of-speech tagging, sentence breaking, and punctuation [21]. HanLP 2.0, on the other hand, integrates a lot of language models and algorithm toolkits for Chinese and possesses strong extensibility [22]. This research further refined and corrected the algorithms based on this foundation (Fig. 3).

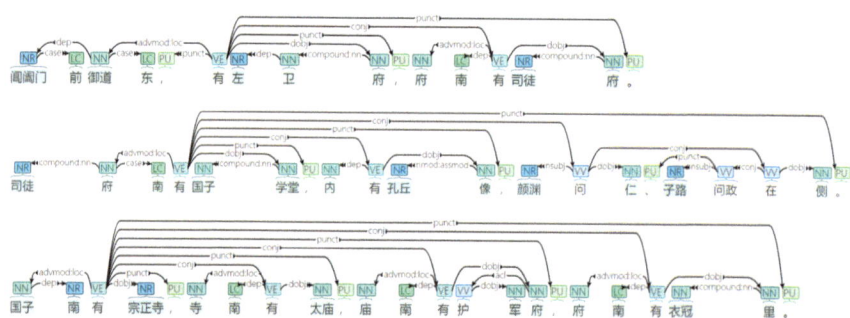

Fig. 3. Semantic analysis of "Luo Yang Qie Lan Ji" by HanLP2.0 [22].

3   Urban Spatio-Temporal Modelling

3.1   Urban Spatial and Temporal Information Archiving

Through the analysis and processing of historical text data, the research preliminarily achieved the capability to extract basic urban spatio-temporal information from textual materials within historical documents. Building on this foundation, the study archived spatial information from the text corpus from the perspective of architectural urban theory.

Spatio-temporal information in historical texts always exists in the states between precise and vague, which needs more references to generate and cross-corroborate the final city image. There is one specific example to show how to extrapolate ancient city spatial forms based on historical texts:

(1) In " *Chang'an Zhi Tu*", there is one description of "*Xing Ning Fang* has one cross-road in the middle with only three gates on the walls except the north one. The southeast part of southeast district is *Quan Nansheng's Residence*, while *Qing Chan Temple* was built on the southwest side. *Yao Yuanchong's Residence* was located in the southwest district"[1]. Thus, as shown in Fig. 4 (1), *Quan Nansheng's Residence* and *Qing Chan Temple* are located in the southeast and southwest districts respectively with clear relative positions, while *Yao Yuanchong's Residence* is only recorded in the southwest without pointing out the specific location, and placed in the center for the time being;

(2) Based on another sentence of " *Princess Taiping's Residence* was located on the eastern side of *Yao Yuanchong's Residence*"[2], the spatial form comes to Fig. 4 (2);

(3) According to another passage in "*Chang'an Zhi*", "(*Princess Taiping's*) Residence's north is *Wang Maozhong's Residence*"[3], we can further determine the relationship between *Yao Yuanchong's Residence*, *Princess Taiping's Residence* and *Wang Maozhong's Residence*, and get the illustration of Fig. 4(3);

(4) However, as in Fig. 4(4), due to the uncertainty of the scope of the ancient descriptions, we cannot identify whether the so-called "north direction" was due north, or east-north, or west-north, or farther north. Also, we have not yet associated the official ranks of these figures, and thus cannot confirm how much their mansions should have been, so we cannot determine the final form of *Xing Ning Fang* based on the above text alone. Therefore, based on the above texts alone, we cannot determine the final form of *Xing Ning Fang*, and the system needs to be supplemented with more corpus.

This study introduces NLP and LLM technologies to deconstruct keywords and identify named entities in textual materials, categorizes words according to their nature within the city. It then explores the implicit relationships between textual materials based on multiple temporal and spatial clues, further completing the archiving of urban spatio-temporal information. The logical composition and deductive rules of ancient cities,

---

[1] "兴宁坊，坊内有十字街，但无北门。十字街分坊为四部;东南部东南侧泉男生宅,西南侧清禅寺;西南部姚元崇宅。"——《长安志图》

[2] "(姚元崇宅)东本太平公主宅"。——《长安志》

[3] "(太平公主)宅北特进王毛仲宅。"——《长安志》

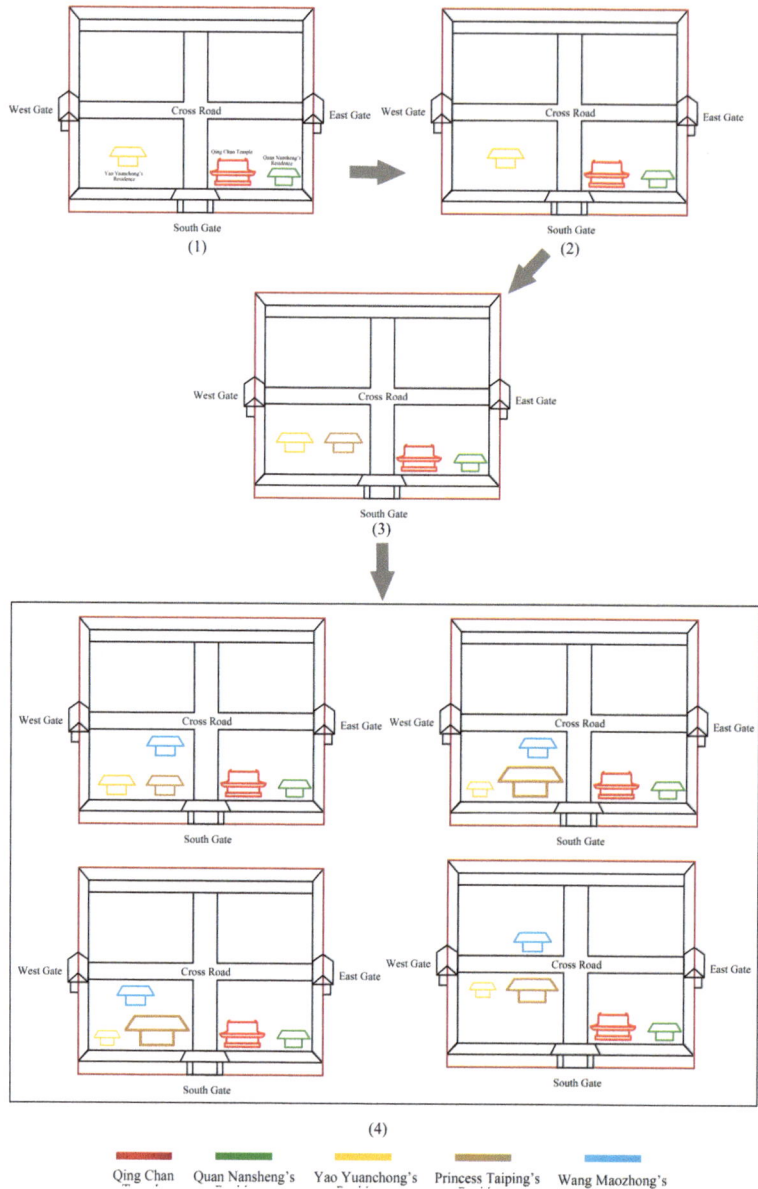

Fig. 4. The process of deducing city form based on historical texts.

along with the diachronic/synchronic relationships between human events and spaces constitute important constraints for historical research. By concentrating fragmented information into tangible and three-dimensional urban forms, the algorithm can generate massive amounts of data on possible spatio-temporal logic relationships, which contributes to solve the urban historical problems that previously could not be addressed with an isolated and local perspective. As the number of historical texts increases, the temporal and spatial information of the city will overlap and validate each other, thereby continuously advancing the updating and calibration of the relational database. This content will run through the entire research process (Fig. 5).

Fig. 5. Diagram for Chang'an city form deduction.

3.2 Object-Oriented Programming for Urban Information

After utilizing the program to analyze the spatio-temporal information of the city, this study also constructed program classes, for urban elements like *Li Fang*, *building*, and *person*. Each class contains its own properties and methods. For example, the *Li Fang* class properties include geometric information such as east-west distance, north-south distance, rectangular outline, diagonal lines, boundary lines, corner coordinates, center coordinates, width of central/peripheral roads, and wall thickness, as well as historical information like a list of *Li Fang* names, related person lists, building lists, internal area lists, and event lists. Its methods include geometric operations such as internal area split, moving, scaling, boundary modification, shape drawing, and historical information operations like updating information and reading/writing information tables.

Through above object-oriented programming, researchers can conveniently handle and manage a large amount of scattered historical information with the help of computer programs. They can also save the information in associated locations, and updating them in real-time as the text increases. For instance, if the program system firstly creates a *Li Fang* class named *"Ju De Fang"*, the contents of this class will gradually be updated and added with expansion of textual readings:

(1) the person list includes: *Xianyu Zunyi, Yuchi Le, Liu Xiangdao, Du Yuanwei, Zhe shi, Su Yuhua, Cao Mingzhao*;

(2) the building list includes: *Yifa* Temple, *Bao'an* Temple, *Ningguan* Temple, *Han Yuanqiu* remnant site, *Xiantian* Temple, *Baoguo* Temple, *Baochang* Temple, *Xianyu Zunyi* House, *Yuchi Le* House, *Liu Xiangdao* House, *Du Yuanwei* House, *Zhe Shi* House, *Su Yuhua* House, *Cao Mingzhao* House, etc.

Moreover, the above person and building lists are also another types of class, which contains various pieces of information related to persons or buildings for the program to modify.

After constructing the classes for urban elements, this study also set up data storage methods suitable for different carriers. Among them, basic storage formats represented by CSV, EXCEL, and TXT, which are convenient for researchers who are not familiar with computer process. While relational databases represented by MySQL are suitable for exploring the establishment of an application layer stage. With these methods, the study generates a "Urban Spatio-temporal Information Correlation Database" where each piece of information in the database mainly includes some or all aspects such as time, place name, orientation data, height, characters, objects, events, landmark buildings, and site features (Fig. 6).

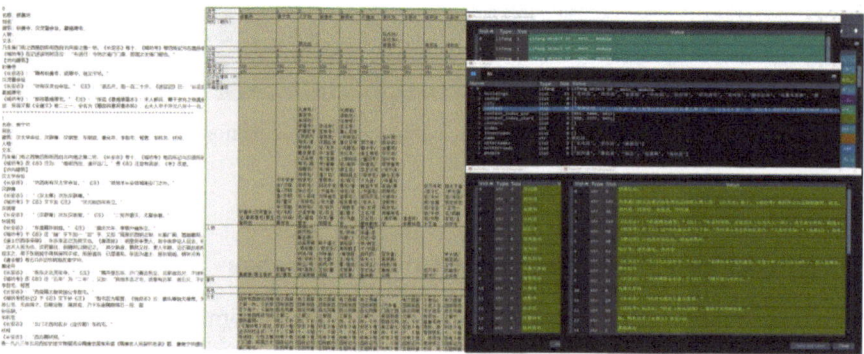

Fig. 6. The interfaces of Chang'an information database.

4   Interactive Applications

4.1   Historical Text Analysis Training Platform

This study also developed an online historical text analysis and training platform on the web based on Streamlit toolkit [23]. We encapsulated python code and elements such as large-scale language pre-training models into the online server, allowing users to interactively input their own urban texts through a simple interface for named entity recognition and information classification. The interactive web page also provides a user modification function to manually correct calculation results in a timely manner, and the modification process will also be retained as an important reference for further correcting and fine-tuning the language model.

As shown in Fig. 7, after the user has completed loading the built-in Chang'an city block information, they can input the text in the dialog box. For example, after entering part of the content about "*Xuan Yang Fang*" which reads "To the south of the West Gate is the residence of *Gao Xianzhi*. To the north of the East Gate is the residence of *Li Qiwu*. To the northwest of the street is the residence of *Li Hai*.", the program can automatically identify the relevant area names within the blocks (south of the West Gate, north of the East Gate, northwest of the street) and buildings (residences of *Gao Xianzhi*, *Li Qiwu*, *Li Hai*) and correspond them accordingly. After the user confirms the information, the program will update the new data into the previously loaded Chang'an city block information database for the "*Xuan Yang Fang*" class, and at the same time, it will also instantly update the local EXCEL spreadsheet content.

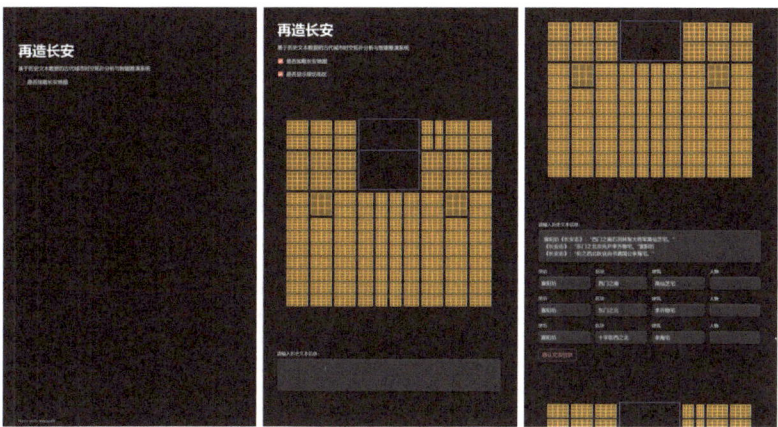

Fig. 7. The interfaces of online historical text analysis training platform.

4.2 Interactive Visualisation of Urban Statistics

In addition to the interactive web pages, this study offers numerous functions for visualizing and speculating on ancient city information. Based on the forementioned database of *Chang'an* city, this research retrieves the relationships between typical historical figures, buildings, and neighborhoods from historical texts, and predicts the ranges of activities in *Chang'an* city using a self-developed urban spatial topology network algorithm. By integrating architectural information across time and space dimensions, one can observe the distribution and evolution of different functional buildings in *Chang'an* city, add some information visualization interactive functions to the related database using tools like pyecharts [24] (Fig. 8), and simulate scenes of *Chang'an* city blocks using proxy models (Fig. 9), etc.

Simultaneously, this research has also compiled the classic *Chang'an* city planning theory into an interactive program that allows users to input parameters using the Pygame toolkit [25]. Additionally, this interactive interface has been integrated with the related database, employing a large language model to interpret the historical text entered by users and dynamically display the results on the map of *Chang'an* (Fig. 10).

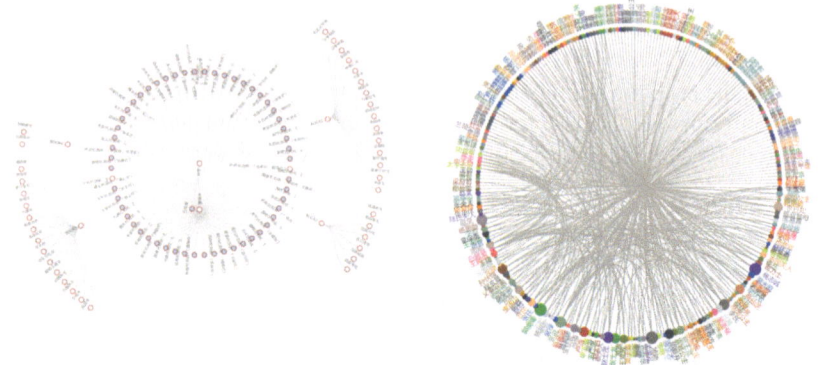

Fig. 8. Pyecharts-based interactive web page for Chang'an data.

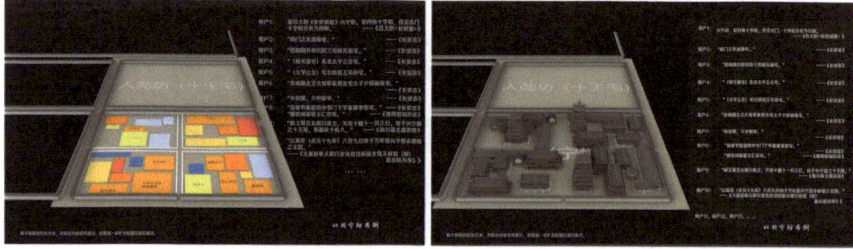

Fig. 9. Li fang modelling for Chang'an City.

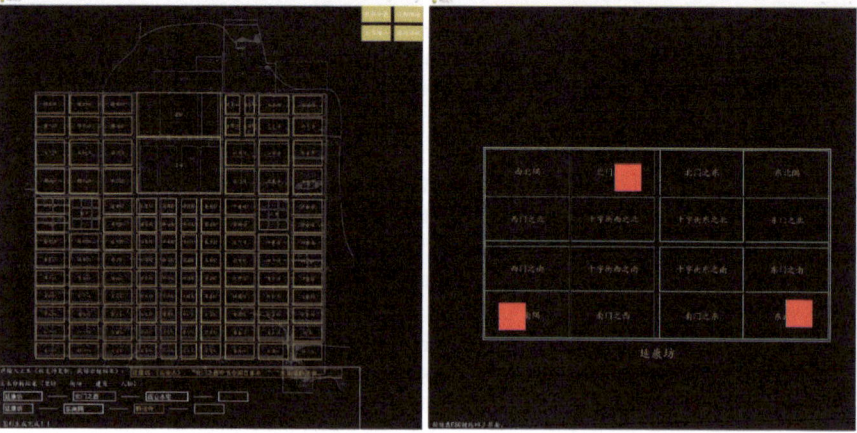

Fig. 10. Prototype of the interaction application program.

5 Conclusions

This research combines digital information technology with relevant architectural theories to generate a system of "possibilities" for historical cities, bringing out the implicit temporal and spatial elements and urban forms from historical texts. The value of this research lies not only in providing ancient urban scholars with efficient research tools for quickly obtaining synchronic and diachronic visual information but also in offering architectural practitioners a simulation platform for historical districts. This helps them dynamically understand the evolutionary patterns of historical districts and design neighborhoods that fit within specific urban formal logics based on similar constraints. Furthermore, it can support the related industries of urban cultural heritage, collaborating with the metaverse, gaming, film, virtual communities, and other online ecosystems to create urban scenes with historical significance. This expands the cultural influence of historical cities, promoting the contemporary transformation and living inheritance of cultural heritage.

References

1. Shusheng, W., Kai, C., Kai, W.: Study on supplementary drawing of Lv Dafang's Chang'an city map. City Planning Review (2016)
2. Guoqiang, G.: Study on Buddhist temples in Chang'an city of Sui and Tang dynasty. Cultural Relics Publishing House (2006)
3. Pingfang, X.: Essays on Chinese urban archaeology. Shanghai Ancient Books Publishing House (2015)
4. Ling, Q.: A preliminary study on the analysis of potential surviving possibilities of the Chang'an city Square site in Sui Daxing Tang. Xi'an University of Architecture and Technology (2017)
5. Institute of Archaeology, Chinese Academy of Social Sciences: Sui and Tang dynasty Chang'an city site Archaeological data compilation. Cultural Relics Press (2017)
6. Luoyang Institute of Cultural Relics and Archaeology: Study on the gate sites of Sui and Tang dynasty Luoyang city. Sanqin Press (2022)
7. Xinjiang, R.: Sui and Tang dynasties: gender, memory and beyond. Sui Tang Chang'an: Gender, Memory and Beyond (2010)
8. Ailing, X.: Study on the historical texts of the conservation plan for the site of Chang'an city in Sui and Tang dynasties. Science Press (2014)
9. Hongnian, Y: The genealogy of the neighbourhoods of the two capitals of Sui and Tang dynasties. Shanghai Ancient Books Publishing House (1999)
10. Hongnian, Y.: An examination of the two capitals of the Sui and Tang dynasties. Wuhan University Press (2005)
11. Lu, G.: Area design Qin-Han, Sui and Tang Chang'an area regional spatial order construction architectural design. China Architecture & Building Press (2019)
12. Tatsuhiko, M.: Sui and Tang dynasty Chang'an and the comparative metropolitan history of East Asia. Northwestern University Press (2018)
13. Xi-nian, F.: Exploration of the planning methods of Chang'an and Luoyang city of Sui and Tang dynasties. Cultural Heritage **03**, 48–63 (1995)
14. Xi-nian, F.: Study on Ancient Chinese urban planning, layout of building groups and architectural design methods. China Architecture and Building Press (2015)

15. Hui, W.: Discussion on the planning methods of Sui and Tang dynasty Luoyang Li Fang. J. Chin. Architectural Hist. **1**, 17 (2008)
16. Wang, H., Cao, K.: Re-consideration of the planning method of Chang'an neighbourhoods in Sui and Tang dynasties. Urban Planning
17. Xiao, C.: New interpretation of design modelling in Chang'an Sui Tang dynasties. Urban Plann. **10**, 9 (2017)
18. Caiqiang, W.: Digital reconstruction of Tang Chang'an. China Architecture Industry Press (2006)
19. Lu, W., Yuan, Y., Wei, D.: Spatial display system of Sui and Tang dynasty Chang'an city site. Popular Archaeol. **12**, 8 (2019)
20. Renhe, J., Lian, Q.: Echoes of the past: a three-dimensional recreation of the Loudangshan Grottoes. Huaxia Geogr. **11**, 8 (2011)
21. Jiajie, Y.: Jiayan. (2019). https://github.com/jiaeyan/Jiayan
22. Han, H., Jinho, C.: The stem cell hypothesis: dilemma behind multi-task learning with transformer encoders. arXiv preprint arXiv:2109.06939. (2021)
23. Streamlit: Streamlit. (2022). https://streamlit.io
24. Pyecharts: Pyecharts. (2023). https://github.com/pyecharts/pyecharts
25. Pygame: Pygame. (2023). https://github.com/pygame/

Optimizing Ionic Style Facade Creation by Integrating Shape Grammars into Stable Diffusion

Yichao Shi[✉] and Chunlan Wang[✉]

Georgia Institute of Technology, Atlanta, GA 30332, USA
{yshi431,cwang932}@gatech.edu

Abstract. Within the domain of AI-driven architectural design, the task of faithfully depicting specific architectural styles, especially in the design of building facades, continues to be a substantial difficulty. This work introduces a novel method that combines shape grammars, which is known for representing design rules, with stable diffusion models to tackle this problem. The specific focus of this study is on the Ionic style. The research pioneers a way to generate building facades by merging the technological capabilities of AI models like LoRA and Dream-Booth with the image-generating abilities of Stable Diffusion. This entails a demanding procedure of creating specialized datasets, training AI models using these datasets, and doing a thorough comparative analysis to assure the accuracy and visual authenticity of the designs in the Ionic style. The main contribution of this study is its illustration of how shape grammar may direct AI models to generate architectural facades that are of excellent quality and consistent in style. The revised model shows the ability to create Ionic-style facades with higher accuracy and AI that follows conventional architectural rules. This study highlights the capacity of AI to enhance the visual elements of architectural design, closing the divide between contemporary computational methods and conventional architectural sophistication.

Keywords: Artificial Intelligence · Shape Grammars · Stable Diffusion · Ionic Style

1 Introduction

The integration of Artificial Intelligence (AI) in architectural design marks a significant shift, offering the potential to enhance creativity, streamline design processes, and foster innovative solutions in facade development [1]. This technology merger blends traditional architectural styles, like the Ionic, with modern, data-driven methods, maintaining aesthetic and functional integrity while pushing the boundaries of design exploration. This hybrid approach underscores the timeless relevance of classic architecture with-in modern practices [2].

Facades play a vital role in determining the aesthetic and functional attributes of buildings. The synthesis of classical designs with modern technological advancements

H. Chai et al. (Eds.): CDRF 2024, *Symbiotic Intelligence*, pp. 297–306, 2025.
https://doi.org/10.1007/978-981-96-3433-0_26

not only enhances the aesthetic appeal and functionality of buildings but also indicates the ongoing evolution of architectural methodologies [3]. The advent of Large-scale Language-Image (LLI) models, particularly stable diffusion (SD), has introduced groundbreaking tools to architects, revolutionizing the process of text-to-image generation and significantly altering design approaches [4].

However, despite the capabilities of AI models like stable diffusion in automating image generation, they often fall short in capturing the intricate details and historical essence of varied architectural styles, particularly in facades. As shown in Fig. 1, the stable diffusion text-to-image fails to generate the Ionic capitals and shafts. This gap is evident in the limited practical use within the architectural sector. Moreover, the absence of substantial empirical research on the use of AI in architecture, particularly for concept generation, points to a gap in understanding these models' real-world implications [5].

Fig. 1. The façade image created within SD Web UI (Prompt: Architectural elevation, Ionic style, six columns, classical building, Greek temple, architectural drawing)

This study suggests using shape grammar (SG) and the Shape Machine (SM) to fill this gap, enabling the integration of architectural languages into computational models, and generating designs consistent with historical styles [6, 7]. This method emphasizes the importance of high-quality datasets for AI training to ensure detailed and precise architectural renderings, merging traditional elegance with modern innovation.

2   Literature Review

The facade of a building is vital for its aesthetic and represents different architectural styles and eras [9]. As shown in Fig. 2, these styles are distinguished by their form, details, and facade elements [10], which significantly affect a building's image [11]. SG plays an essential role in architecture, allowing for the creation of geometric patterns and understanding of architectural languages, proving especially effective in designs like Frank Lloyd Wright's prairie-style homes [12]. Çağdaş highlights its value in the architectural design process [13].

Fig. 2. The facade best represents the design languages of various architectural design styles.

SG reflects the design of facades by incorporating historical and cultural contexts, benefiting from a systematic approach that adds depth to the design process [14]. Martinovic and Murphy underscore its role in design synthesis and digital heritage [15]. Eilouti's work on SG for iconic facades shows its adaptability and alignment with Vitruvius' classical principles [16], while the SM demonstrates advances in its application, improving design accessibility [10, 11, 17].

In conceptualization, architects face challenges with AI tools including better controlling the results, leading to discrepancies [18]. Optimizing AI input can enhance image generation accuracy, ensuring alignment with the original design concepts. AI advancements, like SD, offer customization opportunities through community support [19]. Innovations like LoRA and DreamBooth introduce efficient text-to-image personalization, showcasing significant advancements in AI-driven design personalization [20].

3   Methods

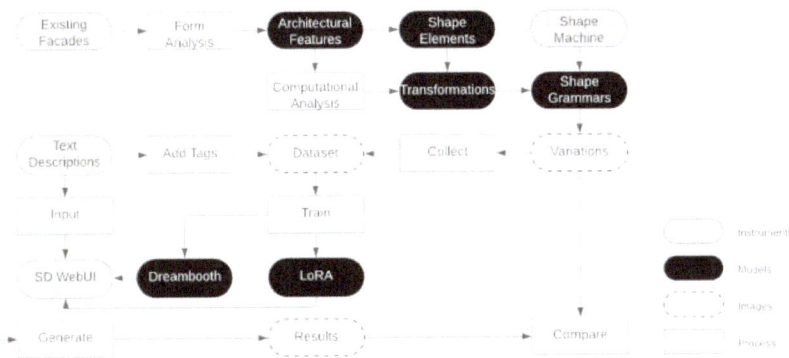

Fig. 3. The overview workflow of the methods

This study explores the use of SG and SD AI to improve the precision, quality, and aesthetics of AI-generated architectural facades, aiming to blend modern technology with classical design elegance. We evaluate designs before and after applying these tools, aiming for facades that merge technological innovations with historical architectural richness. Our approach includes dataset creation, model training, testing, and detailed comparative analysis. We employ tools such as Sd, SM, and the language of classical architecture for a comprehensive examination and analysis as shown in Fig. 3.

3.1 Parametric Shape Grammars of Classical Architecture

In the initial phase of our study, we aim to develop a comprehensive set of SG to generate various images. To achieve this, we must first pinpoint the specific architectural components to include in this grammar. Take Ionic architecture as an example. it's essential to incorporate its unique elements, such as special form of columns, eaves, and foundations as shown in Fig. 4. Variations of the Ionic façade can be represented with various combinations of these elements. For example, the façade of Distyle with Antis has three type types of roofs, including shed roof (A1), Pediment roof (A2), and hipped roof (A3) as shown in Fig. 4. As the count of columns increases to 4 or six, there will also be more variations in the Ionic language.

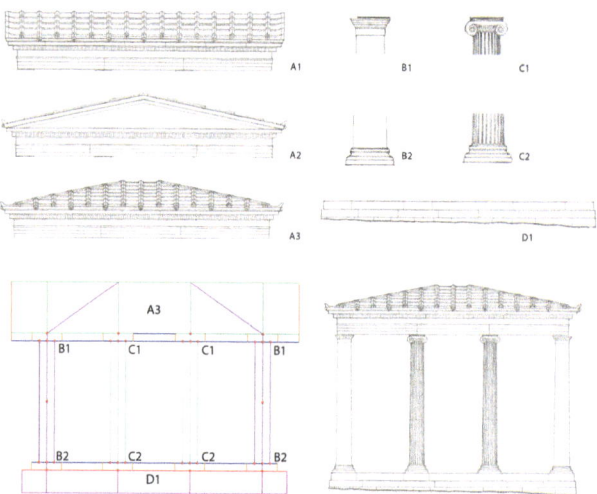

Fig. 4. The architectural elements of an Ionic façade. A1. Shed roof. A2. Pediment roof. A3. Hipped Roof. B1. Anta capital. B2. Anta basis. C1. Column capital. C2. Column basis D1. Stylobate. And then the Façade of Distyle with Antis and hipped roof can be represented with the schema.

Then, we incorporate these components presenting the Ionic elements into the SG framework with a series of shape rules for the schema of the Ionic facades as shown in Fig. 5. With SM we can set loop statement or condition statement to apply the rules.

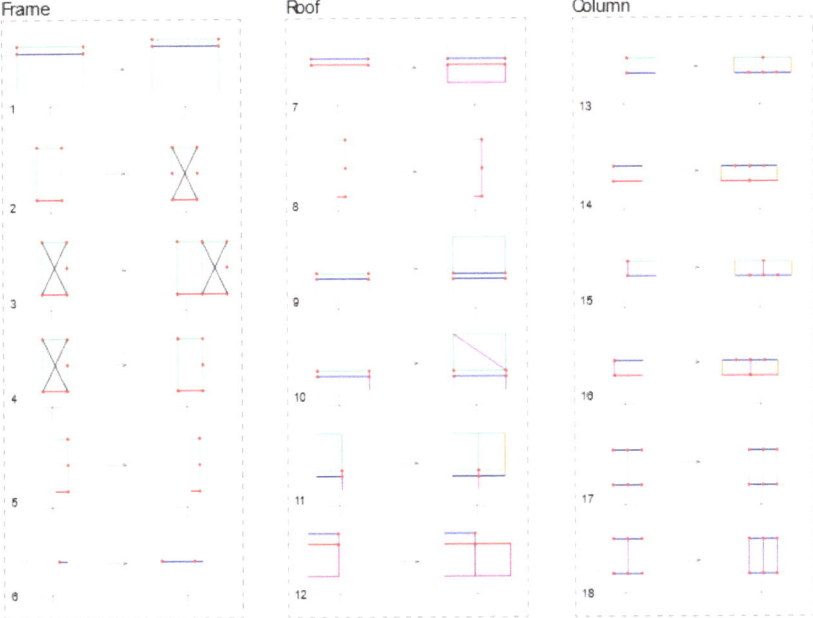

Fig. 5. The shape grammars for the schema of Ionic façade variations

For example, we can apply the rule 1 for specific time to change the hight of the shaft and the building. We can also apply the rule 3 two times for five intercolumniations or three times for seven intercolumniations. By applying the rule 5 for multiple times, we can change the width of the intercolumniations. After these rules are applied, the initial shape can be computed as a specific derivative of the Ionic façade schema (Figs. 6 and 7)

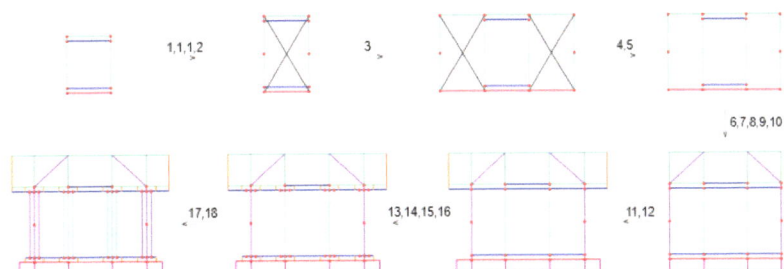

Fig. 6. The production of the schema of the Façade of Distyle with Antis and hipped roof.

To finalize the schema of a specific Ionic façade, we can directly put the corresponding elements on its position based on the schema. Since we can generate a variety of schemas with the provided rules and replace the symbols with the complete drawings of the elements, finally we have amounts of the façade to build the dataset.

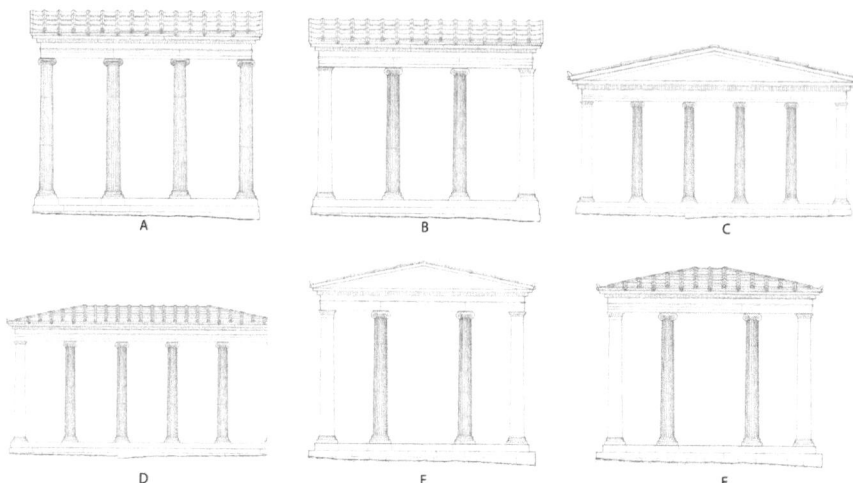

Fig. 7. The variations of the Ionic facades to be used as the dataset.

3.2 Dataset Construction and Processing

With the various imaged produce in 3.1, we still need to tag them with some text descriptions. Each image is tagged with relevant terms like "facade," "Ionic," and "classical architecture." A corresponding text document will be created for each image, ensuring each image-text pair is added to the training dataset.

When using LoRA for training, it can act as a supplementary network to the Checkpoint model of SD. LoRA models can be trained with a limited number of quality images and their descriptions. By applying shape rules, we can alter details in the images, such as swapping out building elements, enabling us to produce 30 image-text pairs for the training set. For the primary SD model, which is the Checkpoint model, we use the DreamBooth training method. This approach requires a significant number of images. We aim to produce 100 datasets initially and will enhance our dataset with Ionic-style drawings from online sources and those generated by the SM. All drawings are tagged accordingly.

3.3 Model Training and Result Analysis

After evaluating 100 variations from the initial selection, we chose 20 images for LoRA's training, capitalizing on its rapidity and adaptability. Conversely, DreamBooth requires a more extensive dataset due to its complexity; therefore, we utilize the entire dataset for its training. Additionally, we supplemented our database with existing, correctly depicted Ionic facade images sourced online to enhance the variety. Upon completing the training of both models, we merged their capabilities within SD web UI. Furthermore, while applying distinct datasets and training approaches for different models, we consistently refine the SD model by adjusting various parameters, such as image size and learning rate. We maintain a consistent image size of 500×500 pixels across all datasets to ensure uniformity.

4   Results

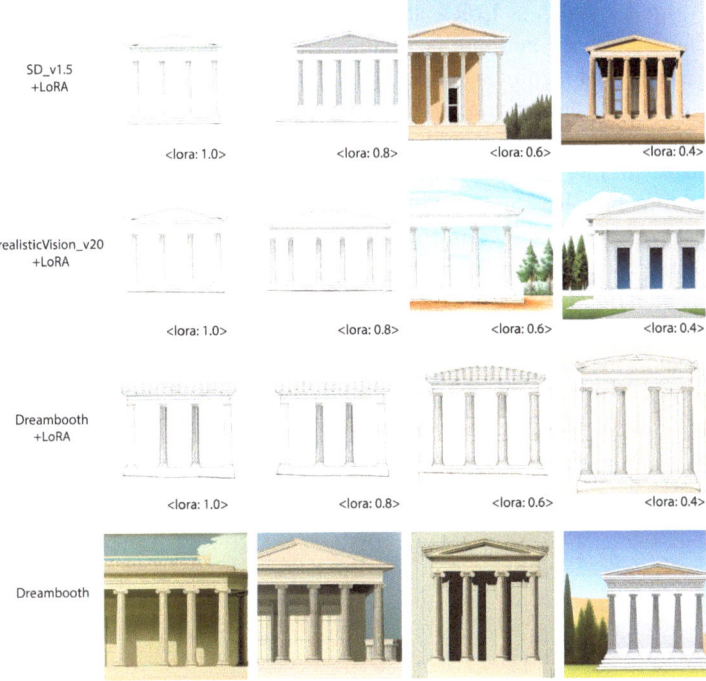

Fig. 8. The generation results by SD with the trained models. Prompt: Ionic facade, Tetrastyle, Greek Temple, classical architecture, Hipped roof, beautiful landscape, colorful background, high quality, best quality. Steps: 30, Sampler: Euler a, CFG scale: 7, Seed: 1531537168, Size: 500 × 500

Ultimately, we gained two SD models. One is a LoRA model that can be utilized in conjunction with checkpoint models such as SD v1.5 or realisticVision V20 as an attached network. As shown in Fig. 8, we experimented with four distinct matches. It is evident that the LoRA model can increase generation accuracy by ensuring that the elements and proportions are accurate. The produced results, however, are prone to overfitting for the training set if the LoRA model's weights are set too high. The outcome will remain similar with the training set even if we alter the prompt or add new descriptions, like "beautiful landscape." When we combined the LoRA model with the DreamBooth model, this phenomenon became especially clear. Combining these two methods yields SD results that are difficult to vary and are highly consistent with the training set, even when the LoRA weights are lowered to about 0.4. It is evident that there is more room for variation in the SD results when we limit our analysis to the ckeckpoint model that was developed during DreamBooth training. Furthermore, these variations are fundamentally incorporated into the Ionic style design language. It seems to have a more balanced representation of the Ionic design language and shape variation than when depending solely on LoRA.

5 Discussion

This study explores the integration of Shape Grammars (SG) with advanced Stable Diffusion (SD) models, such as LoRA and DreamBooth, to generate Ionic-style facades. The intricate nature of Ionic motifs, characterized by classical elements like volutes and dentils, presented significant challenges for the SD model. These challenges highlighted the need for a deeper understanding of shape grammar complexities and a richer architectural vocabulary to capture the full spectrum of architectural designs.

The successful merging of SG with SD's training processes not only optimized the design workflow but also showcased the potential for AI applications within complex architectural contexts as shown in Fig. 9. This approach allows AI to circumvent the labor-intensive process of manual rule definition, providing a quicker, user-friendly solution that maintains high fidelity to design intentions. Additionally, the role of probabilistic models in assessing design suitability without requiring extensive user intervention was explored, demonstrating their potential in enhancing AI's predictive capabilities.

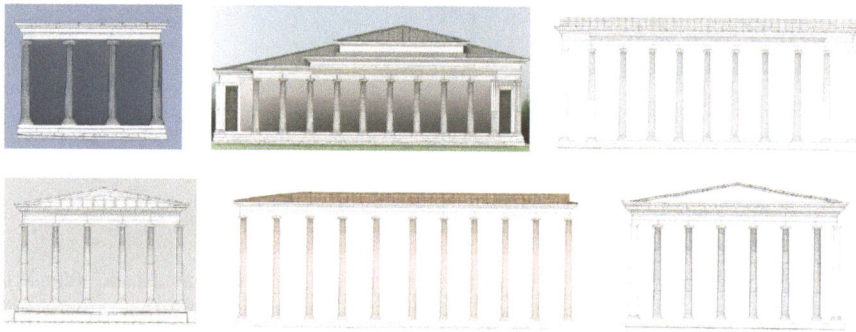

Fig. 9. More generation results with SD

Our research emphasizes the importance of accurately guiding AI systems to produce creative yet precise architectural designs. By refining how AI, specifically the SD model, interprets and utilizes architectural terms, we have enabled the generation of varied outcomes that transcend the rigid constraints of rule-based systems like SG. The goal is to harness both data-driven AI capabilities and rule-based methodologies to improve problem-solving and usability in design processes.

Despite the advantages, integrating SG with SD involves notable challenges, primarily the complexity involved in creating effective training datasets and the high computational demands of model tuning. Nevertheless, the prospect of scalable and efficient design generation that maintains architectural integrity presents a substantial benefit, promising significant advancements in how architectural designs are conceived and realized.

6 Conclusion

This study has provided a framework for improving the AI-assisted architectural design process, particularly through the integration of Shape Grammars with the Stable Diffusion model. By focusing on the generation of Ionic facades, we have demonstrated that AI can be trained to interpret and generate complex architectural styles with greater accuracy and detail. The results show a marked improvement in the model's ability to produce architecturally accurate representations based on complex stylistic inputs.

Further, our research has revealed the broad applicability of this approach, suggesting that similar methods could be employed to master other architectural languages and styles. This capability could make the design of specific style of architecture more accessible and less time-consuming and could also foster more experimental and innovative design practices.

However, the success of this integration is not without its challenges. The creation of effective training datasets and the computational demands of model tuning are significant hurdles that require ongoing attention. Future research should focus on refining these processes and exploring the integration of additional Shape Grammars and advanced probabilistic models to enhance the model's autonomy in design decision-making.

In conclusion, this study not only advances our understanding of AI's potential in architectural design but also sets the stage for future innovations that could further transform the field. The methodologies developed here serve as a blueprint for others looking to harness the power of AI in architecture, promising a future where design is more integrated with technology, leading to faster, more efficient, and creatively unbounded architectural practices.

Referencess

1. Bengio, Y., Courville, A., Vincent, P.: Representation learning: a review and new perspectives. IEEE Trans. Pattern Anal. Mach. Intell. **35**(8), 1798–1828 (2013). https://doi.org/10.1109/tpami.2013.50
2. Lu, Q., Zhu, L., Xu, X., Whittle, J.: Responsible-ai-by-design: a pattern collection for designing responsible artificial intelligence systems. IEEE Softw. **40**(3), 63–71 (2023). https://doi.org/10.1109/ms.2022.3233582
3. Maksoud, A., Al-Beer, H., Mushtaha, E., Yahia, M.: Self-learning buildings: integrating artificial intelligence to create a building that can adapt to future challenges. Iop Conf. Ser. Earth Environ. Sci. **1019**(1), 012047 (2022). https://doi.org/10.1088/1755-1315/1019/1/012047
4. Raviv, T., Park, S., Simeone, O., Eldar, Y., Shlezinger, N.: Online meta-learning for hybrid model-based deep receivers. IEEE Trans. Wireless Commun. **22**(10), 6415–6431 (2023). https://doi.org/10.1109/twc.2023.3241841
5. Sze, V., Chen, Y., Yang, T., Emer, J.: Efficient processing of deep neural networks: a tutorial and survey. Proc. IEEE **105**(12), 2295–2329 (2017). https://doi.org/10.1109/jproc.2017.2761740
6. Stiny, G.: Introduction to shape and shape grammars. Environ. Plann. B. Plann. Des. **7**(3), 343–351 (1980)
7. Economou, A., Hong, T.C.K.: Back to the drawing board: shape calculations in shape machine. In: Gero, J.S. (ed.) Design Computing and Cognition'22, pp. 549–567. Springer International Publishing, Cham (2023). https://doi.org/10.1007/978-3-031-20418-0_33

8. Economou, A., Hong, T.C.K., Ligler, H., Park, J.: Shape machine: a primer for visual computation. In: Lee, Ji-Hyun. (ed.) A New Perspective of Cultural DNA, pp. 65–92. Springer Singapore, Singapore (2021). https://doi.org/10.1007/978-981-15-7707-9_6

9. Fontenele, A., Campos, V., Mesquita, E.: Damage to historic facades: a case study in the historic center of viçosa do ceará. (2022). https://doi.org/10.4322/cinpar.2022.012

10. Shan, L., Zhang, L.: Application of intelligent technology in facade style recognition of harbin modern architecture. Sustainability **14**(12), 7073 (2022). https://doi.org/10.3390/su14127073

11. Askari, A., Dola, K., Soltani, S.: An evaluation of the elements and characteristics of historical building façades in the context of malaysia. Urban Des. Int. **19**(2), 113–124 (2013). https://doi.org/10.1057/udi.2013.18

12. Stiny, G., Mitchell, W.J.: The palladian grammar. Environ. Plann. B. Plann. Des. **5**(1), 5–18 (1978). https://doi.org/10.1068/b050005

13. Çağdaş, G.: A shape grammar model for designing row-houses. Des. Stud. **17**(1), 35–51 (1996). https://doi.org/10.1016/0142-694X(95)00005-C

14. Eilouti, B.: A formal language for Palladian Palazzo Façades represented by a string recognition device. Nexus Netw. J. **10**(2), 255–273 (2008). https://doi.org/10.1007/S00004-007-0068-4

15. Martinovic, A., Gool, L.: Bayesian grammar learning for inverse procedural modeling. (2013). https://doi.org/10.1109/CVPR.2013.33

16. Eilouti, B.: Shape grammars as a reverse engineering method for the morphogenesis of architectural façade design. Front. Archit. Res. **8**(2), 191–200 (2019). https://doi.org/10.1016/j.foar.2019.03.006

17. Okhoya, V.W., Bernal, M., Economou, A., et al.: Generative workplace and space planning in architectural practice. Int. J. Archit. Comput. **20**(3), 645–672 (2022). https://doi.org/10.1177/14780771221120580

18. Dortheimer, J., Schubert, G., Dalach, A., Brenner, L., Martelaro, N.: Think AI-side the box! exploring the usability of text-to-image generators for architecture students. J. Title (not provided) **2**, 567–576 (2023)

19. Rombach, R., Blattmann, A., Lorenz, D., Esser, P., Ommer, B.: High-resolution image synthesis with latent diffusion models. arXiv:2112.10752. (2021). https://github.com/CompVis/stable-diffusion

20. Hu, E.J., et al: LoRA: Low-rank adaptation of large language models (arXiv:2106.09685). arXiv. (2021). http://arxiv.org/abs/2106.09685

Deep Semantics – Design Semantic Universes

Daniel Bolojan[✉], Arie Chocron, Alyssa Scherger, and Thomas Tucker

Florida Atlantic University, Boca Raton, USA
{dbolojan,achocron2018,ascherger2019,tuckert2017}@fau.edu

Abstract. The integration of Generative AI in architectural design offers unparalleled opportunities for design innovation, characterized by the creation of complex, nuanced semantic universes. This paper explores the convergence of AI with various design-based disciplines, aiming to reimagine the design process through a multi-layered strategy. It proposes a unique AI-native workflow for the seamless integration of diverse design concepts, with the primary goal of investigating the capabilities of Creative AI in architectural design and its potential to combine various design fields into a cohesive workflow that addresses the limitations of current generative AI models when applied in architectural design [1]. By employing a trifold methodology consisting of sequential city sections, semantic encoding, and crafting semantic universes, the project utilizes contextual 3D data from Chicago and conducts a comprehensive synthesis across nine design domains, thereby creating a rich 'semantic universe' that serves as the foundation for AI-driven design generation.

Keywords: Creative AI · Architectural Design · Semantic Universe · Multimodality · Encoding Semantics

1 Introduction

The integration of Creative AI in architectural design represents a transformative shift, challenging traditional boundaries and redefining our interaction with the creative process. AI models like Generative Adversarial Networks (GAN) [2], Diffusion models [3], and language models [4], can enhance creative agency and broaden design possibilities [5]. These models enable a transition from static to adaptive semantic models, where formal language is encoded in the AI's latent space, allowing for exploration and adaptation [3]. This research explores the potential of AI models to empower creatives in developing their unique formal language by leveraging the flexibility and generalization capabilities of generative AI. Integrating AI tools in architectural design requires delineating the levels of engagement, from gestalt to organizational and technical aspects, aligning with Alexander's systems theory, which posits that architecture, and the design process are intricately interwoven with multiple interrelated systems [6], and Gero's insights on design complexity, highlighting its multifaceted and nuanced nature [7]. The research focuses on the organizational and gestalt aspects, aiming to merge AI with design disciplines to craft a "semantic universe" through a multi-faceted AI strategy.

© The Author(s) 2025
H. Chai et al. (Eds.): CDRF 2024, *Symbiotic Intelligence*, pp. 307–316, 2025.
https://doi.org/10.1007/978-981-96-3433-0_27

It introduces an AI-native workflow for integrating diverse design concepts, moving beyond replication of computational processes, to uncover the unique affordances [8] offered by these novel AI models.

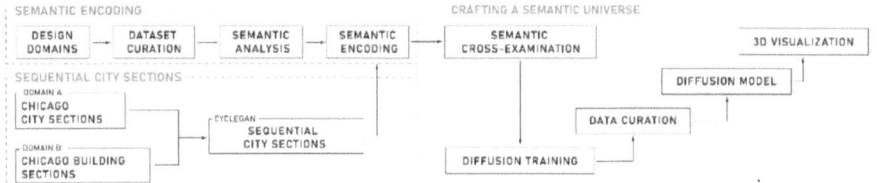

Fig. 1. Workflow diagram of the employed trifold methodology

The methodology is trifold (Fig. 1). First (1) the "*Sequential City Sections*" phase involves extracting and interpreting contextual sectional data from Chicago by deconstructing the Chicago River 3D model into sequential sections to serve as a foundation for architectural inspiration. Second, (2) the "Semantic Encoding" phase focuses on synthesizing information across nine design domains, generating samples for each domain, using an image-to-text (I-2-T) Clip Interrogator model to extract textural semantics, and generating accurate prompt descriptions that encapsulate the observed features and ensure the essence of the visual data is faithfully translated into textual descriptors. Third, (3) in the "*Crafting Semantic Universes*" phase, the curated prompts are used as descriptors for fine-tuning a Stable Diffusion (SD) Model by pairing the descriptors with corresponding images to create a dataset, employing heatmaps and semantic cross-examination processes to refine the models' output and ensure accurate representation of the intended design features. Finally, sections of the Chicago River are input into the fine-tuned SD models to infuse the semantic DNA, which are then reassembled into a cohesive 3D point cloud model embodying the designer's intent.

2   State-Of-The-Art and Background Literature

2.1   Deep Learning Neural Networks

The research employs various neural networks. CycleGAN [9] is used for image-to-image translations and semantic transfer using cycle-consistency loss. StyleGAN [10], manipulates styles at different levels of the generation process, allowing more control over sample synthesis. Stable Diffusion (SD) [3], generates high-quality images from textual descriptions. The CLIP Interrogator uses the CLIP [4] architecture for I-2-T transformations, creating detailed textual descriptions from images via contrastive learning, facilitating accurate caption generation and image analysis. The LAION Aesthetic Predictor model assesses visual samples based on aesthetic principles (i.e. composition, color harmony, subject matter, perspective, etc.), using CLIP ViT-L/14 embeddings to provide an aesthetic score. Diffusion Attentive Attribution Maps [11] generate pixel-level attribution maps for T-2-I SD models, using cross-attention scores between words and pixels to show how specific textual inputs influence image generation, offering insights into semantic segmentation and syntax's role in shaping the image.

2.2 Background

Recent research has explored the integration of AI in various design disciplines, show-casing AI-driven tools' potential to augment the creative process and generate novel solutions. Langenhan et al. [13] highlighted the importance of semantic representation in design by investigating strategies for accessing knowledge in semantic models. Kim et al. [14] demonstrated AI's potential to capture and generate domain-specific design language through the Text2Form Diffusion Framework. Gal et al. [15] introduced Textual Inversion for personalizing T-2-I generation. Bolojan et al. [1] proposed interconnected processes between GANs and LLMs to address design complexity and limitations of T-2-I models. Our research distinguishes itself by proposing a strategy for creating a semantic universe through a trifold methodology combining (1) sequential city sections, (2) semantic encoding, and the (3) crafting of a semantic universe. Our approach aims to develop a custom semantic DNA, that is later encoded into a semantic universe, high-lighting AI's capacity to function as an integral part of the creative process, synthesizing diverse design influences into architecturally relevant outputs.

3   Methodology

3.1   Sequential City Sections: The Subject of the Experiment

3.1.1   Dataset Curation: Sections Through Chicago

The process involved three main steps. First, representative sections of Chicago buildings were combined using MidJourney's (MJ) blend option to generate a comprehensive dataset of sectional qualities for training custom AI models. Second, sequential sections spanning the Chicago River were created from a 3D model of Chicago, with building cuts *poche* and elevations unaltered. The sectional dataset was fine-tuned for better feature translations between domains, and its sequential format allows for accurate 3D reconstruction of the resulting samples. Finally, the building sections were blended with the city sections using MJ blend option, combining architectural features with the urban context to capture the intricate relationship between buildings and their surrounding urban fabric (Fig. 2).

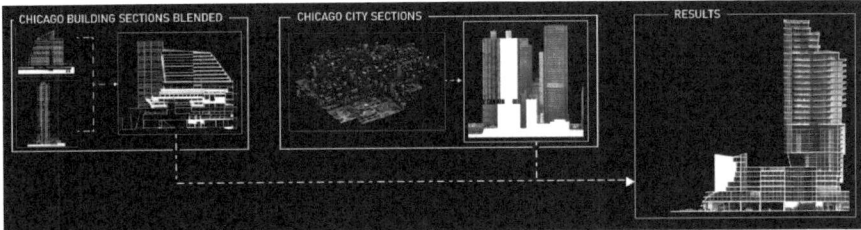

Fig. 2. Blending of MJ generated building sections with city sections of Chicago.

3.1.2   Transferring Features from Building Sections to the City Dataset

After assembling our two datasets—the sequential city sections and the enhanced building sections—we utilize CycleGAN to transfer features from the blended sections of Chicago buildings to the buildings depicted in the sequential sections of the city. Cycle-GAN enables the translation of features from one domain (Building Sections) to another (City Sections) while preserving the structure and composition of the target domain. The approach is designed to provide a foundational architectural structure for the *"poche"* sequential city sections so that future feature transfers have a consistent and more relevant structure be transferred onto (Fig. 2).

3.2   Semantic Encoding

Fig. 3. Feature encoded results from the nine chosen design fields.

3.2.1   Dataset Curation: Visually Defining Nine Design Domains

To create a cohesive dataset for each design domain, a method was developed to capture the field's essential features while accommodating various artists' distinctive characteristics. First (1), five designers were selected for each of the nine domains. Then, ten images showcasing their most iconic designs were chosen to encapsulate their unique stylistic features. In the second step (2), the ten selected images from each designer were blended using MJ to enrich the initial samples and highlight the recurring features observed across their works. In the third step (3), following the creation of blended samples for the five designers across all nine domains, the samples from the designers within each domain were cross-blended, resulting in a dataset representing the main features for each design domain (Fig. 4). This approach allowed the analysis and the encoding of the semantic features of each domain across three levels: (A) individual designer's unique work, (B) designer's overall work features, and (C) prevalent features across all designers, representing the specific design domain.

3.2.2   Translating the Datasets into Semantics and Analyzing the Results

Once a dataset representing the main features of each of the nine domains is generated, a new semantic extraction process is developed. Using CLIPInterrogator, and an I-2-T model, semantic information was extracted from each image for each domain. Recurring words and descriptions are identified to capture the most consistent semantic representations within each design domain. By ranking the extracted semantics based on their frequency and relevance, the results are indexed, and extraneous noise is filtered out, distilling the fundamental gestures prevalent across the diverse design domains (Fig. 5).

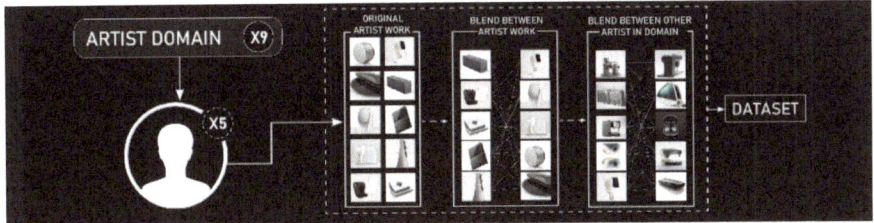

Fig. 4. Typical workflow for dataset curation repeated for each of the nine design domains.

The neural network provides semantics based solely on detectable visual features without subjective bias, consistently yields high-ranking semantics that define the actual subject rather than its constituent features. For instance, in the domain of car design, the most recurrent word was "car," highlighting the significance of unbiased semantic representation. Recognizing that hypernyms like 'car'—broad categories that encompass more specific terms (cohyponyms) [12]—could skew results, we crafted strategies to omit such broad terms across each domain, such as 'shoe' for shoe design and 'table' for furniture design. The selection process focused on retaining high-indexed results, that formed the basis for prompt curation. This method ensures that the semantic descriptors, leveraging cohyponyms, accurately capture the intricate details and subtle distinctions of each design domain, avoiding overly broad terms.

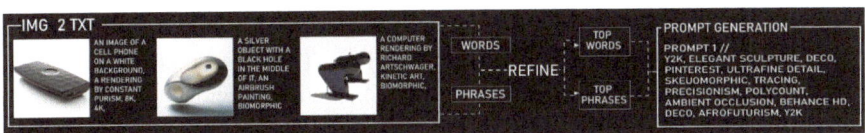

Fig. 5. Refining semantic results and curating prompts based on I-2-T results.

3.2.3 Embedding the Semantic DNA into the Site Setting

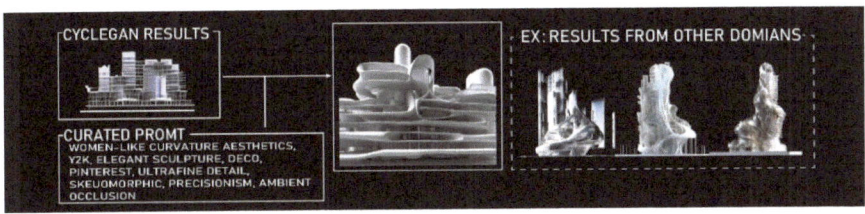

Fig. 6. Combination of the results of the two parallel phases leading to a newly generated dataset.

After analyzing samples from all design domains and extracting accurate semantic descriptors unique to each designer and domain, the embedding process was initiated. The process starts by synthesizing the most relevant words and phrases into carefully crafted prompts (Fig. 5). These prompts, designed to capture the essence of the highest-indexed semantics, are inputted into MJ alongside the manipulated Sequential City sections results as a reference image (Fig. 6). Following curation, only the consistently effective prompts with minimal aesthetic variation, are retained. From this process, three prompts are derived for each domain, each subtly distinct and emphasizing various blended typologies from the encoded semantic DNA. After refinement process, a dataset is produced to explore the potential and versatility of the semantic features within each image sample. Once the semantic DNA of each domain is embedded into our sample images, they are subjected to an Aesthetic Prediction evaluator model, to further refine the dataset and the relationship between semantic features and their cohyponyms descriptors.

3.2.4 Semantic Cross-Examination

Employing the methodologies outlined in the I-2-T analysis, the Semantic Cross-Examination phase extracts semantics from the results of encoding design domain semantics onto reference sectional images. These results undergo an indexing process, consistent with the initial methodology, but are further subjected to a rigorous validity assessment. This assessment entails a semantic cross-examination with the original texts from the design domains. Semantic features identified by the I-2-T model in this iteration are meticulously documented and preserved for future training of an SD model tasked with creating the semantic universe. Unlike the initial text analysis, the I-2-T model uncovers new architectural features in this phase. These new architectural semantics are merged with the design semantics obtained through cross-examination, forming a unique DNA for each generated image sample (Fig. 7). This semantic DNA, along with its corresponding image sample, constitutes the foundational elements for training the SD model, ultimately crafting a detailed semantic universe.

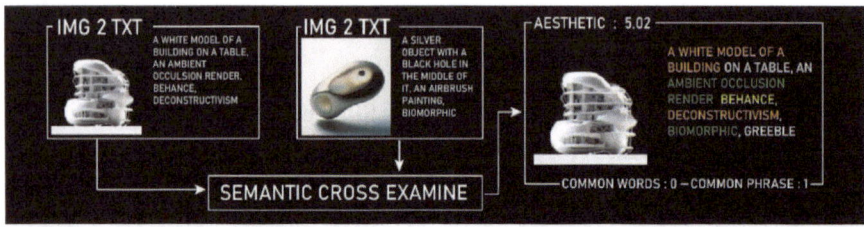

Fig. 7. Semantic cross-examination of the generated texts acquired using I-2-T

3.3 Crafting a Semantic Universe

3.3.1 Translating the Semantic DNA Back into the Sequential City Sections

In the final phase of the trifold methodology, an SD model is fine-tuned using the semantic DNA derived from each design domain (Fig. 3). Additionally, it incorporates the cross-examined semantics and their corresponding image samples that encapsulate architectural features, further refining the SD model (Fig. 7). After fine-tuning, the model effectively separates the connections between the input image and text, creating semantic associations. These associations are encoded into the SD model tailored for each design domain, facilitating the fine-tuning of parameters and textual inputs. The model utilizes the entire Semantic DNA from all domains, ensuring equitable use of text from each to prevent bias in the outcomes.

3.3.2 Allowing AI to Refine Our Results Based on Feature Detection

Building on the fine-tuned model, various text combinations for each domain were evaluated to identify those encapsulating the desired semantic DNA features. After selecting and curating the most effective prompts using test images, sequential sections of the Chicago River, processed through CycleGAN domain transfer, were used as references for the SD model (Fig. 8). This method aimed to transpose the selected features onto the resulting image samples, setting the stage for their transformation into 3D geometry in the next phase.

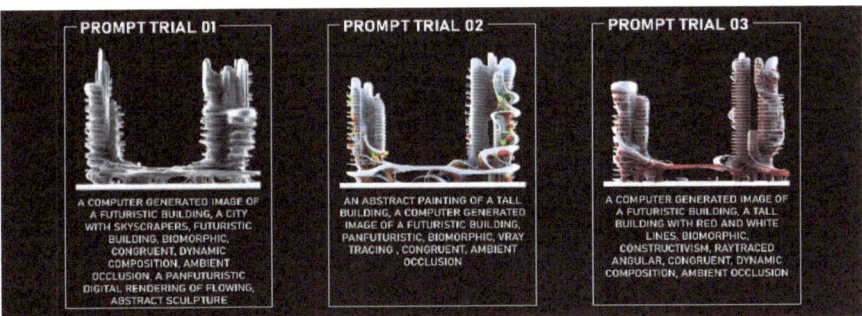

Fig. 8. Testing of semantic combinations with the trained SD model.

3.3.3 Exploring the Results in Three-Dimensional Forms

The final phase translates the results back into the original 3D space (sequential sections of Chicago). Each section in the sequence describes a distinct section of the city. When combined, they describe a larger 3D structure, such as a building or a city block. To facilitate the transition from a sequence of images to a 3D structure, point clouds were generated for each series of image sequences (Fig. 9). This approach efficiently converts an image sequence into discrete points, bypassing the need for complex solid geometries. A Python script was created to streamline stacking the final sequential outcomes into a point cloud. The script set, indexes them with their corresponding values, and imports them into Blender to generate stacked points. This allows for examining the results in the context of all individual images, consolidating them into a 3D structure (Fig. 10).

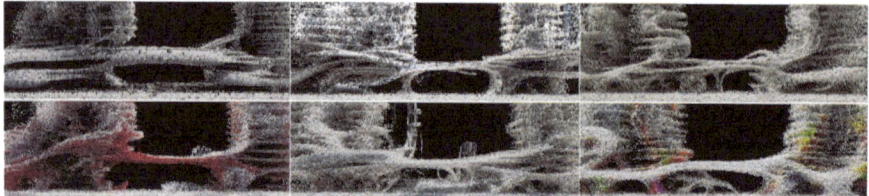

Fig. 9. Point cloud examinations of the design domain encoded city sequences.

4 Discussion and Future Work

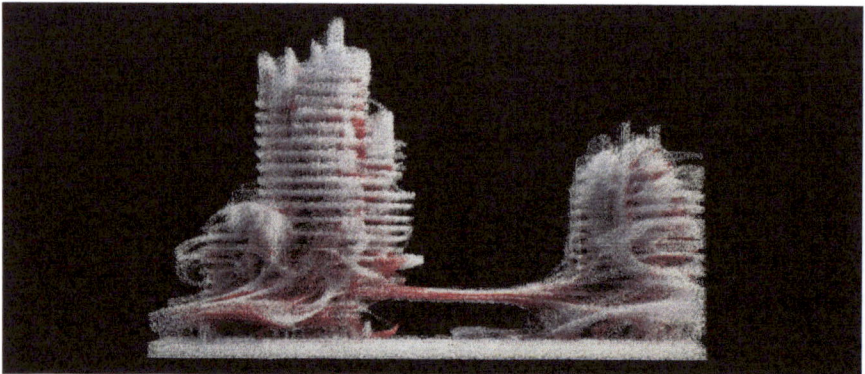

Fig. 10. 3d Point cloud of the design domain encoded for one city sequence.

The methodology showcases the transformative potential of integrating Creative AI into architectural design through a trifold approach of sequential city sections, semantic encoding, and crafting semantic universes, showcasing AI's ability to synthesize diverse design elements cohesively. The research highlights the generation of complex, nuanced semantic universes combining features from various design domains. The semantic encoding process distills essential principles and distinctive features of each domain, creating a rich semantic DNA embedded into the architectural design context. The use of sequential city sections as the foundation for the experiment demonstrates the potential for AI-driven design generation to be grounded in real-world contextual data, while the cross-examination of semantics ensures that the semantic universe accurately captures the desired features and maintains consistency with the initial design intent. Future work will focus on refining the methodology to enhance the integration of AI tools in the architectural design process. This will involve the development of more sophisticated semantic encoding techniques and the exploration of additional design domains. Future work will focus on a more direct real-time interaction with the semantic universe. As the field of Creative AI continues to evolve, it is essential to embrace its potential while remaining cognizant of its limitations and the need for human oversight and guidance in the design process.

Acknowledgements. We extend our sincere gratitude to NVIDIA for their generous support through the NVIDIA Academic Grant Program, which provided us with the essential hardware resources necessary for conducting our research.

References

1. Daniel, B., Emmanouil, V., Shermeen, Y.: Is language all we need? a query into architectural semantics using a multimodal generative workflow. pp. 353–362 (2022). https://doi.org/10.52842/conf.caadria.2022.1.353
2. Goodfellow, I., et al: Generative adversarial nets. Advances in Neural Information Processing Systems. (2014)
3. Rombach, R., Blattmann, A., Lorenz, D., Esser, P., Ommer, B.: High-resolution image synthesis with latent diffusion models. In: Proceedings of the IEEE/CVF Conference on Computer Vision and Pattern Recognition pp. 10684–10695 (2022)
4. Radford, A., et al.: Learning transferable visual models from natural language supervision. In: International Conference on Machine Learning PMLR. (CLIP) pp. 8748–8763 2021 Jul 1
5. Bolojan, D.: Creative AI: augmenting design potency. Archit. Design **92**, 22–27 (2022). https://doi.org/10.1002/ad.2809
6. Alexander, C.: Systems generating systems. Archit. Des. **38**, 605–610 (1968)
7. Gero, J.S.: Ten problems for AI in design. Workshop on AI in Design, IJCAI-91 (1991)
8. Gibson, J.J.: The ecological approach to visual perception. In: Lawrence Erlbaum Associates, United Kingdom, (1986)
9. Zhu, J.Y., Park, T., Isola, P., Efros, A.A.: Unpaired image-to-image translation using cycle-consistent adversarial networks. In: Proceedings of the IEEE International Conference on Computer Vision pp. 2223–2232 (2017)
10. Karras, T., Aittala, M., Hellsten, J., Laine, S., Lehtinen, J., Aila, T.: Training generative adversarial networks with limited data. Advances in Neural Information Processing Systems. (2020)
11. Tang, R., et al.: What the daam: interpreting stable diffusion using cross attention. arXiv preprint arXiv:2210.04885. 2022 Oct 10
12. Dagobert, S.: WordNet. An Electronic Lexical Database. MIT Press, United Kingdom, (1998)
13. Langenhan, C., Haß, S., Weber, M., Petzold, F., Liwicki, M., Dengel6, A.: Investigating research strategies for accessing knowledge stored in semantic models. https://papers.cum incad.org/data/works/att/ecaade2011_014.content.pdf
14. Kim, F., Johanes, M., Huang, J.: Text2Form diffusion framework for learning curated architectural vocabulary. https://papers.cumincad.org/data/works/att/ecaade2023_197.pdf
15. Gal, R., Alaluf, Y., Atzmon, Y., Patashnik, O., Bermano, A., Chechik, G., et al.: An image is worth one word: personalizing text-to-image generation using textual inversion https://arxiv.org/pdf/2208.01618.pdf

Digital Design Theory, Method
and Education

Construction and Analysis of BIM Semantic Graph—Take Revit Model as an Example

Q. Ye and Z. Tong[✉]

Nanjing University, Hankou Road 22, Nanjing, China
tzy@nju.edu.cn

Abstract. With the rapid development of information technology, the construction industry has gradually realized that the traditional two-dimensional design and construction methods can no longer meet the needs of complex projects. BIM, an integrated approach to design and management, is revolutionizing the construction industry through digital modeling, collaboration and information sharing. In this context, Revit has become one of the representative software in the field of BIM with its powerful functions and ease of use. Taking revit as an example, after a series of building prototype modeling experiments, ifc files exported from revit model are further transformed into graphic data structures, and the constructed BIM semantic graph is compared and analyzed, so as to understand revit from the perspective of the graph and enable it to organize and transmit information more comprehensively and reasonably, thus helping to build BIM information semantic graph. Facilitate the sharing and exchange of information.

Keywords: BIM · semantics · graph · Revit

1 Introduction

In recent years, in order to solve the problems of information island, data inconsistency and poor coordination in traditional architectural design and management. BIM technology, an integrated design and management method, supports the management and application of the whole life cycle of buildings with its integrated, collaborative and visual characteristics, changes the traditional architectural design and construction methods, and greatly improves the efficiency and quality. The key to realize building information sharing and collaborative management is that BIM model contains rich semantic information in addition to the three-dimensional presentation of the model. Specifically, it refers to material information, mechanical information, functional classification, location information, cost information, energy consumption information, spatial topological relationship and aggregation relationship, etc. [1]. Based on this, BIM has been widely used and promoted in the field of construction, but there are still some problems in the application process in terms of collaborative communication and data integration.

IFC (Industry Foundation Classes), as a common standard for building information models, plays a key role in the implementation of BIM. However, their interpenetration and complex hierarchical structure of multi-layer references bring trouble to directly

H. Chai et al. (Eds.): CDRF 2024, *Symbiotic Intelligence*, pp. 319–326, 2025.
https://doi.org/10.1007/978-981-96-3433-0_28

obtain architectural information, and the format conversion process will lead to the loss of semantic information [2]. The graph database can store and process various types of data, and can use the semantic association between building elements in the form of a graph to reveal the hidden relationship in building data, and present the internal structure and content of complex and rich data sets in a more comprehensive and intuitive way. They can provide similar functions to BIM technology in some aspects. As a supplement and expansion of BIM technology. Therefore, "map" has received more and more attention in the construction industry, and scholars at home and abroad have conducted a lot of research on it.

Graphs have long been used in computer science to model complex data structures. By classifying graphs, studying their structure and features, and developing algorithms based on these features, graphs have been widely used to solve problems in the field of AEC. Such as hospital layout design, IFC model compression and consolidation, BIM/GIS integration, indoor navigation, building model simplification, building structure identification, IFC model change detection and rule inspection. Langenhan et al. used diagrams to represent the topological relationship between rooms and architectural elements [3]. Chaoyi Jin et al. explored BIM data by using graph-based unsupervised learning [4]. Touffs et al. use Triple graph syntax (TGG) to formally associate IFC with CityGML (Urban Geographic Markup Language) to realize BIM/GIS data integration [5]. X. Zhou et al. proposed a method to generate graphs from IFC, which was used as the first step of an algorithm to compute shapes of IfcProduct instances in parallel [6]. Zhang Yuemei et al. established metadata knowledge base and instance knowledge base, analyzed the relationship between the two knowledge base models, and explained how to store IFC data model and IFC instance file in graph database [7]. However, the current research generally focuses on the modeling and solving of specific problems, and proposes many feasible paths for generating diagrams from BIM software, while relatively little research is done on the construction of BIM semantic diagrams with the help of software.

Based on this, this paper takes the mainstream BIM software Revit as an example to sort out the semantic information of buildings from the perspective of diagrams, explore the current status of IFC, the general standard of building information model, to store and transfer information, and summarize the potential defects of revit software in the BIM environment, so as to better understand the building information model and build a more complete and comprehensive BIM information semantic map. Promote data exchange and collaboration.

2 Construct BIM Semantic Map from Revit Model

2.1 Modeling Experiment

In view of the lack of BIM semantic information in the process of format conversion of building information model, the author takes Revit as an example to explore the problem. The research is mainly divided into three steps: (1) Select the mainstream BIM software Revit, conduct a series of building prototype modeling tests in Revit, and export it into IFC format files; (2) Transform IFC model into graph data structure; (3) Comparative analysis of BIM semantic graph to understand the current situation of information storage and transmission between revit and IFC.

He experiment was carried out by changing the room layout, wall correlation, building components, building space complexity, etc. A total of 13 simple building models were built in revit in 3 groups. Figure 1 shows the graphic design of the 13 models.

Fig. 1. Schematic diagram of architectural prototype modeling experiment scheme

Models 1–5 are the comparison items of the room layout, in which models 1–3 change the room nesting logic by changing the position of the door, and models 4 and 5 are the rich version of model 3. Model 4 adds a second floor on the basis of model 3, and Model 5 further improves the spatial complexity by introducing double-height space, in order to further explore whether the definition of information will change when the number of building floors increases and the spatial complexity increases. Models 6–9 are the comparison items of building components. Based on models 1 and 2, the definition of building components is explored by increasing or decreasing the number of doors and Windows and changing the layout of doors and Windows. Model 10–13 is a comparison item of the association relationship of building walls. All the four models are based on model 3. By changing the definition in the top constraint column of wall attributes, the adhesion relationship between wall and roof in the building is changed, and the possible situations such as no attachment, partial attachment and full attachment between wall and roof are comprehensively listed.

2.2 IFC Data is Represented by Graphical Data Structures

After multiple ontologies of IFC pattern were proposed successively, an ifcOWL ontologies approved by buildingSMART was produced [8]. After that, the researchers developed an IFC to RDF conversion tool that converts IFC to RDF Abox graphs based on the ifcOWL ontology. The converter is an open source project written in Java and is available on Github.

With the IFC -- RDF conversion tool, the author exported 15 simple architectural models built in revit into ifc format, further converted them into ttl format files, and then imported them into Ontotext GraphDB. With the help of the graph data engine, the data could be stored and analyzed, and the complex correlation between entities could be displayed. Draw BIM semantic map. Figure 2 shows two of these semantic graphs. Each node in the diagram represents an entity in the building (wall, column, floor, door and window, etc.), and the edge represents the logical relationship between these two objects.

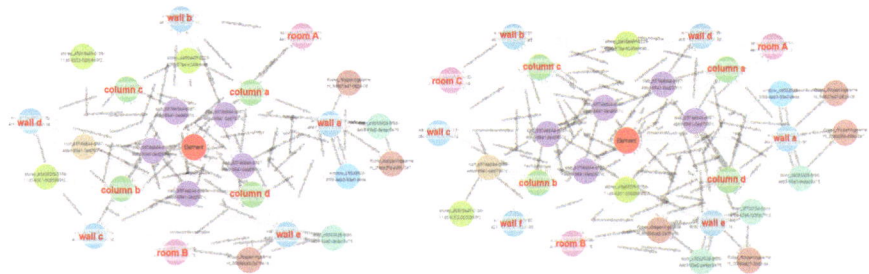

Fig. 2. BIM semantic diagram (left: Model 1, right: Model 4)

3 BIM Semantic Diagram Analysis Based on Revit Model

3.1 Room Layout

Models 1–5 are comparison items of room layout. For convenience of subsequent comparison and analysis, columns and external walls are numbered clockwise (as shown in Fig. 3). Comparing and analyzing 5 BIM semantic diagrams, no matter how to change the room layout logic, the relationship between each column, door, window and wall is accurately defined. For example, door a and window a are always associated with wall a, column b is always associated with walls b and c, and wall f is always associated with walls b and e. However, there are frequent errors in the cognition of the relationship between entity and space, and the description of the topological relationship of space is lacking. Taking the wall as an example, for example, in model 1, the association of room A lacks wall a, and the association of room B lacks walls a and c; In Model 2, the association of room C lacks wall a; In model 3, the association relationship of room C is missing wall a (as shown in Fig. 4), and the association relationship of other rooms is correct. With the improvement of spatial complexity, the confusion of the cognition of the relationship between entity and space becomes more and more serious. In Model 4, the association relation of room C lacks wall a, and the association relation of room D lacks wall a, but there are more walls e and f. In Model 5, the association relation of room C lacks wall a, and the association relation of room D lacks wall a and has multiple walls f; in the relation of room E, walls a, c and g are missing, while walls f and e are extra. Through comprehensive comparative analysis, the gaps and omissions in the cognition of the relationship between entity and space are irregular for the time being, but the principle of projection is followed in the spatial cognition, so the second-floor space will be more associated with the first-floor space wall under the vertical projection.

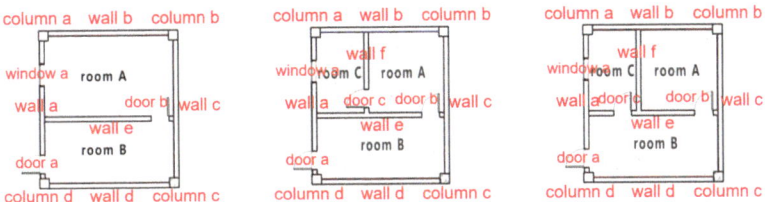

Fig. 3. Building component labels

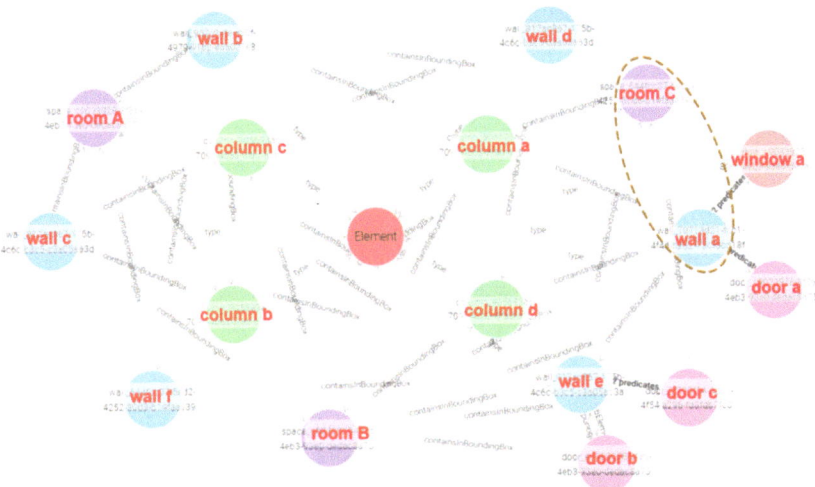

Fig. 4. Model 3 Local semantic map and comparative analysis results (brown dashed line marks the missing semantic association between wall and room)

3.2 Door and Window Components

Models 1, 2, 6, 7, 8, 9 are the comparison terms of doors and Windows, wherein models 6 and 7 have the same spatial topological relationship with model 1, and models 8 and 9 have the same spatial topological relationship with model 2. a comparative analysis of the semantic graph shows that all entity semantic associations of model 1 and model 6 and model 7 are the same except variable window A, door c and door d, and model 2 and model 8 and model 9 are the same. Throughout the 13 BIM semantic maps, no matter how the layout of doors and Windows is adjusted, it will not change the relationship between other entities in the building, and the attachment relationship between doors, Windows and walls is always accurate, but there is a lack of understanding of the relationship between doors and Windows and space and the topological relationship of doors and Windows. For example, the correlation between the door and the room is comprehensive and accurate, but the correlation between the window and the room is lacking (as shown in Fig. 5), which can provide convenience for the indoor navigation of the robot with the door as the carrier, but will reduce the accuracy of the simulation analysis of building energy consumption.

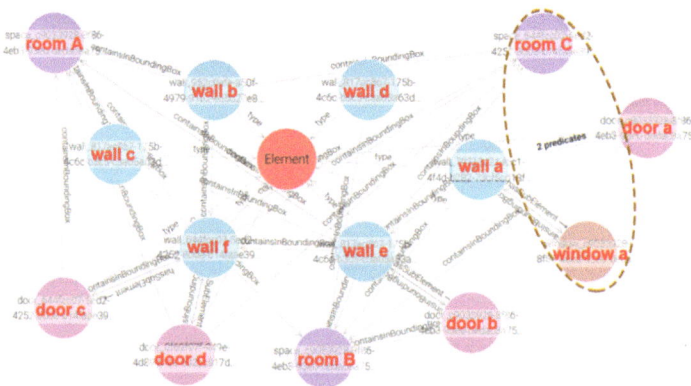

Fig. 5. Model 9 Local semantic map and comparative analysis results (brown dashed line marks the missing semantic association between window and room)

3.3 Relation of Wall

Model 10–13 takes the adhesion relationship between wall and roof in the building as the variable, and the BIM semantic diagram of model 3, 10, 11, 12 and 13 is compared and analyzed. It is found that the different adhesion relationship between wall and roof will affect the definition of entity by the software. For example, wall e of model 11 and 12 is set in the experiment, only the eastern wall is attached to the roof. Therefore, when the internal information of wall e of models 11 and 12 is defined, the eastern wall of wall e is considered to belong to two families, so the semantic graph of models 11 and 12 has one more "wall" element than the other three. Moreover, the cognitive confusion of the association between entity and space still exists. For example, in Model 10, there is more column A in the association relationship of room a, and there is less wall a in the association relationship of room C. In Model 12, there is A lack of wall f in the association relation of room a, and a column A is added, a lack of wall e in the association relation of room B, and a lack of wall A in the association relation of room C. (As shown in Fig. 6).

In the face of the multi-column situation that did not appear before, it is assumed that it is related to changing the adhesion relationship between the wall and the roof, but the conclusion is not conclusive after comprehensive comparison and analysis of

5 semantic maps. For example, when wall f is not attached to the roof, there is more column A in the association relationship of room a in models 10, 12 and 13, but the association relationship between room C next to wall f and column is very accurate. In addition, no matter how the adhesion relationship between wall e and the roof is changed, the association relationship between room B and the column with wall e as one of the boundaries has never been wrong. Therefore, the change of the adhesion relationship between wall and the roof will indeed affect the accuracy of the cognition of the relationship between entity and space, but the degree of influence is uncertain.

In addition, no matter how the adhesion relationship between the walls and the roof changes, the relationship between the pillars, doors, Windows and the entity of the wall is still accurately defined. For example, door c is always associated with wall e, column b is always associated with walls b and c, and wall b is always associated with wall f.

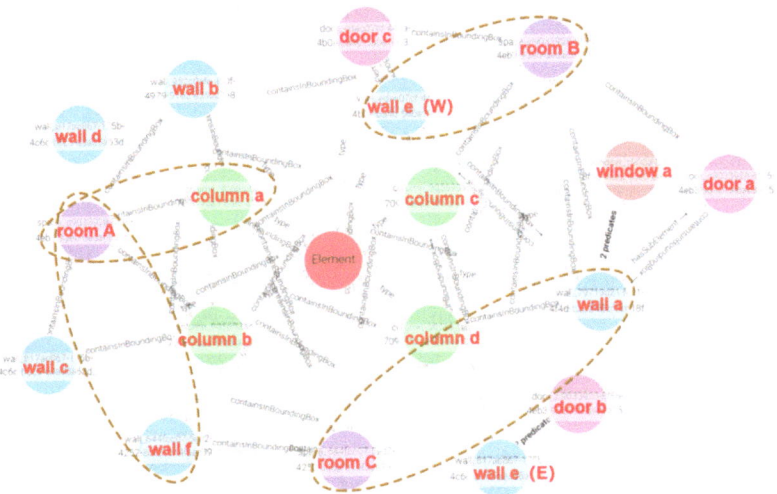

Fig. 6. Model 12 Local semantic graph and comparative analysis results (brown dashed lines show the semantic association fallacy)

4   Conclusions

This paper selects the mainstream BIM software Revit, conducts a series of architectural prototype modeling experiments in Revit, exports all kinds of Revit models into IFC format files, and draws BIM semantic maps with the help of IFC-RDF conversion tool and Ontotext GraphDB. Compare and analyze various semantic graphs to further explore possible problems in Revit software and explore the current situation of IFC information storage and transmission.

After a series of experiments, it is found that Revit pays attention to the definition of entity attributes and the relationship between entities, but has limitations in the cognition of the relationship between entities and space and the topological relationship between space and space, such as lack of expression and inaccurate definition. As a BIM software widely used in the AEC industry today, the shortcomings of Revit in spatial topological relationships may have a certain impact on the life cycle of the building. For example, in architectural analysis and simulation, accurate spatial topology description can improve the simulation accuracy; In terms of operation and maintenance management, a correct understanding of spatial topological relationships helps to improve the efficiency of operation and maintenance and the sustainability of facility management. The article is a useful exploration of BIM technology, helps to understand Revit, facilitates data exchange and collaboration, helps to build BIM semantic maps, and further promotes information integration with Cim.

References

1. Shaohua, J., Bo, Z.: Overview of Semantic Rich Research and application of BIM. Inf. Technol. Civil Architect. Eng. **13**(03), 24–29 (2021). https://doi.org/10.16670/j.cnki.cn11-5823/tu.2021.03.04
2. Lei, S.: Research on object-oriented semantic mapping of architectural structures. Shandong Jianzhu Univ. (2023). https://doi.org/10.27273/d.cnki.gsajc.2023.000451
3. Langenhan, C., Weber, M., Liwicki, M., Petzold, F., Dengel, A.: Graph-based retrieval of building information models for supporting the early design stages. Adv. Eng. Inform. **27**(4), 413–426 (2013)
4. Chaoyi, J., et al.: Exploring BIM data by graph-based unsupervised learning. In: International Conference on Pattern Recognition Applications and Methods (2018)
5. Stouffs, R., Tauscher, H., Biljecki, F.: Achieving complete and near-lossless conversion from ifc to citygml. ISPRS Int. J. Geo Inf. **7**(9), 355 (2018)
6. Zhou, X., Zhao, J., Wang, J., Guo, M., Liu, J., Shi, H.: Towards product-level parallel computing of large-scale building information modeling data using graph theory. Build. Environ. **169**, 106558 (2020)
7. Yuemei, Z., Ge, G., Cheng, P., et al.: Research on IFC model storage technology based on knowledge base. Civil Eng. Inf. Technol. **12**(01):1–7 (2020). https://doi.org/10.16670/j.cnki.cn11-5823/tu.2020.01.01
8. Junxiang, Z., Peng, W., Xiang, L.: IFC-graph for facilitating building information access and query. Autom. Constr. **148**, 104778 (2023). https://doi.org/10.1016/j.autcon.2023.104778

The Research Logic Behind the Surreal 'Voids' – Knowledge Construction and Design Summaries from UCL's MArch Urban Design

Xueyang Miao[1](✉) and Philippe Morel[2](✉)

[1] Tongji Architectural Design (Group) Co., Ltd. No. 1230, Siping Rd., Shanghai, China
laomao0318@126.com
[2] The Bartlett School, 22 Gordon Street, Saint Giles WC1H 0QB, UK
p.morel@ucl.ac.uk

Abstract. With the increasing complexity of urban challenges, enhancing the quality of urban spaces has become a global priority. Urban Design, as a key discipline, has garnered significant attention. In China, Urban Design education is still in its infancy, whereas internationally, it has been evolving for nearly 70 years. Among global institutions, University College London (UCL) is renowned for its forward-thinking approach to design education. However, there is limited literature detailing the methodologies employed in teaching Urban Design at internationally acclaimed universities. This paper explores how UCL's MArch Urban Design program cultivates students through a research-driven approach, drawing on the author's firsthand experience. The study aims to provide insights for reforming Urban Design education in China by examining knowledge construction, teaching methods, and curriculum structure at UCL.

Keywords: UCL · Urban Design · Knowledge Construction · Teaching. Algorithm

1 Introduction

Urban Design is a discipline that pays attention to the layout, appearance, and function of a city. It focuses on urban public space, and its core is to enhance people's mental health by improving the quality of the urban environment (Transick, 1986). In China, Urban Design has been a branch of architecture for a long time, and its teaching style has inherited the mature tutoring system of architecture, that is, paying attention to current hot issues and putting much emphasis on practicality. Students' design of urban form is standardized by carrying out the design courses around the existing theories, such as form aspects like functional layout and traffic networks should conform to urban morphology and typology. Future urban environment is not often talked about in Chinese architectural design courses.

Nowadays, architects are expected to have research skills to deal with the unknown future urban problems, which is different from Chinese teaching direction that focusing on real construction. Some design programmes are prompted to start reforming from being design-oriented to combining design and research. Chinese universities such as

© The Author(s) 2025
H. Chai et al. (Eds.): CDRF 2024, *Symbiotic Intelligence*, pp. 327–337, 2025.
https://doi.org/10.1007/978-981-96-3433-0_29

Tongji University are trying to integrate new technologies in their design programmes. College of Architecture & Urban Planning (CAUP) at Tongji University is ranked 12 in the 2023 QS ranking, which better represents the cutting-edge level and direction of education in architecture and its related disciplines in China. Some of the CAUP design courses to some extent share similarities with current international Urban Design programmes. However, unlike Chinese universities, which are still in the initial stage of the discipline, the Urban Design discipline in European and American architectural colleges have a history of nearly 70 years, and University College London (UCL) from the UK enjoys a global reputation for its future-oriented design thinking. UCL's Bartlett School of Architecture has been ranked in the top three of the QS rankings all year round, and was ranked No. 1 in 2023. It has been exploring the teaching of Urban Design since the 1970s, making various interdisciplinary approach, and finally forming a unique teaching system. The author has had the honour of studying architecture at Tongji University for five years and has obtained the master's degree in Urban Design at UCL, receiving two different styles of education. The paper will take the development of UCL Urban Design as a clue, and use the actual UCL Urban Design programme that the author has experienced as an example to discuss the teaching targets and methods in different stages of its design course, and to show the method of its knowledge construction.

2   Existing Research and Problems with IT

2.1   Current Research on the Teaching of Urban Design at UCL Bartlett School of Architecture

In terms of Bartlett's teaching, students are encouraged to use research to drive design, and construct their own theoretical systems.[1] Existing literature has researched Bartlett's approach to student development through its student work, finding that Bartlett's students are taught to question the accepted qualities of architecture and the urban environment (Chen, Ren, 2015). In addition to outside research on the teaching styles of Bartlett, the school's own faculty have written articles to show their research directions. Professor Pearson from Bartlett found that young people tend to develop their own feelings and understanding of urban environment through video games (Pearson, 2019), so he set up the 'Game Studio' in Bartlett Urban Design. Professor Pasquero has attempted to transform urban environments to reintegrate them with nature by referring to the living habitats of cyanobacteria (Pasquero, 2020), and her studio keeps exploring the inspiration of urban evolution from plants growth. Similar stylistically strong topics can be seen in several studios at Bartlett, and some articles are calling for design classes in Chinese architectural colleges to explore the innovativeness shown in these topics to inform future design curriculum reformation (Jiang, 2019).

---

[1] https://issuu.com/bartlettarchucl/docs/the_bartlett_school_of_architecture_pg_guide_2024.

2.2 Existing Research on Teaching Methods and Related Practices in Urban Design Classes

Some studies argued that it is appropriate for China's Urban Design programme to be set up based on architecture like what North American colleges have done (Jin, 2018). Similar views are also found in the article by Professors Zhuang and Ye of Tongji University, and they found that research-based topics are common in Urban Design programmes of internationally renowned universities, and they should be gradually introduced into Chinese Urban Design classes (Zhuang, Ye, 2017). Some European Urban Design programmes encourage innovative thinking and allowing for some shortcomings in design ability (Dang, Lei et al., 2023). One research discovered that Chinese students lack training in thinking and creativity compared to European students (Ding. 2015). In terms of China's future Urban Design curriculum arrangement, Professor Sun of Tongji University gave a teaching advancement plan for students to quickly master the basic Urban Design ability (Sun, Xu. 2021).

Current Urban Design teaching practice reflects the mainstream teaching methods and innovative approaches in China. Professor Cai of Tongji University has elaborated on the use of urban morphology in Urban Design courses, which to some extent can reflect today's teaching methods of Urban Design (Cai, Jia, Xu, 2021). In terms of curriculum innovation, Professor Cui of Tongji University has for the first time offered a design class on the theme of digitally built environments, encouraging students to develop new design strategies.[2]

2.3 Problems with Existing Research

Although there has been a great deal of research on Urban Design training methods, most of it remains at the level of exploring the teaching framework, without in-depth discussion of the teaching content. There is still a lack of research on the detailed teaching process of international design courses. Current innovative design class is often limited to one semester, which is difficult to cultivate students' ability to think completely. The current conventional Chinese design course is still a practical education mode, lacking the ability to research future urban problems, and there is a big gap between it and the international research-based design teaching. The paper will describe the actual learning experience in a studio of UCL Bartlett's Urban Design, provide some lessons on the experience of research-based design classes, and show a specific perspective as a student when facing such courses.

3   UCL's Approach to Teaching Urban Design

3.1 The Development of UCL Bartlett's Way of Students' Cultivation

Bartlett encourages students to think creatively, and its teaching mode was refined over a protracted period of experimentation and adjustments. Bartlett School of Architecture was founded in the 1910s. In the 1960s, under the influence of new technologies

---

[2] https://www.scla.com.cn/Cn/Index/pageView/catid/16/id/572/ip/12.html.

and ideas, Professor Richard Llewellyn Davies of Bartlett led a reorganisation of the College, introducing studio-based design courses and combining science and art with them. Bartlett began to make interdisciplinary endeavours and to expand the research area.[3] The situation was further boosted by Professor Peter Cook, the Chair of Bartlett from 1991–2005, and during his tenure he introduced a number of young architects to Bartlett, including many from multi-disciplinary backgrounds. Cook sparked new thoughts within the School like sustainability and technology, and the tutoring focus was changed from engineering to design. Lots of new studios were set up and research areas were expanded (Cook, 2000). The thinking about cutting-edge fields was gradually integrated into Bartlett's tutoring system, being refined and expanded over a period of nearly 30 years, and eventually evolving into the current mode. The development of Bartlett's teaching is actually a process of exploring how to incorporate multidisciplinary thinking and a process of deepening thinking about architecture and the built environment (Borden, 2009).

3.2   Curriculum and Assessment Criteria of Bartlett Urban Design

Bartlett's MArch Urban Design was one of the first Urban Design programmes in the world to be established following Bartlett's reorganisation of the discipline in the 1960s. It is a one-year postgraduate programme with three semesters. The programme has five compulsory modules, and 'Urban Design Thesis Final Projects' is the highest valued. At the end of the first semester, students are required to confirm their research direction and submit it in the form of a 1,000-word thesis in English, while at the end of the second semester, students are required to present their latest design results. At the end of the third semester, students should submit not only the final design, but also a final report, and to design a panel for the Bartlett's Summer Show, which takes place in September each year. It is an exhibition that faces to the social public, though not worth any marks, it is still very important as it is possible to meet architects from famous studios like Foster + Partners and, if you are lucky, discuss employment possibilities with them. The final judgement of the project is based on the degree of completion of the work, the degree of innovation of the design concept and the logic behind the design concept (Fig. 1).

3.3   Characteristics of the Bartlett Urban Design's Teaching Mode

Studio instruction is the teaching mode of the design course of Urban Design. Teachers in each studio will add appropriate theory classes according to the needs of the research topics. Each workshop is staffed by 2–3 teachers with different research backgrounds. Currently there are 5 studios, with about 15–20 students per studio. When students first enter the school, there will be a two-week Introduction Week, during which studios' teachers will conduct trail classes to show their research directions. After the trial lectures, students need to submit an order of the studio he or she wants to join in online, and the final choice is determined by an online system. The studio the author attended had two tutors, one of whom had a dual undergraduate background in architecture and computer science, while the other was running a 3D printing company and had a wealth

---

[3] https://www.ucl.ac.uk/bartlett/about/our-history.

Fig. 1. Bartlett MArch Urban Design's curriculum and purposes of each course

of experience in model representation. It proves that the disciplinary backgrounds and practical experiences of the instructors of Bartlett Urban Design are diversified.

4 Learning Experience At Bartlett's Urban Design Studio

4.1 Introduction of Studios

The official name of the studio at Bartlett is Research Cluster (RC), and the author attended RC11. The topic of RC11 is "Deconstructing Central Business Districts (CBDs) and designing their future with the help of computer algorithms". The project was based on La Défense, the largest CBD in Paris. Developed in 1958, La Défense was one of the first CBDs in the world, and is now home to the majority of Paris' high-rise office buildings. However, La Défense is now facing problems such as an increase in the vacancy rate of office space, a decrease in the quality of space, and a homogenous way of life. The target of RC11 is to envision the future of life in La Défense (Fig. 2).

	Topic	Research Content	Tutors' research background
RC11	*Hidden Dimensions*	Speculating AI's impact on the future CBDs	*Philippe Morel:* Architecture; 3D-printing *Julian Besems:* Architecture; Computer Science
RC12	*Videogame Urbanism:* *Welcome to the Metaverse*	Explore urban building logic in videogames	*Luke Pearson:* Architecture; Videogames *Sandra Youkhana:* Architecture; Videogames
RC14	*Machine Thinking Urbanism -* *Cities Beyond Recognition*	Based on specific global issues, designing with very large datasets	*Roberto Bottazzi:* Architecture; Urban Design; Big Data *Tasos Varoudis:* Architecture; Computer Science
RC16	*DeepGreen City*	Explore the growing patterns of specific plants, and use them on design	*Claudia Pasquero:* Architecture; Biology *Filippo Nassetti:* Architecture; Computer Science
RC18	*Relational Urbanism: From the* *Molecular to the Planetary*	Design a new urban model that could correct negative environmental externalities	*Zachary Fluker:* Architecture; Big Data *Enriqueta Llabres-Valls:* Architecture; Economy

Fig. 2. Information about each RC of Bartlett MArch Urban Design in 2021–2022

4.2   Target in Different Phases

The teaching purposes of RC11 in 3 semesters were: constructing a knowledge system, building a design framework, and deepening design. RC11 allowed students to conduct research on the site and to explore the desired research direction for one semester, during which students would not do any design to avoid the expected design results affecting the research process. There were 16 students in RC11, and the project was a cooperative one with 4 students a group. Students would first establish groups of 2. Each group chose 1 topic from the tutors' given 8 to do relevant research for half a semester. The eight topics included traffic, logistics, carbon emission, carbon neutrality, landscape, economy, IT, and big data. After the communication of research results by each group in the half-semester presentation, each group would combine with one other group. The new groups integrated the research contents and expanded them. At the end of the Semester 1, every group determined the research direction that would be moved forward in the Semester 2. In addition to research, students would take complementary technical courses in Semester 1, and in RC11 would teach Python. On-site survey of La Défense was also part of the theoretical research.

The second semester was the process of refining the research direction and eventually developing a design framework. Tutors helped find out the unexpected aspects of each group's proposals. Although tutors would let students broaden their minds, they would not allow abstract topics to appear, so each group was required to describe the theoretical logic behind the proposals in detail. When the logic of the proposal did not make sense, tutors would decisively question it. Through this teaching mode, groups were able to produce a workable design framework in the second semester.

Semester 3 was about deepening the design and reaching to a result. The form should be generated based on the research results got in Semester 2. RC11 was more concerned with the process of generation than the outcome. Whereas the traditional design mode is "I imagine about this form and then approach my design to this form", in RC11 it is "I have worked out a complete design system and am looking forward to seeing what this system will lead to". The only certainty in the entire project was the theoretical system and design logic constructed by the group itself (Fig. 3).

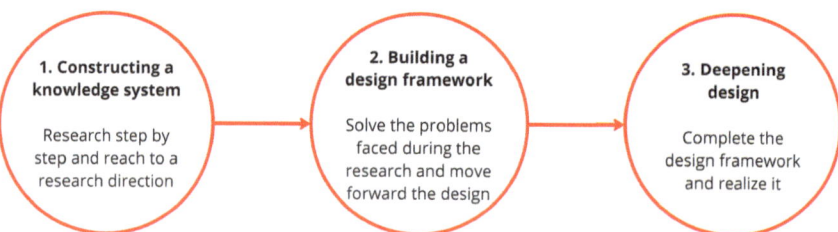

Fig. 3. Three phases during the design

4.3   Research Process and the Inspiration

The author's group chose traffic as the research aspect in the first semester. The first thing learnt during the half-term research was the way of researching from point to surface. The research started with the development of the transportation ways and the road network in La Défense, and then added the study of the commuting ways to La Défense from the Parisian districts at the suggestion of the tutors, which revealed the problem of the excessively long commuting time. Then the research led to the query of why employees did not want to live in La Défense, and finally discovered that the development of the road network of La Défense had the tendency to make the area isolated and thus led to a low-quality living environment. The research content was expanded and finally a relatively complete set of results was formed. After the mid-term, the traffic group formed a new group with the logistics group, because the two groups had come to some coincidental conclusions, such as current function proportion in La Défense was not properly.

At the end of the first semester, the group initially decided the research direction–the optimisation of the supply chain in La Défense. By the second semester, tutors, on the one hand, found it interesting to inverse the urban form from the supply chain, but on the other hand, they pointed out that the supply modes needed to be focused. In addition, one of the tutors' remarks struck the author, *"Efficiency is not something that design is responsible for, as it's a matter of technological upgrading. Design ultimately comes back to urban space and urban life."* The words could be indicative of the approach to design in Bartlett Urban Design–designing at UCL is not about showing off technology, but rather it is based on real-world environments, and thinking seriously about the future of urban space and urban life.

The group discovered the past agricultural background of La Défense through the research of its history. Taking into account the concerns of the Paris people about the quality of their food during the COVID-19, the group focused the design on the food supply chain and improved the design framework, finally settling on the topic to be done in the third semester–La Défense's Plantation Transformation.

All the research would be used for designing in the third semester. In this stage, the elements of the previous design framework were transformed into variables, which were then used to move forward the design by designing computer algorithms. However, the implementation of the code is based on trial and error, so some students would make compromises and simplify a certain logical aspect in order to make the code run successfully. To avoid the situation, tutors helped a lot with the technical aspects, and they would also pay attention to the progress of the project every week and judge whether there was any point could be optimised based on the generated results. The final design outputs included a portfolio describing the research process, a site model and a section model. The portfolio should include all outputs from Semester 1 to Semester 3. Though there is no requirement for the number of pages of the portfolio, it is usually more than 80 pages.

5   How Computer Algorithm is Used in the Design Course

5.1   Purpose of Using Algorithms

Algorithms are used to deal with the complex relationships among variables. Tutors mentioned that algorithms would not be used until the current design framework was completed, and the depth of the research should be sufficient. In addition, they said that algorithms should to be designed by ourselves and they must be progressed step by step through the logic of the design framework. Therefore, the prerequisite for using algorithms was to have sorted out the thread of logic that runs through the steps of the design.

5.2   Conditions of Using Algorithms

The problem identified in the previous research was that the current traffic and logistics flow was determined by the functional layout and ratio of La Défense. La Défense was currently dominated by offices, with a high vacancy rate of office space and residential areas, so the first thing that needed to be done was to change the ratio and layout of functions. In addition, the group wanted to increase the mix of functions in order to cope with the problem of the limited active hours in La Défense. Proportion, layout and degree of mixing interacted with each other and were difficult to deal with by usual way of design, but if they became variables and the rules for function transformation were set, it is possible that they could be moved forward to create unexpected results. In this case, algorithms began to be used.

The functions of La Défense were first classified into five categories, after which the buildings of it were deconstructed into functional units. Each unit followed a transformation rule, which could be expressed as "when there are X identical functional modules in the periphery of a functional module A, A will be left as it is/converted into a functional B/converted into a 'Void' ". 'Void' was a special type of module that represented no function. Then the transformation logic was input into the deconstructed model by Python module of Grasshopper. The algorithm was named after 3D-Cellular Automata because of the similarity of functional transformation logic with Cellular Automata.

5.3   Unexpected 'Void' Space

The final result reached to a relatively stable state after several times of running of the programme. It was found that the proportion of office functions had dropped considerably, while the proportion of residences had increased. The functions were also integrated with each other to some extent, so it could be considered that a good base had been provided for the subsequent design. Furthermore, 'Void' formed clusters in different areas and were wrapped by other functions. If viewed in the language of architecture, the state of 'Void' space was similar to the atrium in a building. However, from an algorithmic point of view, void means "empty" and therefore does not refer to any kind of space (Fig. 4).

3D-Cellular Automata is an algorithm that embodies the urban evolutionary process, so the generated model needed to be further interpreted before they were translated

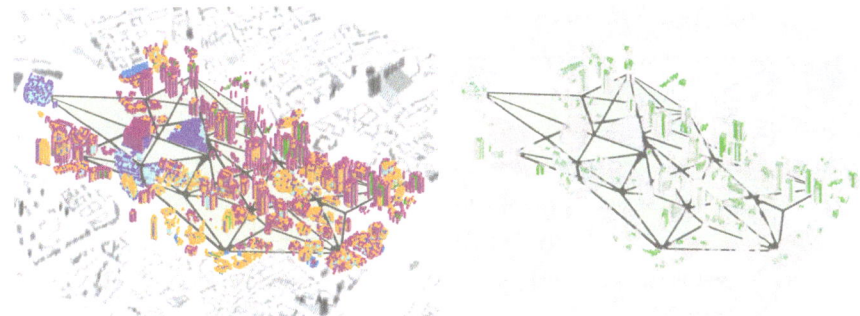

Fig. 4 The new deconstructed La Défense and the 'void' space (green colour)

into the architectural language. According to the logic behind the algorithm, 'void' was generated after the saturation of the same type of function, meaning the space wastage. Therefore, the new deconstructed La Défense transformed by 3D Cellular Automata could be interpreted in such a way: La Défense adjusted its functional ratio and distribution state during the generation-by-generation evolution process and discharged the wasted space.

5.4 Influence Brought by the Appearance of 'Void' Space

The target of 'La Défense's Plantation Transformation' was to realise the supply of some crops from production to consumption in order to increase the food security within La Défense, and to increase the variety of activities through planting. 'Void' was wasted space and therefore could be part of the food supply chain. The appearance of the 'Void' and the interpretation of 'Void' space was favoured by tutors for they were 'unexpected results'. In subsequent steps, 'Void' space was reused as a function related to indoor plantation, and the proportion and distribution of functions were generated by the rewritten 3D Cellular Automata algorithm. During the generation process, the type of crop, suitable planting time and location, and growth cycle were also added to the algorithmic variables, which acted on the void while controlling the planting conditions in the outdoor environment. The results were promising: the modules represented by the crops showed organic variations from month to month, showing the process from growth to harvest, and new crop proportions were added at the beginning of the next planting month, creating varied planting programmes. It means that La Défense's Plantation Transformation could bring about a variety of planting plans, dynamic landscapes and functional proportions that can be adjusted. The emergence of 'Void' space inspired the group that each step of the design process might have its meaning (Fig. 5).

5.5 Impact of the Process of Using Algorithms on the Way of Thinking

The first thing the author learned is that the process of designing algorithms was visualising the research in 3 semesters, and the project design became a system design. Different from the past form-first design mode, system design pays more attention to the process of changes in each stage. System design is rooted in the actual research results and

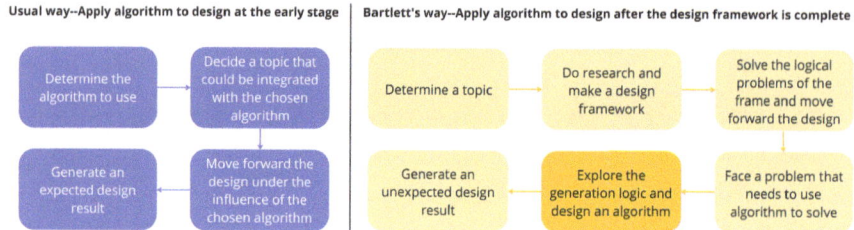

Fig. 5. How algorithm is applied to design in usual way and in Bartlett's way

not limited by the form. In the process of deepening the design, the idea has also gone from being apprehensive at the beginning to being full of anticipation in the follow-up. Although the form was unknown, it must be rigorously logical, and even though it might not be accepted by the mature experience of the present, it could represent the future to some extent (Fig. 6).

Fig. 6. Generate functions in the 'void' space and the final result of the project

6   Conclusion

It can be seen that behind the divergent thinking of Bartlett Urban Design's projects is the focus on students' thinking as well as the generation process. The fact that students' work is influenced by the strong personal style of the instructor is one of the problems in Chinese design courses. Future Chinese Urban Design education could try to get rid of the prospection of the "correct form" in order to realize a more in-depth research-based design. In addition, the cultivation of students' research ability is a long-term process, so there should be a coherent research-based design teaching system in Chinese Urban Design education. The author hopes that the paper could provide a reference for the educational reform of the Urban Design discipline in China and serve as a window to generate an understanding of UCL Bartlett School of Architecture.

References

Trancik, R.: Finding Lost Space: Theories of Urban Design. John Wiley & Sons (1986)

Chen, K., Ren, Z.: Future-oriented architectural education and the cultivation of innovative thinking–a case study of the bartlett school of architecture at UCL. Architectural J. **03**, 95–100 (2016)

Pearson, L.C.: A machine for playing in: exploring the videogame as a medium for architectural design. Des. Stud. **66**, 114–143 (2020). https://doi.org/10.1016/j.destud.2019.11.005

Pasquero, C.: Deep Green. TOPOS **112**, 24–30 (2020)

Jiang, B., et al.: Development and application of "research by design" in contemporary architectural education: at top architectural schools as exemplified in four institutions. Architect **02**, 80–88 (2019)

Jin, G.: Urban design education: the northern America experience and the future road map of China. Architect **01**, 24–30 (2018)

Ye, Y., Zhuang, Y.: Professional education programs in urban design: a comparison among multiple internationally renowned universities. Urban Plann. Int. **32**(1), 110–115 (2017)

Dang, Y., et al.: Architectural education in response to industry transformation-practicing the pilot reform of architectural training programs in Xi'an university of architecture and technology. Architectural J. **06**, 109–114 (2023)

Ding, W.: Transition and transformation-reflections on the knowledge system of architectural education. Architect. J. **05**, 1–4 (2015)

Sun, T., Xu, K.: On the development of urban design specialty in terms of the core elements of the architectural discipline. time + architecture (2021)

Cai, Y.: Outline for concise operation of urban design. Urban Environ. Des. **04**, 340–346 (2021)

Cook, P.: Bartlett Book of Ideas. Bartlett Books of Architecture, London (2000)

Allen, L., Borden, I.: Bartlett Designs: Speculating with Architecture. John Wiley & Sons, Chichester, West Sussex, U.K. (2009)

Adjacencies of the Real: Scholastic Construction in the Twenty First Century

Sean Pickersgill[✉], Damian Madigan, Andrew Lymn-Penning, and Darcy Holmes

University of South Australia, Adelaide, Australia
{sean.pickersgill,damian.madigan,
andrew.lymn-penning}@unisa.edu.au

Abstract. This paper provides an extended understanding of the categorical qualities of intentionality in Don Ihde's Intentionality Matrix (1990) and in P.P. Verbeek's subsequent inclusion of Hybrid and Composite Intentionality. By examining the interrogative posture of Ihde and Verbeek's categories we find that there is insufficient understanding of the dynamic characteristics of these forms of experience. Through an explanation of a pedagogical simulation project undertaken with tertiary architecture and construction students, we provide a more accurate designation of the complex existential modes of attention in engaging with digital twin simulations which rely on authenticity of the modelled environment, and authenticity of the learning experience.

Keywords: architecture · simulation · digital twin · authenticity · intentionality

1 Introduction

Since 2015, the OnSite team at the University of South Australia have been experimenting with the implementation of game-engine software into the digital twin space of construction education. Though initially explored as an expedient alternative to in-person site visits, the project has evolved to encompass a deeper analysis of the role of emulation software and the mediation of the real within architecture.

In constructing and implementing a legitimate digital twin version of a construction site, our team recognised that a core aspect of its pedagogical success rested on the standard of fidelity the simulation model displayed with respect to real-world qualities. A constant within the work program was the accurate modelling of construction components, realistic lighting and texturing effects, entourage items and a user-experience that mirrored a 'real-life' site inspection. To achieve this, we employed Epic Games' Unreal Engine to make use of the contemporary game-engine advancements in immersivity, data management and importantly interactivity. These are key requirements within the contemporary game industry for AAA games and it was logical to borrow this technology for the purpose of the digital twin.

Reflecting on this process, and undertaking empirical observations of user behaviour, we recognised that the fidelity to recognisable construction served three purposes: 1. It validated the technical accuracy of the model and served as a mode of review of the

H. Chai et al. (Eds.): CDRF 2024, *Symbiotic Intelligence*, pp. 338–347, 2025.
https://doi.org/10.1007/978-981-96-3433-0_30

proposed construction process, 2. It validated to student observers that this model had legitimacy as a real-world artefact and that 3. The mode of engagement with the model provided legitimate phenomenological experiences relative to scale, material, lighting, and proprioceptive experience.

Considering Verbeek's notion of composite intentionality, the exercise of implementing digital twins into an educational context reminded the production team of the legitimate value of fidelity when it is directed towards the purpose of defining both an objective state of affairs (how a building is constructed) and that this process (architectural design) is purposive in implementing specific phenomenal values. Moreover, the technological architecture of this world-making is designed to present a non-real location as extensive with 'a' reality. In short, constructing a digital twin of a construction site served the user's fascination with a 'reality' that was delivered by a medium acknowledged to be fully mediated. The purpose of the process was to illuminate the adjacency of the indicators of the 'real world' (accurate phenomenal experience) within a palpably artificial setting (accurate false consciousness).

In this respect the ethics of these worlds becomes startlingly clear. To maintain credibility and functional viability, the model has to maintain an adjacency to reality and, as a consequence, be 'true' to the general test of real-world construction practice – even if the OnSite environment is inherently synthetic. In the true sense of a digital twin emulating a real-world counterpart, the test is not to assess it against an as-built reality, but whether there are world coherences between ours and the model's. As employed, the OnSite model exists for the purpose of asking what are the procedural rules for undertaking a construction process, and in what ways do the components of that world reveal the world-forms (National Building Regulations, vernacular trade practice, professional accreditation/registration standards) from which they emerge? Any indication that these standards are relaxed or missing invalidates the commitment to fidelity and the strength of claims to a version of Verbeek's composite intentionality.

This paper will present the evolution of critical thinking that emerged in the design and construction of the digital twin project, OnSite. It will discuss the importance of fidelity and phenomenal immersivity in the user experience and provide commentary on the effectiveness of understanding the tension between 'real' and 'non-real' markers within the pedagogical experience.

2 The OnSite Experience

OnSite was originally developed in 2015 by Sean Pickersgill and Jennifer Harvey as a protype for experiencing a domestic construction site via an early version of the Unreal game engine.[1] In its current form it is a digital twin environment that replicates a domestic-scale construction site with exacting precision. The construction methods replicate contemporary practice from the conventions of site preparation and set out, through the various phases of the construction process, to practical completion and handover. The highly detailed model is the product of professional input at all stages of the

---

[1] Pickersgill, R. S., Rameezdeen, R., & Harvey, J. (2020). OnSite: the virtual site visit as an environment for construction learning. In *OnSite: the virtual site visit as an environment for construction learning*. IGI Global. https://doi.org/10.4018/978-1-5225-8452-0.ch009.

construction sequence, with the acceptable exception of nails, glue and screws, ensuring that the fidelity of the digital model is paramount. The detailing and construction methodologies are consistent with statutory Australian Standards (the model is based within the context of the Australian construction industry), and the material choices reflect a logical sample of current choices within contemporary suppliers. Allied to this is the capacity within the digital twin to view various stages of the construction process at will. A core governing principle of this staging was to emulate likely moments when appropriate scrutineering of the construction would take place in a real-world scenario.

Users of the model are able to switch between different construction phases in order to develop a visual understanding of the complexities of a building form – from the deployment of monolithic mass to frame and cladding solutions. In short, the user can build up or peel back the layers of construction and do so repeatedly. In addition, the modelling of the environment generally includes entourage elements consistent with the stages of construction. Pivotal in this approach is the recognition that the didactic goals of OnSite require the user to feel that there is an 'authentic' element to the experience. OnSite, as a fundamentally synthetic experience, strives to mask its artificiality through its technical accuracy along with its situatedness in a 'world' that is consistent with expectations of construction processes. It is the argument of this paper that the general trajectory of OnSite's use of a 'realistic' environment is fundamental to its success and constitutes an extensional aspect of the human-technology relations discussed by Verbeek (Table 1).

Table 1. Ihde's Matrix

embodiment relation	(human – technology) $\rightarrow$ world
hermeneutic relation	human $\rightarrow$ (technology – world)
alterity relation	human $\rightarrow$ technology (- world)
background relation	human (- technology – world)

3 Ihde's Matrix

In terms of Don Ihde's (1990) schedule of human-technology relationships, OnSite engages a number of aspects of his matrix.[2] Ihde, of course, distinguishes between those forms of technology that mediate our experience of the world, and those that constitute the mediated experience of the world itself. This is a particular characteristic of games in general, particularly those that place emphasis on the emulation of real-world environments. There is considerable scholarship on the nature of realism in serious simulation environments, particularly those designed to emulate real-world behavioural situations. For example, Neo et al (2021) detail a number of VR experiences in which indexical content identifies the context of the test environment, assisting in ensuring a

---

[2] Ihde D. (1990). *Technology and the lifeworld: from garden to earth.* Indiana University Press.

level of 'presence' for the experience.[3] Because these tests are based on conventional scientific methodologies, they require a level of consistency that a synthetic environment shows. Arguably, however, when reviewing screen images, they are consistently low in quality in comparison with contemporary forms of digital environments, particularly those associated with the commercial games industry. For this reason, there is never a question that the level of world representation is inherently synthetic even though it semantically functions as a signifier of the real world.

Nevertheless, following Ihde's matrix of relations, it is clear that the mode of engagement between participants and their environment in even the simplest form of VR environment is more complex than it first appears and that the forms of immersion that are fundamental to digital twins are a hybrid of the original classifications. Verbeek (2008) proposes an addition to the matrix that includes the designation of 'composite intentionality' which comes after he has made an argument for a form of 'cyborg intentionality' that addresses the emergent reality that prosthetic devices of all kinds have generally disappeared below the horizon of consciousness and can be considered coeval with human experience. His examples include behavioural interventions via neurological implants to pharmacological regimes and associated medico-technological interventions in the human condition. There is a legitimate reservation regarding the broad generalisations of 'cyborg' or human enhancement here, but for the purposes of reviewing the creation of game worlds in general and OnSite in particular, we can acknowledge Verbeek's general strategy.

4   Verbeek on Composite Intentionality

To get to this point, Verbeek's speculation on a further stage of human and technology relations represents itself by referring to the original actants in the matrix, including the dash and the directional arrow. He says:

> Ihde's schematic representations of human–technology relations do not only contain arrows, indicating intentionality, but also dashes, indicating a relation between entities which is not specified further. If we limit ourselves to the embodiment relation and the hermeneutic relation – which are the most relevant relations in the context of intentionality since they ultimately involve relations with the *world* – these dashes indicate a relation between humans and technology or between technology and world.[4] (Verbeek, 2008)

For the purposes of this essay, we would like to concentrate on the term 'world' that Verbeek uses, in particular in the context in which he is assuming that the world itself is a stable entity that is somehow 'out there'. In general, his definition is stable as he is referring to the states of affairs that exist externally to the individual, but it is worth also remembering Verbeek's earlier assertion that the world is always 'for'

---

[3] Neo, Jun Rong Jeffrey, et al. "Designing Immersive Virtual Environments for Human Behavior Research." *Frontiers in Virtual Reality*, vol. 2, 2021, https://doi.org/10.3389/frvir.2021.603750.

[4] Verbeek, PP. Cyborg intentionality: Rethinking the phenomenology of human–technology relations. *Phenom Cogn Sci* **7**, 387–395 (2008). https://doi.org/10.1007/s11097-008-9099-x.

something or is constituted via the apperceptions of conscious experience. The 'world' hence, for Verbeek, we take to mean: 'the world as it is mediated by technology', with the expectation that these mediations are non-trivial.

Verbeek's model of composite intentionality is intended to capture the array of information gathered through the capacities of technology to present (and represent) aspects of the external world that are otherwise hidden to human perception. So the radio telescope that gives a representation of cosmological events beyond human experience nevertheless provides access to a domain of information that is legitimately novel, insofar as it is not simply a representation to us but evidence of empirical realities beyond conventional experience. Verbeek characterises it in this way (Table 2):

Table 2. Verbeek's Amendment

composite relation	human $\rightarrow$ (technology $\rightarrow$ world)

We are unsure that this form of intentionality really extends the scope of intentionality as Verbeek intends it, since the translation function of representing the data from, in this case a radio telescope, effectively returns it to the hermeneutic mode of relations. Within the system design of radio telescopes, the capacity to harvest and recognise aspects of the invisible spectrum of waveforms present in the 'world' is baked into its representational capacities.

Verbeek (2008) goes on to give examples of this form of experience, including examples of long-exposure photography and synthetic 3D image making.[5] These examples, in our view, are the weakest elements on the paper as they require additional discussion for their operations to be credible proofs of Verbeek's larger thesis of composite intentionality. That said, it is clear that the general methodology for adjudicating between the human/technology/world triumvirate has merit and describes aspects of human and technological experience that are grounded in the general ambition to provide clarity in the discussion. In the alternative to Verbeek's composite intentionality, and to move away from the examples he has chosen, we think there is more to be discerned in the unique aspects of world-making within computer 'game' environments in general and the phenomena of digital twins in particular. For the purposes of this paper, we will call this Unstable Intentionality.

5   Unstable Intentionality

Ihde's initial matrix of intentionality identifies differing relations between the human subject and world and between different instances of technology. This, of course, reflects the general purpose of technology – to externalise and reproduce certain human functions, and to engineer a solution that demonstrates necessary and sufficient utility. Ihde's matrix covers single instances but does not consider situations in which multiple relations exist simultaneously. For example, if a person wearing glasses (embodiment),

---

[5] Verbeek, 2008, p.394.

approaches a computer (alterity) that is providing a live feed from a security camera (hermeneutic), then what is the nature of that experience? Is it the sum of its instances, or some other condition? What if one aspect of technology cancels out, or neutralizes, one or more of the others? Is there a procedural priority that should be observed – much as there is within modal logic?

In our view the discussion is, as yet, an unfinished enterprise. The OnSite project outlined above demonstrates some of the complexities of experience that our concern regarding multiple intentionalities raises. In the OnSite Synthetic World (SW) example, the commitment to model the site construction environment with a fidelity to Real World (RW) examples is based on the pedagogical need to encourage authentic experiences for students which we have detailed and discussed in previous publications.[6]

6 Authenticity

Authenticity, in this respect, is a significant term because it attaches to the general pedagogical methodology for e-learning. We relied on, and confirmed, a definition of 'authentic learning' that came from Herrington and Parker's identification of the crucial nexus between learning and engagement in the development of online tools (emerging technologies as they characterise it).[7] Authenticity, in this respect refers to 'Authenticity of Context', 'Authenticity of Tasks and Activities' as well as 'Authenticity of Assessment'. There are other conditions that refer to the pedagogical scaffolding of the process, but for our purposes the issue of context, tasks and assessment are crucial.

The OnSite project addresses the challenge of authentic context by emphasising the comparative realism of the model. A core aspect of the authenticity is the accurate modelling of the projected construction site (Figs. 1 and 2).

Fig. 1. (left): Limestone House, Kew (Photo: Wardle Studio, 2021)

Fig. 2. (right): OnSite Digital Twin, (Screen capture: Authors, 2024)

The accuracy of the model was critical in order to achieve the complex pedagogical task of convincing the student that the synthetic digital twin contained the *exact* form of information that would have been obtained by a site visit. While there is no question for

---

[6] Pickersgill, R. S., Rameezdeen, R., & Harvey, J. (2020), op.cit.

[7] Herrington, J., Parker, J., (2013), 'Emerging Technologies as Cognitive Tools for Authentic Learning', British Journal of Education Technology, Vol44 No4, pp.607–615.

the student that the OnSite model was, in some confused ontological sense, the real thing (or indivisibly RW engaged), it was crucial that this level of uncertainty be developed and maintained to sustain engagement (Fig. 3).

Fig. 3. OnSite Digital Twin 2024, (authors)

In essence, this suspension of disbelief, reproduces the forms of aesthetic engagement typical of the aesthetic attitude towards, for example, theatre or film.

Discussion of this form of aesthetic attitude predates the technology we have been discussing since it refers to levels of engagement within a theatrical performance and the audience's legitimate emotional reaction to events on a stage – a familiar discussion that dates to Aristotle.[8] Nevertheless, there is clearly established thinking on the encounter with representations of reality, whether it is image making specifically or simulation generally.

Ihde, in a text subsequent to the embodied intentionality argument, describes this as 'perceptual isomorphism', pointing to the evolution of simulation technologies that close the gap on the phenomenological experience of certain activities such as flight simulation.[9] Ihde qualifies the proposition that these technologies are moving towards seamlessness with human experience, he describes it as a naivete, with the observation that the entry to this experience is itself a 'framed' encounter.[10] The pilot enters the flight simulator and in so doing deprives themselves of the multi-sensory reality of actually flying an aircraft, or an astronomer observes a planet via a telescope and in so doing is engaged with a technologically enhanced version of observable phenomena. In these examples Ihde is concerned with the posture of embodiment the phenomenologically aware individual is experiencing in the observation of a simulation. Importantly, though this point is not made by Ihde, we can understand that the status of the simulation and its validity as being real is in essence a question of semantic clarity.

---

[8] Aristotle., Baxter, John., Atherton, Patrick., & Whalley, G. (1997). *Aristotle's Poetics*. McGill-Queen's University Press. https://doi.org/10.1515/9780773566606.

[9] Ihde, Don. (2012). *Experimental Phenomenology Multistabilities* (2nd ed.). State University of New York Press., p.135.

[10] Ihde (2012), p.136.

7 Semantic Stacking

If one imagines a representation of a planet obtained via radio astronomy, say Pluto, then the image observed has been mediated by technology through a number of representational strategies – size, shape, detail, focus etc. Each of these representational 'improvements' answers a specific interrogative need of the observer and are, of course, modes of abstraction that answer the semantic question: What does Pluto look like? In this instance the more sophisticated the stack of representational strategies, the closer the representation appears to be 'real'. In its simplest form, this representation might simply be a circle, and then as the sophistication of the semantic stack improves the experience moves towards our expectations of what a real observer may see from their spacecraft in Pluto's orbit (should this ever occur). Our argument is that this example, like the discussion of Verbeek's composite intentionality above, relies on a static posture towards information when a more fulsome description would acknowledge that consciousness of the virtuality of this system is always and necessarily present.

Ultimately, Ihde decides that all forms of simulation are compromised by the very technological architecture that supports them – their externality from human physiognomy. If one's aim, and we argue it is Ihde's, is to search for and value the phenomenologically rich, then all forms of interventionist technology are lacking. This attitude is present in his original matrix and that of subsequent commentators such as Verbeek. However, and this is our argument, the state of deliberation between aspects of the real that come from highly immersive simulations are in and of themselves tangible and meaningful existential encounters.

8 Summary

The deployment of the OnSite model and the accompanying effort to demonstrate heightened levels of reality require the user, in our case the student populace, to deliberate on the reality status of the environment generally and the objects within it. There is a continual, simultaneous debate regarding the status of objects that are recognised to be accurate representations of a complex reality and that their location communicates a legitimate strategic practice of placement (i.e. how buildings are constructed, including this one in particular). The accurate matching of Synthetic World (SW) and Real World (RW) characteristics relies on the complex recognition that the SW's reality is constantly being phenomenally tested. We propose that this form of intentionality can be represented int the following form (Table 3):

Table 3. Semantic Stacking

unstable relation	human $\leftrightarrow$ (technology $\rightarrow$ world)

In this table it should be noted that the two-headed arrow indicates reciprocal relations between phenomenological experience and the recognition of the *question* of synthetic relations. It is a dynamic relationship that anchors and explains our disposition to increasingly sophisticated representations of the real. Our relationship to the quiddity of RW/SW relations is constantly and essentially interrogative since we do not question whether the SW model is synthetic, we are already predisposed to this introductory state, but we do question and reflect on its uncanny sense of reality.

The type of authentic learning that flows from this experience is based on the confidence that seeing and observing the representation of reality is actually a more existentially full problem to be deliberated on than the simple process of observation. More simply, the student has to ask themselves: Is this accurate-appearing representation of a reality both a good representation of the world and is it believable as a legitimate learning experience?

Technology, and in particular the technology of simulation, exists for the purpose of asking these questions of the observer. We have characterised it as 'unstable' because of the acknowledgement that an observer is unable to hold two different perceptions of reality simultaneously. An observer cannot simultaneously believe that their experience is perceptually significant while doubting its reality because of its synthetic mode of production. It is in the very act of switching between these modes, to the degree that an observer senses these differences, that the structural instability of the simulation expresses itself.

As a general principle in developing the learning and teaching strategy for the OnSite project, we have striven for an acute level of representational fidelity, as described above. Our general pedagogical model, based as it is on game-engine technology, is a combination of exploiting the point-of-view characteristics of first-person vision of this medium with accurate modelling, as discussed. By relying on the parallel experiences of immersivity (in the model environment) and estrangement (self-consciousness of mediality) we achieve learning objectives that were originally and organically assumed to only be available to the phenomenological experience of the actual site visit.

References

Aristotle, B.J., Atherton, P., Whalley, G.: Aristotle's Poetics. McGill-Queen's University Press (1997). https://doi.org/10.1515/9780773566606

Herrington, J., Parker, J.: Emerging technologies as cognitive tools for authentic learning. Br. J. Educ. Technol. **44**(4), 607–615 (2013)

Ihde, D.: Technology and the Lifeworld: From Garden to Earth. Indiana University Press (1990)

Ihde, D.: Experimental Phenomenology Multistabilities, 2nd edn. State University of New York Press (2012)

Neo, J.R.J., et al.: Designing immersive virtual environments for human behavior research. Front. Virtual Real. (2021). https://doi.org/10.3389/frvir.2021.603750

Pickersgill, R.S., Rameezdeen, R., Harvey, J.: OnSite: The virtual site visit as an environment for construction learning. In: Mostafa, S., Rahnamayiezekavat, P. (eds.) Claiming Identity Through Redefined Teaching in Construction Programs, pp. 153–176. IGI Global (2020). https://doi.org/10.4018/978-1-5225-8452-0.ch00

Verbeek, P.-P.: Cyborg intentionality: rethinking the phenomenology of human–technology relations. Phenom. Cogn. Sci. **7**(3), 387–395 (2008). https://doi.org/10.1007/s11097-008-9099-x

Verbeek, P.-P.: Some misunderstandings about the moral significance of technology. In: Kroes, P., Verbeek, P.-P. (eds.) The Moral Status of Technical Artefacts, pp. 75–88. Springer Netherlands, Dordrecht (2014). https://doi.org/10.1007/978-94-007-7914-3_5

What Makes a Room a Room for Living?

Sabin-Andrei Ţenea[1]([✉]), Azuka Odiah[2], and Samuel D. Gosling[2]

[1] Ion Mincu University of Architecture and Urban Planning, Strada Academiei, 18-20, 010014 Bucharest, Romania
sabin.tenea@gmail.com

[2] University of Texas at Austin, 108 E. Dean Keeton Street, Austin, TX 78712, USA
azukaodiah@utexas.edu, samg@austin.utexas.edu

Abstract. *What intrinsic attributes constitute the essence of living space, and how are these spaces discerned and defined in the contemporary architectural context?*

This study explores the dynamic relationship between ambiance, spatial design, and human well-being within residential environments. Utilizing a multidisciplinary approach that integrates insights from psychology, architecture, and interior design, the project investigates the dynamic interplay between space structuring, functional organization, and decorative elements in shaping the ambiance and functionality of living rooms. This approach integrates empirical surveys with Generative Algorithms and a Hybrid Recommender System, representing a symbiosis of human intuition and machine precision, not only challenging traditional architectural practices and highlighting the potential of generative AI in design but also offering new insights into designing for user's needs and emotional regulation.

The implications of this research are significant for architects, designers, and stakeholders in the residential design process. It highlights the necessity of a replicable model for consensus on space-ambiance combinations, pointing towards generative design and machine learning as promising tools for bridging the gap between subjective perceptions and objective design goals.

Keywords: Atmosphere · Generative Interior Design · Human-Machine Collaboration · Architectural Perception · Hybrid Recommender System

1 Space and Ambiance

Throughout history, residential spaces have reflected the dynamic interplay between human culture, needs, and technological advancements. From the undifferentiated spaces of primitive huts to the intricate and nuanced living quarters of the 19th century and beyond, our continual reshaping of living environments speaks to an evolving understanding of space as a reflection of societal values, economic conditions, and personal well-being. In contemporary society, where life's increasing complexity and stressors confine individuals indoors, the need to prioritize well-being within these spaces has never been more pronounced. As architects and designers, it is our responsibility to cultivate a deep empathy with the inhabitants of these spaces, ensuring that each space not only meets functional needs but enhances the quality of life and speaks to their needs and aspirations.

H. Chai et al. (Eds.): CDRF 2024, *Symbiotic Intelligence*, pp. 348–359, 2025.
https://doi.org/10.1007/978-981-96-3433-0_31

Gosling et al (2013) proposed the manipulation of one's space can serve three broad functions (Gosling, et al., 2013). First, the activities likely to be performed in the space (the physical features, like the layout, etc., of a living room might make it well suited for watching TV or gathering with family, for example). Second, the items in a space and their arrangement could be used to convey impressions to others. A 2008 study comparing the living spaces of politically liberal and conservative occupants in the United States found that liberals had objects that indicated diverse interests in arts, literature, travel, and other cultures, whereas conservative counterparts tended to have relatively conventional decor (Carney, et al., 2008). Third, features of the space can affect what people think about and how they feel when in that space. For example, people may use photos of loved ones and other reminders of their connections to others ("social snacks") to fend off feelings of loneliness and social isolation (Gardner, et al., 2005).

A 2015 study by Graham et al. further argued that the home provides an informative context for a wide variety of studies examining how social, developmental, cognitive, and other psychological processes play out in a consequential real-world setting, linking it to situation selection in emotion regulation (Graham, et al., 2015). The study demonstrated that not only do people seek out different ambiances, but they also seek out different ambiances across different spaces of their homes—such as associating the Living Room with relaxation and togetherness.

These studies underscore the significance of considering layout and other physical features within living spaces when assessing their psychological impact. The findings also point to ambiance as a crucial component in shaping the quality of a space and the well-being of the people in these spaces. While these are only some of the few works that explore how the layout and physical features of living spaces influence activities and social interactions taking place within that space, none addresses what these layouts and physical features are that evoke ambiance within these spaces, and to what extent.

The concept of "Space + Ambiance" as a composite construct is central to understanding how individuals perceive and interact with their living environments. The terms "Romantic Master Bedroom" and "Cozy Living Room" serve as shorthand for a complex amalgam of sensory experiences, emotional responses, and functional affordances. This condensation of information into succinct labels is invaluable in the design process, allowing for efficient communication and decision-making. However, this efficiency comes at the cost of oversimplification, potentially obscuring the rich, multi-layered experiences that spaces offer.

Furthermore, the evolution of naming conventions for spaces within a home— from size (e.g. hall, salon, chamber) or adjacency-based systems (e.g. antechamber, vestibule) in the 18th century to today's activity-based (e.g. kitchen, dining room, living room) or occupant-based (master bedroom, children's rooms) (Kärrholm, 2019) highlights a move toward personalization and functionality in living spaces, reflecting changing demographics, family structures, and cultural norms. The nomenclature of spaces and ambiances becomes a lens through which the evolving psychology, priorities, needs, culture, and economic status of inhabitants are viewed. Yet, this lens is subject to distortion due to the dynamic nature of language and culture.

Semantic shifts in the names we attribute to our spaces and their ambiances reveal a fluid landscape where meanings and associations are constantly in flux. This fluidity, while reflective of societal evolution, introduces challenges in achieving a shared understanding of what space is meant to convey and facilitate. The disparity in interpretations among designers, residents, and other stakeholders can lead to misalignments in expectations and outcomes, underscoring the importance of developing a common language or model for space-ambiance conceptualization.

The challenge of semantic relativism—where the meaning of words and phrases diverges significantly across individuals and groups—further complicates the endeavor to harmonize design intentions with user expectations. This variability in interpretation can lead to confusion, dissatisfaction, and iterative cycles of redesign, pointing to a critical need for mechanisms that bridge the gap between subjective perceptions and objective design goals.

Recognizing these challenges, our research posits the necessity of a replicable model for establishing consensus on space-ambiance combinations. Such a model would not only aim to streamline the design process but also ensure that the resultant spaces genuinely reflect and accommodate the diverse needs and preferences of their users. To accomplish that, we explore the potential of generative design and machine learning as tools for uncovering and articulating latent preferences in spatial design.

2 Methodology

Spatiophysical Coding System:

Generative AI holds significant promise in architectural design, but the lack of specific guidelines hinders its effectiveness. In exploring living room designs, we face a complex mix of tangible and intangible factors that shape the space's ambiance and functionality. These elements include room dimensions, architectural features, furniture, personal items, activities, the qualities of materials, colors, and light.

To bridge this gap, a 2024 study sought to assess the proficiency of Generative AI models in depicting spaces conveying different ambiances depicted in images (Odiah & Gosling, 2024). In this comparative study, they explored using AI models to recreate spaces conveying intended ambiances, against real photographs conveying similar spaces and intended ambiances. They found that, regardless of how realistic the image, people were always able to better identify spaces and ambiances generated by AI than they were with real photographs. These findings set up Generative-AI as a proficient method for depicting spaces like the living room to further explore the layouts and physical features that could evoke different ambiances within the space. (Fig. 1).

This leads to the principal research question of our project:

What intrinsic attributes constitute the essence of living space, and how are these spaces discerned and defined in the contemporary architectural context?

To unravel this complexity, we draw upon the theoretical framework proposed by Bonnes et al. (1987), which offers a nuanced approach to deconstructing the anatomy of living spaces through a spatiophysical coding system. This system segments living spaces into three distinct but interrelated subsystems, each contributing uniquely to the overall ambiance and functionality of the space.

The first subsystem, the space-structuring code, is foundational, comprising the architectural elements that delineate physical boundaries and dictate the flow within

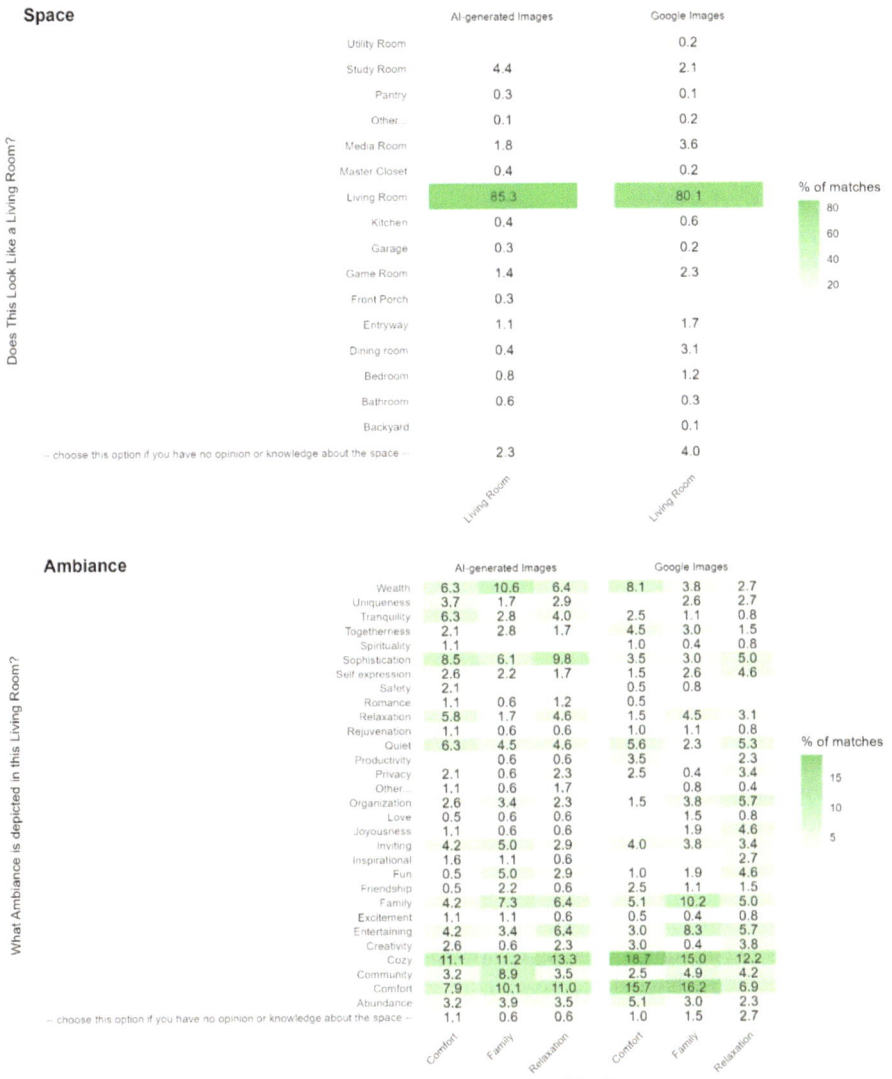

Fig. 1. AI-generated Living Rooms replicate real spaces effectively (top) but struggle to capture intended ambiances (bottom). (Odiah & Gosling, 2024)

the space. Elements such as walls, floors, ceilings, stairs, and ramps define the possibilities for occupancy and movement, influencing how inhabitants experience and engage with their environment. The design and placement of these elements can dramatically affect perceptions of size, openness, privacy, and connectivity within the home.

The second subsystem, the functional organization code, pertains to the layout and configuration of furniture and other functional items within the space. This includes seating arrangements, storage solutions, tables, and appliances, among others. The organization of these elements is pivotal in defining the usability and comfort of the living space, facilitating, or hindering daily activities. The functional organization speaks to the pragmatic aspects of living spaces, aiming to balance aesthetic appeal with practical utility.

The third subsystem, the decoration code, encompasses the objects and elements introduced into the space for decoration, personal expression, psychological impact, or symbolic meaning. This includes lighting fixtures, artworks, books, plants, textiles, and personal memorabilia. These items inject personality into the space, reflecting the identities, tastes, and values of its inhabitants. The decoration code plays a crucial role in transforming a space from a mere functional area into a home, imbuing it with warmth, character, and life.

General Relational Room Model:

Employing these subsystems, we developed a General Relational Room Model, a flexible framework that allows us to experiment with various spatial configurations. This model recognizes that a single spatial structure can support a multitude of functional organizations, which in turn can accommodate numerous decoration scenarios. This multiplicity underscores the vast potential for diversity in living spaces, with each configuration offering a unique experience based on the interplay of its components.

The comprehensive nature of this model, however, introduces a level of complexity that necessitates a focused approach in the initial phases of our research. We therefore elected to concentrate on living rooms, which typically serve as the heart of the home. Living rooms are inherently multifunctional, hosting a range of activities from social gatherings to solitary relaxation. This versatility makes them an ideal subject for studying the dynamic relationships between space-structuring, functional organization, and decoration codes.

By commencing our exploration with living rooms, we aim to gain insights into the fundamental principles that govern the design and perception of living spaces. This focused inquiry will enable us to extrapolate and apply our findings to other areas of the home, advancing our understanding of how to create spaces that resonate with the needs and desires of their inhabitants. Through this research, we endeavor to articulate a coherent narrative of contemporary living space design, illuminating the intrinsic attributes that make our living environments not only functional but truly habitable.

Generative AI-Powered Feature Aggregation:

To overcome the limitations of traditional data collection methods, such as a restricted number of participants and human bias, and to broaden our analysis of living room features, we used a Large Language Model, specifically GPT-4o. To compile a comprehensive list of features found in living rooms, we used prompt engineering to create a verbal picture of living rooms portraying different ambiances. This integrated main categories as well as detailed subcategories, specifically architectural elements (space structure), common furnishings (functional organizations), decorations (decoration scenarios), lighting conditions, and scents. To get a broad range of simulated user preferences, we reiterated the same prompt one hundred times for each ambiance. This approach produced multiple comprehensive descriptions that accurately reflect living rooms with varying ambiances and layouts.

Prompt Utilized:

"As an expert in interior design and architecture, structure your response to detail the material attributes in a residential living room that convey a specific ambiance. Use the following list-like format with detailed attributes for each category: < color, texture, material, feature >. Make sure to adhere to this order in your response:\n – Layout: Outline the spatial arrangement.\n – Furniture: Detail the types of furniture used.\n – Decorative Items: List decorative items.\n – Lighting: Describe the lighting setup.\n – Scents: Note any scents that reflect the specified atmosphere.

Help me understand the material attributes that convey a (AMBIANCE) ambiance in residential living rooms. I'm looking to design a living room that reflects a (AMBIANCE) atmosphere. Please describe, in meticulous detail and in the specified order, the living room focusing on layout, furniture, color schemes and textures, and decorative items. Emphasize spatial qualities, sensory details such as color schemes, textures, lighting, and scents. Use the following list-like format with detailed attributes for each category: <color, texture, material, feature>. The information should be presented in a concise list-like format, adhering to the structure:\n – Layout: Outline the spatial arrangement.\n – Furniture: Detail the types of furniture used.\n – Decorative Items: List decorative items.\n – Lighting: Describe the lighting setup.\n – Scents: Note any scents that reflect (AMBIANCE). Avoid extraneous commentary."

With hundreds of detailed descriptions for different ambiances in the living room, we then used LIWC (Linguistic Inquiry and Word Count) software to run a text analysis using the Meaning Extraction Method (Pennebaker, et al., 2015). Using this method, we captured each word used in the description of the living room for each ambiance. These words represent objects and descriptors – features that make up each unique response. Figure 2 illustrates the features captured in a living room, arranged from most to least common, for each designated ambiance.

This carefully selected compilation serves as a substitute for direct user contributions, summarizing an agreement on the essential components of living room design. By incorporating such diversity, we aimed to capture a wide spectrum of design preferences and functional requirements, ensuring our model is not biased toward any specific group or region.

Visualization:

Addressing the challenge of achieving controlled and consistent outputs in spatial prompting, our study evaluated several AI image generators including Dall-E, Midjourney, and Stable Diffusion. The lack of fine control in these models necessitated a more sophisticated approach for applications in architectural design, environmental psychology, and interior design. To overcome this limitation, we incorporated additional conditional inputs that directly specified the desired room composition through segmentation maps. These maps acted as a conditioning mechanism within the image generation process, ensuring greater fidelity to design intents.

We harnessed Python to perform a quasi-random selection of feature sets from the generated list of features, resulting in fifteen distinct feature groups, each containing between four to twenty features (e.g.: *"wall, sideboard, plant, wall art, sofa, floor lamp, photograph, ottoman, wooden flooring"* or *"sideboard, sofa, carpet, wooden*

Fig. 2. Word cloud for each mbiance showing the frequency of features mentioned in living room descriptions through LIWC analysis.

flooring"). This process was aimed at capturing a comprehensive spectrum of possible living room configurations. Subsequently, these configurations were operationalized into detailed three-dimensional models through the General Relational Room Model, utilizing Rhinoceros3D and Grasshopper3D for precise rendering. This approach facilitated a rigorous analysis of the spatial dynamics and potential user experience impacts of diverse living room layouts.

Ambiance Creation with Stable Diffusion, Realistic Vision, and ControlNet:
After model creation, the configurations were rendered into detailed segmentation maps. These maps serve as control images, functioning as foundational canvases for the subsequent incorporation of physical attributes. By devoiding these images of explicit ambiance indicators, we established a neutral basis, facilitating the controlled integration and assessment of atmospheric elements in later stages. This underscores the study's commitment to a systematic and replicable approach to evaluating the influence of specific design elements on spatial perception and user experience.

The control images, derived from the segmentation maps, were subsequently inputted into Stable Diffusion, an open-source latent diffusion text-to-image model, specifically utilizing the Realistic Vision Model Version augmented by Conditional Control using ControlNet (Zhang, et al., 2023).

We employed a structured prompt-generation strategy to guarantee the stability and consistency of the model's outputs. Each control image was appended with a quasi-random embedding-style prompt that detailed the desired ambiance, informed by the list of objects present within each image. Beginning with the initial embedded list of features that defined the 3D model of the living room, we added detailed descriptors from our previously aggregated inventory of potential materials, colors, textures, and design specifics to each object. Examples of such prompts are:

"light wood long sideboard, textured teal wall, black potted plant with lush foliage, modern abstract wall art with black and white geometric shapes, sleek white and teal sofa with bold lines, bright yellow throw pillow, minimalist design black floor lamp, large framed lone tree photograph, teal ottoman, rich wooden flooring, natural and artificial light", and in the case of control images with fewer elements and features: *"dark pink textured accent wall, metallic long sideboard, a baby pink sofa, a lush dark carpet, rich wooden flooring"*.

This detailed approach ensured that each object in the segmentation map had corresponding visual, consistent, and identifiable physical properties in the generated images. This procedural application resulted in a series of photographs, each embodying the intended atmosphere through the diffusion-driven addition of textures, colors, and lighting. The foundational control images served as precise blueprints for the text-to-image model, guiding the accurate and realistic incorporation of design elements to align with each specified ambiance. This approach facilitated a nuanced exploration of how various atmospheric conditions affect spatial perception, with the generated images providing a visually compelling representation of these effects (Fig. 3).

Evaluation and Further Generation:
The series of images generated through the application of Stable Diffusion, augmented with descriptive physical qualities, constitute the initial living room image database presented to study participants. Participants are tasked with evaluating these images, focusing on the degree to which each visual representation aligns with their conceptualization of a living room, particularly in terms of the ambiance conveyed.

To capture the participants' assessments, the study employs a dual feedback mechanism. Firstly, a Likert scale is used to gauge ambiance valence, with endpoints ranging

Fig. 3. Space-Ambiance Combinations: The upper row displays the segmentation input maps created using Grasshopper 3D. The leftmost column shows the prompts for Stable Diffusion XL with Control Net. The accompanying images represent the outcomes of these combinations.

from "Uncomfortable" to "Comfortable" or "Agitating" to "Relaxing," enabling quantifiable measurement of participants' reactions to the ambiance each image portrays. Secondly, participants are encouraged to provide open-ended responses, offering a platform for more nuanced insights into their perceptions.

To further refine the evaluation process of living room images, our study integrates a Hybrid Recommender System (HRS), which combines the strengths of content-based filtering and collaborative filtering techniques (Burke, 2002). This system is pivotal in analyzing participant feedback by mapping their assessments to custom user profiles. These profiles are constructed by interpolating the data collected from the participants, particularly linking their quantitative ambiance valence scores to the descriptive elements of the living room images.

Upon aggregating and analyzing the participant data, if the HRS identifies statistically significant trends—such as a specific layout or a color combination correlating with higher "Comfort" scores—it triggers the generation of new images. These images are adjusted to emphasize the identified parameters and to test the hypothesis that certain design elements contribute significantly to the perceived ambiance.

This iterative process, illustrated in Fig. 4, is designed to be adaptive. As the system receives additional feedback and usage data over time, it refines its recommendations to better match users' changing preferences and needs. This adaptability is crucial for maintaining the relevance and accuracy of the model across different contexts.

Limitations:

Participant Selection   Generated Recommendations

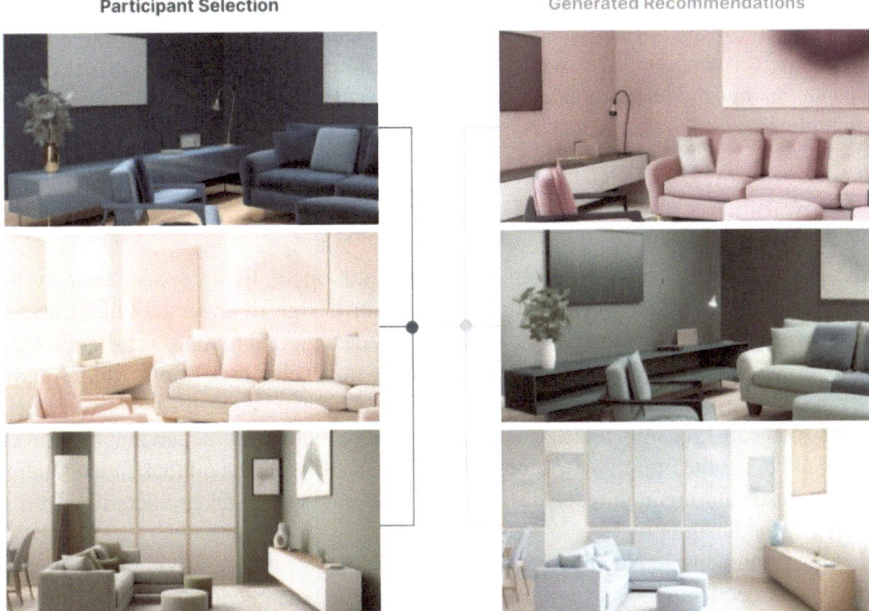

Fig. 4. Recommender Generations: This illustrates a participant's interaction with the hybrid recommender system. The participant's selection of certain space-ambiance combinations prompts the system to produce additional room examples based on its identification. The left column shows the participant's selections, while the right column displays the corresponding outcomes generated by the system

Our research into the integration of AI tools for empathetic residential space design, despite its successes, encounters several limitations that need acknowledgment to ensure a balanced perspective on the system's capabilities and areas for enhancement. Key areas of concern include the inherent biases in AI training data, the primarily interpolative nature of our system, and the challenges in data acquisition due to subjective feedback.

The generative models employed, including ChatGPT, Stable Diffusion, and ControlNet, depend on extensive datasets that often carry cultural, aesthetic, and socioeconomic biases. This can adversely affect the universality and acceptance of the designs across diverse user groups, making careful calibration of these models essential for their effective application.

Our system primarily functions as an interpolative tool, generating designs from existing data on structures, layouts, and decor. This reliance on pre-existing data restricts the system's ability to innovate and address unique design challenges independently. To overcome this, we empower designers to introduce new elements and machine learning models manually, enhancing the system's flexibility and ensuring that AI supports rather than replaces human creativity.

Additionally, our evaluation process is heavily influenced by participant feedback, which is intrinsically subjective. Factors like emotional state, personal biases, and cultural background can skew perceptions, introducing variability that makes it difficult to draw universally valid conclusions. To address these challenges, we plan to conduct comprehensive cross-validation studies in various demographic settings. Analyzing the system's performance in diverse environments will help us identify and correct biases, ensuring that our model remains effective and relevant across a broad spectrum of users, ultimately providing reliable space-ambiance recommendations.

3 Conclusions and Future Directions

This research embarked on a multifaceted exploration of living spaces, integrating insights from psychology, architecture, and interior design to understand the intricate relationship between ambiance, residential spatial design, and human well-being. Central to our findings is the development of a General Relational Room Model and a machine-learning augmented workflow that bridges the gap between subjective perceptions and objective design goals. This symbiosis of human intuition and machine precision underscores the importance of empathetic design, fostering a mutual understanding and consensus among architects, designers, clients, and other stakeholders involved in the residential design process.

Our comprehensive data aggregation included a wide variety of living room configurations sourced from diverse cultural, socio-economic, and geographical backgrounds, capturing a broad spectrum of design preferences, functional requirements, and ambiance qualities. The Hybrid Recommender System (HRS) we developed is designed to be adaptive, continuously learning from new data inputs and adjusting its recommendations to better match evolving user preferences and needs.

The model's framework is inherently scalable and flexible, allowing it to be adapted for different types of residential spaces beyond living rooms. This scalability is facilitated by the General Relational Room Model, which can be customized to accommodate various spatial configurations and user requirements. The flexibility of our model ensures it can be applied to a wide range of design scenarios, making it a versatile tool for architects and designers globally.

Looking ahead, we aim to broaden our research to encompass the entire home and incorporate emerging technologies such as employing Mixed Reality (MR) and other technologies to investigate how digital immersion might influence previous results. Furthermore, by analyzing commonalities across user profiles, we seek to devise design strategies that accommodate collective preferences, thereby enhancing the utility and emotional resonance of shared spaces. The insights and data accrued hold promise for developing text-to-space machine learning models that prioritize empathy, enhancing evidence-based design practices. This endeavor not only enriches our understanding of living spaces but also sets the stage for future innovations in residential design that prioritize human well-being.

References

Bonnes, M., Giuliani, M.V., Amoni, F., Bernard, Y.: Cross-cultural rules for the optimization of the living room. Environ. Behav. **19**(2), 204–227 (1987)

Burke, R.: Hybrid recommender systems: survey and experiments. User Model. User-Adap. Inter. **12**, 331–370 (2002)

Carney, D., Jost, J., Gosling, S., Potter, J.: The secret lives of liberals and conservatives: personality profiles, interaction styles, and the things they leave behind. Polit. Psychol. **29**(6), 807–840 (2008)

Gardner, W., Pickett, C., Knowles, M.: Social snacking and shielding: using social symbols, selves, and surrogates in the service of belonging needs. In: Williams, K.D., Forgas, J.P.. Hippel, W.V. (eds.) The social Outcast: Ostracism, Social Exclusion, Rejection, and Bullying. s.l.: Psychology Press, pp. 227–241 (2005)

Gosling, S.D., Gifford, R., Mccunn, L.: The selection, creation, and perception of interior spaces: an environmental psychology approach. In: Brooker, G., Weinthal, L. (eds.) The Handbook of Interior Architecture and Design, pp. 278–290. Bloomsbury Publishing Plc (2013). https://doi.org/10.5040/9781474294096.ch-020

Graham, L.T., Gosling, S.D., Travis, C.K.: The psychology of home environments: a call for research on residential space. Perspect. Psychol. Sci. **10**(3), 346–356 (2015)

Kärrholm, M.: The life and death of residential room types: a study of Swedish building plans, 1750–2010. Architectural Histories **8**(1), 1–18 (2019)

Pennebaker, J.W., Booth, R.J., Boyd, R.L., Francis, M.E.: Linguistic Inquiry and Word Count: LIWC2015. Pennebaker Conglomerates, Austin, TX (www.LIWC.net) (2015)

Zhang, L., Rao, A., Agrawala, M.: Adding Conditional Control to Text-to-Image Diffusion Models. arXiv e-prints, arXiv (2302.05543) (2023)

Odiah, A., Gosling, S.D.: Laying the foundations for using generative AI images in architectural research: do images convey the intended spaces and ambiances? Arch. Intell. **3**(1), 35 (2024). https://doi.org/10.1007/s44223-024-00076-x

Advancing Prototyping Pedagogy—A Design-Build Studio Approach

Darcy Zelenko[1]([✉]), Michael Minghi Park[2], and Rochus Hinkel[2]

[1] Design and Architecture, Monash University, Building 4.0 CRC, Monash Art, Clayton, Australia
darcy.zelenko@monash.edu

[2] Faculty of Architecture, Building, and Planning, University of Melbourne, Parkville, Australia

Abstract. This paper investigates the advantages of integrating a prototyping-based pedagogy into a design-build studio program. A specific design-build studio is analysed as a case study to discuss the opportunities associated with prototyping playing a more vital role in architectural education. The case study presents a unique integration of prototyping with digital fabrication, thereby enabling students to rapidly iterate designs and realise complex geometries not feasible with traditional methods. The study discusses the compatibility of prototyping-based pedagogy with digital fabrication and computational design in the educational setting. Through an analysis of the studio's structure, the paper assesses the educational benefits associated with engaging in prototyping activities, whilst drawing comparisons to established prototyping theories from other fields. Finally, the study advocates for further research to evaluate the efficacy of prototyping-based pedagogy and develop a prototyping theory specific to architectural education.

Keywords: Prototyping · architectural education · design-build

1 Introduction

This paper examines the efficacy of a pedagogical approach to design-build education that heavily utilises prototyping, involving graduate architecture students in a research-led education context. Students from the Melbourne School of Design were involved in the delivery of a dendriform structure (Fig. 1), that serves as a housing for an evaporative cooling unit (ECU). The ECU is constructed from solid timber and utilises a predictive algorithm to tailor the cooling effect to environmental conditions, and occupants (Stojanovic et al., 2023). It was developed using computational design methods and produced with digital fabrication equipment. As part of the subject, students were involved in the fabrication and assembly processes, and were required to propose a revised design utilising lessons learned.

This paper explores the educational benefits of engaging students in such a design-build studio environment, where they use prototyping as an experience to guide their learning (Pelman, 2022). It examines the role that prototyping plays in fostering a symbiotic relationship between students' design endeavours and advanced computational

H. Chai et al. (Eds.): CDRF 2024, *Symbiotic Intelligence*, pp. 360–370, 2025.
https://doi.org/10.1007/978-981-96-3433-0_32

tools. Additionally, it provides critical reflection on running a research-led subject for Master's architecture students.

An overview of the ECU is first outlined as background to the paper. Following, a review of relevant literature from design-build studio pedagogy, and prototyping provides context. Further literature review of digital fabrication and computational design is used to delineate the novelty of approach from traditional design-build settings. Next, the case study is presented that dissects the students' involvement with the full-scale ECU prototype (referred to as Prototype 1) and reviews the students' observations of the digital fabrication and assembly processes. The case study is extended to also focus on how these lessons learned were used to inform the design of a revised structure (Fig. 2). In the discussion a critical review of student work is presented and connected to prototyping theory. This is used to make the case for a more considered use of prototyping to augment to guide educational practices in design-build studio context. It suggests that this is particularly relevant to the teaching of computational design, and digital fabrication. It also raises the need for a built environment-specific prototyping theory to inform both education, and industry practice (Zelenko & Maxwell, 2023). This theory is poised to enhance architectural education, particularly in design-build contexts, by offering a structured approach to guide prototyping activities. Future work aims to apply and evaluate this theory in a university setting, assessing its impact on fostering innovative and symbiotic design solutions.

2 Background

Stojanovic et al. present the conceptual design for an evaporative cooling system, comprising multiple ECUs, that utilises a predictive algorithm to tailor its cooling effect according to local weather conditions, and tailored to the number of current occupants (Stojanovic et al. 2023). Alongside the research was the development of a 1:2 scale prototype of an ECU, created from 3d-rpinted nylon. As a development of this research, the ECU was re-designed to utilise timber as the majority material, as opposed to 3d-printed nylon. All parts needed to be re-detailed to accommodate the shift to timber, an anisotropic material, and facilitate the required subtractive digital fabrication process.

Alongside the realisation of Prototype 1 at full scale to test feasibility, the project was undertaken as part of a research-led teaching initiative in the Architecture, Building and Planning Department of the University of Melbourne in 2021. This was through the subject; DF_Lab: Designing Making (DF_Lab),—a six-week long summer elective subject. DF_Lab is part of the Digital Design and Fabrication Electives at the University of Melbourne, that enables students to critically engage with and apply contemporary design tools. DF_Lab allows students to prototype using digital design and fabrication techniques, enabling the re-conceptualisation of architectural components to demonstrate how digital technologies can innovate tectonic, aesthetic, and performative aspects. In the iteration of DF_Lab that ran in 2022, students were involved in the fabrication process of parts, participated in the post-production process, and assisted with the assembly of the Prototype 1. Alongside this prototyping endeavour, as part of their learning students were directed to design a revised version of the ECU (referred to as Prototype 2). The design for Prototype 2 was required to be informed by the experiences gained from Prototype 1. This design exercise was undertaken in small groups in a competition setting, where a single design was selected to be progressed as an entire class.

Fig. 1. ECU Prototype (Ransome, 2022)

3 Methodology

This paper adopts a qualitative research methodology, focusing on the observation of architectural students engaged in the DF_Lab: Designing Making course, a component of the Digital Design and Fabrication Electives at the University of Melbourne. The methodology is grounded in the principles of action research, allowing for the iterative exploration of design and fabrication processes within a prototyping-based pedagogical framework. Observations were systematically recorded, documenting the progression from initial design concepts to the realization of physical prototypes. This approach facilitates an in-depth understanding of the students' learning experience, the pedagogical impact of prototyping, and the integration of digital tools in architectural design and fabrication. The observation method is supported by Schön's (1983) concept of the reflective practitioner, emphasizing the importance of reflection in and on action as a fundamental learning process in design education. Additionally, the method aligns with Cross's (2001) advocacy for design thinking as an iterative process of problem-solving, which is central to the development and refinement of prototypes.

4 Literature Review

4.1 Design-Build Pedagogy

Design-build pedagogy in architectural education is a teaching methodology that bridges theoretical knowledge with practical application through the design and construction of real-world projects. This pedagogical approach enriches students' learning experiences by engaging them directly in the process of creating architectural works, from conceptualisation to execution. Design-build pedagogy allows students to apply theoretical concepts in real-world scenarios, fostering a deep understanding of architectural design, materials, and construction techniques.

By participating in the creation of a physical structure, students can directly observe the outcomes of their design decisions, enabling a profound learning experience that emphasises the practical aspects of architecture (Mohareb & Maassarani, 2018). The design-build approach promotes a holistic educational experience by requiring students to consider various aspects of architectural projects, including cost management, spatial form exploration, and material properties. This comprehensive approach prepares students for the multifaceted nature of architectural practice, equipping them with the skills needed to navigate complex design and construction challenges. By integrating digital design technologies with traditional construction practices, design-build pedagogy encourages students to explore innovative solutions within architectural design. This fusion of technology and craftsmanship fosters creativity and experimental thinking needed to progress the field of architecture (Colopy, 2019). Engaging in design-build projects prepares students for the realities of architectural practice by providing hands-on experience with the planning, development, and execution of projects. This experience not only enhances students' technical skills but also develops their ability to work collaboratively, manage projects, and communicate effectively with clients and stakeholders, key competencies for a successful architectural career (Khalifa & Hefnawi, 2020). By combining theoretical learning with practical application, this approach not only enhances students' understanding of architecture but also prepares them for the challenges and opportunities of professional practice.

Symeonidou (2022) demonstrates an application of design-build pedagogy leveraging digital fabrication to assist students to understand the complex process of thermoforming and achieve refined design outcomes. Reither and Wit (2015) suggest that design-build studios utilising digital fabrication should be research-driven from at inception, rather than focusing on a conventional architecture brief. This allows novel ideas for projects to emerge that can later be incorporated into an architectural design proposition. Abdulghany et al. (2023) suggest that adequate scaffolding is needed to improve opportunities for cognitively effortful and effectively meaningful learning, especially when introducing digital fabrication and making procedures to novice trainees.

4.2 Prototyping

A prototype can be described as the initial, full-scale implementation of a new design, featuring some degree of functionality (Zelenko and Maxwell 2023). Prototyping is a widely used tool in product development across many disciplines, each developing its own unique methods. It provides critical information about design performance and acts as a learning tool to inform subsequent improvements (Mitomi et al. 2022). Prototyping is a key strategy to enable the adoption, adaptation, or development of new technologies. It provides valuable information about design performance and is used as a learning tool to guide refinements. Camburn et al. synthesise four objectives to prototyping from literature;

- Refinement - gradually improve a design's quality over time.
- Exploration - generate, and narrow-down, new design concepts, where out of the box thinking is encouraged.
- Learning - acquisition of knowledge—tacit knowledge in particular—regarding the performance of a design.
- Communication - the sharing of design information through prototypes to internal, and external stakeholders or users (Camburn et al. 2017).

The objectives of Refinement and Exploration are diametrically opposed as they both concern the interaction with a design concept and can be grouped together. Similarly, Learning and Communication, are linked as they refer to the generation, and dissemination, of knowledge, through prototyping.

Houde and Hill (1997) propose an important model of prototyping used in software development, which includes three types: Look and Feel, Implementation, and Role, all of which can combine to create a fourth type: Integration. In this model, "Role" relates to the functional aspects of a product and its impact on the user's life. "Look and Feel" addresses the tactile and visual sensory experiences of the user while interacting with the product. "Implementation" involves the specific technologies or programming methods used to fulfil the product's intended Role. "Integration" prototypes combine these three aspects to present an initial version of the final product, used to understand the overall design comprehensively. David G. Ullman's classes of prototyping connect the production of prototypes to stages of the design process as proof of Concept, Product, Process, and Production (Ullman 2010). Prototyping theory provides a framework to contextualise the inherent complexity of such projects, allowing for the discussion of critical aspects of different kinds of prototypes.

Using Ullman's model as a classification tool, this paper extensively describes the development of a proof-of-product prototype (Prototype 1) of the ECU described by Stojanovic et al. In this context, the geometry, materials, and manufacturing processes of the ECU are advanced to suit a full-scale implementation of an individual unit. The algorithm and its simulation proposed by Stojanovic et al. can be described as an implementation prototype according to the model by Houde and Hill. This is because it clearly demonstrates the system's capabilities but lacks the specifications required for real-world application. The 1:2 3D-printed prototype, which didn't include the cooling system or algorithm, is more accurately considered a Look & Feel prototype. It demonstrates a potential concrete experience of the ECU artefact. According to Ullman's classification, this prototype serves as a proof-of-concept, created as a learning tool to communicate the overall product and hint at its function. Using Ullman's model as a classification tool, this paper extensively describes the development of a proof-of-product prototype (Prototype 1) of the ECU described by Stojanovic et al. In this context, the geometry, materials, and manufacturing processes of the ECU are advanced to suit a full-scale implementation of an individual unit.

4.3 Digital Fabrication/Computational Design

The term "computational design" emerged as a comprehensive label encompassing various approaches that use digital technologies in the design process. Caetano et al. make a significant contribution by defining "computational design" as a distinct term, clarifying its applications in design. These applications include automation of design processes, optimizing parallel execution of unique design tasks, responding flexibly to design changes, and assisting designers with feedback (Caetano, Santos, and Leitão 2020).

The term "digital fabrication' describes the interface between computer-aided-design (CAD) and computer-aided-manufacturing (CAM) processes that facilitate the translation of digital models into physical structures (Kolarevic 2003). In architectural education, the utility of digital fabrication lies in 'rapid prototyping,' enabling designers to optimise their concepts through physical iterations (Burry 2005). Students directly engage with material, structural, manufacturing, and assembly parameters, enhancing their understanding of project complexities.

In the context of DF_Lab, students are offered a hands-on experience with tools that directly translate digital designs into physical artifacts. They are introduced to the complete file-to-factory process for the project, provided with copies of the working files used, and are provided with an environment to experiment with CNC machinery—the primary digital fabrication equipment utilised on the project. This exposure gives them a practical education about best-practice techniques for complex project, that they are encouraged to augment in their own design work.

5 Case Study: Dissection of Prototype 1

Observing the process from redesign to realisation of Prototype 1 was a valuable learning experience for the students involved in DF_Lab. This case study outlines the major activities the students engaged with, and lessons learned that were used to inform the design of Prototype 2.

5.1 Timber Nesting Strategy

The design of Prototype 1 sought to integrate an evaporative cooling unit within a solid timber housing and employed computational design methods to optimise the lengths of timber required. Students were involved with nesting of parts to understand the CNC fabrication process. Students identified that the design neglected to consider timber width in part nesting, and that they could achieve greater efficiencies to reduce material wastage.

5.2 Milling and Digital Fabrication Challenges

The milling process adopted a combination of 3-axis and 4-axis strategies, requiring considerable time dedicated to the tool pathing process, resulting in instances of human error (Fig. 2). Students encountered several challenges in this phase, and identified a need for tighter tolerances in the milling of node connections and the inefficiencies of deep drilling operations. Students identified that the large number of parts added considerable production time to the project. It was concluded that a subsequent prototype could utilise fewer, larger parts to achieve a similar result, and would require less milling time, due to the reduced number of transfer movements the CNC would have to make across multiple parts.

5.3 Post-processing and Assembly Insights

Post-processing revealed the critical role of fine-tuning milled parts in the project, from removing bridges and sanding to addressing minor issues in CNC milling that affected fitment. Students highlighted that the assembly process underscored the importance of accuracy in component fabrication, with instances of misalignments in node connections observed. They also highlighted the labour-intensive nature of manual post-processing and the need for strategic planning in assembly to mitigate stresses on the structure.

Fig. 2. Slight material shifts during fabrication process were rectified by hand

5.4 Reflections and Learning Outcomes

The dissection of Prototype 1 was instrumental in identifying key areas for improvement in design rationalisation, fabrication efficiency, and assembly accuracy. The process illuminated the intricate relationship between design intent and material reality. Students gained firsthand experience in the challenges of bringing a conceptual design to physical form, understanding the value of precision, planning, and collaboration in architectural fabrication.

5.5 Towards Prototype 2

Students in DF_Lab developed the design of Prototype 2 (Fig. 3) using the lessons learned from Prototype 1. Key improvements included the adoption of a more efficient nesting strategy, the simplification of node connections for ease of assembly, and the integration of a resolved misting system. These enhancements were complimented by a unique formal aesthetic. The analysis of both prototype projects demonstrates that the student group was able to achieve enhancements on the initial design. Students achieved this through the use of group design thinking and problem-solving skills, paired with the lessons learned from their involvement with the first prototype. Figure 4 shows that Prototype 2 used 30% less parts that Prototype 1. This was a key consideration for the project and was a design metric that was used to drive the design process.

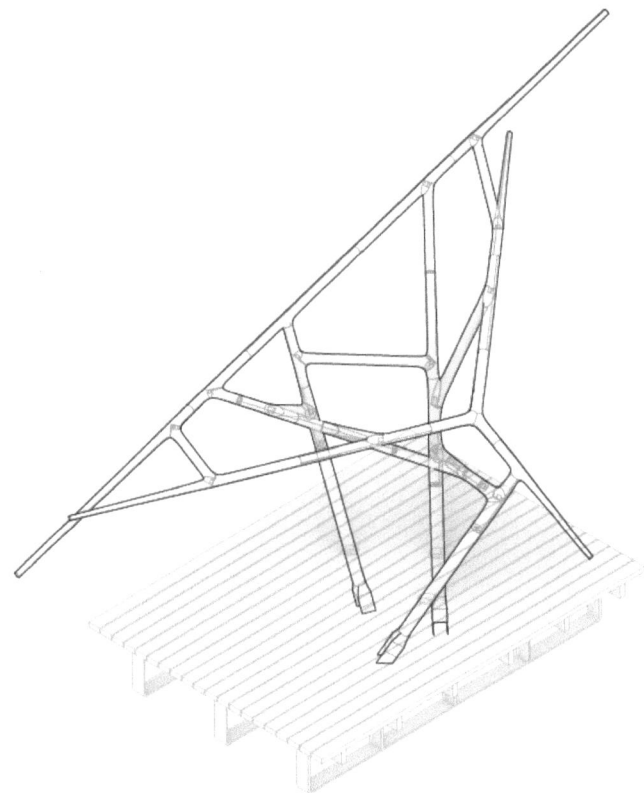

Fig. 3. Student-designed Prototype

6 Discussion: Development of a Prototyping-Based Pedagogy

The prototyping-based pedagogy provided students with valuable learning experiences and challenges, and enabled them to produce a higher quality design outcome. It is characterised by active involvement in a prototyping endeavour that enables them to obtain hands-on, experiential learning. Students undertaking he design exercise—to propose a design for revised ECU—were directed to maintain a similar material system and fabrication method, so as to make the most of the prototyping knowledge generated in the cohort. Students are provided with buy-in to share in the benefits from achieving the prototyping objectives outlined by Camburn et al.

At the core of this pedagogy is the integration of design, technology, and fabrication, that students are enabled to engage with as active participants in the prototyping process. In the context of six-week elective, this participation allowed participating students to obtain a deeper understanding of the delivered computational design, and digital fabrication concepts. The hands-on experience with digital fabrication tools and techniques significantly enhances students' technical skills. The transition from Prototype 1 to Prototype 2 highlighted the importance of mastering digital fabrication technologies, from 3D modelling software to CNC milling. Furthermore, confronting and solving the myriad of challenges encountered during the design-build process strengthens problem-solving abilities, a critical skill for future architects.

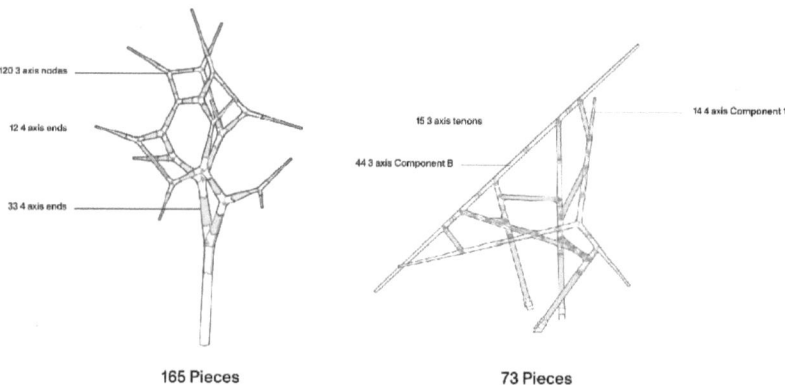

Fig. 4. Efficiency comparison between Prototype 1 and Prototype 2

This research contributes to developing built environment specific prototyping theory, and highlights a symbiosis between digital fabrication and architectural creativity. It also points towards the use of a structured approach leveraging prototyping theory to facilitate innovative design education. The insights gleaned from the prototyping process undertaken by students suggest the need for a more comprehensive prototyping theory in architectural education. Such a theory would formalise the role of prototyping in the curriculum, outlining methodologies for integrating design, fabrication, and analysis within the educational framework. It would also address the pedagogical implications of prototyping, including its impact on creativity, technical proficiency, and collaborative learning.

7 Conclusion

This paper identifies a pedagogy for design-build education that uses a prototyping exercise at the beginning of student engagement, to enable students to achieve richer design outcomes. This is enabled through a collaboration with an active research project. By embedding prototyping at the heart of the curriculum, architectural education can better prepare students to navigate the complexities of the modern built environment, equipped with the skills, knowledge, and mindset necessary for innovation and success. The work is limited by a lack of empirical evidence to support the efficacy of the pedagogy. Further research should be directed at generating further data such as anonymous student surveys to capture the responses of students to the prototyping activities, and also study its applicability to other educational contexts.

Acknowledgement. This research is supported by the Monash Cox Scholarship in Architecture and Building 4.0 CRC. The support of the Commonwealth of Australia through the Cooperative Research Centre Programme is acknowledged. The researchers acknowledge the contribution of Mitchell Ransome towards the delivery of the project.

We would also like to acknowledge the contributions of students of DF_Lab 2022: Leelee Chea, Yekun Shi, Haoyu Chen, Jessica Broad, Angus Grant, Theodore Lehrer, Christine Chen, Jack Halls, Andrew Lee, Nadine Alirani, Kester Wen Zhe Cheong, Mingjie Zhang, Min Ghi Park, and Ling Tian.

References

Abdulghany, R., Youssry, M., ElKhateeb, S.: Digital fabrication as an approach for innovative architecture education. MSA Eng. J. **1**(4), 79–97 (2022). https://doi.org/10.21608/msaeng.2022.280424

Burry, M., Burry, J.: Prototyping for Architects. Thames & Hudson (2016)

Caetano, I., Santos, L., Leitão, A.: Computational design in architecture: defining parametric, generative, and algorithmic design. Front. Architectural Res. **9**(2), 287–300 (2020)

Camburn, B., Viswanathan, V., Linsey, J., Anderson, D., Jensen, D., Crawford, R., et al.: Design prototyping methods: state of the art in strategies, techniques, and guidelines. Design Sci. **3**, e13 (2017)

Cross, N.: Design cognition: results from protocol and other empirical studies of design activity. In: Eastman, C., Newstetter, W., McCracken, M. (eds.) Design knowing and learning: Cognition in design education, pp. 79–103. Elsevier (2001)

Houde, S., Hill, C.: What do prototypes prototype? In: Helander, M.G., Landauer, T.K., Prabhu, P.V. (eds.) Handbook of Human-Computer Interaction, pp. 367–381. Elsevier (1997)

Kolarevic, B.: Architecture in the Digital Age: Design and Manufacturing. Spoon Press (2003)

Mitomi, K., Ikenoue, T., Takizawa, K., Mitsuhashi, T.: A proposal for a prototyping method focused on communication that increases economies of prototyping (2022)

Mohareb, N., Maassarani, S.: Design-build: an effective approach for architecture studio education. Int. J. Architectural Res. **12**, 146–161 (2018)

Ransome, M.: ECU Prototype (2022). https://disposablemitch.org/Ambient-Cooling-Prototype

Riether, G., Wit, A.J.: Redefining the parametric pedagogy. In: Anais do XIX Congresso da Sociedade Ibero-americana de Gráfica Digital 2015, pp. 713–718 (2015). Editora Edgard Blücher. https://doi.org/10.5151/despro-sigradi2015-110215

Schön, D.A.: The Reflective Practitioner: How Professionals Think in Action. Basic Books (1983)

Symeonidou, I.: Form D-Form – A design studio on digital fabrication and thermoforming (2022)

Ullman, D.G.: The Mechanical Design Process: Product Discovery, Project Planning, Product Definition, Conceptual Design, Product Development, Product Support, 4th edn. McGraw-Hill (McGraw-Hill series in mechanical engineering) (2010)

Zelenko, D., Maxwell, D.: Built environment prototyping for design-value. In: Integration of Design and Fabrication. International Association for Shell and Spatial Structures (IASS), Melbourne, Australia (2023). https://www.ingentaconnect.com/contentone/iass/piass/2023/00002023/00000006/art00015

Performance-based Design, Analytics & Optimization

Digital Design of Artificial Reef with Computational Fluid Dynamics and Topology Optimization

Jiacheng Yu[1], Dan Luo[1(✉)], and Ding Wen Bao[2(✉)]

[1] The University of Queensland, Brisbane, QLD, Australia
d.luo@uq.edu.au
[2] RMIT University, Melbourne, VIC, Australia
nic.bao@rmit.edu.au

Abstract. This paper explores the development and optimization of artificial reefs by introducing a novel generative design method incorporating Computational Fluid Dynamics (CFD) and Optimization (BESO). Since the 1950s, efforts to create artificial reefs have been pursued to improve marine ecosystems. Our study first surveyed the existing design of artificial reefs. Addressing the limitations in existing design methods, this research employs a topological optimization strategy, focusing on optimizing space allocation for polyphony expansion within these reefs. By analyzing fluid dynamics and connecting it with an iterative optimization process, we assess the effectiveness of material exchange in these artificial structures. This is critical for the design process, considering constraints from advanced manufacturing to allow for quick production with natural-like geometries. To advance the design of artificial reefs and explore new possibilities, we introduce a novel generative design approach, where the design of the reef emerges from the interaction between CFD and BESO, through an iterative process of material removal and addition in response to the external loading condition. The artificial reef generated is compared against benchmarks of current artificial design to assess its material efficiency, structure performance, and geometrical characteristics in complex underwater conditions.

Keywords: Artificial Reef · Topological Structural Optimization · Computational Fluid Dynamics (CFD) · Generative Design

1 Introduction

Over the course of history, estuaries and coastal seas have served as central hubs for human settlement and marine resource utilization. However, the extensive periods of overexploitation, habitat alteration, and pollution have obscured the complete extent of degradation in estuarine ecosystems, leading to biodiversity loss and compromising their ecological resilience (Lotze et al., 2006). And the reduced carrying capacity of estuarine ecosystems is primarily attributed to the depletion of natural habitats, encompassing rocky reefs, seagrasses, and mangroves (Elliott et al., 2016).

© The Author(s) 2025
H. Chai et al. (Eds.): CDRF 2024, *Symbiotic Intelligence*, pp. 373–384, 2025.
https://doi.org/10.1007/978-981-96-3433-0_33

Recent years, despite extensive interest in developing submerged man-made structures known as Artificial Reefs (AR) aiming to mimic certain characteristics of a natural reef to increase marine biodiversity and density in targeted undersea environments, the design of the Artificial Reefs remained simple. Most of the existing artificial reef designs are made from repetitive elements based on simple rectilinear geometries. Such AR design confronts a few common challenges, such as poor hydrodynamic performance and vulnerability to drag and hydrodynamic forces. The complex flow dynamics are only applied to analysis the performance of the pre-designed AR solutions but are not actively engaged in the design and optimization process structure and geometric geometrical. Also, rigorous geometry sets itself apart from organic geometry in the natural environment, creating homogenous spaces that are different from the diverse spaces that are preferred for a natural habitat with a diversity of species.

Addressing those challenges, this research proposed a novel generative design method for artificial reefs via an interactive interaction between Computational Fluid Dynamics and Topology Optimization. This generative design flow aims to explore a design process that can develop an optimized artificial reef structure in reaction to the dynamic of the local current, as well as create the structure of organic geometry with diversified void space, mimicking the formal characteristic of natural reefs. The design outcomes will be assessed against existing industrial benchmarking artificial reef design regarding key structure performance and spatial diversity criteria.

2 Background

In recent decades, artificial reefs (ARs) have gained increasing attention and popularity worldwide. Nevertheless, there is a notable number of these artificial reefs have experienced destruction from storms or corrosion, displacement or burial in sediment and other issues, resulting in their failure to achieve the intended objectives (Baine, 2001; Reed et al., 2006; Xue et al., 2023). The limited success of these projects can be attributed to the numerous uncertainties associated with constructing ARs in the complex marine environment. Consequently, assessments are crucial for ensuring the effectiveness of ARs in attaining their intended objectives.

Artificial reefs play a crucial role in restoring and enhancing coastal ecosystems and their allied ecological services. This involves the association of sessile organisms with the surface, structure and surrounding water column of artificial reefs, ultimately leading to an increase in biomass at the site (Seaman, 2000). Through previous studies, a link between the reef and its ecological effect has been demonstrated. The exchange and mixing of water between layers can be promoted by the change of flow field generated by the interaction between the structure of artificial reef and the water column, which will affect the attachment of biomass and aggregation of fish (Jiang et al., 2020; Xue et al., 2023).

With the evolution of CFD, numerical simulations have progressively gained significance in the hydrodynamic analysis of ARs. Which have undergone validation to quantitatively characterising the flow field, providing comprehensively evaluation of the hydraulic characteristics. Considering that the structures and configurations of ARs have a significant impact on hydrodynamics, extensive engineering-related works for the design, installation, and management of ARs have been carried out.

Interdisciplinary efforts have opened new avenues in structural design, particularly in analyzing biological structures for application in topology optimization. Scholars have demonstrated the effectiveness of methods like the adjoint technique and the bidirectional evolutionary structural optimization (BESO) algorithm, which systematically eliminate inefficient material while introducing additional material to evolve structures toward an optimal state (Lin et al., 2018; Zhao et al., 2018). For example, BESO has been successfully applied to optimize artificial coral structures, minimizing resistance to matter interchange, and hindering harmful substance deposition (Lin et al., 2023). Challenges in biological evolution mechanisms can be addressed through multiobjective optimization (Lin et al., 2023). A hybrid generative design method integrating Computational Fluid Dynamics (CFD) and BESO has been proposed to predict and optimize artificial reef performance (Wang et al., 2022). A modified version of this method, involving parametric design facilitated by CFD and BESO, allows for iterative optimization of artificial reef designs, enabling early-stage assessment of structural configurations and environmental impact.

3 Methodology

To overcome the limitations of current artificial reef design, an iterative process combining Computational Fluid Dynamics (CFD) analysis and topology structure optimization is developed. This process involves computing surface pressure under flow at each iteration to provide dynamic feedback for generating the design of artificial reefs. The void ratio of the optimized structure is continuously assessed until it reaches a preset target ratio determined by a benchmark artificial reef (AR). Bi-directional Evolutionary Structural Optimization (BESO) is employed iteratively in this process. BESO is a topology optimization method widely used in structural design, guiding the evolution of a structure towards an optimum by removing inefficient material while adding materials as needed (Huang & Xie, 2010). By incorporating BESO into the design process, a preliminary structural configuration meeting predefined criteria based on finite element analysis can be achieved.

As an advanced form-finding method in digital design (Lin, 2020; Zhao and Chen, 2016), topology optimization holds the capacity to refine material layouts and explore novel forms within contemporary design frameworks (Bao, 2022). The BESO method has garnered attention in artificial stone design research due to its ability to generate diverse organic forms with superior structural performance. Throughout the design process, a hybrid generative design approach has been first employed, integrating CFD and BESO techniques to optimize stone design considering both environmental and structural factors (Feng et al., 2022).

Via the incorporation of CFD and BESO into a iterative generative design process (Fig. 1), this method addresses the following limitation of the current method in AR design:

- The force and impact of flow identified by CFD directly informs the structure and geometrical design and optimization for AR.
- An iterative process allows for incremental development of AR responding to the interplay between the geometrical variation and the subsequent impact on the CFD.

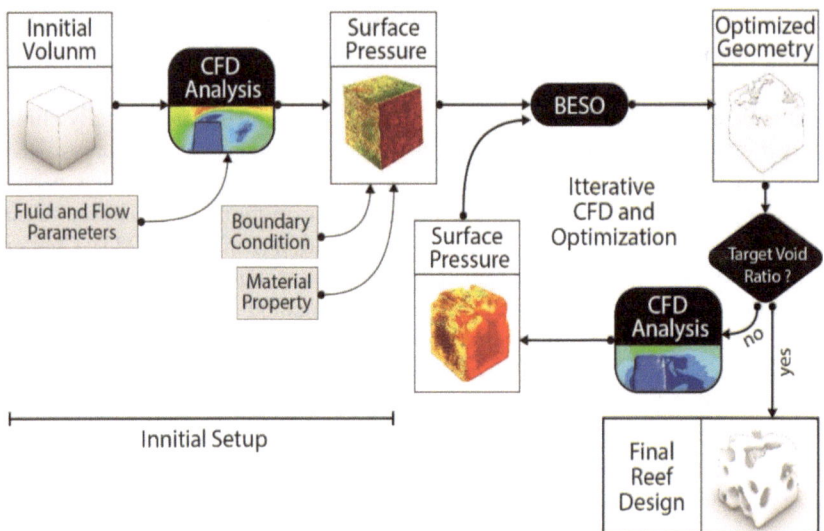

Fig. 1. Workflow for generative artificial reef design based on CFD and BESO

- Key criteria such as drag, force, and void ratio are validated at each incremental iteration of design and optimization to ensure the overall performance of the AR.
- The outcome generates organic geometries with topologically complex spatial qualities that aesthetically blend in and better support the natural habitat.

3.1 Computational Fluid Dynamics Parameters and Conditions

ANSYS-Fluent, which is a finite volume method (FVM), is employed to analyse the flow fields around optimised artificial reef models in this study. The Reynolds-Averaged Navier-Stokes (RANS) is employed as the governing equation, and the turbulence model is modelled with the RNG k-ε model as it offers the capability to incorporate the influence of swirl or rotation by adjusting the turbulent viscosity accordingly (Jiao et al., 2017).

   In this study, there are several appropriate hypotheses are introduced for the flow analyses using ANSYS-Fluent to decrease the complexity of numerical calculation for the three-dimensional turbulent flow fields around artificial reefs: The water is an incompressible, viscous, and Newtonian fluid; Isothermal flows exist in the water, regardless of the heat exchange; The inflow is in a steady state; The water depth sufficiently deepens to disregard the influence of wind and waves on the near AR flow field; The seabed is regarded as rigid bottom without considering the sediment scouring.

   The structure of the original artificial reef model is shown in Fig. 2 below. In order to verify the validity of the CFD and BESO combined design method, the optimised artificial reef obtained from the previous study (Xue et al., 2023) has been employed as a benchmark for this study. And the hydrodynamics of the optimised AR were compared to optimised ARs from this study. Therefore, the original artificial reef height is 3 m, the bottom side length is 3 m, the slope angle between the side of the reef body and the bottom surface is. As it has been concluded that the optimal performance is achieved by AR with slope angle close to (Xue et al., 2023).

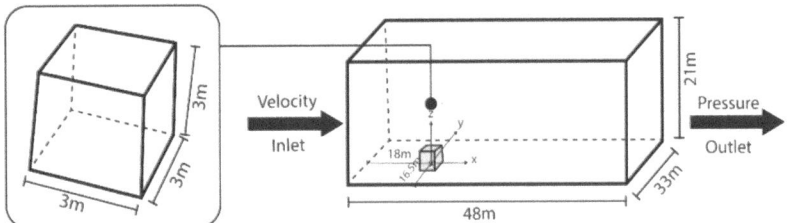

Fig. 2. Left:Original artificial reef model and definition of slope angle; Right: Computational domain (Xue et al., 2023)

The boundary conditions are categorised as inlet, outlet, and wall. All boundary conditions are described as follows: At the inlet, the mean velocity of inflow is 0.8 m/s; The pressure, which is the boundary condition, is uniform and relative at the outlet; The seabed is assumed to be rigid without considering the sediment, thus the stationary non-slip boundary condition is employed for the side wall; The free surface is treated as a 'moving wall', which has zero shear force and the same speed as incoming flow fluid.

In this study, A 3D pressure-based solver and second-order implicit unsteady formulations are adopted for simulation. The least-square cell-based method is employed to calculate the gradients of the governing equations. The RNG k-ε turbulent model was selected as the viscous model. The constants in this study are set as default values. The SIMPLE scheme is used to iterate the equations. Additionally, a second upwind discretization scheme is used for the convection term of momentum, turbulent kinetic energy, and energy dissipation rate equations. The solution is assumed convergent when all residuals fall below 1e-6, and a maximum of 200 iterations per time step was considered if the residuals failed to pass these thresholds.

3.2 Topology Optimization and Iteration Parameters

The mesh in each AR optimization iteration is divided into approximately 2000 quadrilateral faces. 6,000 to 10,000 surface points are sampled for pressure in CFD, informing subsequent BESO iterations (Fig. 3). External loads are applied to the center of each face based on pressure and area. Supports are positioned at the AR's lower corners, with translation constrained on the XYZ axis. The material of the AR in this study is defined as concrete, the most common material for exiting AR, with Young's Modulus of 28.60 GPa, Poisson's ratio at 0.15 and density of 2300 kg/m^3.

As illustrated in Fig. 3, the current iteration of artificial reef (AR) undergoes bi-directional evolutionary structural optimization using the Ameba plugin in Rhino 3D. Minimum compliance is targeted, with a 5% evolution ratio and a 3-filter radius. The final AR volume target is 40%, similar to benchmarks in research literature, but each iteration's volume target varies between 60–80%. Typically, less than 20 iterations are needed, with a maximum set at 100. Three iteration processes analyze integrating Computational Fluid

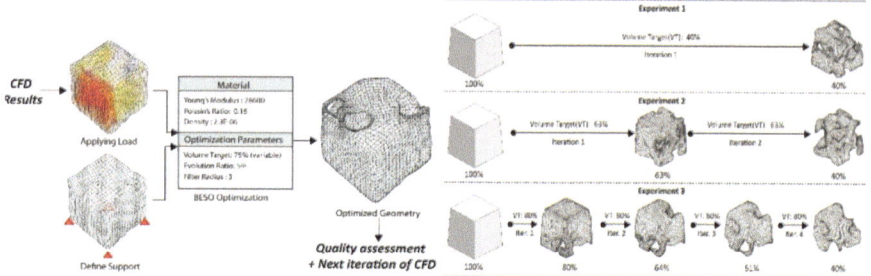

Fig. 3. Left: BESO Set-up; Right: Experiment with Volume Targets and # of iterations

Dynamics (CFD) into topology optimization, aiming for a 40% final volume ratio. Three experimental setups with volume targets at 40%, 63%, and 80% explore the impact of CFD feedback. Material removal ratios indicate 1, 2, and 4 iterations are needed to achieve the volume target. Increasing CFD iterations provide feedback on optimization interim states, allowing adaptation to fluid dynamics changes.

4   Outcome and Assessment

A series of geometrical and structure analysis is conducted to compare the design outcomes of the integrated BESO and CFD against the conventional AR design by Xue (Xue et al., 2023) as the benchmark. The benchmark AR with a slope angle of 85.2° generates different flow filed effect from the hollow cubic AR from previous validation part. It can be observed that the upwelling and back eddy volume increased dramatically. And according to the study conducted by Xue et al. (Fig. 4), from the aspect of flow field effct, the optimal slope angel is 85.2° for the pyramidal reefs, which can produce the largest volume of upwelling and back eddies by per unit reef volume (Xue et al., 2023). Therefore, the pyramidal reef with a slope angle of 85.2° is chosen to be the benchmark AR for this study. For this benchmark design, it's drag coefficient is 1.77, hydrodynamic drag is 7124.03 N and overturning moment is 9797.12 N*m.

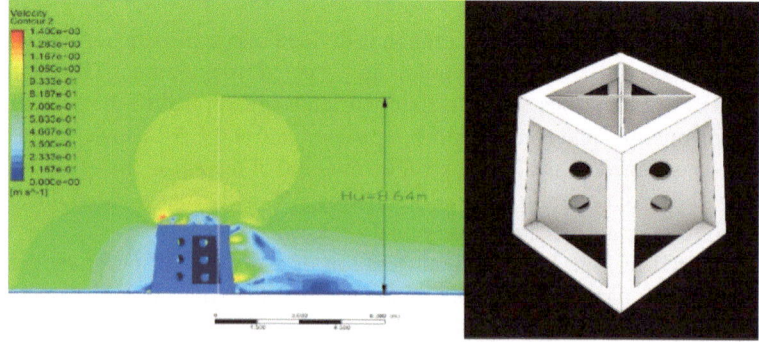

Fig. 4. Velocity contour and vectors on the symmetry plane of the Benchmark AR

4.1   Geometrical Evaluation

The geometry of the generated artificial is compared against the benchmark to assess its geometrical features, as well as discuss adaptivity serving habitat for AR. Recent studies have shown that reducing open spaces while enhancing the intricacy of a reef can impact the composition of fish communities (Eklund, 1996). Numerous investigations have focused on the relationship between the complexity of the environment and the fish species it supports. Typically, these investigations have identified a favorable link between a reef's structure's complexity and the number and variety of fish species present (Spieler et al., 2001). Thus, this research, the ARs are evaluated in terms of the complexity of surface curvature and the complexity of spaces created and voids.

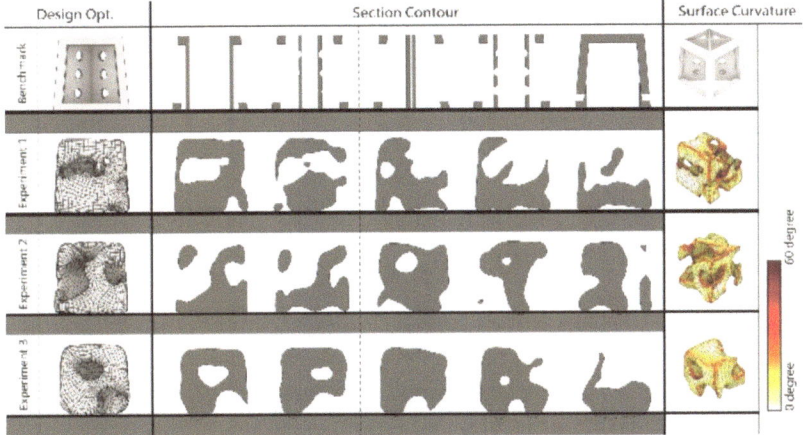

Fig. 5. Surface curvature analysis and sectional analysis of void cavities.

Compared with the rectilinear benchmark AR design, the outcome of all 3 design experiments has demonstrated an increasing level of complexity in surface curvature where topologically complex geometries are created by smooth transitioning of concave and convex surfaces (Fig. 5). The comparison with the benchmark AR design reveals that the three experimental designs exhibit increased surface curvature complexity, creating diverse spaces and cavities suitable for marine species. These variations range from small tunnels to larger semi-protected areas, providing habitat diversity. Each design iteration produces unique topographical features optimized for structural performance under water flow conditions. Refining the design process with feedback from computational fluid dynamics (CFD) can further enhance structural integrity, supporting a diverse range of reef species.

4.2   Performance Evaluation

In this study, three sets of iteration processes are established to analyse and compare the effects of integrating Computational Fluid Dynamics into the iterative process of topology optimization. Figure 6 below illustrates the iterative calculation process of the third experiment with 4 iterations with controlling variables for the above-mentioned main parameters. It can be observed that with the process of iteration, the surface that faces the inflow, the surface pressure tends to become lower. And as illustrated from the contour of eddy viscosity (Fig. 7), the original artificial reef has a larger region of back eddy with a higher pressure. And the back eddy region becomes smaller with the process of iteration while the pressure becomes smaller.

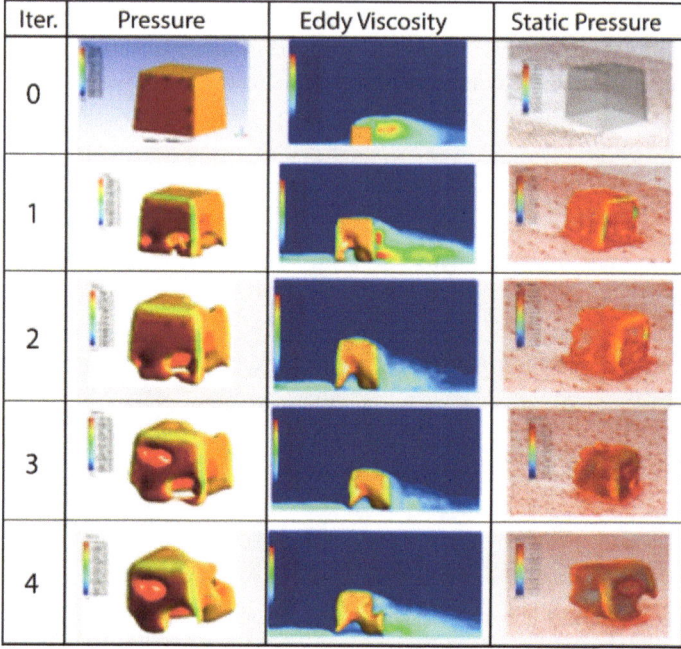

Fig. 6. Iterative calculation process in CFD for Experiment 3

The comparison between benchmark and optimized artificial reefs (ARs) shows differences in eddy viscosity, with the benchmark AR exhibiting higher viscosity and a larger back eddy region (Fig. 8). Lower viscosity around optimized ARs allows for smoother water flow, reducing destabilizing forces and turbulence, enhancing stability critical for AR effectiveness. While significant interaction between ARs and water can lead to pronounced flow field effects, achieving both favorable flow effects and stability is challenging. Therefore, instead of solely focusing on flow field effects, examining AR performance and hydrodynamic parameters is essential.

The hydrodynamic parameters of optimized artificial reefs (ARs) from each iteration are summarized in Table 1. The simulation results show a trend of increasing drag coefficient after each iteration. In the context of artificial reefs, a higher drag coefficient indicates increased resistance to water flow. This results in higher hydrodynamic drag and overturning moment after each iteration. Some resistance can be beneficial for promoting water circulation and enhancing habitat complexity. Slower water flow around the reef can provide shelter for fish and other marine species, concentrate plankton and

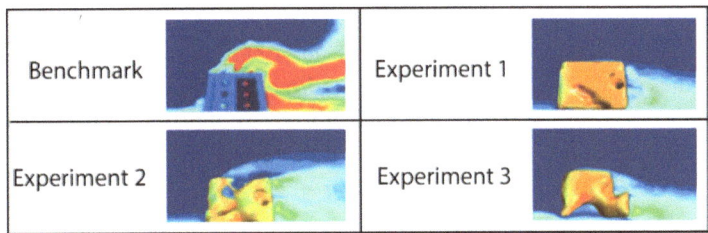

Fig. 7. Eddy viscosity contours on the symmetry plane of the Benchmark AR and optimised AR

other organisms, and promote the settlement and growth of coral larvae and invertebrates. However, excessively high drag coefficient can lead to issues such as increased turbulence, slipping, overturning, and potential damage to reef structures, negatively impacting marine life and the effectiveness of ARs as habitat and coastal protection. Therefore, checking the physical stability of ARs is crucial to avoid slipping or overturning.

Table 1. Drag coefficient, hydrodynamic drag, overturning moment, anti-slipping and anti-overturning coefficient of ARs

Experiment Number		Hydrodynamic Parameters				
		C_d	Hydrodynamic Drag (N)	Overturning Moment (N*m)	Anti-Slipping coefficient	Anti-Overturning coefficient
1	Original	1.14	3273.30	4503.29		
	Iteration 1	3.53	9936.24	14889.03	8.24	8.80
2	Original	1.14	3273.30	4503.29		
	Iteration 1	1.49	4201.09	7513.29	19.63	17.56
	Iteration 2	2.55	6909.03	10239.30	8.42	22.73
3	Original	1.14	3273.30	4503.29		
	Iteration 1	2.85	8028.99	13492.14	11.00	10.48
	Iteration 2	1.94	5351.48	8938.92	14.91	14.28
	Iteration 3	1.78	4685.83	8494.20	15.25	13.46
	Iteration 4	4.59	11650.30	19233.78	5.21	15.77

The safety factors of anti-slipping coefficient and anti-overturning coefficient are often employed to check the AR stability (Xue et al., 2023). According to the ARs stability studies connected by Zhong et al., and Yu et al., (Yu et al., 2019; Zhong et al., 2006), the anti-slipping coefficient and the anti-overturning coefficient are defined as:

$$S_F = \frac{F_{fmax}}{F_D} = (\mu(\sigma - \rho)V_g)/F_D$$

$$S_M = \frac{M_1}{M_2} = ((\sigma - \rho)V_g l_w)/M_2$$

where F_{fmax} represents the maximum frictional force, μ denotes the maximum static friction coefficient, which is chosen to be 0.5 in this study. Additionally, σ and ρ signify the densities of the artificial reef and seawater. Which are 2000 kg/m^3 and 1025 kg/m^3 respectively. V is the volume of the AR, F_D is the hydrodynamic drag force. M_1 is the resultant moment of gravity and buoyancy on the overturning center, l_w is the horizontal distance from the centre gravity to the overturning center. And M_2 is the hydrodynamic moment of the flow on the overturning center. And g is the gravity acceleration which is equal to 9.81 m/s^2.

The anti-slipping and anti-overturning coefficient of different ARs is shown in Table 1 above. It can be observed that both coefficient of optimised ARs is much higher than those of benchmark AR. Therefore, ARs optimised from iterative process based on CFD analysis and topology structure optimisation are able to fulfill these purposes effectively, artificial reefs must remain in place and intact despite dynamic marine environments.

5 Conclusion and Discussion

With employment of the generative design method for artificial reefs via an interactive interaction between Computational Fluid Dynamics and Topology Optimisation, there is an opportunity to develop an optimised artificial reef structure which has diversified void space, mimicking the formal characteristics of natural reef. And thereby enable AR to provide sufficient spatial diversity in the geometries to serve as habitat for marine species and provide shelter to median reef species from swift currents and predator.

The generative design method proposed in this study can be the potential solution to the challenge that notable number of previous artificial reefs have experienced destruction from complicated underwater environment, resulting in failure to achieve intended objectives. The comprehensive CFD simulations conducted in this study provided valuable insights into the performance and effectiveness of optimised AR. By comparing with the benchmark AR, optimised AR can generate the unique flow field effect and actively react to the dynamic of the local current. By comparing hydrodynamic parameters including drag coefficient, anti-slipping coefficient and anti-overturing coefficient with benchmark, optimised AR cis able to avoid issues such as increased turbulence, slipping and overturning which can negatively impact its effectiveness and stability.

However, only the stability of the AR was considered in this study to evaluate the performance of the optimised structures. In fact, the biological process of coral growth, underwater adaptability and practicality is complex, influenced by various factors. However, accurate mathematical descriptions of these processes remain elusive. Utilising universal methods like evolutionary optimization algorithms could offer a promising approach to analyse these challenges.

References

Baine, M.: Artificial reefs: A review of their design, application, management and performance. Ocean wCoast. Manag. **44**, 241–259 (2001)

Bao, D.: Performance-driven digital design and robotic fabrication based on topology optimisation and multi-agent system (2022)

Dod, J.: Effective substances. In: The dictionary of substances and their effects. Royal Society of Chemistry (1999)

Eklund, A.: The effects of post-settlement predation and resource limitation on reef fish assemblages. Dissertation. University of Miami, Miami, Florida, USA (1996)

Elliott, M., et al.: Ecoengineering with Ecohydrology: successes and failures in estuarine restoration, p. 176. Estuarine, Coastal and Shelf Science (2016)

Jiang, Z., Liang, Z., Zhu, L., Guo, Z., Tang, Y.: Effect of hole diameter of rotary-shaped artificial reef on flow field. Ocean Eng. **197**, 106917 (2020)

Jiao, L., Yan-xuan, Z., Pi-hai, G., Chang-tao, G.: Numerical simulation and PIV experimental study of the effect of flow fields around tube artificial reefs. Ocean Eng. **134**, 96–104 (2017)

Huang, X., Xie, Y.: Evolutionary topology optimization of continuum structures. Evolutionary Topology Optimization of Continuum Structures: Methods and Applications (2010)

Le, Q.T.N., Jung, S., Na, W.B.: Wake region estimates of artificial reefs in Vietnam: effects of tropical seawater temperatures and seasonal water flow variation. Sustainability **12**(15), 6191 (2020). https://doi.org/10.3390/su12156191

Lin, S., et al.: Design and fabrication of artificial brain coral: evolution principle, turbulent hydrodynamics and matter interchange. Comput. Struct. **276**, 106955 (2023)

Lin, S., et al.: Shell buckling: from morphogenesis of soft matter to prospective applications. Bioinspir. Biomim. **13**(5), 051001 (2018). https://doi.org/10.1088/1748-3190/aacdd1

Heike, K., et al.: Depletion, degradation, and recovery potential of estuaries and coastal seas. Science **312**(5781), 1806–1809 (2006). https://doi.org/10.1126/science.1128035

Reed, D., Schroeter, S., Huang, D., Anderson, T., Ambrose, R.: Quantitative assessment of different artificial reef designs in mitigating losses to kelp forest fishes. Bull. Mar. Sci. **78**, 133–150 (2006)

Seaman, W.: Artificial reef evaluation: with application to natural marine habitats. CRC. (2000)

Spieler, R.E., Gilliam, D.S., Sherman, R.L.: Artificial substrate and coral reef restoration. Bull. Mar. Sci. **69**(2), 1013–1030 (2001)

Woo, J., Kim, D., Yoon, H.-S., Na, W.-B.: Characterizing Korean general artificial reefs by drag coefficients. Ocean Eng. **82**, 105–114 (2014)

Xue, D., Wang, C., Huang, T., Pan, Y., Zhang, N., Zhang, L.: Flow field effects and physical stability of pyramidal artificial reef with different slope angles. Ocean Eng. **283**, 115059 (2023). https://doi.org/10.1016/j.oceaneng.2023.115059

Yu, D., Yang, Y., Li, Y.: Research on hydrodynamic characteristics and stability of artificial reefs with different opening ratios. Periodical Ocean Univ. China **49**(4), 128 (2019)

Zhao, B., Chen, T.Y.: Application of topology optimisation to architectural design. Archit. Cult. **11**, 104–105 (2016)

Zhao, Z.-L., Zhou, S., Feng, X.-Q., Xie, Y.M.: On the internal architecture of emergent plants. J. Mech. Phys. Solids **119**, 224–239 (2018)

Zhong, S., Sun, M., Zhang, S., Zhang, S.: Study on the design and stability of the artificial steel prism reef. Mar. Fish. **28**(3), 234–240 (2006)

Research on Office Building Facade Design Based on Visual Comfort of Daylit Office

Xi Huang, Dagang Qu, Cheng Sun(✉), and Shi Sun

School of Architecture and Design, Harbin Institute of Technology; Key Laboratory of Cold Region Urban and Rural Human Settlement Environment Science and Techmology, Ministry of Industry and Information Technology, Harbin 150006, China
suncheng@hit.edu.cn

Abstract. This paper presents a model for predicting visual comfort of natural lighting office, considering both the horizontal paper work and vertical VDT (Visual Display Terminal) work mode, based on field experiments and questionnaires. By using the prediction model, an intelligent prediction platform system for visual comfort is built. The model and platform are applied to investigate the prototype office space model in Harbin, China. By simulation and control variable method, the influence of various building facade form on the visual comfort of natural lighting office space is analyzed. Finally, this paper summarizes and proposes office building facade design strategies based on visual comfort, to enhance the overall visual comfort level of office space, and guide office building design.

Keywords: visual comfort · office space · building façade · design strategy

1 Introduction

Nowadays, the average daily working hours of urban office population in China are 8.66 h, and about 90% of the population in developed cities spends time in the office. A large number of relevant studies have pointed out that indoor light environment has an important impact on human physical and mental health and work efficiency. With the increase of work pressure, the office population has increasingly improved the quality requirements of spatial light environment. Insufficient light comfort in office space will directly lead to low work efficiency of people, resulting in additional environmental intervention behaviors, such as closing curtains and so on, which will eventually increase building energy consumption, which is not conducive to building operation cost control, and it is difficult to achieve the goal of energy saving and emission reduction.

In recent years, both building light environment design optimization and adaptive skin related research are gradually from light environment analysis to user visual comfort. The influencing factors of visual comfort in natural light environment are complex, which are related to illumination [4], glare [1], brightness [6], view quality [3] and other factors, and affected by psychological and physiological factors of people [5]. The facade of office buildings not only affects the building aesthetic, but also plays an important role in the visual comfort of indoor space. The contemporary office mode mostly adopts the hybrid

H. Chai et al. (Eds.): CDRF 2024, *Symbiotic Intelligence*, pp. 385–393, 2025.
https://doi.org/10.1007/978-981-96-3433-0_34

horizontal and vertical office mode. Compared with the past horizontal office mode, the visual comfort of the contemporary office mode is more affected by the facade of the office building. Due to the insufficient research on the visual comfort prediction method of office space in the hybrid horizontal and vertical office mode, the automatic prediction platform system has not been proposed, and the influence law of office building facades on the visual comfort of indoor space has not been revealed. Therefore, architects lack effective office building façade design strategies to guide office building design, resulting in the problem of increased operation and maintenance energy consumption caused by low visual comfort. Based on the above content, the author, based on the research results obtained, applies the established prediction model of visual comfort of office space to study the influence law of typical window form of office building facades on the visual comfort of space, so as to guide the architect to design office building facade based on visual comfort.

2 Methodology

The study of the impact of office building façade form on indoor visual comfort is divided into two steps.

Firstly, the author applied the self-developed visual comfort prediction platform to simulate the visual comfort of typical office building facades with different window forms. The simulation experiment focuses on two types of windowing forms: vertical windowing and transverse windowing.

Secondly, the visual comfort simulation results under different façade window forms were compared and analyzed, the influence laws of façade window forms on indoor visual comfort were summarized, and the façade design strategies of office buildings based on indoor visual comfort were proposed.

The visual comfort simulation experiments in this study were carried out by applying the space visual comfort prediction platform. The visual comfort prediction platform was further constructed based on the author 's research on "Construction of the visual comfort prediction model for daylit office based on deep learning".

In 2022, we selected typical office spaces in Harbin to carry out comfort research experiments. The experimenters used the 7 Likert scale scoring table (1 = very discomfort, 7 = very comfort) to score the visual comfort, and the equipment was used to measure and calculate the horizontal illumination of the desktop, the vertical illumination of the human eye, and the vision quality, and finally obtained 432 sets of effective experimental data. On this basis, the author constructed the visual comfort prediction neural network and completed the network training.

The author 's team studied and analyzed the limitations of the visual comfort prediction neural network. In order to facilitate the comparison of the spatial visual comfort under different window morphologies in this study, the author proposed two indicators: "spatial visual comfort time ratio" and "high visual comfort space ratio".

The spatial visual comfort time ratio refers to the ratio of the annual cumulative working time length of the simulation grid at or above the visual comfort level i ($i \in (1,2,3,4,5,6,7)$) to the sum of the annual working time of all the simulation grids, expressed as $\Lambda_{(S \geq i)}$. This indicator can macroscopically describe the time ratio of the center point of each simulated grid in the predicted space under different visual comfort levels with the specified orientation.

The high visual comfort space ratio is the ratio of the spatial grid with $\Lambda_{(S\geq4)} \geq 0.7$ to the total number of simulated spatial grids, expressed as P. This indicator describes the proportion of high visual comfort space in the simulated space, which is used to measure the uniformity of the visual comfort distribution of the office space.

The two parameters are obtained by the mathematical statistical calculation of the output results of the visual comfort prediction neural network, which are the output results of the visual comfort prediction platform in this study. In order to clearly and intuitively express the distribution of spatial visual comfort, the spatial visual comfort prediction platform in this study also outputs the visual comfort map. The drawing method of the visual comfort map is the same as that used in the author 's previous research [2].

3   Simulation Experiment Research of Office Building Façade Based on Visual Comfort Simulation

This paper studies the relationship between office building facade design and spatial visual comfort by using visual comfort prediction platform and control variable method. This study respectively carries out simulation experiments on the distribution of visual comfort in the indoor space of office buildings with different windowing forms and unilateral lighting. This simulation experiment did not study the relatively complex building skin.

In this study, experiments are carried out on two types of facade window forms, vertical form and transverse form. The methods and steps of the simulation experiments of the two types of windowing forms are the same. The experiment is divided into three steps, and Fig. 1 shows the workflow of the experiment.

The first step is to establish an office building model, conduct fine modeling, material labeling, and divide the simulation grid in the modeling software Rhino.

The second step is to simulate the annual vertical illuminance and horizontal illuminance and view quality values of the center point of the simulated grid through ClimateStudio to obtain the numerical set.

In the third step, the values of vertical illumination, horizontal illumination and View Quality are automatically sorted and input into the prediction model to obtain the instant comfort value, and the annual spatial comfort metric value and visual comfort map are output (Fig. 1).

Each point is carried out the simulation in the east and west directions, which are the sight line parallel to the window, that is, the simulation is carried out.The height of the building space is 4.5 m, the southern side of the building is the lighting interface, and the other three sides have no lighting holes. The reflectance of the floor, wall and ceiling of the office space is 80%, 80% and 20% respectively, the landscape outside window is the same. The windows of the building are all double-layer hollow Low-E ultra-white glass. No shading system is used in all buildings. The window area of each facade form model is 45 m^2, the window wall area ratio is 0.5, and the window ground area ratio is 0.225. In the simulation, the grid is divided with a step length of 1.5 m, and 88 points

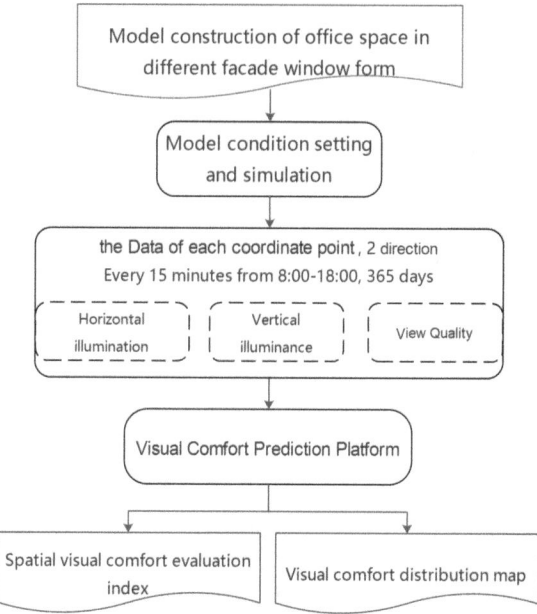

Fig. 1. Visual comfort of office building façade experiment workflow

are distributed in the simulation space. The height of the simulated plane for horizontal illumination is 0.75 m, and the height of the simulated plane for vertical illumination is 1.2 m. The simulation is carried out from 8:00–18:00 for 365 days, and the time step is 15 min. The visual comfort of east and west orientation at each point is simulated.

3.1 Vertical Windows Facade Forms Experiment

Due to the influence of the building structure and the pursuit of the aesthetic trend of towering buildings, vertical windows are widely used in office buildings. The study on the spatial visual comfort of the vertical window facade studied two kinds of conditions, including uniform distribution of fenestrated facade and non-uniform distribution of fenestrated facade. A total of six vertical facade form models were used to conduct visual comfort simulation experiments.

3.1.1 Evenly Distributed Vertical Window Facade Form

Three evenly distributed vertical window Facade form models are used to study the influence of facade and indoor visual comfort. The window height is all 4.5 m, and the window width of Facade A, Facade B, and Facade C is 1.3 m, 0.9 m, and 0.5 m, with evenly window distribution (Fig. 2). The annual spatial visual comfort metric value (Table 1) and visual comfort maps (Fig. 3) are output by the spatial visual comfort prediction platform developed by the author.

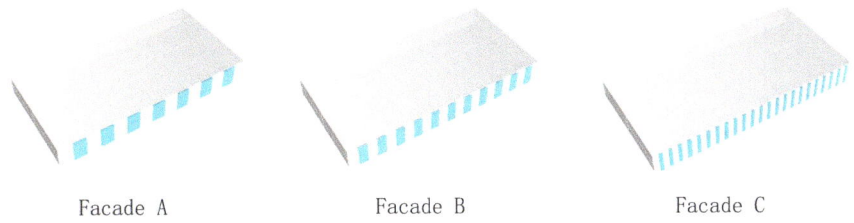

Facade A   Facade B   Facade C

Fig. 2. Evenly distributed vertical windows facade forms

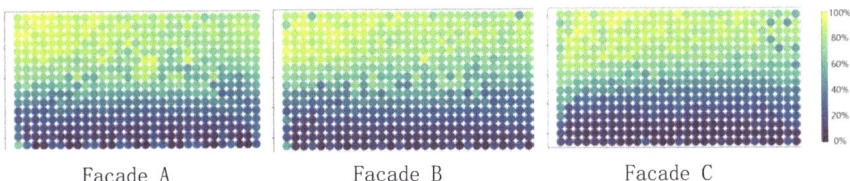

Facade A   Facade B   Facade C

Fig. 3. Visual comfort distribution map of evenly distributed vertical window facade forms

Table 1. Spatial visual comfort evaluation metric value of evenly distributed vertical window facade forms

Facade A		Facade B		Facade C	
$\Lambda_{(S\geq4)}$	55.56%	$\Lambda_{(S\geq4)}$	54.08%	$\Lambda_{(S\geq4)}$	51.03%
$\Lambda_{(S\geq5)}$	41.44%	$\Lambda_{(S\geq5)}$	39.80%	$\Lambda_{(S\geq5)}$	37.92%
$\Lambda_{(S\geq6)}$	22.72%	$\Lambda_{(S\geq6)}$	21.56%	$\Lambda_{(S\geq6)}$	21.48%
$\Lambda_{(S=7)}$	3.99%	$\Lambda_{(S=7)}$	2.99%	$\Lambda_{(S=7)}$	1.53%
P	31.96%	P	31.75%	P	32.06%

3.1.2 Unevenly Distributed Vertical Windows Facade Forms

Three unevenly distributed vertical window facade models are used to study the influence of facade and indoor visual comfort. The window height is 4.5 m, Facade D and Facade E are unevenly facades generated randomly, and the window of Facade F is on one side of the facade (Fig. 4). The annual spatial comfort metric value (Table 2) and visual comfort map (Fig. 5) are obtained through the spatial visual comfort prediction platform.

3.2 Horizontal Windows Facade Forms Experiment

Three cases of horizontal window mode, the window width is 20 m for the overall length of the facade, and Facade G, Facade H, and Facade I are facade with 2, 3, and 4 strip windows respectively, with the window heights of 0.9 m, 0.65 m, and 0.5 m respectively (Fig. 6), with uniform window distribution without partition, and the horizontal line of sight of 1.2 m height of the indoor personnel 's sitting eye height is constructed without blocking under the three facade forms. The comfort index values (Table 3) and the

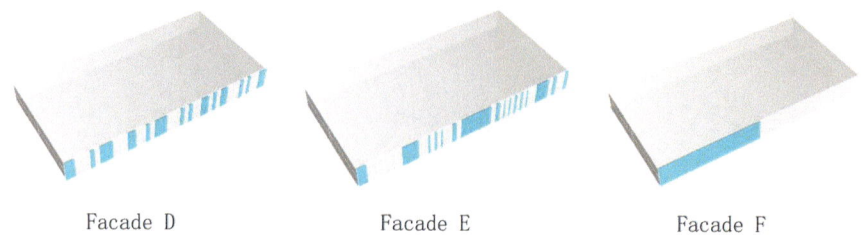

Facade D   Facade E   Facade F

Fig. 4. Unevenly distributed vertical windows facade forms

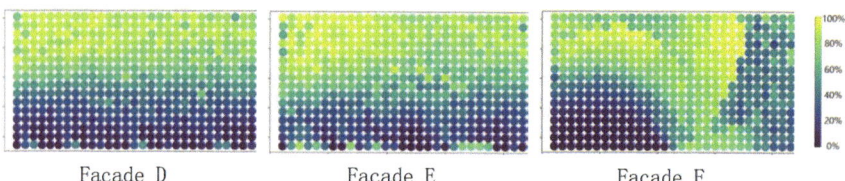

Facade D   Facade E   Facade F

Fig. 5. Visual comfort distribution map of unevenly distributed vertical windows facade forms

Table 2. Spatial visual comfort metric value of unevenly distributed vertical window facade forms

Facade D		Facade E		Facade F	
$\Lambda_{(S \geq 4)}$	52.67%	$\Lambda_{(S \geq 4)}$	56.27%	$\Lambda_{(S \geq 4)}$	59.62%
$\Lambda_{(S \geq 5)}$	38.14%	$\Lambda_{(S \geq 5)}$	41.39%	$\Lambda_{(S \geq 5)}$	49.56%
$\Lambda_{(S \geq 6)}$	20.11%	$\Lambda_{(S \geq 6)}$	22.87%	$\Lambda_{(S \geq 6)}$	30.00%
$\Lambda_{(S = 7)}$	1.70%	$\Lambda_{(S = 7)}$	2.57%	$\Lambda_{(S = 7)}$	6.97%
P	30.95%	P	33.57%	P	33.77%

comfort distribution map (Fig. 7) of the space throughout the year are obtained through the simulation and prediction platform.

The author also used three facade models to study the influence of horizontal window form on spatial visual comfort. The window width is 20 m, and the Facade G, Facade H, and Facade I are 2, 3, and 4 horizontal strip windows, with window heights of 0.9 m, 0.65 m, and 0.5 m respectively. The windows are evenly distributed. The three facades do not block people 's view (Fig. 6). The annual spatial comfort metric value (Table 3) and visual comfort map (Fig. 7) are output by the prediction platform.

4   Analysis and Discussion

According to the above experimental results, the building facade form has an impact on spatial visual comfort. Through the study, it is found that under the condition of the same window-wall ratio, the visual comfort of the office space of the horizontal window is higher than that of the vertical window. $\Lambda_{(S \geq 4)}$ of Facade G is 18.13% higher than

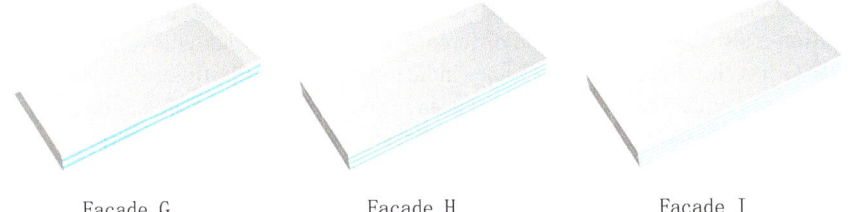

Facade G   Facade H   Facade I

Fig. 6. Visual comfort distribution map of horizontal strip window facade forms

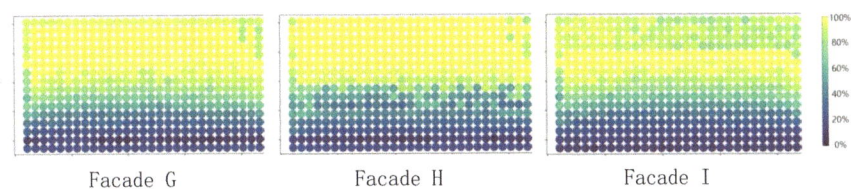

Facade G   Facade H   Facade I

Fig. 7. Visual comfort distribution map of horizontal strip window facade forms

Table 3. Spatial visual comfort evaluation metric value of horizontal strip window facade forms

Facade G		Facade H		Facade I	
$\Lambda_{(S \geq 4)}$	69.15%	$\Lambda_{(S \geq 4)}$	67.48%	$\Lambda_{(S \geq 4)}$	65.31%
$\Lambda_{(S \geq 5)}$	53.35%	$\Lambda_{(S \geq 5)}$	49.37%	$\Lambda_{(S \geq 5)}$	48.15%
$\Lambda_{(S \geq 6)}$	37.24%	$\Lambda_{(S \geq 6)}$	31.29%	$\Lambda_{(S \geq 6)}$	28.73%
$\Lambda_{(S = 7)}$	14.26%	$\Lambda_{(S = 7)}$	13.14%	$\Lambda_{(S = 7)}$	11.28%
P	34.88%	P	35.38%	P	34.68%

that of Facade C, whose is the minimum. The proportion of high visual comfort space of Facade G is 2.82% higher than that of Facade C. It can be seen that the facade of the horizontal window is better than that of the vertical window. If the design permits, the priority can be given to the design of the building facade with horizontal windows.

In the case of vertical window facade, the Facade F with centralized window arrangement has the best visual comfort, and the area furthest away from the window also have good visual comfort. The area with low comfort is small and concentrated, which can be designed as office auxiliary functional areas with low visual comfort requirements. Under the condition of evenly distributed form, the facade A with the largest window width is the best, and the facade C with the smallest window width is the worst. To sum up, the layout form of Windows in the office building facade with vertical Windows has a great impact on the visual comfort. It is easier to ensure a higher visual comfort in the interior space by adopting a centralized window layout mode or expanding the width of the window as much as possible.

In the case of the building facade with a horizontal window, the visual comfort evaluation metric value of the three different heights of the window is relatively close, Facade G has the highest $T_{(S \geq 4)}$ value, indicating that its total time of comfort level greater than or equal to 4 is the longest. Facade H has the highest P value, indicating that it has the highest ratio of high visual comfort space. From the visual comfort map, it can be seen that the visual comfort level of the single window height of the Facade I in the near window area is lower than that of other forms, and the comfort level $S \geq 6$ area is relatively small. The effect of window height on visual comfort is relatively small in the horizontal window facade but the window height should not be too small.

5   Future Outlook

Based on the engineering design application, this study carried out a systematic study on the prediction method of visual comfort of natural lighting office space. Through the visual comfort prediction platform constructed by the existing research, through a variety of vertical, horizontal, uniform distribution, non-uniform distribution of the facade window form simulation experiments and comfort prediction results comparison, the design strategy of building facade window form was summarized and put forward. When the designers combine the specific geographical conditions, creative ideas, energy saving requirements and other factors to reasonably choose the building facade form, the quality of office visual comfort will be improved to a certain extent. Limited by the research time and length of the paper, this paper mainly discusses the unilateral lighting building facade form based on visual comfort. The same experimental research method can be used in the future to explore the facade form of office buildings with multiple lighting surfaces, different plane forms, curtain walls or shading systems.

Based on the application requirements of engineering design, this study applies the visual comfort prediction platform developed in the research to conduct simulation experiments on the spatial visual comfort under a variety of facade windowing modes, studies the effect of the facade windowing form of office buildings on the spatial visual comfort, and summarized the design strategy of office building facade. Limited by the research time and the length of the paper, this paper mainly discusses the facade forms of one-sided lighting buildings based on visual comfort. The subsequent research will study the facade design strategies of office buildings with complex windowed facades and different plane forms, so as to better assist architects to complete the facade design of office buildings and improve the quality of building space.

References

1. Daich, S., Saadi, M.Y., Piga, B.E., et al.: A combined method for an exhaustive investigation of the anidolic ceiling effect on improving indoor office daylight quality: an approach based on HDR photography and subjective evaluations. J. Daylighting **8**, 149–164 (2021)
2. Huang, X., Qu, D., Sun, C., et al.: Construction of visual comfort prediction model for daylit office space based on deep learning (in Chinese). Architect. J. **659**(10), 50–54 (2023). https://doi.org/10.19819/j.cnki.ISSN0529-1399.202310008

3. Jamrozik, A., Clements, N., Hasan, S.S., et al.: Access to daylight and view in an office improves cognitive performance and satisfaction and reduces eyestrain: A controlled crossover study. Build. Environ. **165**, 106379 (2019). https://doi.org/10.1016/j.buildenv.2019.106379

4. Kaushik, A., Arif, M., Ebohon, O., et al.: Effect of indoor environmental quality on visual comfort and productivity in office buildings. J. Eng. Des. Technol. **21**(6), 1746–1766 (2023)

5. Lan, L., Wargocki, P., Lian, Z.: Thermal effects on human performance in office environment measured by integrating task speed and accuracy. Appl. Ergon. **45**(3), 490–495 (2014). https://doi.org/10.1016/j.apergo.2013.06.010

6. Rockcastle, S., Andersen, M.: Dynamic annual metrics for contrast in daylit architecture. In: Spring Simulation Multiconference Cambridge 2012, USA (2012)

Effectiveness and Optimization of Passive Design for Climate Adaptation in the HSCW Zone—Taking a High-Rise Apartment Retrofit in Philadelphia as an Example

Zhen Lei[1], Tong Zhang[1(⊠)], and Yue Fang[2]

[1] School of Architecture, Southeast University, 2 Sipailou, Nanjing 210096, China
hytong@seu.edu.cn
[2] College of Engineering, Northeastern University, Boston, MA 02115, USA

Abstract. For architectural design to actualize climate adaptation, it is essential to optimize building energy efficiency, emission reduction, and passive survivability. However, passive design strategies for building retrofit in the hot summer and cold winter (HSCW) zone are limited in current building energy simulation and optimization (BESO) studies, which have not been widely applied in architectural practice due to the lack of a unified standard. This paper aims to explore the effectiveness and optimization methods of passive design for the typical high-rise apartment retrofit in Philadelphia, considering the dynamic effects of energy consumption, thermal comfort, and future climate scenarios. In this study, the developed future weather files were used to plot the Givoni bioclimatic chart (GBC), and building datasets were constructed based on the EnergyPlus model simulation. Meanwhile, the optimal solutions are realized based on the Morris sensitivity analysis (SA) and NSGA-II method. The results indicate solar protection remains the most effective passive design strategy, especially for south-facing room units, while the cooling effect of natural ventilation by window opening will significantly decrease over time. It is expected that in the future, the thermal coefficient (TC) of the wall and window will increase the effectiveness of energy efficiency to 235% and 152% respectively. The combinations of passive parameters in various climatic scenarios for the overall high-rise apartment retrofit can reduce both heating and cooling loads by up to 50%, and improve the duration of passive survivability by over 400 h.

Keywords: Climate adaptation · Passive design · Energy efficiency · Thermal comfort · Building retrofit

1 Introduction

According to the IPCC 2022 report, climate change has become an imperative challenge facing the 21st century. The trend of global warming and the frequency of extreme weather have drawn attention from nations worldwide to research on human habitats and

© The Author(s) 2025
H. Chai et al. (Eds.): CDRF 2024, *Symbiotic Intelligence*, pp. 394–406, 2025.
https://doi.org/10.1007/978-981-96-3433-0_35

building performance [1]. In developed countries like Europe and North America, buildings account for approximately 40% of total energy consumption. Energy conservation and emission reduction efforts have inducted "climate adaptation" into contemporary topics such as building vulnerability and urban resilience [2]. This study is situated in Philadelphia, USA, which lies in a humid subtropical climate zone characterized by a typical hot summer and cold winter (HSCW). In areas such as City Hall and University City, high-rise apartments are prevalent as a type of residential building, many of which were built in the mid to late 20th century.

In existing green building systems, building layout, envelope structure, geometric form, material properties, and infiltration & airtightness have been identified as critical passive parameters for the HSCW zone [3]. The effectiveness of daylighting, ventilation, temperature, and dehumidification in residential buildings can be verified through building energy simulation optimization (BESO), providing relevant guidelines for the mixed-mode building environment [4]. In recent years, "passive survivability," as a standard within the LEED and RELi rating systems developed by the USGBC, refers to the idea that certain buildings, especially houses and apartments, should be designed and built to maintain habitable temperatures in the event of an extended power outage or interruption in heating fuel. It provides a certain compliance path for effective evaluation through psychrometric analysis [5]. In the current research, many scholars from different regions are examining various aspects of passive design effectiveness in the HSCW zone [6], including district thermal resilience in the United States [7], indoor overheating risks in Hong Kong [8], the standards of green building rating tools in China [9], and bioclimatic design visualization in Iran [10].

The previous studies have revealed limitations in data-driven passive design methods, with most relying on a typical multi-time optimization [11]. Many researches indicate that the minimum number of simulations required for effective optimization of building energy models ranges from 1400 to 1800 times [12], involving cumbersome steps and long calculation times, which fail to provide conclusive insights into coupling effects during the decision-making process. To fill the gap, this study aims to explore an effectiveness and optimization method for passive design based on adaptability to climate change, guiding the high-rise apartment retrofit in the HSCW zone.

2 Methodology

2.1 The Framework of This Study

This study primarily adopts a three-phase method (pre-processing, optimization, and post-processing) based on BESO, integrating sensitivity analysis (SA) and Genetic Algorithm (GA) for the application of passive design research. Specifically, in the first step, future climate datasets are utilized to generate and compute bioclimatic charts, assessing the dynamic trends of passive design strategies. In the second step, building thermal models and passive design variables are configured, along with the introduction of thermal criteria. Subsequently, the third step employs the Morris screening within the SA method to identify the importance and correlation of multiple parameters with target values. Finally, the NSGA-II model is utilized to provide Pareto optimal solutions in order to validate the feasibility of the decision-making process. In summary, the overall framework of the effectiveness and optimization of passive design in this study is shown in Fig. 1.

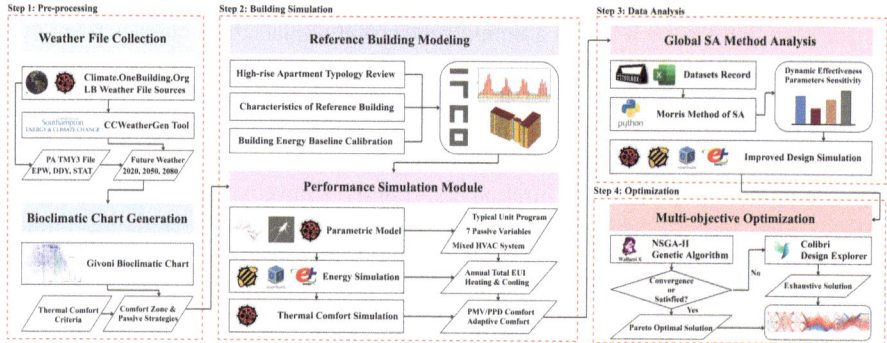

Fig. 1. Methodological framework of the study

2.2   Thermal Model and Bioclimatic Chart

This study investigates Sansom Place West, a graduate apartment constructed in the 1970s, as the sample for simulation, analysis, and optimization. In thermal modeling, the typical floor is divided into thermal zones, while the remaining parts are considered adiabatic geometry to enhance computational efficiency, as shown in Fig. 2. The HVAC ideal model is entirely built by the OpenStudio cross-platform plugin Ladybug Tools based on EnergyPlus, with simulation results closely matching the building's actual situation. The annual baseline energy use intensity (EUI) is 262.25 kWh/m^2, with the predicted mean vote (PMV) of 38.71%.

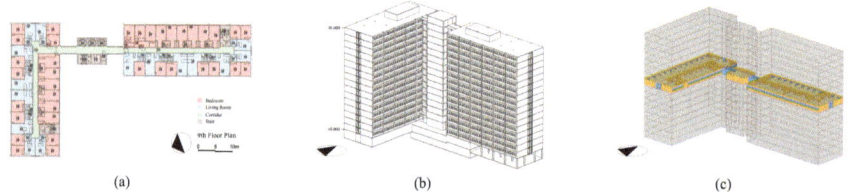

Fig. 2. Typical floor plan (a), the whole building model (b), and simulation thermal model (d) of Sansom Place West

Passive survivability evaluates buildings under off-grid conditions using "extreme caution" thresholds for dry bulb temperature (DBT) and wet bulb globe temperature (WBGT). Specifically, during summer, indoor conditions should not exceed the values of the heat index, i.e., DBT below 32 °C and WBGT below 28 °C. In winter, indoor DBT should be above 10 °C. Referring to the ASHRAE 55 standard and the adaptation level of residents in hot and humid climates due to psychological thermal expectations, the adaptive comfort standard (ACS) model is adopted to set the threshold for natural ventilation conditions at 22–27 °C [13]. The indoor comfort temperature (t_{comf}) of the ACS model is calculated based on the mean outdoor dry bulb temperature ($t_{a, out}$) as follows:

$$t_{comf} = 0.31\, t_{a,out} + 17.8 \tag{1}$$

The Givoni bioclimatic chart (GBC) is widely applied in the research of environmental regulation. This method calculates temperature and humidity ranges through the weather data, visualizing the dynamic impact of climate for thermal comfort [14]. Drawing from the Milne-Givoni bioclimatic chart, this study horizontally divided the hot and cold blocks to facilitate the comparison of passive design strategies suitable for the HSCW zone, as illustrated in Fig. 3. Additionally, using Philadelphia's typical meteorological year (TMY) weather file as the baseline, future data for three time periods in the 21st century (2020/2050/2080) were generated using the CCWorldWeatherGen tool to analyze the duration and dynamic effectiveness of passive design strategies.

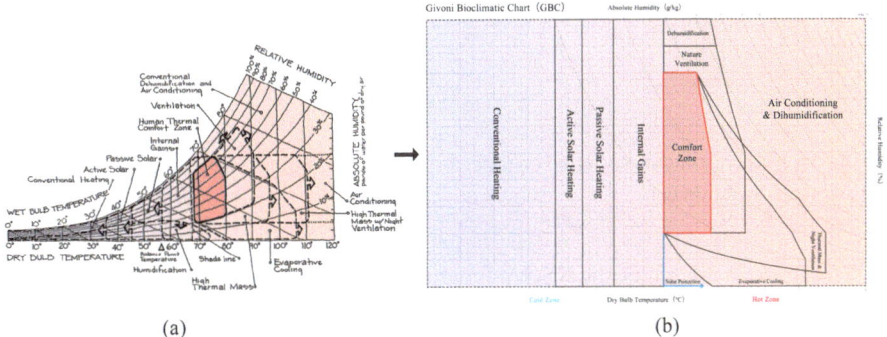

(a)   (b)

Fig. 3. Milne-Givoni bioclimatic chart (a) and horizontal GBC drawn by authors

2.3 Sensitivity Analysis and Optimization Method

This study quantifies the importance of input parameters on target values by calculating sensitivity coefficients. It employs the Morris screening for global SA based on the SALib module in Python, wherein each successive computation, only one parameter changes [15]. The sampling is perturbed by the probability distribution equation to realize the ranking of multiple parameters. SA is based on the ratio of standard deviation σ to absolute mean μ^*, as shown in Fig. 4. Here, μ^* represents sensitivity, with larger values indicating greater influence, while σ represents nonlinearity, with larger values indicating stronger interactions. When the output $\sigma/\mu^* < 0.5$, the parameters exhibit a certain degree of monotonicity. The basic formulas for σ and μ^* in SA are as follows:

$$\sigma_k = \sqrt{\frac{1}{nR}\sum_{r=1}^{nR}\left(EE_k^r - \mu_k\right)^2} \tag{2}$$

$$\mu_k^* = \frac{1}{nR} \sum_{r=1}^{nR} |EE_k^r|$$
(3)

where k represents input parameters, r represents the number of parameter changes, nR represents the number of sampling trajectories, and EE represents the elementary effect. In this study, based on four climate scenarios and drawing from relevant research experiences [16], the number of parameter changes was set to 8, the number of sampling trajectories to 40, and the minimum number of simulations to 1280.

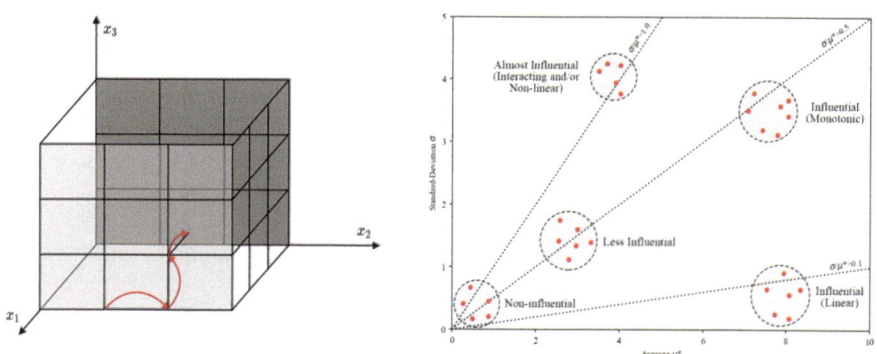

Fig. 4. Morris screening of SA based on trajectory, and SA coordinate with absolute average (μ*) and standard deviation (σ)

Moreover, in this study, the NSGA-II based on the Pareto optimal solution is adopted as a multi-objective optimization algorithm with an elite retention strategy. It can utilize fast non-dominated sorting of population to realize the solution space by selection, propagation, crossover, and mutation [17]. Firstly, the important and linear parameters selected by SA are employed as input values for GA optimization. The WallaceiX tool is utilized to set up fitness functions to determine the combinations of passive parameters. Secondly, for other nonlinear parameters, Colibri's Design Explorer tool is utilized to simulate exhaustively, ultimately achieving the optimal solutions under different climate scenarios.

In this regard, by studying the passive system principles of building envelope structure, the passive factors and variable relationships of the environmental regulation can be constructed (Fig. 5). Among these, the passive system is categorized into three functional modules (daylighting, ventilation, and thermal) from top to bottom, forming a positive interaction with environmental parameters through heat exchange of building layers. Table 1 shows the units and specific ranges of the important parameters in passive design strategies, and parameter variations are set to be evenly distributed.

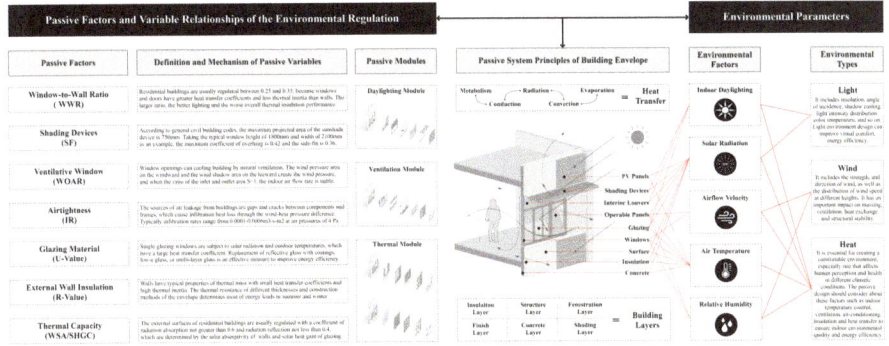

Fig. 5. Passive component factors and variable relationships of the environmental regulation

Table 1. Baseline parameters and ranges of variables under uniform distribution

Parameter	Abbr.	Unit	Default	Range
Window U-value	WinU	W/m^2-K	3.5	0.30–5.90
Solar Heat Gain Coefficient	SHGC	-	0.6	0.20–0.69
Wall R-value	WallR	m^2-K/W	2	1.68–4.48
Window Open Area Ratio	WOAR	%	10	10–80
Window-to-Wall Ratio of BR	WWRB	%	40	20–55
Window-to-Wall Ratio of LR	WWRL	%	90	34–90
Wall Solar Absorptance	WSA	-	0.7	0.10–0.73

3 Results

3.1 Dynamic Trends of Passive Design Strategies in GBC

Based on the GBCs generated from Philadelphia's EPW file, dynamic hours and importance levels are calculated for different passive design strategies, as depicted in Fig. 6. Overall, the color blocks exhibit a significant rightward shift over time, indicating an increasingly hot and humid climate in the region. Numerically, there is a slight fluctuation in the duration of comfort zone hours over the 21st century, decreasing from 1035 h in the TMY to 937 h in 2080. This might be attributed to the weakening of heating demand during winters due to the warming climate in the subtropical region, thus compensating for the comfort zone. Among them, internal heat gains decrease by 220 h, while traditional heating reduces by 1036 h, nearly halving the duration. Besides, the overall solar heating variations are relatively minor, given the limited impact of climate warming on solar radiation.

In contrast to heating, most cooling exhibits significant fluctuations in duration throughout the 21st century, as illustrated in Fig. 7. Among them, natural ventilation maintains a high and stable proportion. However, the duration of solar protection almost

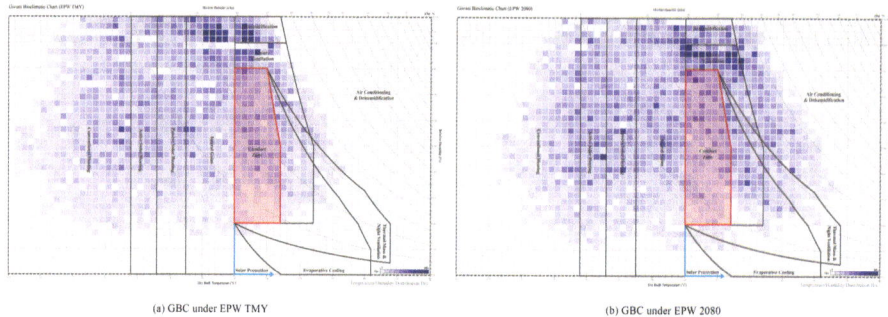

Fig. 6. GBCs of Philadelphia under climate scenarios of TMY (a) and 2080 (b)

doubles, indicating a more pronounced demand for buildings to mitigate sunlight and radiation in future climates. This also underscores the crucial role of window-wall thermal coefficients in ensuring thermal comfort under future overheating conditions.

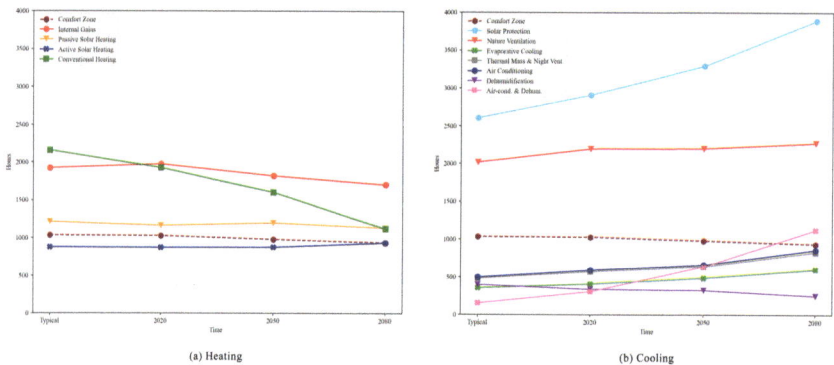

Fig. 7. The number of hours for heating (a) and cooling (b) strategies from TMY to 2080

3.2 Dynamic Evaluation of Passive Parameters in SA

In this study, Morris SA was conducted on the simulated dataset to plot multi-objective radar charts of the μ^* values for each passive parameter, as shown in Fig. 8. Throughout the climate change of the 21st century, the thermal properties of building envelope materials have a significant impact on building energy consumption. Parameters such as WinU, WallR, and WWRB exhibit high sensitivity, while SHGC and WSA show rapidly increasing sensitivity. Relatively, the impact of each passive parameter on thermal comfort is minor, and the effectiveness of parameters between the unconditioned ACS model and the mechanical PMV model is different.

A comprehensive sensitivity ranking is shown in Fig. 9, which could be a more intuitive assessment of the significance of each parameter to target values. Particularly for cooling, SHGC, WSA, and WWRB exhibit high sensitivity and stable timeliness, which

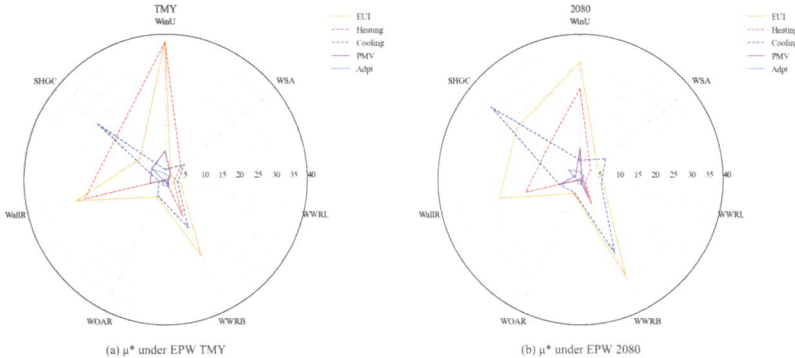

Fig. 8. The radar charts of μ^* under climate scenarios of TMY (a) and 2080 (b)

become the key variables for enhancing building performance in the future. However, WOAR shows a decreasing trend across multiple target values, indicating a noticeable weakening of the dynamic effectiveness of natural ventilation over time.

Objective	EUI				Heating				Cooling				PMV Comfort				Adaptive Comfort			
EPW	TMY	2020	2050	2080	TMY	2020	2050	2080	TMY	2020	2050	2080	TMY	2020	2050	2080	TMY	2020	2050	2080
WinU	1	1	1	1	1	1	1	1	6	6	7	6	1	1	1	1	5	2	2	2
SHGC	4	4	4	4	3	3	3	3	1	1	1	1	2	3	4	4	1	1	1	1
WallR	2	3	3	3	2	2	2	2	7	7	6	5	2	2	2	2	6	6	6	6
WOAR	5	5	7	7	7	7	7	7	4	4	5	7	7	7	7	7	3	5	5	5
WWRB	3	2	2	2	4	4	4	4	2	2	2	2	4	4	3	3	2	3	3	3
WWRL	6	6	5	5	6	6	6	6	5	5	4	4	6	6	6	5	7	7	7	7
WSA	7	7	6	6	5	5	5	5	3	3	3	3	5	5	5	6	4	4	4	4

Most Sensitive ◄――――――――――――――――――――― Least Sensitive

1	2	3	4	5	6	7

Fig. 9. The heat map of sensitivity ranking of passive parameters to five target values

According to the definition of the SA scatter plot, the distribution of parameters σ/μ^* was identified by comparing TMY with 2080 (Fig. 10). In the TMY, SHGC exhibits a monotonic and significant influence on energy consumption, while WOAR remains effective for the current climate. In 2080, SHGC, WSA, and WWRB demonstrate a monotonic trend for cooling loads. Besides, for comfort models, only WinU's value approached 0.5 and right, indicating its significant impact on passive survivability.

3.3 Dynamic Optimization of Decision-Making by GA

In the multi-objective optimization phase, parameters with high sensitivity and mono-tonicity were selected as the main variables in the BESO. Specifically, for TMY, WinU, WallR, SHGC, and WOAR were chosen, while for 2080, WinU, WallR, SHGC, WSA, and WWRB were selected. In the GA results, the standard deviations of the annual EUI and PMV shifted significantly to the left and gradually converged (Fig. 11). Additionally, the distribution frequency of genetic variables exhibited certain regularity in the final generations of solution space (Fig. 12). Hence, the simulation results align with the expectations of dynamic optimization of decision-making by GA.

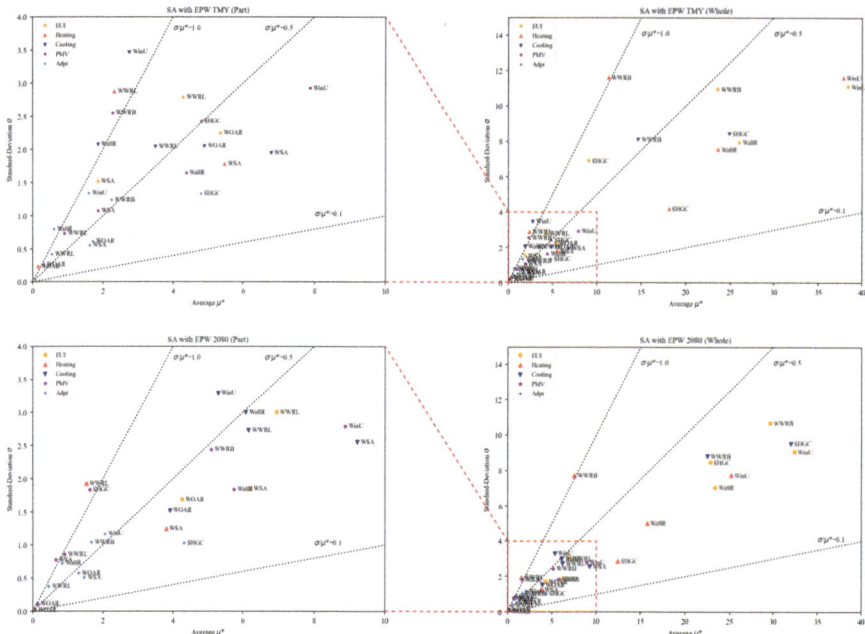

Fig. 10. The scatter plots of σ/μ^* under climate scenarios of TMY and 2080

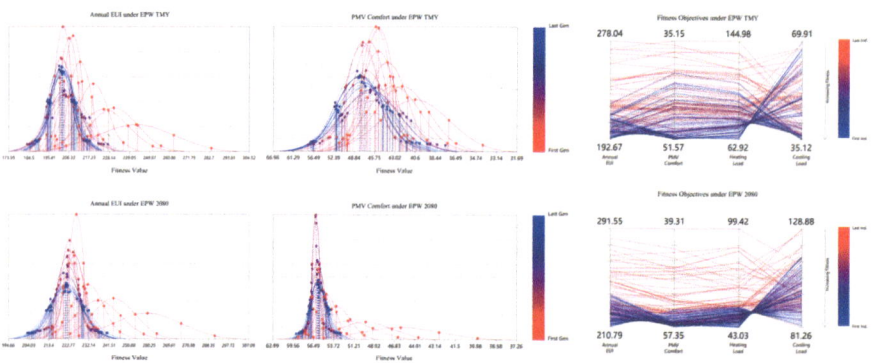

Fig. 11. Standard deviations and parallel coordinates of fitness objectives under the climate scenarios of TMY and 2080

The selected parameter combination with the highest average fitness rank underwent optimization by the exhaustive method. Solar protection was added this time, and WWRB was additionally included for TMY. By comparing the results (Fig. 13), specific measures and ranges for passive design strategies targeting the high-rise apartment retrofit under current and future climates were summarized, as illustrated in Table 2.

In local optimization, assuming the above passive approaches are adopted, three shading strategies (overhangs, side-fins, and louvers) were applied to south-facing units most affected by the HSCW zone. The remaining passive parameters were included

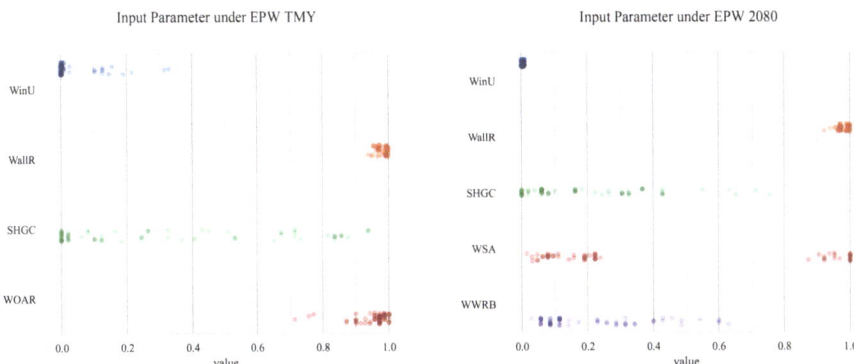

Fig. 12. Distribution frequency of passive parameters in all Pareto front solutions

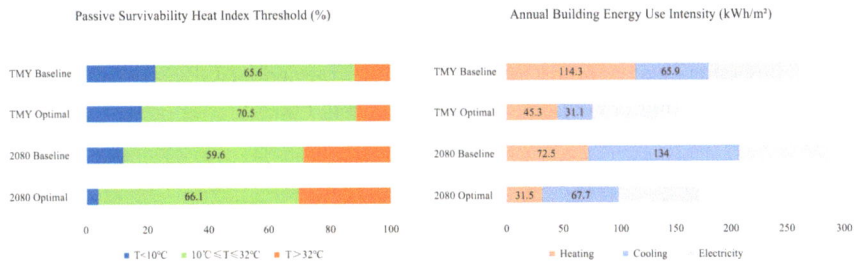

Fig. 13. Comparison of simulation results under the climate scenarios of TMY and 2080

Table 2. Passive measures and ranges under the climate scenarios of TMY and 2080

Scenario	Shading Device	Window & Glazing	Envelope Structure
Retrofit Measures in TMY	Adding short external overhangs and side fins; Adding the operable louvers	Replacing single clear glazing with Low-e or double glazing; Increasing sliding window opening area	Adding 10 mm inner expanded polystyrene or composite PCM layer; Adding external Dry hanging stone or ceramic panels
Retrofit Measures in 2080	Adding long external overhangs and side fins; Adding the operable louvers	Replacing single clear glazing with Low-e coated grey glazing; Replace sliding window to top-opening window	Adding 10 mm inner expanded polystyrene layers; Adding external dry hanging light-colored tiles; Adding external vertical greening

(continued)

Table 2. (*continued*)

Scenario	Shading Device	Window & Glazing	Envelope Structure
Parameter Range in TMY	Overhang length = 350–500 mm Side fins width = 0–350mm	WinU = 0.3–1.5 W/m²-K SHGC = 0.2–0.6 WOAR = 60–80%	WallR = 4–4.5 m²-K/W WSA = 0.4–0.7 WWR = 0.35–0.5
Parameter Range in 2080	Overhang length = 500–750 mm Side fins width = 0–350 mm	WinU = 0.3–1 W/m²-K SHGC = 0.2–0.4 WOAR = 40–60%	WallR = 4–4.5 m²-K/W WSA = 0.1–0.3 WWR = 0.2–0.35

in a secondary optimization. In particular, indoor useful daylight illuminance (UDI), PMV thermal comfort, and the distribution frequency of cooling and heating loads in summer and winter were considered, and the unconditional simulation results of the indoor temperature and humidity were drawn into the indoor GBCs to compare the changes in the duration of comfort zones, as plotted in Fig. 14.

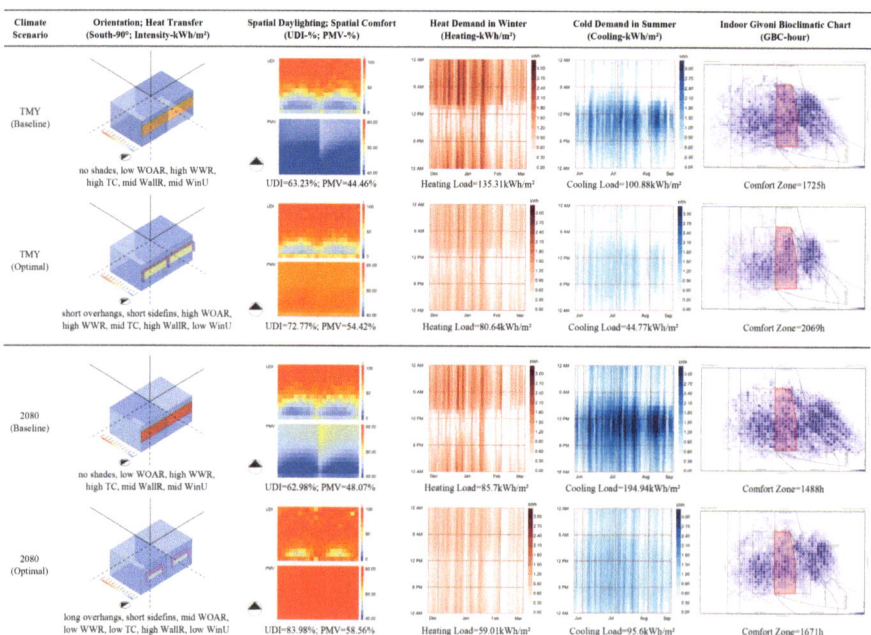

Fig. 14. The effectiveness comparison of passive design strategies for south-facing units under the climate scenarios of TMY and 2080

4   Discussion and Conclusion

In the optimization results, by applying the above passive design, the high-rise apartment can effectively and significantly save energy and continue to maintain passive survivability. In particular, the south-facing units demonstrate the increase in the intensity of heat transfer through the building envelope and the reversal of the energy demand for heating and cooling as a result of climate change. In the optimization of passive design to ensure daylighting, thermal comfort, and energy efficiency, combing shading devices can control the total number of hours in indoor GBCs in the middle, avoiding the impact of extreme temperatures in both winter and summer.

The study shows that the built environment in the HSCW zone will face the issues of current thermal insulation and future overheating risks. In this case, WinU and WallR have stable effectiveness on the overall building performance, while the effectiveness of WSA and SHGC on energy efficiency will be significantly increased to 235% and 152%. In the current, adding WOAR to adopt natural ventilation can effectively mitigate thermal discomfort. Meanwhile, replacing facade materials and adding shadings can significantly improve building performance. It is expected that the cooling and heating load will be reduced by 52.9% and 60.4% respectively, and the duration of passive survivability will be increased by 431 h. In the late of this century, multi-shading and window-wall control will become the key factors. It is expected that the heating and cooling load will be reduced by 49.5% and 56.6% respectively, and the duration of passive survivability will be increased by 561 h. Admittedly, this study only discusses the effectiveness and optimization of passive design for the built environment in the HSCW zone, while other environmental factors have significant impacts and challenges for building retrofit. Therefore, further research that takes into account the comprehensive analysis will be more instructive for climate adaptation.

References

1. Intergovernmental Panel on Climate C, PöRtner, H.O., Belling, D., et al.: Climate change 2022: impacts adaptation and vulnerability: Working Group II contribution to the Sixth Assessment Report of the Intergovernmental Panel on Climate Change. Cambridge University Press, Cambridge (2023)
2. Rajkovich, N., Holmes, S.H.: Climate adaptation and resilience across scales: from buildings to cities. In: Routledge, New York, NY, p. 1 online resource (2022)
3. Tian, Z., Zhang, X., Jin, X., et al.: Towards adoption of building energy simulation and optimization for passive building design: a survey and a review. Energy Build. **158**, 1306–1316 (2018)
4. Chen, X., Yang, H.: A multi-stage optimization of passively designed high-rise residential buildings in multiple building operation scenarios. Appl. Energy **206**, 541–557 (2017)
5. Holmes, S.H., Phillips, T., Wilson, A.: Overheating and passive habitability: indoor health and heat indices. Building Research & Information **44** (2016)
6. Wu, H., Zhang, T.: Multi-objective optimization of energy, visual, and thermal performance for building envelopes in China's hot summer and cold winter climate zone. J. Build. Eng. **59**, 105034 (2022)
7. Wijesuriya, S., Kishore, R.A., Bianchi, M.V.A., Booten, C.: Enhancing thermal resilience of US residential homes in hot humid climates during extreme temperature events. Cell Reports Physical Science, 101986 (2024)

8. Liu, S., Kwok, Y.T., Lau, K.K.-L., et al.: Effectiveness of passive design strategies in responding to future climate change for residential buildings in hot and humid Hong Kong. Energy and Buildings 228 (2020)
9. Li, X., Feng, W., Liu, X., Yang, Y.: A comparative analysis of green building rating systems in China and the United States. Sustain. Cities Soc. **93**, 104520 (2023)
10. Roshan, G., Farrokhzad, M., Attia, S.: Defining thermal comfort boundaries for heating and cooling demand estimation in Iran's urban settlements. Build. Environ. **121**, 168–189 (2017)
11. Konis, K., Gamas, A., Kensek, K.: Passive performance and building form: an optimization framework for early-stage design support. Sol. Energy **125**, 161–179 (2016)
12. Hamdy, M., Nguyen, A.-T., Hensen, J.L.M.: A performance comparison of multi-objective optimization algorithms for solving nearly-zero-energy-building design problems. Energy Build. **121**, 57–71 (2016)
13. Albatayneh, A., Jaradat, M., Alkhatib, M.B., et al.: The significance of the adaptive thermal comfort practice over the structure retrofits to sustain indoor thermal comfort. Energies **14**(2946), 14 (2021)
14. Morillón-Gálvez, D., Saldaña-Flores, R., Tejeda-MartíNez, A.: Human bioclimatic atlas for Mexico. Solar Energy **76**, 781–792 (2004)
15. Mcculloch, A.: Sensitivity analysis in practice: a guide to assessing scientific models. Journal of the Royal Statistical Society Series A: Statistics in Society 168 (2005)
16. Campolongo, F., Cariboni, J., Saltelli, A.: An effective screening design for sensitivity analysis of large models. Environ. Model. Softw. **22**, 1509–1518 (2007)
17. Wang, S., Yi, Y.K., Liu, N.: Multi-objective optimization (MOO) for high-rise residential buildings' layout centered on daylight, visual, and outdoor thermal metrics in China. Building and Environment 205 (2021)

Analysis of Campus Crowd Behavior Based on Location Data and Physical Environment Data: A Case Study of Southeast University Wuxi Campus

Ye Tang[✉], Junqiang Sun, Guangjin Wang, Wenjin Hong, and Li Li[✉]

Southeast University, Si-Pai-Lou 2#, Nanjing, Jiangsu Province, China
{220230214,101012053}@seu.edu.cn

Abstract. The study on the behavior of on-campus individuals provides valuable insights for campus management, resource allocation, and planning layout. The application of multi-source data offers more objective and in-depth opportunities for exploring behavioral phenomena. Focusing on the Wuxi campus of Southeast University, this research utilized Wi-Fi probe positioning technology combined with a physical environment sensor system to comprehensively collect 28.87 million positioning data points and 340,000 environmental data points over a period of 14 days. After cleaning redundant, missing, abnormal, drifting, and ping-pong data, both types of data underwent visual analysis, and their correlations were studied. Additionally, trajectory feature extraction was conducted using a convolutional autoencoder neural network. The study revealed the temporal distribution of pedestrian flow, the spatial distribution of stopover behavior, and the spatiotemporal characteristics of pedestrian trajectories. This provides a reliable basis for guiding crowd behavior by improving specific campus areas and the physical environment.

Keywords: Wi-Fi probes · Physical environment sensors · Data visualization · Characterization · Data correlation

1 Introduction

High school campus spaces belong to the observational scale in urban spaces and feature comprehensive and concentrated urban facilities. Faced with the challenge of limited resources in urban spaces, various universities in China opt to expand their scale through new campus development, imposing higher demands on management, resource allocation, and planning. The characteristics of crowd behavior have become a focal point in urban design, reflecting both the spatiotemporal features of urban areas and the behavioral needs of users. Analysing crowd behavior on university campuses helps obtain concise and diverse research samples, assisting designers and administrators in understanding the environment and providing site-specific design guidance. Currently, research on crowd behaviour characteristics is mature, with diverse data collection methods such as

H. Chai et al. (Eds.): CDRF 2024, *Symbiotic Intelligence*, pp. 407–417, 2025.
https://doi.org/10.1007/978-981-96-3433-0_36

Ultra-Wideband (UWB), base station positioning, Wi-Fi positioning, and GPS positioning. However, due to the concentrated nature and moderate scale of campuses, a single data source may not comprehensively capture crowd behavior.

Taking Southeast University's Wuxi campus as an example, this paper combines the visualization and feature analysis of on-campus crowd behavior using positioning and physical environment data. The research provides essential references for campus management, resource allocation, and planning, offering valuable insights for the preliminary research of expansion plans for other universities.

2   Data Collection Tools and Methods

The study covers the detailed collection, cleaning and visualization of pedestrian positioning data and environment perception data (Fig. 1). Obtaining sufficient and valid data is a critical step in this study before proceeding with data processing. This chapter aims to provide a comprehensive overview of the methods used to acquire positioning data and physical environment data in the related study, with an in-depth discussion of the instrumentation employed, the data collection procedures, and the pre-processing steps used to obtain the dataset, with the aim of establishing a solid foundation for subsequent data analysis.

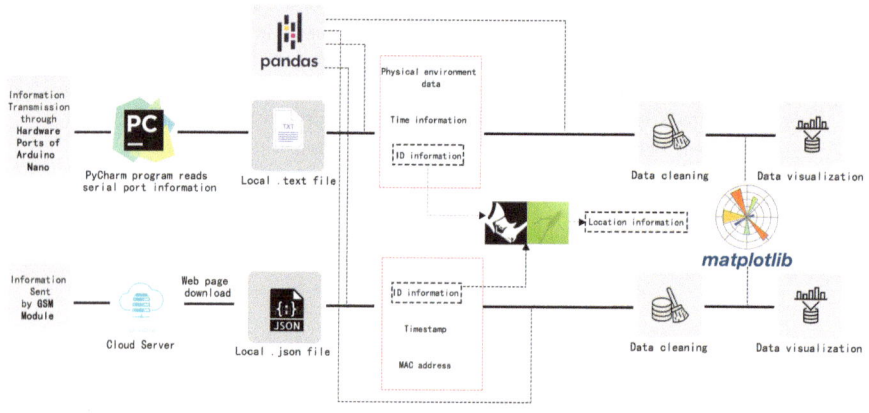

Fig. 1. Flowchart for data collection, cleansing, and visualization

2.1   Existing Equipment and Sensor Construction

Traditional crowd tracking methods suffer from significant manpower and resource input, as well as subjective influences. In 2011, Cyriac et al. introduced Wi-Fi probe technology, utilizing MAC addresses for anonymous tracking of Wi-Fi devices. Scholars like Hongyu Wan [1] extended the analysis of positioning data to architectural design using low-precision Wi-Fi probe devices. Yizhou Wu et al. [2] created layered user mobility models with Wi-Fi access point data. In this study, a self-developed Wi-Fi probe device integrates GSM, Wi-Fi, and power modules on a circuit board.

Traditional methods for obtaining building environmental data include satellite remote sensing, building information model simulations, and GIS for regional meteorological information. Yizhou Wu et al. [3] analyzed the impact of the physical environment on human behavior by simulating sound, light, heat, and landscape. Wang W et al. [4] compared datasets from EPW data, urban climate stations, and local microclimate stations. Purchasing meteorological station products has been a common method in past research, but these products are expensive, especially when monitoring multiple points.

For studies of this nature, a more suitable setup for physical environment sensors was implemented (Fig. 2): cost-effective and compact DHT11 temperature sensors, light-sensitive sensors, and three-cup ABS material anemometers were selected. A local area network communication system was constructed using LoRa modules to cover the entire campus, avoiding the use of expensive 4G modules. Sensor integration involved connecting the main control board Arduino Nano, LoRa module, and various sensors using a breadboard and dedicated jumper wires to enhance LoRa signal stability. Waterproofing measures were applied to the encapsulated sensors, ensuring normal operation under various weather conditions.

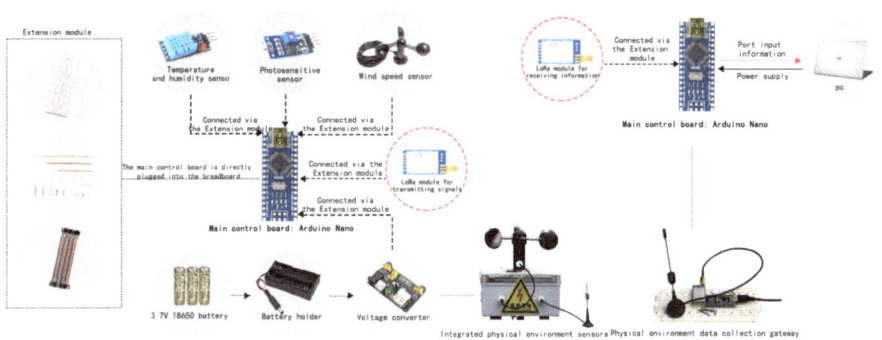

Fig. 2. Physical data sensors construction using embedded system technology

2.2 Data Collection

This study focuses on the Southeast University Wuxi Campus, which is organized in a circular layout with various functional zones surrounding the central landscape area. To comprehensively capture representative trajectories of human behaviour, the campus is divided into eight regions: Student Dormitory Area, Academic Experiment Zone, Professional Teaching Zone, Life Service Area, Library, Central Landscape Area, Public Teaching Zone, Sports and Fitness Zone, and Faculty Dormitory Area. Wi-Fi probe monitoring points are strategically placed at key intersections and public spaces within each region, following principles such as positioning them at high-frequency usage locations, ensuring unique paths between monitoring points, keeping them away from buildings to minimize indoor device impact, maintaining distance between monitoring points to reduce ping-pong data, and establishing points at campus entrances and exits.

After a week of preliminary data collection, 19 monitoring points were finalized, including critical traffic nodes and public space nodes like 202, 204, 205, 208, 213, 210, 215, 214, 211, 299 (Fig. 3). The information from these 20 points is directly uploaded to a cloud server via a 4G module. Over the course of a 14-day winter data collection period, a total of 28.87 million positioning data points were collected.

Physical environment monitoring points are placed in the main public spaces within the Wi-Fi probe coverage area. Sensors are positioned at heights corresponding to human standing behavior to ensure that monitoring accurately reflects people's perceptual conditions. Information from the 10 monitoring points is transmitted via LoRa modules to two gateways located on the 5th floor of the campus library and the 1st floor of the Two Rivers Building. Over the 14-day winter data collection period, a total of 3.4 million pieces of physical environment information were collected.

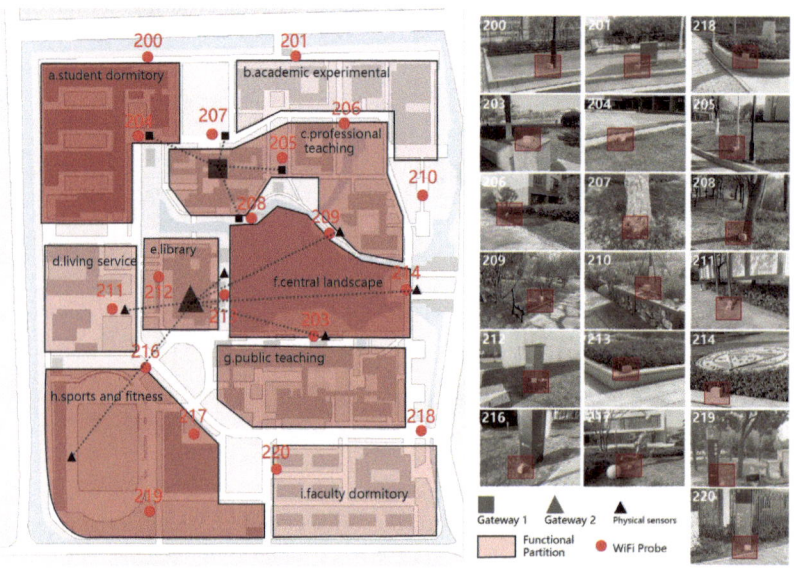

Fig. 3. Layout of monitoring points for collection devices

2.3 Data Cleansing

Data cleansing aims to detect and clean the dirty data and noise in the original data to ensure that the cleaned dataset reflects the nature of the data more realistically and improves the data visualization.

The data pre-processing stage requires judgment based on the specific project context. After analysis, five types of data requiring pre-processing were identified: ① Anomalous data: data lacking certain critical information; ② Time restrictions: removing data between 23:00 at night and 6:00 in the morning to reduce computation; ③ Drift data: data representing positions outside the possible reach within a certain time due to smart device or equipment malfunctions; ④ Redundant data: MAC addresses appearing continuously or intermittently within a monitoring range over a period; ⑤ Ping-pong data: devices repeatedly detected between two probe monitoring points. Simultaneously, a cleaning framework was established to sequentially clean these five types of data, improving cleaning efficiency. Ultimately, 400,000 valid data points were obtained, achieving a cleaning rate of 96%. This provides a high-quality data foundation for subsequent analysis.

The characteristics of physical environment data are simpler, and its data cleaning framework mainly includes two parts: data import and data preprocessing. The information collected by the gateway is written into a local.TXT file through the serial port of the computer, which is converted into a.CSV file or a.JSON file for subsequent cleaning. Information transmission through the Lora module will lead to a part of the missing information, the two parts of the extracted keywords need to be merged into a complete information. Some outliers need to be filtered out by setting thresholds for special values.

3   Behavioral Characterization of Crowd Activities

Physical environmental data exhibit variations in microenvironments within the same region, primarily influenced by campus layout. Therefore, conducting a correlation analysis between the dwell rate in campus public spaces and the physical environment is a crucial aspect of an in-depth exploration of campus layout. By determining the minimum dwell time of samples on the same device within a specified period in positioning data, it can be assessed whether a sample has lingered at a device point. In this study, the dwell rate (J) is calculated as the ratio of the number of unique MAC addresses (W) to the count of recorded MAC addresses (S), indicating preferences in people's dwellings at different monitoring points in space (Fig. 4).

In addition, visualizing preferences in individuals' walking paths is achieved by restoring Wi-Fi probe data. However, due to the limited number of Wi-Fi probes, initial data can only be transformed into "point-to-point" line graphs, making it challenging to accurately represent actual trajectory situations. To optimize trajectory maps, the Dijkstra algorithm is employed to find the shortest paths between different monitoring points (Fig. 5). The trajectory of a specific MAC throughout the day is presented on a two-dimensional plane, where point size indicates the duration of stay, and line thickness represents the frequency of walking (Fig. 6).

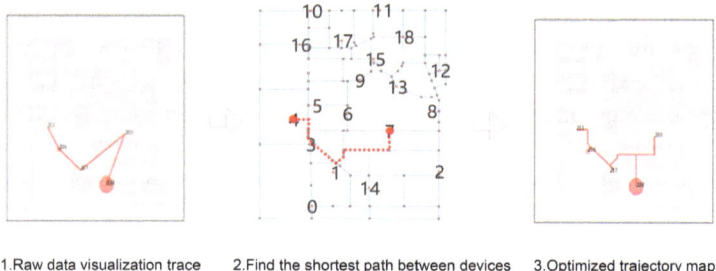

1.Raw data visualization trace   2.Find the shortest path between devices   3.Optimized trajectory map

Fig. 4. Schematic diagram of optimized trajectory using Dijkstra algorithm

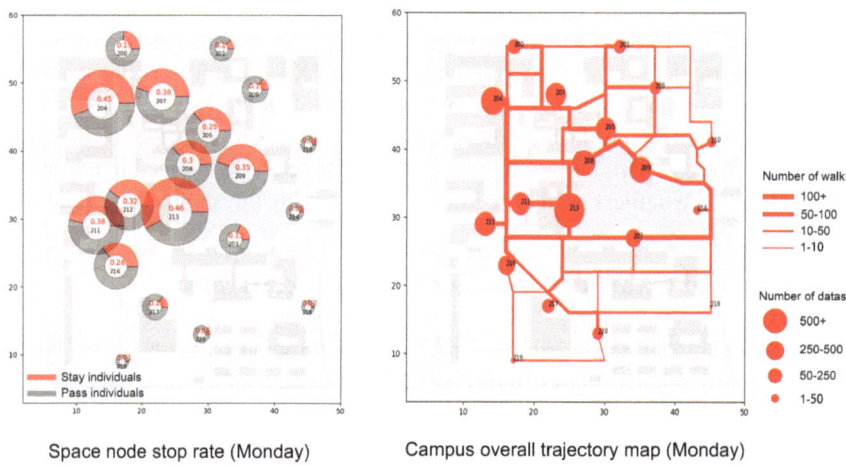

Space node stop rate (Monday)   Campus overall trajectory map (Monday)

Fig. 5. On-campus Personnel Stopover Rate Plan

Fig. 6. Layout of monitoring points for collection devices

3.1 Spatial Visualization of Physical Data

By distributing the physical environment data spatially in the form of heat maps, it is possible to visualize the differences in the physical environment of different public spaces on campus (Fig. 7). The following characteristics were observed when observing the thermograms: ①The monitoring points around the centre lake, such as 213, 214, and 215, exhibited higher wind speeds. It is worth noting that point 204 has an unusually high wind speed despite being in an area surrounded by buildings. This area is a windy plaza that is usually underutilized despite the presence of public facilities. ②215, 205, 202, and 204 have relatively high temperatures. Among them, 202, 204, and 205 are in the dormitory area and workstation complex, which have higher temperatures relative to other public spaces. ③209, 214, 215 and 213 at the lake shore have significantly higher humidity than the other monitoring points, while point 299 located in the playground has the lowest humidity and is the driest public space. ④213 on the east side of the lake and

215 on the south side of the lake have the highest illuminance. They share the common characteristics of being open environments, away from buildings, and shaded by fewer trees.

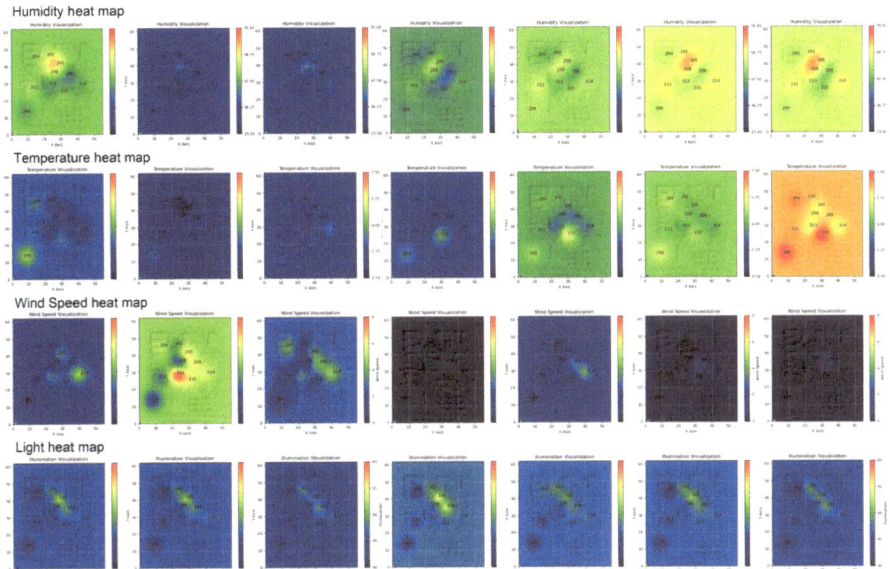

Fig. 7. Four visual views of the physical environment for each day of the week

3.2 Fitting of Stopover Rates to Environment Data Indicators

Upon visualizing the environmental data, particular attention was directed towards the dwell rates at ten key public space nodes. Calculations were performed at three time points during the day—morning, noon, and afternoon. The results from both aspects were integrated and simulated, yielding line graphs illustrating the data fluctuations (Fig. 8).

From the indicator line chart analysis, it is evident that the stay rates at all monitoring points are negatively correlated with wind speed and unrelated to illumination. The relationship between stay rates and temperature-humidity varies across different monitoring points, for examples, the stay rates at 208, 214, 213, and 215 are more closely related to humidity, while those at 209 and 205 are more closely related to temperature. In conclusion: ① In the campus under study, people's choice to stay in a particular public space is most affected by the wind environment. ②Due to fewer obstructions and wider roads on campus, the wind environment can easily cause discomfort to individuals during winter. ③Temperature and humidity are secondary factors influencing people's choices for staying activities, with individuals more likely to linger in leisure-oriented public spaces during the winter when humidity and temperature are higher. ④The stay rates at some public spaces serving as transportation hubs show no apparent correlation with temperature and humidity.

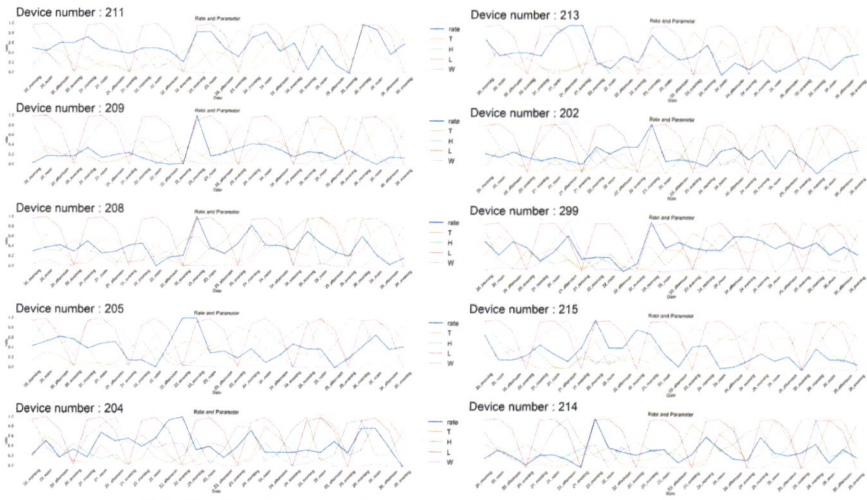

Fitting line chart of stop rate and physical data indicators at different monitoring points

Fig. 8. Fitting line chart of stop rate and physical data indicators

3.3 Behavioural Trajectory Visualization and Clustering

Analyzing the overall trajectories across the campus for seven days reveals crucial public space nodes, such as the library, central landscape area, and dormitory areas. However, as two-dimensional trajectories cannot capture time information, this representation introduces time as the third dimension. If a MAC experiences a stationary behavior, the trajectory grows along the z-axis. This approach allows the observation of the trajectory's starting and ending points, facilitating the inference and categorization of activities.

Based on the cleaned data, this study obtained over 9,000 trajectory images within a week (Fig. 9). The diverse coordinate patterns of each trajectory make it challenging to intuitively extract patterns from numerous images. To address this issue, a Convolutional Auto-Encoder (CAE) was employed for trajectory clustering. In this study, a set of RGB images with pixel dimensions of 88 × 88 extracted over the week was used as the training base. During the clustering analysis, feature extraction was performed using the pre-trained ResNet-50 model. After a series of convolutions and pooling, the feature dimensions were compressed, and the K-means ++ clustering algorithm was applied. The clustering results were then compressed into three-dimensional visualizations using t-SNE. Both two-dimensional and three-dimensional trajectory clustering yielded four distinct results (Fig. 10).

From the two-dimensional trajectory clustering results, campus trajectories can be categorized into four types: ① stationary activities around the central landscape area, ② activity trajectories centered around dormitories, cafeterias, and workstations with

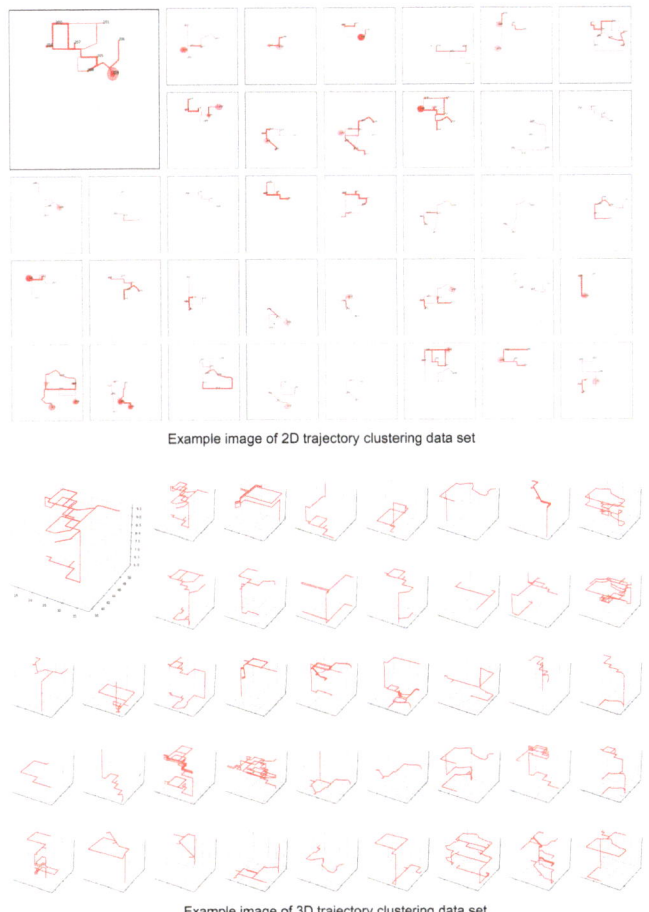

Example image of 2D trajectory clustering data set

Example image of 3D trajectory clustering data set

Fig. 9. Parts of 2D-3D trajectory map used as a training set for clustering

weak mobility, ③ more active trajectories centered around dormitories, cafeterias, and the library, ④ activity trajectories centered around professional classrooms. In the three-trajectory clustering results, people's trajectory ranges were generally singular and limited, mostly consisting of closed-loop trajectories around specific points, with a few non-closed trajectories differing in their starting points.

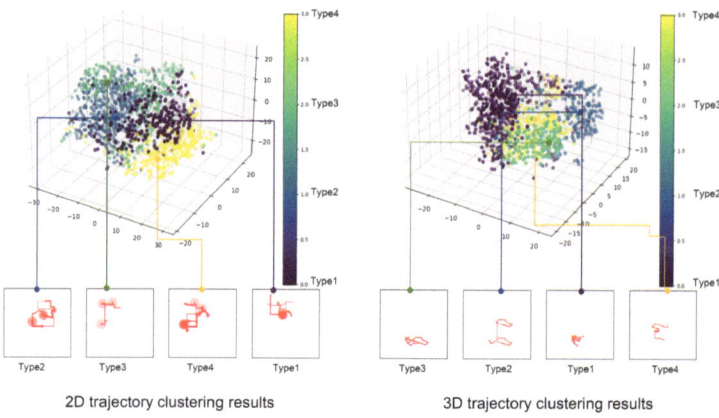

Fig. 10. Clustering results for 2D-3D trajectories

4   Summary and Discussion

This study combines Wi-Fi probe technology and physical environmental sensors to comprehensively analyze pedestrian flow and environmental conditions at the Southeast University Wuxi campus. The selection of effective data collection techniques forms the foundation of the research, offering insights for studying urban spaces at various scales. From preliminary investigations to the analysis of current campus conditions, this data-driven research process holds practical significance for assessing and enhancing urban environments and validating design proposals.

Despite achieving initial outcomes, the study faces limitations. Challenges include the insufficient sample size affecting the comprehensive representation of crowd behavior in trajectory clustering. The sensors based on LoRa modules are influenced by buildings and trees, resulting in signal losses. The data collection period spans only two weeks in winter, reflecting the impact of seasonal variations on human activity.

Future research endeavors could optimize data cleaning processes, increase sample sizes, enhance sensor deployment strategies, and extend data collection durations. This would contribute to a deeper understanding of dynamic urban spatial features, ultimately improving the practicality and applicability of the research.

References

1. Wan, H., Wang, S., Du, J., Li, L., Zhang, J.: Trace analysis using Wi-Fi probe positioning and virtual reality for commercial building complex design. Automation in Construction (2023)
2. Cohen, A., Cohen, B.: Integrating Large-Scale Additive Manufacturing and Bioplastic Compounds for Architectural Acoustic Performance. In: Proceedings of the 28th CAADRIA Conference (2023)

3. Wu, Y.Z., et al.: Integrated Evaluation Method of the Health-Related Physical Environment in Urbanizing Areas: A Case Study from a University Campus in China. Frontiers in Public Health, 10, Article 801023 (2022)
4. Wang, W., Li, S.G., Guo, S.Y., Ma, M., Feng, S.H., Bao, L.: Benchmarking urban local weather with long-term monitoring compared with weather datasets from climate station and Energy Plus weather (EPW) data. Energy Rep. **7**, 6501–6514 (2021)

Energy-Efficient Optimization of Digital Twin Air Handling Unit (AHU) Systems Based on Indoor People Counting: Case Study

Yucheng Xiao[1], Zhi Zhuang[1(✉)], Wanlin Zhang[2], and Tao Yu[1]

[1] College of Architecture and Urban Planning, Tongji University, Shanghai, China
zhuangzhi@tongji.edu.cn

[2] Sino-German Applied Sciences College, Tongji University, Shanghai, China

Abstract. Precision maintenance and operation of building systems are crucial for energy efficiency and carbon reduction in building sector. Fluctuations in people flow play a significant role in building cooling load. Traditional air handling unit (AHU) operating under constant air volume tends to waste energy, as it lacks the capability to dynamically adjust to varying demand. This study introduces an intelligent digital twin AHU system considering the people flow variation, and the energy-saving potential is evaluated by a commercial building in Shanghai. The IoT technology is used to monitor the parameters of the AHU, using MobileNetSSD and NMS algorithms to extract people flow information from a camera. Through sensitivity analysis of the collected data, the key parameters for load prediction are identified, forming the basis for establishing a dynamic load prediction model using ANN. A digital model of the AHU is set up by TRNSYS, and the optimal control strategy is determined to minimize fan and pump energy consumption under different cooling loads. The digital twin system has been successfully implemented in a shopping mall in Shanghai, and 29.5% reduction of fan energy consumption can be achieved by dynamically adjusting air supply rate in response to the changing cooling loads.

Keywords: Digital Twin · Internet of Things · AHU · Data driven · Interactive Environments

1 Introduction

The national dual-carbon development strategy has placed higher demands on building energy efficiency and low-carbon development. Specifically, 21.7% of the national total carbon emissions come from the operational phase of buildings, of which 40% comes from the heating, ventilation and air conditioning (HVAC) [1]. Therefore, the optimization of HVAC operation and maintenance emerges as a key factor in achieving building energy efficiency and reducing carbon emissions.

In commercial buildings, cooling loads fluctuate due to variations in people flow. Traditional constant air volume (CAV) air-handling-unit (AHU) systems lack dynamic

H. Chai et al. (Eds.): CDRF 2024, *Symbiotic Intelligence*, pp. 418–427, 2025.
https://doi.org/10.1007/978-981-96-3433-0_37

adjustment, resulting in a large amount of energy waste. Numerous scholars have conducted studies to address this problem [2, 3]. For instance, Sultan M. Alghamdi [2] implemented PID control of CAV-AHUs to compare energy-saving and cost-reduction options. Retrofit studies highlight people flow's crucial role in cooling load. Advances in computer technology enable video-based people flow recognition. Haifeng Lan [5] applied image recognition, using YOLOv5 in classrooms for retrofitting air conditioning systems, showing the potential of intelligent solutions in optimizing energy efficiency.

Digital twins are becoming increasingly important as a tool for equipment maintenance. Proposed by American professor Michael W. Grieves, it is widely applied for tracking product lifecycles and has gained research interest in the AHU field [6–8]. Hosamo [6] created a digital twin model to study cooling coils. The study focused on heat transfer and fluid dynamics and used a PID feedback controller to optimise fluid temperature and flow rate to improve system performance and energy efficiency. Nevertheless, implementing digital twin model for AHU requires addressing practical issues like data availability and compatibility.

Research indicates that integration of digital twin technology with building equipment is in initial stages. Traditional retrofit methods lack real-time response in AHU system upgrades, hindering flexibility and impacting energy efficiency. Retrofit technologies with DT faces implementation challenges, requiring extensive real-time data collection and integration. In this study, a digital twin system was constructed for AHUs in a commercial building in Shanghai, and an on-site IoT sensor system was deployed for related analyses. The digital model in TRNSYS software offers a new perspective on AHU retrofitting and intelligent control.

2 Method

2.1 Digital Twin System Architecture for Air Handling Unit

Digital twin is a digital model built based on physical entities, which can be divided into two key parts: the physical and the digital system. For the physical system, data monitoring via IoT is required, which includes video image-based monitoring of people flow. By analysing the correlation between the monitored parameters and the load, the best parameter combination is selected. The digital system mainly consists of load prediction through ANN and models in TRNSYS to determine the optimal control method for the AHU (Fig. 1).

2.1.1 Monitoring Method of AHU Based on IoT

IoT uses sensors and technologies for real-time data collection, connecting physical objects to a network via Internet. Monitoring the AHU requires a system with sensors, communication devices, PLC, and upper computer. Data is transferred via RS485 to the end LoRa communication module, which communicates with the upper LoRa module. The data is then sent to the upper computer, which can be accessed by external users via the 4G module. This interconnected system ensures efficient AHU monitoring and data retrieval through network connection (Fig. 2).

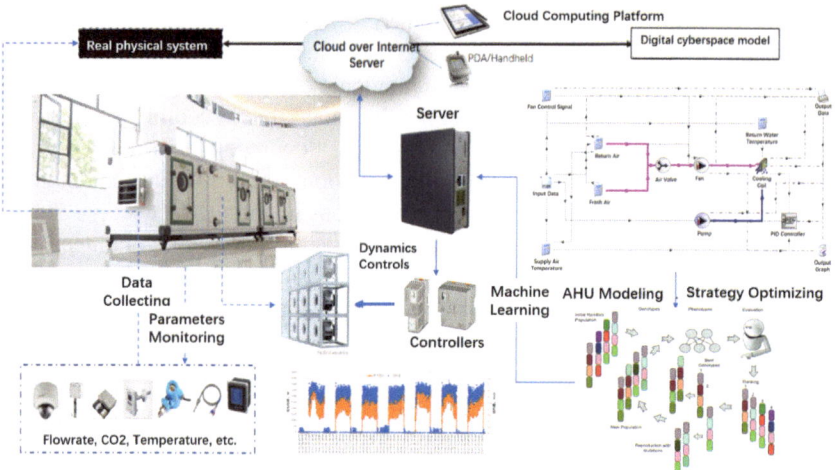

Fig. 1. Digital Twin system architecture

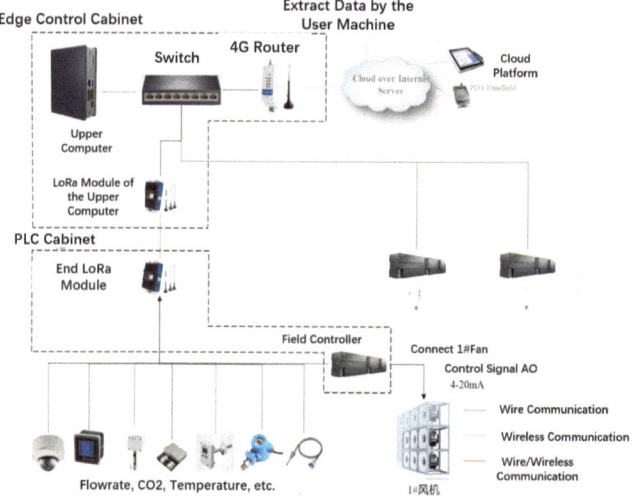

Fig. 2. IoT System architecture diagram

2.1.2   Methodology for Monitoring the People Flow

Rapid socio-economic progress increases population density, especially in public places like shopping malls, which impacts indoor air temperature, and thus it is crucial to monitor it. This study focuses on a mathematical model and algorithm for dynamically measuring pedestrian flow in commercial buildings. A camera and RaspberryPi 4B via RS485 are used to conduct detection, image recognition, and analysis, with results transmitted to the upper computer through the LoRa module.

Various algorithms, including MobileNet, are used for human flow detection. Image recognition involves decomposing images into parts and identifying them through convolution. MobileNet divides traditional convolution into depth and point-by-point convolution, reducing computation for scenes with limited equipment conditions. SSD (Single Shot Multibox Detector) utilizes a multi-scale feature map for target detection, applying different convolution kernel sizes and ratios on various feature map levels. The MobileNet-SSD algorithm efficiently combines two network structures for resource-constrained devices, addressing the need for lightweight target detection models in pedestrian flow monitoring.

2.1.3 Load Forecasting Method

CAV to VAV retrofit enables accurate indoor load forecasting by adjusting the fan to dynamic load changes. Parameter monitoring, including wind and water sides, load and energy consumption related, supports the prediction model. While more parameters can improve accuracy, cost constraints limit data collection. Balancing scientific validity and economic feasibility is important, which can be achieved by analyzing the correlation coefficient analysis between parameters and load [9].

This paper adopts the artificial neural network (ANN) for load prediction, applying the BP algorithm. BP algorithm is the most widely used ANN learning algorithm, which is divided into two learning processes: the signal forward propagation and the error backward propagation. Usually, the cooling and heating load imposed by the conditioned air can generally be calculated using the following formula:

$$Q = m * c * (T_{hf} - T_{sf}) \tag{1}$$

where Q is the indoor cooling load; m is the mass flow rate of air; c is the specific heat capacity of air; T_{hf} and T_{sf} are the temperature of the return air and the temperature of the supply air.

After load forecasting using ANN, the accuracy of the forecasting results needs to be assessed with the help of evaluation metrics such as root mean square error and correlation coefficient.

2.1.4 Digital Twin Modelling and Optimisation Method

TRNSYS is a platform for transient systems and multi-area buildings that allows users to build physical equipment by connecting component models. In this paper, the software models AHU components, including the evaporator, inverter fan, air mixing device, PID controller, etc. Input parameters include water supply temperature and air supply volume, and independent variables like return air temperature and fresh air volume. The total energy consumption of the fans and pumps can be determined by calculating the return air temperature through the evaporator at the next moment, as in Fig. 3.

The system aims to minimize total energy consumption of fans and pumps. Optimization variables are supply water temperature, pump flow rate, and supply air volume, while independent variables include the load prediction value, return air temperature, return water temperature, etc. The next moment's return water temperature is obtained through the evaporator. Total energy consumption is calculated and various parameter

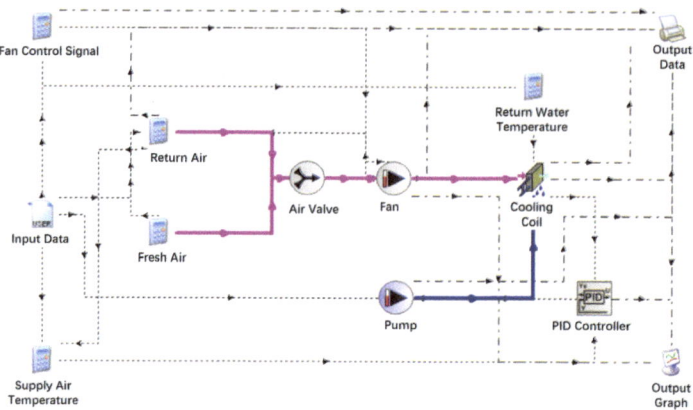

Fig. 3. TRNSYS modelling diagram

combinations are compared to identify the wind and water side parameters combination to minimize total energy consumption.

3   Case Study

3.1   Parameter Monitoring and Data Processing

3.1.1   Case Introduction

The AHU in the use case is a single-cooled unit located in a Shanghai shopping mall. Initially the AHU operated in CAV mode. Recognizing the substantial energy waste associated with this operation, a sensor network and control equipment are implemented for unit monitoring. The aim is to propose a new control strategy involving variable-air-volume modification, intending to mitigate energy consumption.

3.1.2   People Flow Data

For people flow monitoring, a 180-degree infrared camera captures lift images. The video is processed by a Raspberry Pi for edge detection and analysis. The algorithm uses pre-training, combining non-maximal value suppression and center-of-mass tracking for accurate pedestrian flow segmentation. After capturing videos through the camera, the Raspberry Pi performs person recognition on the video. The backend program collects data every minute. The data collection frequency, occurring at one-minute intervals, causes some degree of variation in the detected number of individuals, resulting in a noticeable deviation in the graph due to the real-time fluctuations in people flow within the shopping mall, as in Fig. 4.

The area where the retrofitted AHU is located was monitored during 2 to 8 October. As can be seen from the figure, the flow of people in the area gradually increased with the opening of the mall at 9:00 a.m., reaching a maximum at 4:00 p.m. on a general holiday (2–7 October) and at 13:00 p.m. and 4:00 p.m. on a weekday (8 October), followed by

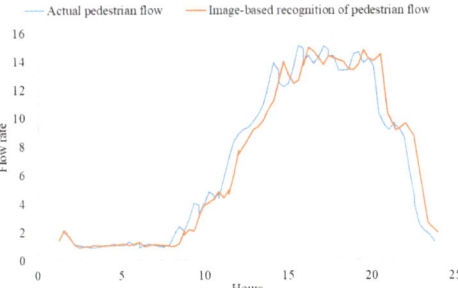

Fig. 4. All day people flow counts (Saturday)

a gradual decrease. The maximum flow reaches 57 persons per hour and varies between 269 and 490 persons throughout the day (Figs. 5 and 6).

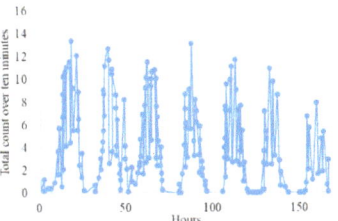

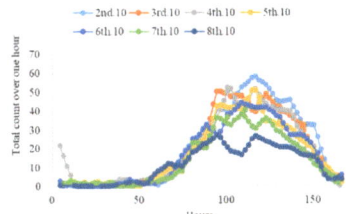

Fig. 5. Patterns of distribution for monitoring the flow of people (left: cumulative number of people in 10 min, right: cumulative number of people per hour)

Fig. 6. Patterns of distribution for monitoring the flow of people (left: cumulative number of people in 10 min, right: cumulative number of people per hour)

3.1.3 Parameters of AHU

Sensors collected data from July 3 to October 15, 2021. Determining the input parameters combination is crucial for load forecasting. In this case, 15 parameters related to the control of the AHU were monitored to find the key parameters affecting the building load.

Condition 1 includes all parameters (Carbon dioxide, supply air volume/ humidity/ temperature, fresh air volume/ humidity/ temperature, return air volume/ humidity/ temperature, chilled water flow rate, supply water temperature, return water temperature, pressure difference, indoor human flow rate). Initial filtering of correlation coefficients considers values above 0.5 as load-correlated. Condition 2 (Supply air volume/ humidity/ temperature, fresh air volume/ humidity/ temperature, return air volume/ humidity/ temperature, water supply temperature, pressure difference, indoor human flow rate) encompasses load-related influence factors.

Condition 3 (Supply air temperature/ humidity, fresh air temperature/ humidity, return air temperature/ humidity, flow rate, water supply temperature, pressure difference, indoor human flow rate) includes parameters which are not autocorrelation. Finally, Condition 4 (Fresh air temperature, supply air temperature, return air temperature, pressure difference, indoor human flow rate) includes fundamental parameters in the analysis.

Next the parameters in the above conditions are substituted into the load prediction model and the load is predicted by applying BP algorithm described above. Three layers of the network are set up with the number of neurons of 128-256-512, the optimizer learning rate is 0.001, and the number of iterations is 1000. The error analyses obtained under different combinations of parameters are shown in the table below (Table 1).

Table 1. Analysis of load prediction errors under different operating conditions

Working condition	RMSE	R
Condition 1	2.387	0.993
Condition 2	2.665	0.991
Condition 3	2.362	0.993
Condition 4	1.946	0.995

By comparison, it can be found that the R-values are all close to 1, indicating that the load prediction using the ANN model has higher accuracy and better fitting effect. Among them, the prediction effect of condition 4 is the best, indicating that there is a certain overfitting phenomenon in the rest of the conditions, and low-cost monitoring can reduce this phenomenon to a certain extent to achieve better load prediction, and it is recommended that this model be used in practical engineering applications.

3.2 Digital Twin Model and Control Scheme

The accuracy of the digital twin model was assessed by bringing the filtered data into the established digital twin model. Since the actual water system is not a single pump, so the fan energy consumption is compared. Figure 7 shows the trend of two folds, revealing a close alignment between the actual and simulated fan energy consumption. The correlation coefficient, reaching 0.8, indicates the accuracy of the digital twin in predicting load and conducting simulations.

Due to the current on-site implementation, the AHU only accepts VAV control. Therefore, during actual control of the AHU, it calculates the predicted supply air volume for the current moment based on the forecasted load value and the set return air temperature difference for the upcoming moment. It then compares this calculated value with the minimum fresh air volume, choosing the greater value as the current required supply air volume.

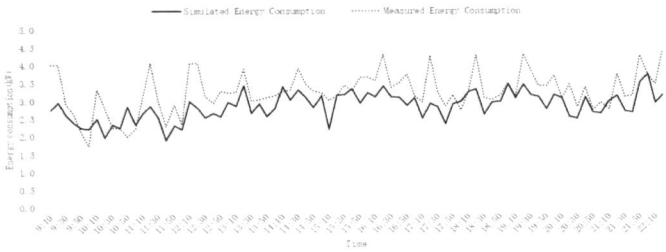

Fig. 7. Comparison of measured and simulated fan energy consumption

3.3 Optimisation of Energy Efficiency Evaluation

Based on the monitored data, a data analysis was conducted on the operational conditions of the selected AHU before and after the implementation of the variable air volume retrofit in this project. The CAV operation of the fan was set as Case 1, and the working period was from 21 July to 24 September. The VAV operation was set as Case 2, and the working period was from 25 September to 15 October (Figs. 8 and 9).

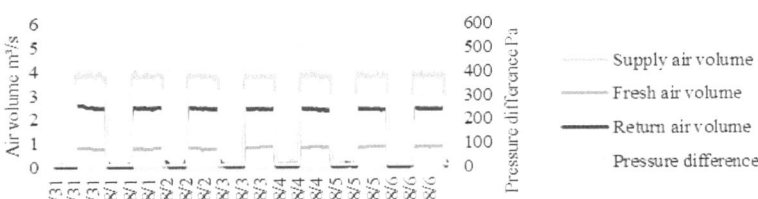

Fig. 8. Case 1 (before modification) fan conditions

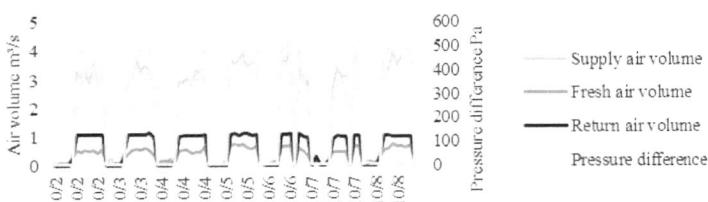

Fig. 9. Case 2 (after modification) fan conditions

From the comparison of the two cases, it can be found that during the operation of case 1, the supply air volume, return air volume and fresh air volume of the AHU were kept in a more stable state with less fluctuation. During the operation of case 2, the supply air volume and return air volume of the AHU are dynamically changed under intelligent control (Table 2).

Table 2. Analysis of energy saving effect of air-conditioning box under different conditions

Category	Case 1	Case 2
Control Mode	CAV	Intelligent VAV
Average value of room load (kW)	40.35	39.76
Average value of supply air volume	3.80	3.37
Average value of fan pressure difference	431.5	423.9
Average value of fan power (W)	3393.2	2393.8
Average value of fan efficiency (%)	48.3	60.6
Energy saving rate (%)	/	29.5

After analyzing the energy-saving effect of intelligent control, it can be found that after intelligent control of the fan, the average value of the supply air volume is reduced, and the power of the fan is reduced accordingly, and at the same time, the efficiency of the fan has been improved greatly, and the energy-saving rate can reach 29.5%.

4 Conclusion

This paper addresses the issue of energy wastage in commercial buildings due to the inability of traditional constant air volume air handling units to dynamically adjust with human traffic. To solve this, we develop a digital twin model for air handling units based on people flow monitoring. We propose optimal combinations of the return air temperature difference, supply-return water temperature difference, and supply water temperature to minimize the energy consumption of the AHU's air and water systems. The accuracy and energy-saving effect of the digital twin model are validated through an application example involving the air handling units of a commercial building in Shanghai.

An IoT network enables dynamic monitoring of AHU parameters. MobileNet-SSD detects human flow in videos, while ANN accurately predicts room loads. The results show that the root mean square error value of load prediction using basic parameters is 1.946 and the R-value is 0.995, proving high accuracy. The digital twin model of the AHU is constructed using the TRNSYS platform and has been validated with historical data, achieving a correlation coefficient of 0.8, thereby demonstrating its practical applicability. By employing an energy consumption minimization strategy, dynamic adjustments to the fan airflow in the air handling unit resulted in a 29.5% reduction in fan energy consumption compared to fixed air volume conditions.

As the actual retrofit project eventually implements variable air volume control, this study only achieves optimal control and energy saving in the fan part, and the future study can be further extended to the water system to achieve comprehensive energy saving optimization of the air conditioning system.

Acknowledgement. This work is supported by the Fundamental Research Funds for the Central Universities, Research Program of Shanghai Municipal Commission of Housing and Urban-Rural Development (No. 2024-005-005) and Excellent Experimental Project of Tongji University.

References

1. China Association of Building Energy Efficiency: 2022 Research Report of China Building Energy Consumption and Carbon Emissions. Chongqing, China (2022)
2. Sultan, M.A., Mohammed, N.A., Nidal, H.A., et al.: Using proportional-integral-derivative controllers and PCM and a new design of building air intake with five scenarios to present a multi-zone CAV-AHU for tackling high energy consumption. Journal of Building Engineering (2022). https://doi.org/10.1016/j.jobe.2022.104764
3. Michael, A., Dominic, T.J.O., Ken, B.: Implementation of the IDAIC framework on an air handling unit to transition to proactive maintenance. Energy and Buildings (2023). https://doi.org/10.1016/j.enbuild.2023.112872
4. Zhihao, R., Jung, I.K., Jonghoon, K.: Assessment methodology for dynamic occupancy adaptive HVAC control in subway stations integrating passenger flow simulation into building energy modeling. Energy and Buildings (2023). https://doi.org/10.1016/j.enbuild.2023.113667
5. Haifeng, L., Huiying (Cynthia), H., Zhonghua, G., et al.: Computer vision-based smart HVAC control system for university classroom in a subtropical climate. Building and Environment (2023). https://doi.org/10.1016/j.buildenv.2023.110592
6. Haidar, H.H., Paul, R.S., Kjeld, S., et al.: A digital twin predictive maintenance framework of air handling units based on automatic fault detection and diagnostics. Energy and Buildings (2022). https://doi.org/10.1016/j.enbuild.2022.111988
7. Kang, C., Xu, Z., Burkay, A., et al.: Digital twins model and its updating method for heating, ventilation and air conditioning system using broad learning system algorithm. Energy (2022). https://doi.org/10.1016/j.energy.2022.124040
8. Lv, Z., Xie, S.: Artificial intelligence in the digital twins: State of the art, challenges, and future research topics. Digital twin (2021)
9. Jihong, L., Na, D., Jincheng, X., et al.: An improved input variable selection method of the data-driven model for building heating load prediction. Journal of Building Engineering (2021). https://doi.org/10.1016/j.jobe.2021.103255

VR, AR and Interactive Technology

Metaverse-Based Evaluation of Passenger Navigation: A Case at Shanghai Pudong International Airport

Mingyan Zou[1], Chengyu Sun[1(✉)], and Shuyang Li[2]

[1] College of Architecture and Urban Planning, Tongji University, Shanghai 200092, China
cy.sun@tongji.edu.cn

[2] College of Design and Engineering, National University of Singapore, Singapore 119077, Singapore

Abstract. The concept of the metaverse, representing a cutting-edge virtual reality paradigm, has gained significant traction as a mode of communication and data acquisition within virtual environments, applied across diverse domains. This study investigates the utilization of a metaverse-based platform in environmental behavior analysis, focusing on wayfinding behavior. We developed a metaverse-based platform for the Satellite Terminal 1 (S1) of Shanghai Pudong International Airport (PVG), offering participants an immersive and intuitive virtual environment. Leveraging the serious game concept, we designed a wayfinding task, through which 2,746 wayfinding trajectory data were collected for performance evaluation purposes. Notably, this metaverse-based platform presents a feasible and efficient approach for studying wayfinding behavior and streamlining data collection processes.

Keywords: Metaverse · Wayfinding · Gamification · Data visualization · Transportation building · Performance evaluation

1 Introduction

Studying wayfinding behaviors of passengers in large public buildings is essential for architectural design, pedestrian traffic management, and emergency response. While laboratory experiments are valuable for examining behaviors, they are often constrained by extended durations, complexity, high costs, and limited behavioral diversity. Meanwhile, as depicted in the movie "Ready Player One", the metaverse is developing into a shared three-dimensional (3D) virtual platform, creating a parallel virtual universe that transcends time, place, and location constraints. This enables users worldwide to engage in a shared virtual environment. Consequently, the metaverse has emerged as a significant Internet application that integrates various advanced technologies, offering new possibilities for behavioral studies.

This paper explores the potential of a metaverse-based platform for studying passenger behaviors, such as wayfinding. The research addresses the following questions:

© The Author(s) 2025
H. Chai et al. (Eds.): CDRF 2024, *Symbiotic Intelligence*, pp. 431–441, 2025.
https://doi.org/10.1007/978-981-96-3433-0_38

RQ1: Is a metaverse-based platform an effective tool for studying passenger behavior?
RQ2: How can a metaverse-based platform be designed to study wayfinding behavior?
RQ3: What are the advantages and disadvantages of using a metaverse-based platform for behavior studies compared to current methods?

To answer these questions, we developed a realistic metaverse-based platform using Unity3D, creating a parallel virtual world for S1 of PVG. This platform includes virtual interior spaces of the building and avatars as surrounding passengers, and employs the Browser/Server (B/S) architecture, allowing remote access via compatible browsers to enable broad user participation. Additionally, we integrated serious games with virtual wayfinding experiments, incorporating game elements to attract and motivate participants. Internet users were recruited to participate in an experiment on the metaverse-based platform, and a wayfinding dataset was collected for performance evaluation.

2 Literature Review

Wayfinding refers to the ability of individuals to reach their destinations quickly and without stress (Peponis et al. 1990). It can be considered a series of continuous problem-solving tasks based on environmental cues (Arthur and Passini 1992). In complex facilities such as hospitals and airports, the efficiency of wayfinding becomes particularly important.

In the context of airports, wayfinding behavior involves the dynamic route selection challenges faced by passengers upon entering airport terminals (Fewings et al. 2001). Due to time pressures and transfer requirements, passengers need to find their destinations as easily as possible. However, the complexity of the interior spaces of airport terminals often makes navigation difficult. Studies have shown that factors such as building layout and signage significantly impact an individual's wayfinding performance (Slone et al. 2015). To enhance navigation efficiency within terminals, methods such as the visibility index (VI) (Churchill et al. 2008), inter-connection density (ICD) (O'Neill 1991), and space syntax (Turner et al. 2001) have been employed to quantify ease of wayfinding. However, these methods only evaluate wayfinding performance on a two-dimensional plane. Additionally, wayfinding studies can be conducted through on-site surveys or VR experiments in the laboratory. Although these methods provide valuable insights, they are limited by the number of participants, and the required time and costs for evaluation are relatively high.

The Metaverse and gamification can attract more participants to actively engage in specific wayfinding tasks at a lower cost, offering new perspectives and innovative solutions for studying wayfinding behaviors. Recently, various metaverse-based applications, such as Roblox and Meta, have garnered increasing interest (Dwivedi et al. 2022). As the concept of the metaverse continues to evolve, numerous definitions have emerged. Generally, the widely discussed metaverse is a shared virtual world parallel to the real world. As a next-generation iteration of the internet, it integrates diverse new technologies, fostering numerous innovations.

Technological advancements have not only enhanced the realism of participant experiences but also made metaverse-based platforms more accessible, distinguishing them from previous applications. With continuous improvements in web technology, web browsers have evolved from simple display tools into highly functional application platforms (Zhao et al. 2019). By leveraging hardware acceleration capabilities of underlying graphics processing units (GPUs) and incorporating standard cross-platform OpenGL interfaces, web browsers no longer require plugins for 3D display. This enables web browsers to render virtual content without any plugin support. Consequently, participants can access metaverse-based platforms locally through web browsers, using common devices such as smartphones, tablets, or laptops. This seamless accessibility facilitates easy and continuous engagement with the metaverse. Some studies have been conducted on metaverse platforms (Gu et al. 2023; Gattullo et al. 2022).

Gamification highlights the application of game design elements in nonentertainment contexts. "Serious games," a term first introduced by Zyda (2005), refer to games primarily intended for purposes other than entertainment. These games combine real-world issues with gaming elements and employ mechanics such as goals, points, leaderboards, and rewards to create engaging experiences that stimulate participants' interest and enhance their experience. Serious games have been applied in crowd behavior studies (Connolly et al. 2012; Li et al. 2017) to record decision-making and behaviors during gameplay, offering valuable opportunities for collecting behavioral data. With the growing maturity of digitization and gamification, serious games are gradually becoming a significant form within metaverse-based platforms. They provide participants with more realistic experiences and motivate them to take action. The high accessibility and continuity of metaverse-based games offer unique opportunities for researching wayfinding behavior.

3 Metaverse-Based Platform

3.1 General-Purpose Framework

Figure 1 illustrates the platform architecture of the metaverse-based system, which employs a B/S architecture. This system comprises two main components: the "web frontend" and the "web server". On the web server side, a lightweight virtual world, including virtual building interior spaces and avatars, is transmitted to the web frontend via the internet. The web frontend utilizes the WebGL graphics engine for scene rendering, while HTML5 and JavaScript are used for user interface design and logical event processing. Participants interact with the virtual world through a web-based interface and participate in a virtual wayfinding experiment hosted by the web server. During this experiment, wayfinding data is transmitted back to the server in real-time and stored in a database. Additionally, the avatars within the virtual environments are driven by an algorithm and recorded data stem from the web server.

3.2 Platform Development

3.2.1 Data Acquisition and Modeling

The purpose of data acquisition and modeling is to create a realistic virtual building environment. Data acquisition involves capturing panoramas with a panoramic camera

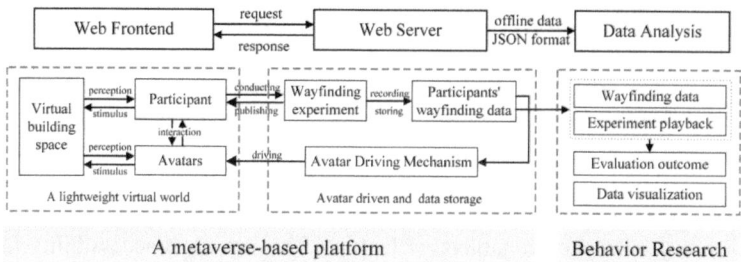

Fig. 1. The framework of the metaverse-based platform

and gathering detailed information about the target building through field research or 3D scanners. Advanced tools like Rhino can generate highly detailed meshes, facilitating the construction of a 3D model based on the acquired data. However, these highly detailed models often lead to slower loading and running speeds for web frontends, resulting in a poor user experience. To mitigate this issue, it is essential to reduce the model's complexity without compromising the realism of the virtual scenes. This can be achieved by applying high-resolution panoramas as textures to a model with relatively low accuracy. A 3D model with relatively low accuracy ensures that only components related to sightline occlusion are considered. Then, panoramas as textures are applied to the models to achieve high appearance quality. Consequently, this method maintains higher loading and running speeds while preserving the details of the virtual environments.

3.2.2   Unity Engine

Game engines aimed at outputting high-frequency renderings efficiently are widely used to build metaverse-based platforms (Choi et al. 2022). Unity engine allows users to build games and experiences in both 2D and 3D, and it offers a powerful scripting API in C#, which has strength in building virtual worlds (Bansal et al. 2022). Furthermore, the Unity engine has started offering ways to export and publish directly to the web. Consequently, the Unity engine is used to develop this metaverse-based platform in this study.

4   An Application Case

The International Departure part of the S1 of Shanghai Pudong International Airport was chosen as a case to be replicated, demonstrating the capabilities and features of the metaverse-based platform (see Fig. 2).

4.1   Building Metaverse-Based Platform

4.1.1   A Realistic Virtual Building Environment

For the International Departure part, 150 decision points were established, and at each point, 360-degree high-resolution panoramas were captured. A 3D model, scaled to match the real building, was created using Rhino software. Due to the increasing complexity of the model, which scales with the size of the building, optimization is necessary

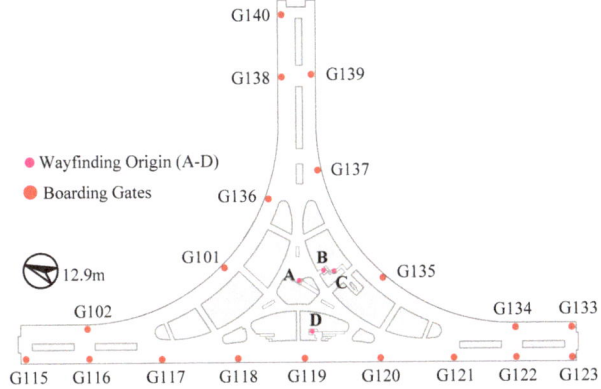

Fig. 2. The layout of the International Departure of the S1

to ensure smooth real-time rendering performance. To address this, we segmented the 3D model into several local models. Each segment allows participants to navigate, with high-resolution panoramas mapped as textures to these local models (see Fig. 3). Thus, each segment comprises a 3D local model and a high-resolution panorama at its center, resulting in a total of 150 segments. This technique balances the need for a highly detailed, realistic virtual environment with the requirement for smooth performance.

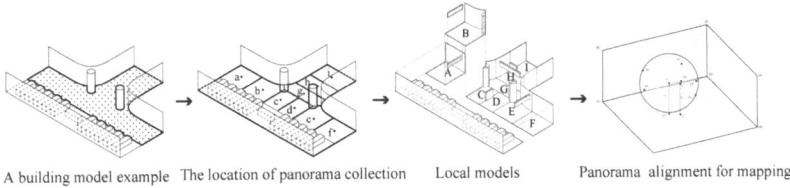

A building model example   The location of panorama collection   Local models   Panorama alignment for mapping

Fig. 3. The procedures for building a realistic virtual environment

4.1.2 Interaction with Avatars

In the real world, people often interact with those around them to gather information on how to find their way. To replicate this high interaction fidelity in virtual environments, the context generation and pointing algorithm are used to simulate real-life inquiring behavior between the participant and avatars. Participants can ask for directions by clicking on any surrounding avatar. When clicked, a dialog box with common wayfinding questions and contextual information appears. Participants select a relevant question, and the avatar generates an arrow pointing towards the destination.

4.1.3   Roaming in the Metaverse-Based Platform

The metaverse-based platform utilized in this study was developed using the Unity engine and can be deployed on both VR devices and web browsers. However, requiring participants to wear HMDs for the wayfinding experiment in non-laboratory settings presents challenges. On the web frontend, participants navigated and explored the virtual environment from a first-person perspective (see Fig. 4). They could rotate the view using a mouse, simulating the head movements in an HMD, allowing them to observe their surroundings. Movement between different decision points was facilitated by clicking on light columns in the virtual environment, enabling jump roaming instead of continuous roaming. Freedom navigation included three operations: forward/backward jump roaming, left/right jump roaming, and left/right rotations. The virtual world loaded in less than two seconds during transitions between decision points on the web frontend.

Fig. 4. The metaverse-based platform

4.2   Serious Game in Wayfinding

The task design incorporates game-thinking principles, integrating knowledge of public buildings from traditional design contexts into a unified virtual spatio-temporal framework. Various design elements, such as competition and game types, are considered. Consequently, wayfinding tasks were designed in an "activity-boarding" mode based on the actual operational situation of the case. Participants undertaking the wayfinding task have individual virtual timelines, which include start time, travel time, time spent on other activities, and departure time. Start and departure times are determined by the task settings, while travel time represents the duration spent on the route. Based on unique login controls and participants' ages, corresponding speed intervals are automatically matched: the elderly ($\geq$60 years) at 0.7–1.1 m/s, adults (18–60 years) at 0.9–1.3 m/s, and adolescents (<18 years) at 0.8–1.0 m/s. Travel time is calculated by measuring the actual distance between spatial nodes in the virtual environment, cumulatively consuming virtual time.

Engaging in activities such as shopping and using the washroom during wayfinding also consumes virtual time. The time spent on these activities is calculated by randomly flipping "leave" cards. The consumed virtual time for these activities is the sum of the base time and the card-flipping time. The base time for shopping and using the washroom is 5 min and 2 min, respectively. Card-flipping time is calculated as 1 s of real-time equivalent to 1 min of virtual time. Participants receive timely feedback on their manipulation through the virtual time and the progress of task completion displayed on the screen. A task scoring mechanism that adds competition is employed, with rewards distributed based on score (i.e., task performance). This incentivizes participants to perform better in virtual wayfinding tasks by fostering a competitive environment.

4.3 Evaluation and Analysis of Wayfinding Behaviour

A total of 370 participants were recruited through social media to experience this metaverse-based platform. A total of 2,746 pieces of data were collected, of which 2,634 were valid, yielding an effective rate of 95.9%. All data were recorded in JSON format, which can be used to analyze wayfinding behavior.

4.3.1 Task Success Rate

A statistical analysis of task completion was conducted based on wayfinding data (see Table 1). The results indicate a significantly lower success rate for the sub-task of using the washroom compared to the other two tasks. This suggests that participants encountered more difficulties during finding the washroom in the virtual environment due to insufficient directional signages about the washroom within this case.

Table 1. Success rate of different tasks

Task Type	Boarding task	To the washroom	Shopping
The total number of tasks	2634	1302	1332
The number of failed tasks	11	32	20
Task success rate (%)	99.58	97.54	98.50

4.3.2 Wayfinding Trajectories

Generally, shorter wayfinding trajectories indicate higher wayfinding efficiency and better-designed directional signage. Additionally, trajectory consistency reflects the similarity of participants' paths during the same virtual wayfinding task. Analyzing the length and consistency of wayfinding trajectories can help designers identify locations with wayfinding challenges more effectively.

Figure 5 shows that when participants navigated from origin A to boarding gates G118-G123, their trajectory lengths were relatively longer. This could be due to the difficulty in identifying the wall-attached illuminated directional signage. A similar issue appeared when participants navigated from origin B to boarding gates G118-G121. For participants starting from origin C to find various boarding gates, it was observed that the trajectory lengths were unsatisfactory. It was primarily because only directional signages of G135 were placed at the elevator (i.e., origin C), thereby limiting the overall wayfinding effectiveness. Furthermore, although the wayfinding trajectory results from origin D to gates G118-120 were suboptimal, it was not due to issues with directional signage but rather because the shops were located farther from origin D, leading participants to take longer paths according to the sub-task of shopping. In contrast, gates G136-G140 were also far from origin D, and the discontinuous placement of directional signage for these gates likely contributed to the difficulty in wayfinding over long distances.

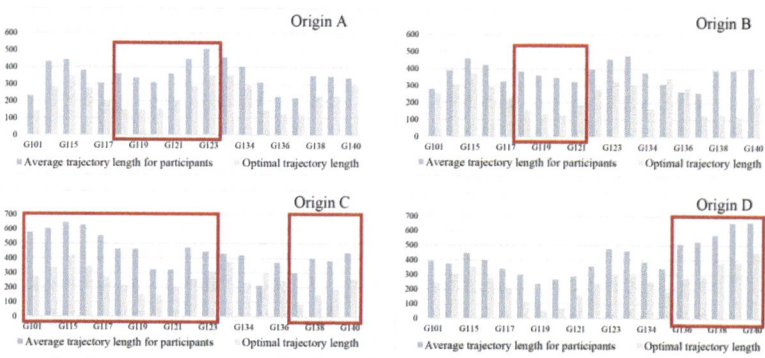

Fig. 5. Analysis of trajectory length

Figure 6 illustrates the trajectory consistency when the destinations are G119 and G140, respectively. The trajectory consistency for G140 was higher. A visual blind spot existed on the route to G119 because the signage was parallel to the passengers' sightline, resulting in inconsistent wayfinding trajectories, suggesting the directional signs must be placed according to passengers' forward direction.

Fig. 6. Illustrations of Trajectory Consistency Analysis (G119, G140)

4.3.3 Decision Time

Decision time at different decision points refers to the duration participants spend thinking and responding at these points. Quantifying decision time can help evaluate participants' wayfinding performance and assist designers in identifying locations that may confuse participants. The average decision time was calculated for each decision point. It was found that decision time exceeded 20 s for 27 decision points (see Fig. 7). These decision points were categorized based on visual range, spatial scale, and the number of available directions for selection.

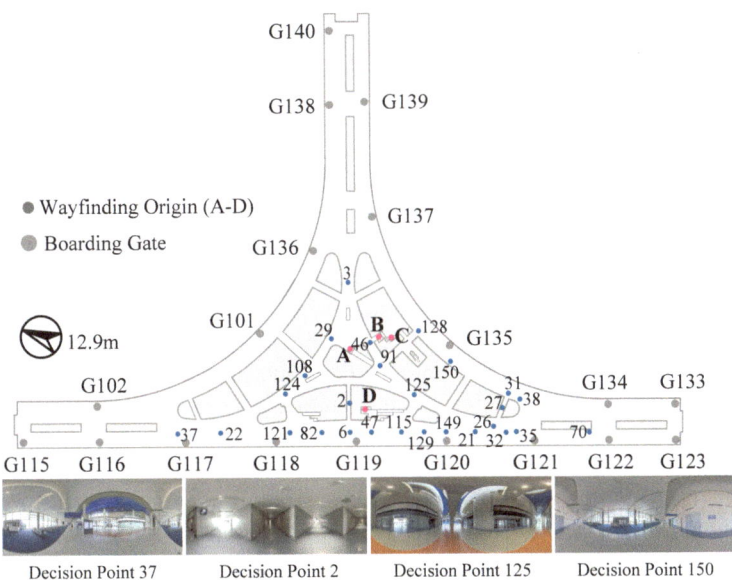

Fig. 7. Decision points with a longer decision time

Among these, the following decision points are classified into a category characterized by multiple directional choices in large-scale spaces with long visual ranges: 6 (38.09s), 115 (32.97s), 128 (32.78s), 82 (28.47s), 121 (28.35s), 35 (26.28s), 32 (25.97s), 47 (24.99s), 26 (22.93s), 38 (22.77s), 31 (22.75s), 149 (21.53s), 37 (21.53s), 22 (21.10s), and 21 (20.94s). Additionally, decision points 27 (28.69s) and 2 (22.62s) fall into another category primarily characterized by limited visibility in small spaces with two directional choices. Decision points 125 (28.11s), 46 (27.85s), 124 (26.75s), 29 (25.92s), 91 (20.92s), 108 (20.72s), and 3 (20.10s) are classified into a category featuring multiple directional choices in large-scale spaces with near visual ranges. Lastly, decision points 129 (27.53s), 70 (21.68s), and 150 (21.67s) are classified into a category characterized by two directional choices in large-scale spaces with long visual ranges.

Decision time significantly increases when participants are confronted with more than two directional choices at a decision point. This suggests that having multiple directional options can heighten the cognitive load for participants, thereby extending the time required to make a decision. To enhance wayfinding efficiency in such scenarios, the design of interior installations and landscape elements can aid participants in grasping the spatial structure more rapidly.

In decision points situated within confined spaces with limited visibility, featuring only two directional choices, participants still encounter difficulty in making prompt decisions due to the constrained visual range. Improvements such as ample directional signage can assist individuals in navigating through such spaces. Furthermore, lighting can direct participants' attention; for instance, employing lights of varying colors or brightness to signify different directions.

In decision points within expansive spaces with elongated visual ranges and two directional choices, participants may need to process a greater amount of information before making decisions. In these instances, well-designed guiding signage should prevent information overload and strategically position directional signage at crucial junctures.

5 Discussion and Conclusion

In this study, we developed a metaverse-based platform as a parallel virtual world for Shanghai Pudong International Airport (PVG) and employed it to gather a total of 2,746 samples for wayfinding evaluation. The findings indicate that the metaverse-based platform serves as an effective tool for both data acquisition and behavioral research purposes. We integrated serious games into the development of the metaverse-based platform. Incorporating gaming design elements into a non-gaming context significantly increased participants' engagement with the platform. This metaverse-based platform offers a more immersive and interactive experience for serious games, thereby enhancing the environment's credibility and participants' involvement in the tasks. Moreover, the adoption of a B/S architecture in the metaverse-based platform notably improves data acquisition efficiency, highlighting the potential of metaverse-based platforms in behavioral studies.

However, despite the advantages presented by metaverse-based platforms in behavioral studies, certain limitations persist. Virtual environments may not fully replicate the complexity and richness of real-world settings. The interactive performance of the current metaverse-based platform still has scope for enhancement. Participant-avatar interaction primarily occurs through dialog boxes (i.e., text), diverging from real-world interaction experiences. Introducing audio features to the interaction between passengers and avatars could be meaningful and enhance the immersion.

Moving forward, we aim to enrich the content and modes of interaction with avatars, broadening its applications across various domains. Furthermore, while the current metaverse-based platform features a single-player mode, the capabilities of the metaverse supporting multiplayer collaboration suggest avenues for future exploration. By delving into the multiplayer mode, we can unlock new dimensions of interaction, thereby augmenting the attractiveness of the metaverse-based platform.

Acknowledgement. This work was supported by the National Natural Science Foundation of China (grant number 51778417).

References

Arthur, P., Passini, R.: Wayfinding: People, Signs, and Architecture. McGraw-Hill, New York (1992)

Bansal, G., Rajgopal, K., Chamola, V., et al.: Healthcare in metaverse: a survey on current metaverse applications in healthcare. IEEE Access **10**, 119914–119946 (2022)

Choi, S., Yoon, K., Kim, M., et al.: Building Korean DMZ Metaverse using a web-based Metaverse platform. Appl. Sci. **12**(15), 7908 (2022). https://doi.org/10.3390/app12157908

Churchill, A., Dada, E., de Barros, A.G., et al.: Quantifying and validating measures of airport terminal wayfinding. J. Air Transp. Manag. **14**(3), 151–158 (2008)

Fewings, R.: Wayfinding and airport terminal design. J. Navig. **54**(2), 177–184 (2001)

Connolly, T.M., Boyle, E.A., MacArthur, E., et al.: A systematic literature review of empirical evidence on computer games and serious games. Comput. Educ. **59**(2), 661–686 (2012)

Dwivedi, Y.K., Hughes, L., Baabdullah, A.M., et al.: Metaverse beyond the hype: Multidisciplinary perspectives on emerging challenges, opportunities, and agenda for research, practice and policy. Int. J. Inf. Manage. **66**, 102542 (2022)

Gattullo, M., Laviola, E., Evangelista, A., et al.: Towards the evaluation of augmented reality in the Metaverse: information presentation modes. Appli Sci. **12**(24), 12600 (2022). https://doi.org/10.3390/app122412600

Gu, J., Wang, J., Guo, X., et al.: A Metaverse-based teaching building evacuation training system with deep reinforcement learning. IEEE Trans. Syst. Man Cybern. -Syst. **53**(4), 2209–2219 (2023)

Li, C.Y., Liang, W., Quigley, C., et al.: Earthquake safety training through virtual drills. IEEE Trans. Vis. Comput. Graph. **23**(4), 1388–1397 (2017)

Peponis, J., Zimring, C., Choi, Y.K.: Finding the building in wayfinding. Environ. Behav. **22**(5), 555–590 (1990)

O'Neill, M.J.: Effects of signage and floor-plan configuration on wayfinding accuracy. Environ. Behav. **23**(5), 553–574 (1991)

Slone, E., Burles, F., Robinson, K., et al.: Floor plan connectivity influences wayfinding performance in virtual environments. Environ. Behav. **47**(9), 1024–1053 (2015)

Turner, A., Doxa, M., O'Sullivan, D., et al.: From isovists to visibility graphs: a methodology for the analysis of architectural space. Environ. Plan. B-Plan. Des. **28**(1), 103–121 (2001)

Zhao, S.L., Jin, S., Ai, C.F., et al.: Visual analysis of three-dimensional flow field based on WebVR. J. Hydroinform. **21**(5), 671–686 (2019)

Zyda, M.: From visual simulatiou to virtual reality to games. Computer **38**(9), 25 (2005)

The Consciousness Printer

Establishing Architectural Environments Through Brain-Computer Interface Technologies

Fanyi Tang[✉] and Shengyu Liu[✉]

Politecnico di Milano, Piazza Leonardo da Vinci, 32, 20133 Milano, Italy
{fanyi.tang,shengyu.liu}@mail.polimi.it

Abstract. This research aims to investigate human perception of architectural spaces through EEG signal recognition and develop an innovative interactive-based creative mode. Artificial intelligence algorithms are increasingly used in architectural design, but they are currently limited to imitating and replicating the neural mechanisms of the biological brain. Architectural space extends beyond mere form and style. This paper explores the dynamic connection between architectural space and human cognition by utilizing brain-computer interface (BCI) technology to obtain a direct brainwave signal source. The focus is specifically on the relationship between architectural space and emotional response. Emotion recognition is achieved through the implementation of advanced deep-learning algorithms. Simultaneously, we utilize VR technology to immerse testers in the architectural environment, studying neural responses to various spaces. Our experiments have preliminary revealed spatial features that are linked to human emotions, as well as differences between the designer's expectations and the audience's experience. The AIGC model converts emotional data into corresponding geometric spatial features and generates new spatial scenes that contain emotional, temporal, and personal attributes. This approach comprehensively understands the interplay and co-creative essence between architecture and human consciousness, as well as externalizes human consciousness as a method in architectural design.

Keywords: Architecture · Brain-Computer Interface · Deep-Learning · Emotional Perception · Neuroscience · AIGC

1 Introduction

Today, architecture is facing numerous practical challenges and crises. Extensive research has been conducted over the long term on various innovative approaches to digitally transform architectural design through artificial intelligence algorithms to alleviate these problems. However, it is important to note that neural network algorithms are derived from the neural mechanisms of biological brains. In essence, artificial intelligence aims to emulate and replicate the operating principles of the brain's neural functions. Architectural design is not just about form and style; generating designs based solely on superficial analysis of visual data is incomplete.

© The Author(s) 2025
H. Chai et al. (Eds.): CDRF 2024, *Symbiotic Intelligence*, pp. 442–456, 2025.
https://doi.org/10.1007/978-981-96-3433-0_39

Architectural spaces interact with and shape people's subjective consciousness and experiences, becoming a projection and extension of them. This is a complex issue involving architecture, neuroscience, and philosophy. Integrating architecture and neuroscience can help us better understand the essence and significance of architectural design, as well as the interplay between architecture and human consciousness, and even the neural mechanisms behind architectural design decisions. It enables us to investigate the correlation between consciousness and materiality, as well as desire and representation.

Arbib (2013) [1] divides the relationship between neuroscience and architecture into three parts: (1) the neuroscience of the design process, which investigates the brain of the architect; (2) neuro-morphic architecture, which studies the 'brain' of the building; and (3) the neuroscience of architectural experience, or individual experiences in architectural environments. This paper will focus on the third part, exploring the ways in which individuals experience architectural environments from a neuroscientific perspective.

1.1 Architecture

In the field of architecture, 'architectural typology' can be understood as architects' systematic categorization of core and essential issues within architecture. It has evolved over time, initially focused on classifications related to specific functions. Architectural typology further developed in the early 20th century with Ernst Neufert's proposals (1936) [2] based on geometric forms and functions. He believed that a building's form and scale should correspond to its function and user requirements. He guided the architectural design process through systematization and standardization.

However, in the latter half of the 20th century, the rise of postmodernism led to a subversion and reinterpretation of traditional architectural types. Aldo Rossi (1966) [3] argued that architecture is not only an accumulation of geometric forms and functions but also an art form rich in symbolism and meaning. Through meticulous design and organization of architectural elements, proportions, light, and materials, architects create captivating spaces that convey emotions, atmosphere, and meaning. Architecture is a medium capable of inspiring emotions and evoking individual and collective memories (Figs. 1 and 2).

Rossi's perspective aligns with cognitive scientist Andy Clark's theory (1988) [4], which proposes that human cognition is not solely reliant on internal brain processes but is a result of interactions and coupling with the external environment. Additionally, French philosopher Bernard Stiegler introduced the concept of the 'Organology of Dreams' and 'Original Cinema' (2016) [5], offering a new perspective for comprehending architectural design. Modern technology and media extend our senses and consciousness, acting as tools through which we interact with and experience the world. Dreams are our response and re-creation to these tools. Original cinema involves projecting consciousness onto objects, weaving perception into a montage. This creates a 'third retention' as a historical component that becomes part of the narrative, affecting the consciousness and perception of the audience and extending their sensory tools. Rather, it is a projection and extension of an architect's senses, consciousness, and cognition, acting as an organ for their interaction and experience with the spatial environment. To endow architectural space with meaning and value, it requires the collective memory of history and culture, serving as the 'third retention.'

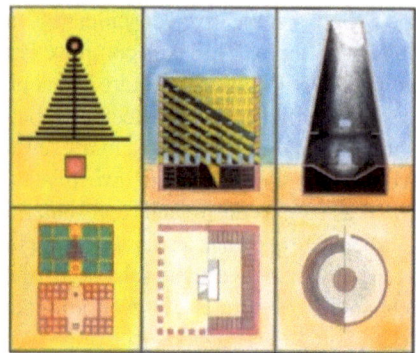

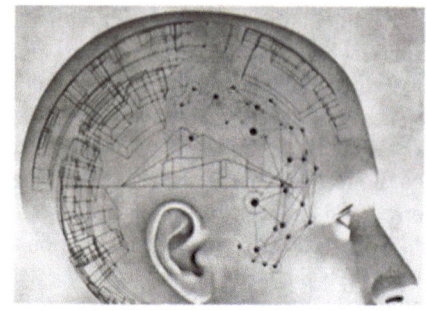

Fig. 2. (Right) Architecture and Brain Neuroscience.

Fig. 1. (Left) Aldo Rossi Architectural Typology

Architecture is not solely defined by physical structures, geometric forms, or machine functions, nor is it merely a style of representation. The spatial organization and functional layout of architectural spaces can impact human cognition. People also shape the meaning and function of architecture through their perception, thought, and behavior. This mutual interaction forms a complex dynamic coupling system between human cognition and architectural environments. To promote progress and innovation in architectural design, it is crucial to understand its emotional and cognitive dimensions. This understanding will better serve human mental and physical well-being, as well as overall quality of life.

1.2 The Brain and Emotions

Consciousness is believed to originate from the brain, making it fundamental to understand how the brain works. The frontal lobe, which accounts for approximately 40% of the cerebral cortex, serves as the executive center of brain functions. The limbic system, located at the top of the brainstem, includes the hippocampus, hypothalamus, and amygdala, among other parts, and is responsible for generating emotions and memories. The prefrontal cortex is primarily associated with higher cognitive functions in humans. It directs attention to specific goals and enables detailed, rational thinking. Additionally, it has a regulatory effect on the limbic system, which generates emotions, and can inhibit impulsive outbursts. Therefore, human behavior is regulated by both systems. (2009) [6].

Emotions can significantly influence and interfere with our thinking and decision-making. Therefore, this article starts by discussing the relationship between space and emotion. To explore the neural mechanisms behind architecture, we can leverage theories and technologies from neuroscience. To reduce the impact of human emotions on judgment and achieve more objective and accurate emotional transition processes, we utilized brain-computer interface technology to directly acquire emotional signals for emotional judgments. Additionally, we utilized VR technology to create architectural spatial environments, study neural responses to different environmental stimuli (including architectural spaces) and attempt reverse engineering.

2 Related Work

2.1 Architecture and Neuroscience

In the field of neuroscience, researchers use devices such as Electroencephalogram (EEG) and Functional Magnetic Resonance Imaging (fMRI) to study the brain's responses to different environments. A comprehensive qualitative review of the literature on the intersection of neuroscience and architecture by Tulay Karakas, Dilek Yildiz, and their colleagues (2020) [7] presents a broad discussion of these concepts.

Kirk et al. (2009) [8] discovered differential brain reactions between experts and non-experts in the process of aesthetic judgment. Vartanian et al. (2013) [9] also employed fMRI to compare straight vs. curved interior spaces. They reported that images containing curved spaces activated aesthetic processing sections of participants' brains. Essawy et al. (2014) [10] utilized electroencephalography equipment to measure participants' reactions to mental space. Choo et al. (2017) [11] claim to have identified neural activity patterns associated with particular architectural styles. Coburn et al. (2020) [12] purportedly identify neural reactions in the brain to attributes of interior design images like coherence (organization and ease of understanding a scene), allure (information richness and intrigue of a scene), and "hominess" (the degree to which a scene reflects personal space).

Avishag Shemesh et al. (2021) [13] explored the experiential measurement and quantification of emotional responses to geometric manipulations within architectural spaces. This research utilized multimodal measurements using physiological sensors like EEG (Electroencephalogram), GSR (Galvanic Skin Response), and Eye-tracking (ET), coupled with VR for spatial rendering. Their findings suggest that emotional responses to spaces, whether positive or negative, can be gauged through proportion variations of curvature (C), protrusion (P), scale (S), height P (H), or width P (W) in virtual spaces. Large symmetrical spaces tend to have a positive impact on users. The more extreme the proportions, the stronger the emotional response. Notably, non-designers exhibited stronger reactions to the geometric shapes of virtual spaces than designers did.

However, most of these concepts have largely been left unverified experimentally, suggesting an array of concepts that require further exploration.

2.2 Brain-Computer Interface and Emotion Recognition

Brain Computer Interface (BCI) refers to the collection and conversion of brain activity signals into output control signals through computers and other electronic devices without involving the peripheral nervous system and muscle tissue. This technology enables communication between humans or animals and external devices by creating a direct connection that facilitates the exchange of information between the brain and the device. The electroencephalogram signal (EEG) is composed of 5 sub-band signals: delta, theta, alpha, beta, and gamma. Each sub-band is associated with different mental states and conditions.

Compared to recognizing facial expressions, body movements, and gestures, electroencephalogram (EEG) signals are more direct and less susceptible to interference. Therefore, EEG is a suitable method for extracting human emotions.

This article will use non-invasive brain-computer equipment and adopt the most objective data values. M. R. Islam et al. (2021) [14] conducted a preliminary statistical analysis of EEG-based emotion recognition algorithms. The study listed the architecture of general emotion recognition and existing algorithm systems, including data management, preprocessing, feature extraction, feature selection, and classification algorithms.

Data Set. For emotion recognition, the first step is to collect or record Electroencephalographic (EEG) data. Using the same raw data across multiple studies allows for comparison of their performance or accuracy. Some scholars have created research-grade datasets that are freely available to the public. Currently, the most widely used dataset is the DEAP (A Dataset for Emotion Analysis using Physiological and Audiovisual Signals) [15]. Figure 3 shows that EEG data were collected from 32 healthy participants, 16 males and 16 females. The participants were both physically and psychologically healthy. EEG signals were captured using electrode caps based on the international electrode placement system '10–20', consisting of 32 leads, as shown in the electrode distribution diagram. Participants watched 40 one-minute videos while their EEG signals were recorded at a sampling frequency of 512Hz. The dataset includes 32 EEG channels and 16 additional channels capturing other common signals such as electrooculogram (EOG) and electrocardiogram (ECG). After watching the videos, subjects rated them on a scale of 1 to 9 for Valence, Arousal, and Dominance.

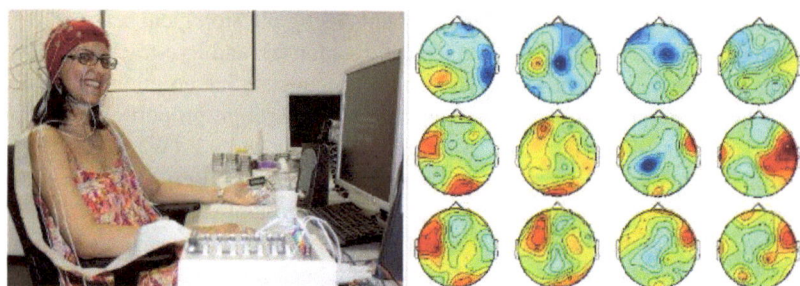

Fig. 3. (Up) Acquisition Scenarios & EEG Signal Distribution in DEAP Dataset. The DEAP dataset contains EEG signals from participants in different video viewing states. Participants were asked to rate their Valence, Arousal, and Dominance levels on a scale of 1 to 9 to indicate the depth of their feelings during the viewing process.

CDCN [16] . Regarding the extraction of Differential Entropy (DE) features, each channel in the raw data undergoes processing to extract DE features across five principal frequency bands. This process transforms the original data into a matrix with both frequency and temporal domain features. For EEG signals with E channels, the dimensionality of each resultant feature matrix is [E,5]. This serves as the input for the CDCN model, as described in the computational formulae mentioned earlier Fig. 4. The dense block comprises a convolutional layer, followed by batch normalization and ReLU (Rectified Linear Unit) activation. This is intended to improve the flow of information and promote feature reusability. The block also introduces direct connections from

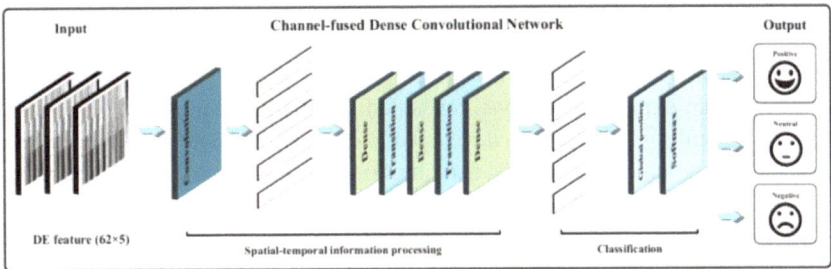

Fig. 4. (Down) CDCN Framework for Emotion Recognition includes DE feature extraction, convolutional layers, pooling layers, dense blocks, and transition blocks.

any layer to subsequent layers to extract higher-level features. The transition block is a combination of batch normalization, convolutional layers, and max-pooling, designed to reduce feature dimensionality. This architecture needs modifications to be suitable for EEG signal applications. In practical implementations, transition blocks comprise batch normalization layers, convolutional layers, and max-pooling layers. The study uses the Backpropagation Through Time (BPTT) algorithm to optimize network parameters until achieving the optimal solution or reaching the maximum number of epochs. The cross-entropy objective function is used as the loss function to enhance model accuracy, and the Adam optimization algorithm is used to train the CDCN framework. The framework achieved an accuracy rate of 90.63% on the DEAP emotion dataset. We chose this algorithmic framework for our current experiment and plan to make specific adjustments based on the distinctive features of our study.

De Feature. In the DE feature process, we obtain EEG signals and sampling frequency from DEAP and Emotive datasets. We define five frequency bands (delta, theta, alpha, beta, and gamma) and initialize a differential entropy matrix of size (number of channels, 5). The function then loops through each channel and computes the power spectral density using the Welch method. For each frequency band, the function finds the corresponding frequency index and computes the differential entropy feature. Finally, the function returns the differential entropy feature matrix. This function can be used to analyze the level of neural activity in different frequency bands of EEG signals, which is useful for investigating the relationship between brain cognition and cognitive dysfunction.

$$h(X) = -\int_{-\infty}^{\infty} \frac{1}{\sqrt{2\pi\sigma^2}} \exp\frac{(x-\mu)^2}{2\sigma^2} \log\frac{1}{\sqrt{2\pi\sigma^2}}$$
$$\exp\frac{(x-\mu)^2}{2\sigma^2} dx = \frac{1}{2}\log 2e\pi\sigma^2 \tag{1}$$

$$x_l = H_l(x_{l-1}) \tag{2}$$

$$x_l = H_l\left([x_0, x_1, \ldots, x_{l-1}]\right) \tag{3}$$

3   Method and Experiments

In the present study, which aims to investigate the relationship between spatial environments and human emotions, we have developed the following methodology: First, an "Emotion Recognition Machine" is developed to establish the correspondence between electroencephalogram (EEG) signals and emotional states. Subjects are then invited to participate in experiments in which spatial experience serves as a stimulus, allowing real-time EEG data to be collected throughout the experience. The emotional fluctuations of the subjects can then be assessed using the "Emotion Recognition Machine" developed in the previous stage. Finally, we conduct further research and data processing based on the emotional fluctuations and other data features obtained earlier. This study uses Russell's Circumplex 2D (1980) [17] model for emotion classification, which uses two-dimensional data composed of Valence and Arousal.

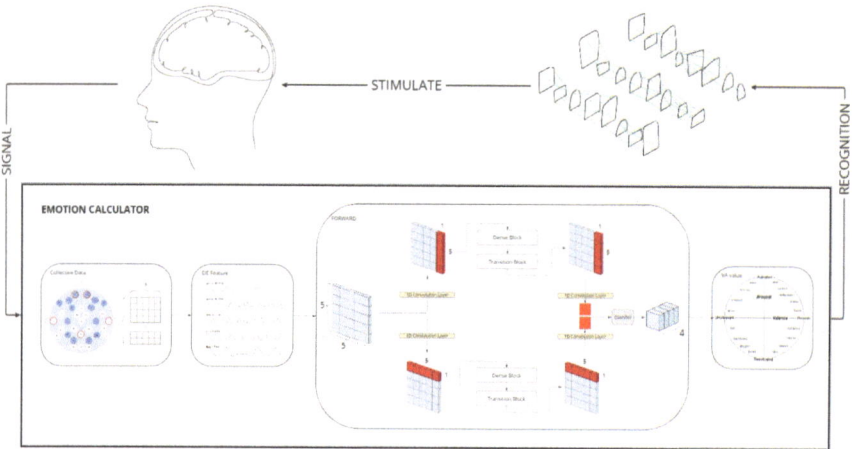

Fig. 5. Experiment Framework: First, train the 'emotion recognition machine' using the DEAP dataset. Then, explore the interactive relationship between space and human experience.

3.1   Emotion Recognition Machine

In this study, we will use the mainstream DEAP dataset as the basis. Since the electroencephalogram (EEG) device used in this experiment is the EMOTIV Insight (a 5-channel EEG device), we will extract the corresponding five signal channels from the 32 channels in the DEAP dataset that have a higher correlation with emotional values. These channels are T7, T8, FP1, FP2, and PZ. At the same time, we will extend the CDCN algorithm framework based on this to ensure that it is better suited to the characteristics of this experiment, as shown in Fig. 5. Experiment Framework.

Feature Extraction. The function takes two parameters: EEG signals and sampling frequency. First, the function swaps the channels and sample size of the EEG signals and then defines five frequency bands: theta (4–8 Hz), alpha (8–12 Hz), low-beta (12–16 Hz), high-beta (16–25 Hz), and gamma (25–45 Hz). The function then initializes a differential entropy matrix with dimensions corresponding to the number of channels and frequency bands. The Welch method is used to compute the power spectral density, iterating over each frequency band. For each band, the function identifies the relevant frequency indices and performs the calculations. Finally, the function returns a differential entropy feature matrix. The result is a 5*5 matrix imbued with channel and frequency domain information.

CDCN Network. CDCN is a deep neural network model consisting of convolutional layers, a series of Dense Blocks and Transition Blocks. The Dense Blocks are designed to improve information flow and feature reuse, while the Transition Blocks are used to reduce feature dimensionality. The architecture aims to ensure the effective extraction and use of features while maintaining the depth of the model.

Given the specific needs of our current experiment, we introduce moderate adjustments to the CDCN neural network. After the DE feature extraction described above, we obtain a 5*5 feature matrix that serves as the input to the neural network model. After passing through the initial one-dimensional convolutional layer, the output is fed into the dense blocks and transition blocks. A global average is taken in both time and frequency dimensions, and then a linear classifier is used to obtain a softmax probability distribution for emotional categories. This process is followed by another convolutional layer. The initial matrix is then transposed, and another forward propagation is performed. Finally, the results of these two branches are combined into a single output matrix representing the Valence and Arousal dimensions. Our approach adopts a lightweight architectural design. The CDCN algorithm is simplified to perform emotion recognition classification tasks. In the terminal part of the network, a fully connected layer transforms the feature vector into the final output classes, the number of which is determined by num_classes. The output layer uses a softmax function, which allows the output of the model to be interpreted as probabilities for each class.

In summary, we have built a neural network based on the CDCN framework. By extracting EEG data from five channels and performing DE feature extraction, we obtain a 5*5 feature matrix. This matrix serves as input to the CDCN neural network model, resulting in a 1*2 output matrix corresponding to Valence and Arousal. With the completion of EEG feature extraction and dataset training, we have preliminarily configured our "Emotion Recognition Machine", which is capable of assessing the emotional states of subjects based on five-channel EEG signals.

3.2 Spatial Drawing

The experiment requires two sets of spatial sequences as stimulus files that are then integrated into VR devices to better capture participants' responses. We prepared two different sets of spatial sequences for this purpose.

As shown in Fig. 6, the first set of sequences consists of seven spatial scenes, each lasting ten seconds. These scenes undergo gradual transformations in their spatial morphology along three-dimensional parameters-namely, height, width, and curvature-while keeping lighting and color conditions constant to control for variables. Episodes 1 to 3 feature square spaces with aspect ratios of 5:1, 1:3, and 1:1 respectively. From episodes 4 to 7, the space boundaries gradually transition from right angles to arcs, resulting in a seamless and edgeless space by episode 7.

The second set of spatial sequences is composed of five segments, each lasting ten seconds, depicting different spatial perspectives of Tadao Ando's Senju Art Museum. The spatial light environment undergoes gradual changes. Direct light sources, such as ep1, create high-contrast and clear-boundary spaces that gradually transition to spaces with a compromise between light and dark and low-sharp boundaries, such as ep5.

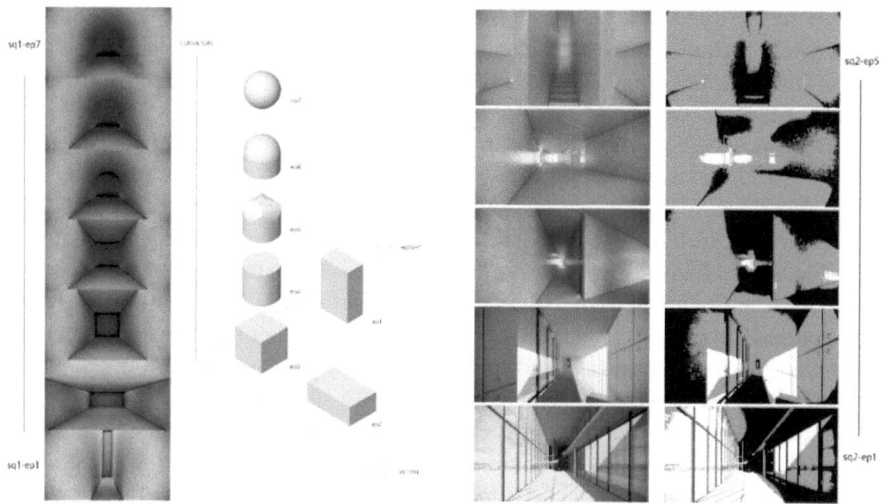

Fig. 6. Spatial Drawing: Contains two spatial sequences. The left sequence consists of seven ten-second episodes. In a constant light source environment, the spatial form changes in three-dimensional parameters (height, width, and curvature) that vary from bottom to top (ep1 to ep7). The second sequence (right) consists of five episodes, ten seconds for each as well: These clips showcase five different spatial perspectives inside the Tadao Ando Narahama Art Museum, with different light environment, arranged from bottom to top (ep1~ep5).

3.3 Experiment

In the experiment, we recruited volunteers of different genders and ages, with 50% being designers and the remaining 50% being non-designers. Participants were asked to wear electroencephalogram (EEG) equipment (Emotiv Insight 1.0) and virtual reality (VR) equipment (PICO Neo3). The VR system used two pre-constructed spatial scenario videos as stimulus files, exposing subjects to changes in spatial conditions over a specified time frame. As the videos played, real-time EEG signals were collected from participants to record their neural responses to changes in the spatial environment. During the second phase of the spatial sequence test, the designer participants were instructed to assign subjective feature values to the scenes, while the EEG signals of the non-designer participants were recorded in different spatial settings.

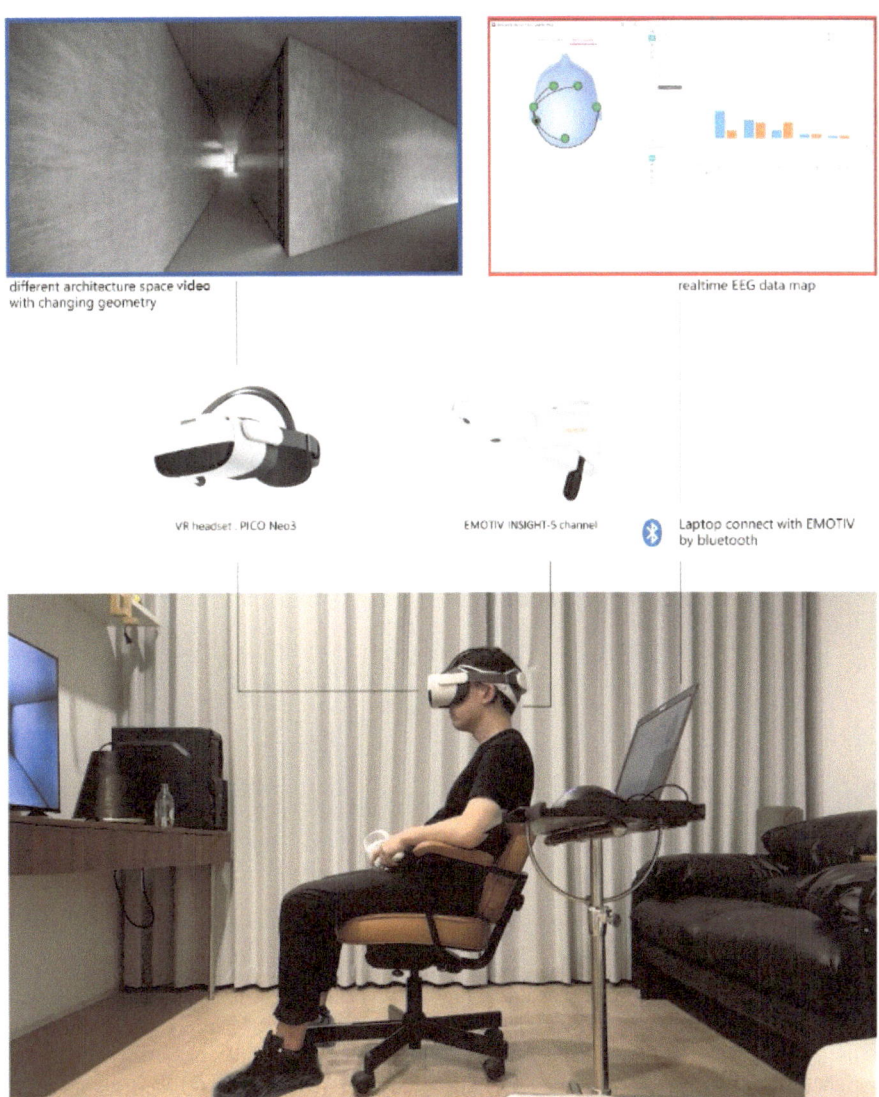

different architecture space video
with changing geometry

realtime EEG data map

VR headset : PICO Neo3   EMOTIV INSIGHT-5 channel   Laptop connect with EMOTIV
by bluetooth

Fig. 7. Experimental Scene. The participants wore an electroencephalogram device (Emotiv Insight 1.0) and a virtual reality (VR) headset (PICO Neo3). The experiment recorded their brainwave signals as they experienced changes in the spatial environment. In the second test space sequence, the designers among the participants assigned subjective feature values to the scenes, while the brainwave signals of non-designer participants were recorded.

The experimental setup is shown in Fig. 7. Subjects were asked to wear an EEG device (Emotiv Insight 1.0) and a VR device (PICO Neo3), and the experiment recorded brainwave signals as subjects experienced spatial and environmental changes. Brainwave signals as the subjects experienced spatial changes in the environment. In the second test space sequence, the designers in the test subjects were asked to assign subjective eigenvalues to the scene, while the non-designer test subjects' brainwave signals were recorded in different spatial environments. Throughout the experimental procedure, participants remained in a relaxed state. The EEG signals were transmitted via Bluetooth to a PC, whereupon a previously developed "Emotion Recognition Engine" was used to identify emotional Valence and Arousal values.

4 Results

4.1 Experimental Conclusions

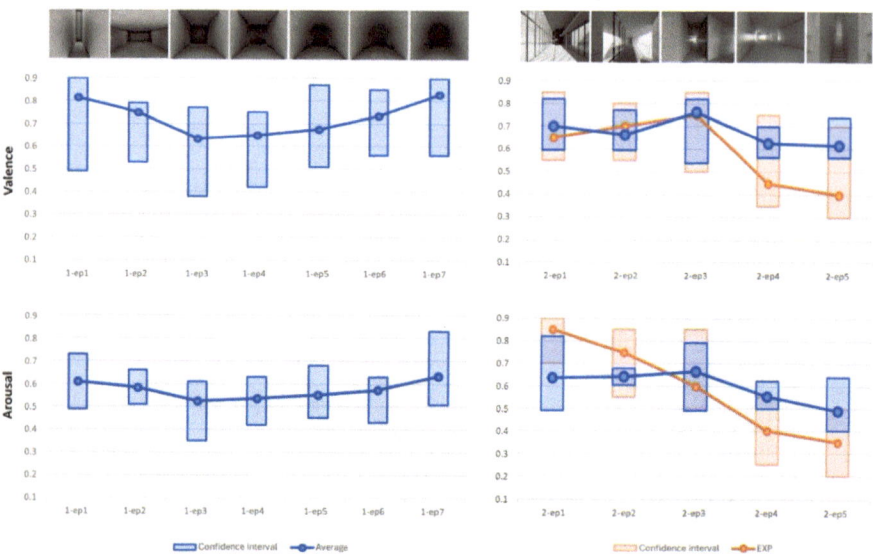

Fig. 8. Experimental data results. The horizontal coordinates represent the geometric space with temporal ordering, and those of vertical correspond to the range of participants' Valence and Arousal values. The left graph displays the sampled values of the first spatial sequence, while the right graph displays those of the second spatial sequence. The graph displays the mean and confidence interval of the detected emotion values in blue, and the expectancy ratings of designers among the participants in orange.

The data results of the experiment are presented in Fig. 8, which illustrates the relationship between different spatial experiences and participants' emotions, as well as the difference between designers' expectations and audience experience. To enhance the fitness, we remapped the final emotional value range to a scale of 0.1–0.9, corresponding to the previous range of 1–9 (from weak to strong).

The left side displays the statistical results of the emotional response to the first spatial sequence modulated by different parameters. Ep1–3 demonstrates that a significant spatial aspect ratio has a much more positive impact on people's emotions compared to when the aspect ratio is close to 1, especially in the Valence value, which dropped from 0.81 to 0.63 on average. Starting from ep4, the visual contours and boundaries blur as the softness of the space increases. The values of Valence and Arousal steadily increase, with the mean Valence rising from 0.63 to 0.82. The change range of arousal is moderate, but the average value also gradually increases from 0.52 to 0.62. According to these results, spaces with larger aspect ratios and higher softness have a more positive effect on people's emotions. These findings preliminary suggest a correlation between the physical characteristics of a space and its impact on emotional response.

The image on the right displays the sampling results from a second spatial sequence. It compares the designer's intended judgment of the space with the actual measured audience experience. The sampling space scenes are arranged in order from strong to weak based on the spatial illumination and contrast. Based on the pre-judgment statistics of the designer subjects, it is predicted that the space with relatively balanced illumination and contrast, ep3, will have the highest positive impact on people's emotions with a mean Valence value of 0.75. In contrast, the strong or soft spaces, ep1, ep2, ep4, and ep5, are less effective than ep3, especially the expected Valence values of ep4 and ep5 are as low as 0.45 and 0.4, respectively. The expected average value for Arousal level steadily decreases from 0.85 to 0.3 as illumination and contrast decrease (from ep1 to ep5), according to the designer participants. However, the actual emotional test results for normal participants showed differences from the designer's expected values. It is obvious that ep1 and ep2, the two spaces with the strongest illumination and contrast, did not arouse the audience as expected. The measured Arousal value is 0.65, which is lower than the expected values of 0.85 and 0.75. Additionally, the Valence and Arousal values of ep4 and ep5 are still higher than the expected average, despite being lower than those of ep1–3.

4.2 Spatial Regeneration

Next, we try to implement novel approaches based on the experimental results: converting emotional data into corresponding feature inputs and autonomously optimizing raw spaces with individual attributes. We attempt to apply these results to the processing of original files and generate new spatial scenes incorporating emotional, temporal, and personal attributes. We use the collected spatial prediction values and translate their basic representational meanings into the LLM (Large Language Model). This allows the model to generate contextually appropriate prompt text based on its analysis of the values. Once the appropriate prompts are received, we use stable diffusion techniques to create emotional spaces that match the collected data.

Finally, we chose the 7 spaces of basic geometry (sq1-ep1 ~ sq1-ep7) to make spatial reconstruction Fig. 9. Two representative sets of test data were selected for each space. We can see that although each group generates a different prompt, the final spatial regeneration reflects the tested values of Valence and Arousal. The higher the Valence value, the greater the color saturation of the reconstructed space; the higher the Arousal, the more detailed the reconstructed space becomes. It can be shown that this method can be effective for spatial regeneration.

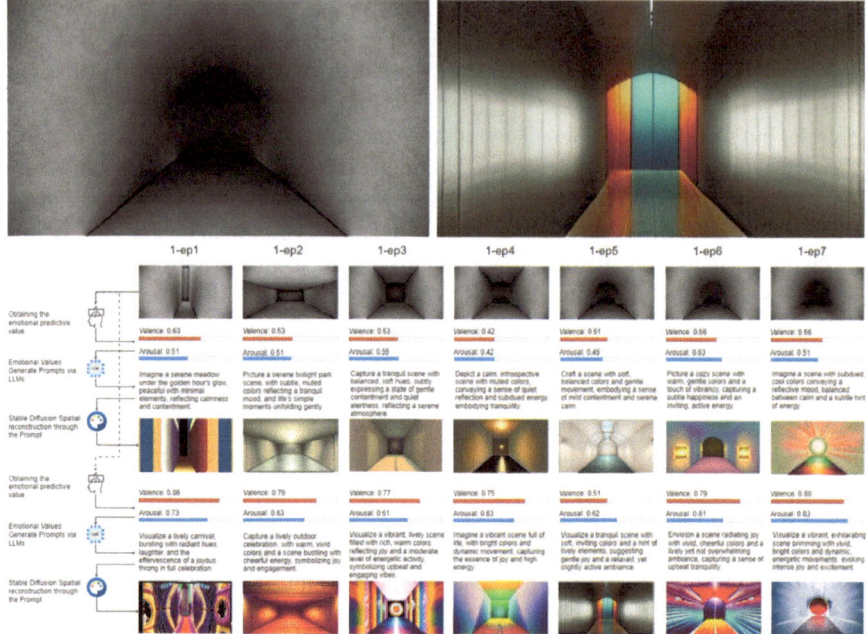

Fig. 9. Spatial Regeneration. Seven geometric spaces were selected, each with two sets of representative predictions.

5   Conclusion

This study aims to integrate architecture and neuroscience, investigating the relationship between architectural spaces and human emotion. EEG data and deep learning emotion recognition algorithms are used to measure human emotional experiences in virtual spatial scenarios. Based on our initial findings, it appears that individuals tend to provide more positive emotional feedback to spaces that have significant aspect ratios, softer boundaries, and a relative balance of illumination and contrast. Additionally, there is a notable distinction between the emotional response elicited by the architectural space and the designer's expectations. Furthermore, by integrating emotional features with AIGC, we achieve spatial regeneration, attempting a methodology that externalizes human consciousness into architectural design.

In summary, this research goes beyond the visual representation of architecture and delves into the intrinsic co-creative and interactive relationship between architectural spaces and human consciousness. It reveals how people construct their cognitive and emotional experiences through interactions with their environment, and how architectural spatial design serves as a manifestation and projection of human consciousness. For future research, the scope and implications of the concept of a 'consciousness printer' can be further expanded. To deepen our understanding of the relationship between form and meaning, we can continue to explore different architectural geometric forms. Additionally, architects, as shapers of space, are an intriguing area of research in terms of the neural mechanisms underlying their design decision-making process. Interdisciplinary collaboration between architecture and neuroscience can provide deeper insights and inspiration for both the theoretical and practical realms of architecture.

References

1. Dougherty, B.O., Arbib, M.A.: The evolution of neuroscience for architecture: introducing the special issue. Intell. Buildings Int. **5**(sup1), 4–9 (2013)
2. Neufert, E., Neufert, P.: Architects' Data. Bauwelt-Verlag (1936)
3. Rossi, A.: The Architecture of the City. MIT press (1984)
4. Clark, A., Chalmers, D.: The extended mind. Analysis **58**(1), 7–19 (1998)
5. Stiegler, B.: Art, Difference, and Repetition in the Anthropocene [Lecture]. China Academy of Art, Hangzhou (2016)
6. Martin, E.I., Ressler, K.J., Binder, E., Nemeroff, C.B.: The neurobiology of anxiety disorders: brain imaging, genetics, and psychoneuroendocrinology. Psychiatr. Clin. **32**(3), 549–575 (2009)
7. Karakas, T., Yildiz, D.: Exploring the influence of the built environment on human experience through a neuroscience approach: a systematic review. Front. Architectural Res. **9**(1), 236–247 (2020)
8. Kirk, U., Skov, M., Christensen, M.S., et al.: Brain correlates of aesthetic expertise: a parametric fMRI study. Brain Cogn. **69**(2), 306–315 (2009)
9. Vartanian, O., Navarrete, G., Chatterjee, A., et al.: Impact of contour on aesthetic judgments and approach-avoidance decisions in architecture. Proc. Natl. Acad. Sci. U.S.A. **110**(supplement_2), 10446–10453 (2013)
10. Essawy, S.M., Kamel, B., Elsawy, M.S.: Timeless buildings and the human brain: the effect of spiritual spaces on human brain waves. Int. J. Architect. Res.: ArchNet-IJAR **8**(1), 133 (2014)
11. Choo, H., Nasar, J.L., Nikrahei, B., Walther, D.B.: Neural codes of seeing architectural styles. Sci. Rep. **7**(1), 40201 (2017)
12. Coburn, A., Vartanian, O., Kenett, Y.N., et al.: Psychological and neural responses to architectural interiors. Cortex **1**(126), 217–241 (2020)
13. Shemesh, A., Leisman, G., Bar, M., Grobman, Y.J.: A neurocognitive study of the emotional impact of geometrical criteria of architectural space. Archit. Sci. Rev. **64**(4), 394–407 (2021)
14. Islam, M.R., Moni, M.A., Islam, M.M., et al.: Emotion recognition from EEG signal focusing on deep learning and shallow learning techniques. IEEE Access. **22**(9), 94601–94624 (2021)
15. Koelstra, S., Muhl, C., Soleymani, M., et al.: Deap: a database for emotion analysis; using physiological signals. IEEE Trans. Affect. Comput. **3**(1), 18–31 (2011)
16. Gao, Z., Wang, X., Yang, Y., et al.: A channel-fused dense convolutional network for EEG-based emotion recognition. IEEE Trans. Cognit. Develop. Syste.. **13**(4), 945–954 (2020)
17. Russell, J.A., Lewicka, M., Niit, T.: A cross-cultural study of a circumplex model of affect. J. Pers. Soc. Psychol. **57**(5), 848 (1989)

Research Review and Prospects of EEG Technology in the Field of Space Emotion Cognition

Hongyi Men[1], Sky Lo[2], and Xiangmin Guo[1(✉)]

[1] School of Architecture. 5, Harbin Institute of Technology (Shenzhen), Shenzhen 18055, China
24904404@qq.com
[2] School of Design. 1, The Hong Kong Polytechnic University, Hong Kong 00872, Singapore
tt-sky.lo@polyu.edu.hk

Abstract. With advancements in biology and sensor technology, neuroscience, particularly electroencephalograph (EEG) technology, has seen significant developments and applications in the field of space emotion perception. EEG technology records brain electrical activity through electrodes placed on the scalp. The signals collected by these electrodes reflect the brain's activity status and are utilized for studying cognition, diagnosing diseases, and more. EEG technology has revolutionized the measurement of human emotions within space contexts by providing preliminary insights into human space cognition. Traditional methods, such as subjective survey questionnaires and field visits, are characterized by high levels of subjectivity, instability, and delays, which significantly undermine the objectivity, accuracy, and real-time nature of emotion recording. In contrast, EEG technology offers real-time and objective feedback on changes in space elements, effectively guiding designers to enhance space quality and improve user satisfaction and well-being within these spaces.

Keywords: EEG · Space Design · Emotion Perception

1 Introduction

The advancement of neuroscience, particularly the emergence of electroencephalograph(EEG) technology, has introduced novel methods for measuring space emotion perception. EEG technology analyzes physiological signals originating from the brain, the source of emotions. Consequently, its application in the field of space emotion perception offers significant advantages over other physiological signal-based methods. EEG technology can quantitatively, in real-time, and objectively measure individual emotion responses to space elements, providing valuable feedback to designers to facilitate truly people-oriented space design. Currently, EEG technology is applied in the field of emotion perception across different space scales. This paper examines these applications, comparing their strengths and weaknesses, summarizing the advantages and limitations of current EEG applications, and proposing future development directions. The study aims to expand the application scope of EEG technology and provide more intuitive guidance for space design.

© The Author(s) 2025
H. Chai et al. (Eds.): CDRF 2024, *Symbiotic Intelligence*, pp. 457–467, 2025.
https://doi.org/10.1007/978-981-96-3433-0_40

2   The Relationship Between EEG Technology and Space Emotion Perception

2.1   The Development of EEG Technology

The brain, as the highest component of the human central nervous system, serves as the foundation of all conscious activities.

EEG technology measures brain electrical activity by recording electrical signals generated by neuronal activity via electrodes placed on the scalp's surface. These EEG data are typically presented in waveforms, classified by their frequency, amplitude, and shape characteristics. Symbolizing the emotion and cognitive responses of individuals to external stimuli, they reveal different functional states of the brain. The origins of EEG research trace back to early human medical studies and have evolved into a widely used method in modern neuroscience. In 1924, German scientist Hans Berger first recorded the human brain's electroencephalogram. Since then, EEG technology has found extensive application in clinical diagnosis and neuroscience research, particularly in diagnosing mental disorders such as epilepsy and depression, as well as in assessing sleep status and brain injuries. It is also employed in conjunction with functional magnetic resonance imaging (fMRI) for emotion recognition, fatigue monitoring, stress recovery, aesthetic research, and other areas within cognitive neuroscience.

In recent years, the integration of EEG technology with machine learning and artificial intelligence (AI) algorithms has significantly expanded its application potential across multiple fields.

2.2   The Development and Mechanism of Space Emotion Perception

2.2.1   The Development of Space Emotion Perception

Emotion, as a crucial medium for individual cognitive expression, has seen continuous development and evolution in both concept and understanding.

From the ancient Greek philosopher Aristotle's assertion that emotions are closely related to rational thinking, to Charles Darwin's 19th-century work, *"The Expression of the Emotions in Man and Animals,"* which posited that emotion expression is universal across cultures and species. There has been a significant shift in emotion research from philosophy to biology. In the 20th century, the study of emotions became an essential branch of psychology, with the James-Lange theory and the Cannon-Bard theory jointly laying the physiological foundation for it. The emergence of cognitive psychology further emphasized the close relationship between emotions and individuals' subjective cognitive assessments of events.

Today, emotions are regarded as a multidimensional structure encompassing physiological responses, cognitive evaluations, emotion experiences, and behavioral expressions. The complex role of emotions in individual adaptation to the environment has also received considerable attention.

2.2.2   The Mechanism of Space Emotion Perception

The mechanism of space emotion perception comprises four key stages: sensation, perception, emotion, and behavior.

Sensation is the initial stage, where sensory organs (such as the eyes, ears, nose, tongue, and skin) convert space information into neural signals. Perception follows, involving the processing and interpretation of these sensory signals. The brain integrates and filters sensory inputs, transforming them into meaningful perceptual objects. Emotion, the most critical component of this process, represents the subjective experience and response to perceived information. Behavior is the final stage, serving as the response to the perceived information and representing the ultimate goal and outcome of the perception process.

Emotions drive space behavior during the perception process. When entering a specific space, individuals subconsciously and continuously perceive their surroundings.[1] If the perceived information meets their current needs, positive emotions are elicited, which subsequently drive space approach behaviors. Conversely, if the perceived information is evaluated as conflicting with their needs, negative emotions arise, leading to space avoidance behaviors. Upon completion of the current behavior, the results are fed back into a new cycle of environmental perception.

In summary, emotions serve as a medium for space perception and expression, playing a crucial role in regulating space behavior.

2.3   The Correlation Between Emotions and EEG Signals

Modern research indicates that the generation of emotions involves the interaction of multiple brain regions, particularly the cerebral cortex.

The cerebral cortex is comprised of the frontal, temporal, parietal, and occipital lobes [2] (Fig. 1). The amygdala, located deep within the temporal lobe, plays a central role in processing emotion information and triggering adaptive emotion responses. The prefrontal cortex regulates the activity of the amygdala and participates in the development of coping strategies for emotion arousal events [3]. As an intuitive graphical representation of electrical activity generated by neuronal activity in the brain, EEG can directly reflect emotion valence, offering unique advantages in monitoring and studying emotion feedback.

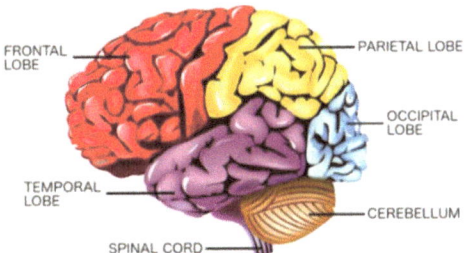

Fig. 1. Schematic diagram of cerebral cortex zoning [2]

Understanding the correspondence between emotions and EEG signals requires the separate classification of emotions and EEG signals. The most commonly used classification method is Russell's two-dimensional emotion classification model. It divides emotions into two dimensions: arousal and valence. Arousal indicates the intensity of emotions, while valence represents the positivity or negativity of emotions. There are two types of indicators in EEG technology: EEG signals and event-related potentials (ERP). Currently, EEG signals are more commonly used in research related to space design, so this article does not explore experiments using ERP signals. Different frequencies of EEG waves are associated with various psychological states and emotion experiences (Table 1). Among these, α(alpha), β(beta), δ(delta), and θ(theta) waves are common standards for space emotion perception experiments.

Table 1. Frequency of EEG Signals and Corresponding Emotions and Cognition

EEG signal	Frequency (Hz)	Emotions and cognition
α	8–13	relaxed and positive emotion
β	12–30	alert and vigilant emotion
θ	4–8	relaxed and calm emotion
γ	≥30	focused and positive experience
δ	0.5–4	sleepy state, no emotion

This relationship provides a solid theoretical foundation for enhancing the measurement framework of space emotion perception.

3 The Application of EEG Technology in the Field of Space Emotion Cognition

Currently, there are three primary theories in the interdisciplinary field of neuroscience and space emotion perception: neuroarchitecture, neurourbanism, and neurolandscape. These theories examine emotion perception in space across various scales. This article explores the application of EEG technology in the field of space emotion perception, drawing on insights from these three theories.

3.1 Application of EEG Technology in Neuroarchitecture

Neuroarchitecture aims to understand how the architectural environment affects the human brain and behavior, thereby enhancing positive emotions in space by altering specific space elements. Currently, EEG technology primarily investigates the impact of changes in three types of indoor space elements on individual emotion perception.

The first type is space proportion features, which include the aspect ratio, window-to-wall ratio, and area of the space. Experimental results indicate that as these elements increase, subjects exhibit higher levels of α and β waves, [4] particularly in larger rooms, which also promote θ waves, enhancing focus [5, 6]. Notably, most relevant experiments have been conducted in classrooms, suggesting that reasonable aspect ratios, window-to-wall ratios, and room sizes can effectively improve students' learning efficiency (Fig. 2).

Fig. 2. Space proportion feature experimental scenario [4–6].

The second type is space aesthetic elements, including the form of space interfaces and windows, colors, and materials of space interfaces. Experiments comparing free curves and regular shapes of windows and walls have shown that curved shapes positively influence individual emotions and alleviate environmental stress [7]. Additionally, colored spaces promote positive emotions, while hard paving can induce alertness [8]. In studies on aesthetic elements, some scholars have employed multimodal perception technology, using both heart rate and EEG signals to measure participants' emotions.

The third type is space functional elements. Experiments have demonstrated that prayer spaces in monasteries can arouse calm emotions in individuals [9]. However, research on such elements is limited by two issues: the narrow range of functional spaces studied and insufficient exploration of the strength of functional impacts on emotions.

3.2 Application of EEG Technology in Neurourbanism

Neurourbanism, proposed by Adli in 2017, aims to improve the urban environment and enhance the mental health of urban residents. Currently, EEG technology has mainly used to explore the impact of two types of functional urban spaces on individual emotion perception.

The first type is resting environment, characterized by relative quietness and greenery. In such environments, participants feel happy and relaxed. Research on this environments often focuses on both adult and elderly groups [10]. The second type is street environment, which is relatively noisy and can make people feel excited but cautious. Two specific environments within the street category are of particular interest. The historical environment, containing historical buildings or elements, can induce a joyful mood and relieve stress [11]. The human environment, where human activities dominate, can evoke feelings of vigilance and even fear [12]. Research has examined human environments both during the day and at night (Fig. 3).

Fig. 3. Human environment experimental scenario [12]

3.3 Application of EEG Technology in Neurolandscape

Neurolandscape focuses on how different landscape environments affect human brain activity, emotion states, and overall health. Currently, EEG technology primarily investigates the impact of changes in the attributes and categories of green landscapes on individual emotion perception.

Green landscape attributes include shape and density. Both free-form and high-density greenery can enhance individuals' enjoyment in the space [13] (Fig. 4). The categories of green landscapes include urban landscapes, natural landscapes, and seasonal landscapes. Urban landscapes encompass park and community landscapes, both of which can increase feelings of pleasure. The less artificial the urban landscape appears, the more relaxed people tend to feel [14]. Some studies have explored the presentation methods of urban landscapes by allowing participants to observe them from windows at different heights. Results indicated that viewing from a 12th-story window had the most positive emotion impact, while both higher and lower floors induced negative emotions [15]. Natural landscapes, such as forests, can stimulate positive emotions and alleviate stress. Seasonal landscapes, comparing summer and winter environments, found that summer landscapes had a more significant relaxation effect on participants [16].

Fig. 4. Green landscape attributes experimental scenario [13]

4   Comparison of EEG Technology Applications in 4 Different Scale Spaces

The elements of space vary across different scales, leading to distinct emotion perceptions. This article classifies the space elements involved in the experiments into three categories: space proportion, space aesthetics, and space function. It also compares the selected subjects and technical methods used in applying EEG technology to different theories. Finally, it summarizes the strengths and limitations of these theories in the application of EEG technology. This study finds that (Table 2):

- In studying space elements, neuroarchitecture focuses more deeply on space proportion and aesthetics. Neurourbanism conducts in-depth research on space function, while neurolandscape maintains a relatively balanced approach. However, these three theories have seldom explored specific data on space proportions and aesthetic elements. Furthermore, they have not constructed a comprehensive system integrating both macro (functional) and micro (proportional and aesthetic) space elements.
- Regarding subject selection and experimental methods, research on neurourbanism is more advanced than that on neuroarchitecture and neurolandscape. However, across all three theories, the application of EEG technology has not included comparisons of different gender groups in specific spaces. Additionally, the studies have not focused on vulnerable groups such as adolescents, children, and post-war trauma survivors.

In summary, the application of EEG technology within these three theories should be improved in the following ways.

- Neuroarchitecture should account for the varying emotion needs of individuals in different functional spaces, establish a macro framework for space elements, and refine the micro framework with detailed data. Additionally, it should increase focus on vulnerable groups (such as the elderly and children) and negative emotions (such as sadness and fear). Combining multiple physiological signals (such as ECG) in experiments can further enhance the accuracy of results.
- Neurourbanism should consider a secondary division of urban spaces based on different functions and extract typical space elements. For example, the existence, transparency, form, and color of commercial interfaces hold significant research value for understanding consumers' emotions. Additionally, quantitative research should be conducted on these elements, such as determining the saturation value range of interface colors that evoke comfort.
- Neurolandscape should increase its focus on water landscapes and conduct comparative research on green and water landscapes. Further studies on the proportion of green landscapes, including the types and heights of greenery, should be undertaken to construct a reasonable element system. Additionally, the selection of subjects and experimental methods should be more complex.

Table 2. Comparison of EEG technology applications in different theories

Theory	Space Scale	Space Aesthetic	Space Function	People	Signal	Positive	Negative
Neuro architecture	Aspect Ratio	Color	Prayer Room	Adults	EEG	Element	Function
	Window-to-Wall Ratio	Form					
	Area	Material					
Neuro urbanism	-	-	Resting Environment	Adults	EEG	Function	Element
			Street Environment (Historic, Human)	The Elderly	HRV		
Neuro landscape	Density	Form	Urban Landscape (Park, Community)	Adults	EEG	Green	Water

5   The Prospects and EEG Technology in the Field of Space Emotion Perception

This article summarizes the application of EEG technology in emotion perception in different scale spaces. Among these space scales, the application of EEG technology has the following highlights.

- Firstly, in the applications discussed, changes in space elements are interrelated with changes in EEG signals, which are linked to emotion changes. This significantly enhances the real-time and objective nature of the experiments, thereby improving the accuracy of the results.
- Secondly, the selection of experimental methods and environments in the aforementioned applications is becoming increasingly realistic. The process has evolved from viewing images in the laboratory to watching panoramic videos, roaming in a Virtual Reality(VR) environment, and ultimately using real-life scenes. This progression better reflects the influence of space on emotions, significantly enhancing the credibility of the experiments.
- Thirdly, the technological methods used in the aforementioned applications are becoming increasingly diverse and sophisticated. Initially relying on a single EEG technique, these experiments have evolved to incorporate multimodal perception techniques combined with heart rate signals, thereby improving the accuracy of capturing individual emotions in space. Additionally, the subjective feedback mechanism has progressed from simple questionnaires to the Self-Assessment Manikin (SAM) scale, enabling participants to express their emotions more effectively.
- Finally, some of the experimental conclusions mentioned above have become increasingly diverse and flexible in their application. From the study of the relationship between EEG and space emotions for academic purposes, to the post evaluation of real space use, and finally to the prediction of design using experimental conclusions. This design guidance based on emotion data can improve the user experience and satisfaction of the space, ensuring that the designed space environment is more in line with people's emotion needs.

However, these applications also have certain limitations.

- Firstly, there are significant physiological differences between individuals. Even with technical processing, EEG signals cannot completely account for these differences, leading to substantial variation in how certain space elements affect individual emotions. Moreover, emotion responses can be influenced by multiple factors beyond the space environment itself.
- Secondly, there are inherent limitations to the use of EEG devices. Due to their non-invasive nature, EEG signals are highly susceptible to external interference, such as environmental noise and user head movement, which can degrade data quality.
- Additionally, the processing of EEG signals is quite cumbersome. Specialized analysis and interpretation of activity across the entire brain region require substantial computing power, necessitating larger physical devices.
- Finally, there is insufficient coherence in application scenarios. Currently, academic achievements in using EEG technology for space emotion cognition primarily aim

to verify conclusions from previous subjective questionnaire experiments. This approach has not been significantly applied in emerging technology scenarios, resulting in relatively limited effects on improving human well-being.

Considering technological advancements and equipment progress, this article suggests that EEG technology can overcome the aforementioned limitations through several future development directions.

- For the first two limitations, brain-computer interface (BCI) technology may offer a solution. BCI is a revolutionary human-computer interaction technology that establishes a direct communication and control channel between the brain and external devices by capturing brain signals and converting them into electrical signals. In 2023, a depression patient in China had chips implanted in their brains to alleviate symptoms, with successful surgical outcomes. On January 29, 2024, Elon Musk announced on social media that his company, Neuralink, had completed the first human brain device implantation surgery, with preliminary results showing effective detection of neuron-related activity.

In the field of space emotion perception, BCI technology can enhance the personalized recognition of individual emotion states by directly detecting brain electrical activity and emotion-related regions. This offers a new method for personalized emotion regulation. Additionally, BCI technology improves the human-computer interaction experience and addresses the issue of non-invasive devices being significantly affected by external factors. By combining EEG signals with emotion intelligence algorithms, this technology is expected to achieve higher levels of emotion recognition and understanding, leading to the development of more intelligent human-computer interaction systems and emotion assistance tools.

- For the latter two limitations, smart cities may offer a solution. A smart city is an urban development concept that leverages advanced technologies such as information and communication technology (ICT) and the Internet of Things (IoT) to intelligently transform and optimize city management. This approach aims to improve urban operational efficiency, resource utilization, and residents' quality of life.

In the future, with the advancement of brain-computer interface technology and the improvement of EEG signal recognition accuracy, the IoT could evolve into the "Internet of Brains". By integrating EEG signal analysis devices within the city's infrastructure, the necessary computing power and device scalability can be ensured. Moreover, providing personalized services, the application of EEG technology can enable smart city systems to intelligently respond to residents' emotion needs. In transportation and public spaces, it can enhance the city's care and support for its inhabitants via offering customized operational strategies. Additionally, this integration can improve the overall happiness and mental health of urban residents while providing new data sources and decision support for urban management and policy making.

6 Inclusion

This article first analyzes the development process of EEG technology and emotion perception, exploring the relationship between EEG signals and emotions. It then reviews and compares the application of EEG technology in emotion perception across different space scales, proposing directions for improvement. Finally, based on existing research findings, the article summarizes the advantages and limitations of EEG technology in the field of space emotion perception and suggests future development directions. The construction of brain-computer interface technology and smart city systems is proposed as a means to further advance EEG technology in space emotion perception. They can provide faster and more intuitive feedback and guidance for space design, ultimately enhancing residents' quality of life (Fig. 5).

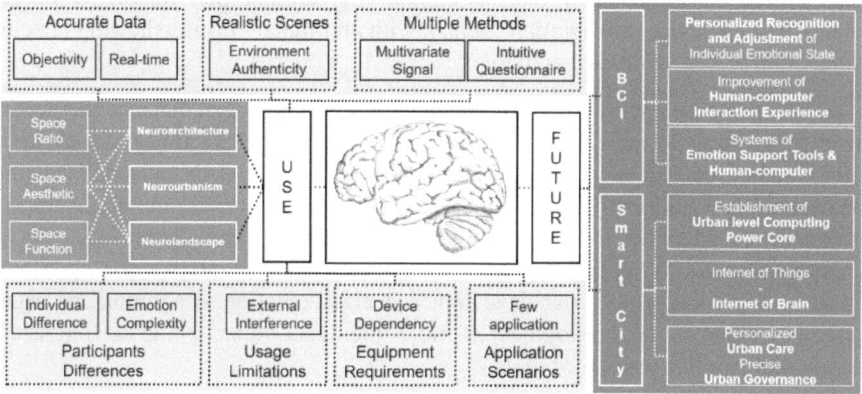

Fig. 5. Current applications and future prospects of EEG technology in the field of space emotion perception

References

1. Tamietto, M., de Gelder, B.: Neural bases of the non-conscious perception of emotional signals. Nat. Rev. Neurosci. **11**(10), 697–709 (2010)
2. Ribas, G.C.: The cerebral sulci and gyri. Neurosurg. Focu **28**(2), E2 (2010). https://doi.org/10.3171/2009.11.focus09245
3. Buhle, J.T., Silvers, J.A., Wager, T.D., Lopez, R., Onyemekwu, C., Kober, H., et al.: Cognitive reappraisal of emotion: a meta-analysis of human neuroimaging studies. Cerebral Cortex **24**(11), 2981–2990 (2013). https://doi.org/10.1093/cercor/bht154
4. Kim, S., Park, H., Choo, S.: Effects of changes to architectural elements on human relaxation-arousal responses: based on VR and EEG. Int. J. Environ. Res. Public Health **18**(8), 4305 (2021). https://doi.org/10.3390/ijerph18084305
5. Cruz-Garza, J.G., Darfler, M., Rounds, J.D., Gao, E., Kalantari, S.: EEG-based investigation of the impact of room size and window placement on cognitive performance. J. Building Eng. **53**, 104540 (2022). https://doi.org/10.1016/j.jobe.2022.104540

6. Banaei, M., Hatami, J., Yazdanfar, A., Gramann, K.: Walking through Architectural Spaces: The Impact of Interior Forms on Human Brain Dynamics. Front. Hum. Neurosci. 11 (2017). https://doi.org/10.3389/fnhum.2017.00477

7. Mirkia, H., Nelson, M.S.C., Abercrombie, H.C., Thorleifsdottir, K., Sangari, A., Assadi, A.: Recognition memory for interior spaces with biomorphic or non–biomorphic interior architectural elements. J. Inter. Des. **47**(3), 47–66 (2022). https://doi.org/10.1111/joid.12224

8. Küller, R., Mikellides, B., Janssens, J.: Color, arousal, and performance—A comparison of three experiments. Color Res. Appl. **34**(2), 141–152 (2009). https://doi.org/10.1002/col.20476

9. Vijayan, V.T., Embi, M.R.: Probing phenomenological experiences through electroencephalography brainwave signals in Neuroarchitecture study. Int. J. Built Environ. Sustain. **6**(3), 11–20 (2019). https://doi.org/10.11113/ijbes.v6.n3.360

10. Neale, C., Aspinall, P., Roe, J., Tilley, S., Mavros, P., Cinderby, S., et al.: The Aging Urban Brain: analyzing outdoor physical activity using the Emotiv Affectiv suite in older people. J. Urban Health **94**(6), 881 (2017). https://doi.org/10.1007/s11524-017-0209-3

11. Reece, R., Bornioli, A., Bray, I., Alford, C.: Exposure to green and historic urban environments and mental well-being: results from EEG and psychometric outcome measures. Int. J. Enviro. Res. Public Health **19**(20), 13052 (2022). https://doi.org/10.3390/ijerph192013052

12. Mavros, P., Wälti, M.J., Nazemi, M., Ong, C.H., Hölscher, C.: A mobile EEG study on the psychophysiological effects of walking and crowding in indoor and outdoor urban environments. Sci. Reports **12**(1) (2022). https://doi.org/10.1038/s41598-022-20649-y

13. Ren, H., Zheng, Z., Zhang, J., Wang, Q., Wang, Y.: Electroencephalography (EEG)-Based comfort evaluation of Free-Form and Regular-Form landscapes in virtual reality. Appl. Sci. **14**(2), 933 (2024). https://doi.org/10.3390/app14020933

14. Herman, K., Ciechanowski, L., Przegalińska, A.: Emotion well-being in urban wilderness: assessing states of calmness and alertness in informal green spaces (IGSs) with Muse—Portable EEG Headband. Sustainability **13**(4), 2212 (2021). https://doi.org/10.3390/su13042212

15. Olszewska-Guizzo, A., Escoffier, N., Chan, J., Yok, T.P.: Window View and the Brain: Effects of floor level and green cover on the alpha and beta rhythms in a passive exposure EEG experiment. Int. J. Environ. Res. Public Health **15**(11), 2358 (2018). https://doi.org/10.3390/ijerph15112358

16. Wang, Y., Xu, M.: Electroencephalogram application for the analysis of stress relief in the seasonal landscape. Int. J. Environ. Res. Public Health **18**(16), 8522 (2021). https://doi.org/10.3390/ijerph18168522

Research on Visual Elements of Spatial Experience in Historical and Cultural Districts Based on Eye Movement Analysis

Yue Cai[1,2], Hui Chen[1,2], Xuange Zhu[1,2], Wenquan Gan[3,4], Bo Liu[1,2], and Haohao Xu[1(✉)]

[1] School of Architecture and Planning, Hunan University, Changsha 410082, China
54879503@qq.com
[2] Hunan International Innovation Cooperation Base on Science and Technology of Local Architecture, Changsha 410082, China
[3] School of Design, Xi'an-Jiaotong Liverpool University, Suzhou 215123, China
[4] School of Environmental Sciences, University of Liverpool, Liverpool L69 3BX, UK

Abstract. Under the background of new urbanization, protection historical and cultural districts is crucial to the urban development. The purpose of this study is to explore the public's visual preference for block space, providing empirical insights and technological innovation for protection and regeneration. A hybrid method of environmental behavior and evidence-based design theory is used to capture the visual behavior of participants in virtual reality through eye tracking technology, revealing perceptual experience and choice preference. Focused on representative historical and cultural districts in Changsha, our investigation delves into the nuanced interplay between design elements imbued with cultural significance such as architectural style, facade materials, color palettes, and street layouts and their capacity to captivate visual attention within these urban enclaves. The analysis highlights the pivotal role of culturally characteristic design elements in engaging visual interest in these districts. Furthermore, data comparison reveals a strong correlation between visual appeal, emotional resonance, and cultural identity. These insights provide a scientific basis for strategies to preserve and regenerate historic and cultural districts. This study quantitatively analyzes the challenges of block space design from a humanistic perspective. In the renewal of historical and cultural blocks, designers need to emphasize the display of historical and cultural elements in buildings and protect historical buildings in order to realize the effective inheritance and activation of blocks.

Keywords: Historical and cultural districts · Eye tracking technology · Mixed-research method · Attractive cultural characteristics · Historical spatial experience

1 Introduction

Historical and cultural districts constitute the core part of urban culture, and also carry the continuation of the value of urban civilization (Xiao et al. 2017). Such blocks have profound cultural heritage, valuable material heritage, unique local customs and exquisite

H. Chai et al. (Eds.): CDRF 2024, *Symbiotic Intelligence*, pp. 468–478, 2025.
https://doi.org/10.1007/978-981-96-3433-0_41

local art (Wang et al. 2012). The protection of historical blocks in the West has been recognized earlier, and has evolved from the protection of a single building to the overall protection of the heritage of urban blocks. In recent years, with the rapid development of urbanization in China, the development of urban space has entered the stage of "stock renewal" from "incremental expansion." In view of the fact that historical and cultural districts are an important part of urban space, the past 'rigid' and 'specimen' protection models need to be optimized. Under the background of new urbanization, the historical and cultural districts have adopted the concept of "people-oriented." Through refined and humanized planning and design, it has become an important research topic in the field of architectural planning to explore how to make the people in the blocks obtain happiness and enhance their sense of identity. The internal cognitive feelings of such individuals are often associated with psychology. In the past, the satisfaction survey of historical and cultural districts was mostly carried out by means of the SD method of psychological experiments (Zhang 2011). However, it is difficult to distinguish the semantic deficiencies in the in-depth study of complex problems. More is still a qualitative analysis, mainly relying on subjective feelings and empirical judgments, lacking objectivity and scientific. With the development of big data and artificial intelligence technology, more scientific emerging technologies have begun to be applied to landscape evaluation research, such as VR, GIS, EEG, eye tracker (Wang et al. 2021). In recent years, eye tracking technology has been widely used by re-searchers in the field of architecture, mainly in urban streets (Chen et al. 2022; Chen et al. 2014), public space (Amati et al. 2018; Fu et al. 2022; Li et al. 2022; Hollander et al. 2020), traditional villages and towns (Ma et al. 2023; Xing et al. 2024), indoor space visual experience (Zhang et al. 2022; Liu et al. 2020; Hamedani et al. 2020), space navigation (Zhou 2016; Sun and Yang 2019). Utilising eye movement data to acquire more precise and scientifically valid research findings contributes to enhancing collective spatial perception, particularly from a "people-oriented" perspective. Nevertheless, it is evident that eye-tracking technology remains relatively underutilised in examining historical and cultural block spaces within academic research. Thus, to fill this gap, this research utilise Chaozong Street in Changsha City as a case study, integrating quantitative analysis employing eye tracking technology. By immersing participants in a panoramic representation of the block space, the study tracks their eye movement indices to extract quantitative data. Subsequently, these datasets are synthesised to derive quantified preferences regarding spatial visual elements within historic and cultural districts.

2 Materials and Methods

2.1 Research Framework

This study uses eye tracking technology to explore the design methods of historical and cultural blocks. Firstly, we collected panoramic photos of historical blocks, and then carried out eye movement experiments. Through eye movement software, the research team collected relevant data such as the total fixation time and the first fixation time of the subjects in the panoramic environment of the block. After that, the results of the eye movement experiment are analyzed, and the conclusion about the preference of specific visual elements in historical and cultural blocks is drawn (Fig. 1).

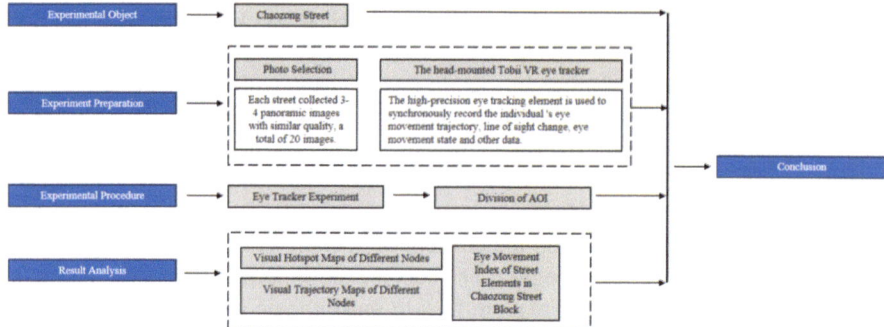

Fig. 1. Research Framework

2.2 Survey Development Process

2.2.1 Photo Selection.

Under the same time, weather conditions and imaging technology, we selected five main streets, and each street collected 3–4 panoramic images with similar quality, a total of 20 images. On the basis of ensuring the consistency of shooting weather and shooting height, the three most representative street images are selected from the sample images. Each photo contains the most representative historical and cultural elements and modern elements of the block. In order to maximize the simulation of the real scene of Chaozong Street seen by the public, the sample uses panoramic images (Fig. 2).

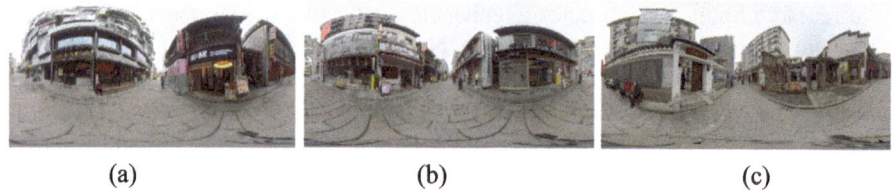

(a)   (b)   (c)

Fig. 2. Chaozong Street Block Selection

2.2.2 Data Collection Method for Eye-Tracking

In this experiment, the head-mounted Tobii VR eye tracker was used to present the main streets of Chaozong Street to the subjects through panoramic photos. The sampling frequency of the eye tracker is 250 Hz, and the high-precision eye tracking element is used to synchronously record the individual's eye movement trajectory, line of sight change, eye movement state and other data. In order to avoid the influence of other factors, the experiment was carried out in the eye movement test laboratory to reduce the environmental noise (less than or equal to 40 dB) and the interference of other personnel. After the participants agree to the experiment, are familiar with the experimental environment and complete the wearing under the guidance, they stand in the experimental area with no obstacles and free movement around them, and the experimenter explains the experimental process and requirements to the participants. Before the beginning of the experiment, the subjects will be calibrated at five points, and then watch a panoramic photo sample similar to the experimental street. At the beginning of the formal experiment, the participants would watch three panoramic photos played in random order through the eye tracker, each of which was watched for 30 s, and 5 s black screen was set in the middle of each picture to relieve eye fatigue and concentrate.

2.3 Data Collection

2.3.1 Research Area Selection

The research object is Chaozong Street, which is located in Kaifu District, Changsha City, Hunan Province. This block is rich in visual landscape elements, including historical and cultural elements and modern elements: commercial buildings, ancient buildings, street design, etc. It is an important place for the coexistence of modern and historical sense. This experiment focuses on the public's preference for visual elements in historical blocks, and selects several important streets with relatively large flow of people. These streets are areas where cultural protection buildings and modern commerce are combined, and the visual elements are rich and complete. It is helpful to explore more attractive visual elements in historical and cultural districts.

2.3.2 Participant Recruitment

In this experiment, 12 undergraduate, postgraduate and doctoral students from Hunan University were selected, including 7 males and 5 females, aged between 22 and 28 years old. There are 5 participants whose majors are related to architecture and 7 non-architectural majors. Participants were the first time to see the experimental samples, with a certain aesthetic ability, and uncorrected visual acuity or vision correction is normal, visual imaging color is normal.

2.3.3 Final Survey Development

Tobii Pro Lab software was used to process eye movement data, and the required data was extracted after dividing the interest area. In order to better analyze the public's visual preference and attraction for spatial elements when they are in historical blocks, we selected 6 eye movement indicators and divided the block elements into 3 categories with a total of 11 visual elements (Table 1).

In eye tracking research, areas of interest (AOIs) are based on eye movement behavior data, which are used to quantitatively analyze objects or regions of interest to users. Through AOI, researchers can deeply understand user concerns, interest points and visual trajectories, and optimize content design. This study defines specific areas and divides street elements into building skins, structures and decorative elements. Architectural skin design plays an important role in historical and cultural blocks, covering ancient architecture and modern cyber-punk style. Structures are secondary in block design, including railings, eaves, etc. Decorative elements are reflected in the store style (Fig. 3).

Table 1. Interpretation of the Specific Meaning of Eye Movement Index

Eye movement indicators	Abbreviation	Specific meaning of indicators
Total duration of fixations	TFD	It refers to the total fixation time in the region of interest (AOI)
Average duration of fixations	AFD	It refers to the average fixation time in the region of interest (AOI)
Number of fixations	FN	It refers to the number of times people view an area
Time to first fixation	TFF	It refers to the moment when you first see the AOI in the picture
Duration of first fixation	FFD	It refers to the duration of the first fixation on AOI
Average amplitude of saccades	ASA	The number of saccades occurring during the interval
Number of saccades in AOI	SNA	The number of saccades occurring during the interval
Average peak velocity of saccades	ASPV	The average peak velocity of all saccades in this interval

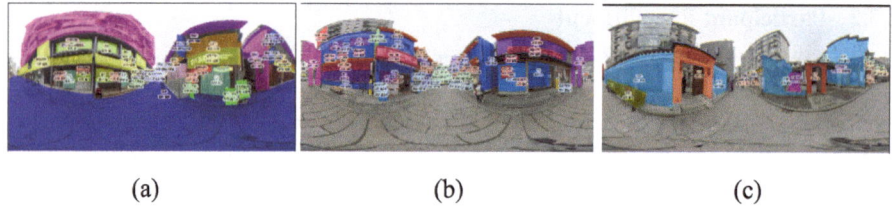

(a)   (b)   (c)

Fig. 3. Elements of Interest Zoning

3   Results

3.1   Description of the Results

According to the divided AOI, we collected data from different streets in the same block, and collected 12 eye movement data from each street. The absolute fixation time of each street is visualized to generate a heat map.

In the heat map, red and green are used to represent the length of the absolute fixation time of the subjects in the street scene. The redder the fixation area indicates that the subjects have a longer absolute fixation time in the area, and the green indicates that the subjects are in this area. The absolute fixation time is relatively short, and the part without color may not reach the recorded threshold or the subject has less interest in this area (Fig. 4).

According to the heat map, it can be seen that the public's visual focus in the street space is mainly on the architectural shop signs and facades. At the same time, the visual

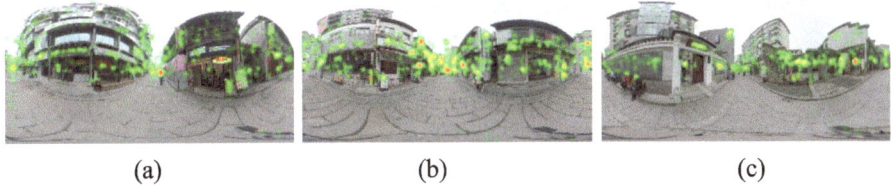

(a)   (b)   (c)

Fig. 4. Visual Hotspot Maps of Different Nodes

trajectory map is generated in the software. The digital logo is used in the trajectory map to mark the attractive area of the street space to the public for the first time, so as to understand the street space elements that the public pays attention to for the first time in the street, which is conducive to the design of the first attractive space (Fig. 5).

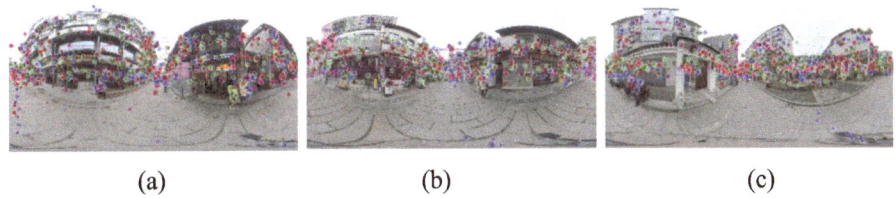

(a)   (b)   (c)

Fig. 5. Visual Trajectory Maps of Different Nodes

In the visual trajectory map, it can be seen that the area that the public first looks at in the space of the historical and cultural block is more part of the store sign and the building facade. Through the heat map and visual trajectory map, the most direct visual preference results can be obtained from the image itself, that is, the elements that the public is more interested in in the historical streets of Chaozong Street are building facades and commercial store signs. However, it is impossible to accurately understand the specific indicators of specific elements only through image results, so it is necessary to export the data of eye movement indicators to obtain specific results.

3.2 Built Environment and Participant Preferences Association

After the eye movement experiment, we integrated the data of 12 participants, selected 6 data types, divided 11 interest areas, and obtained relevant data in the corresponding interest areas. From the well-organized results table, it can be seen that the brick wall is the highest area of interest for TFD, FN, and SNA; the vertical store sign is the highest interest area of AFD and FFD. The building railing is the earliest interest area of TFF (Table 2).

Comparing different elements of the same property, it can be found that:

The TFD, FN, TFF and SNA of brick walls on buildings are larger than those of stainless steel exterior walls. From this point of view, it can be known that the public in historical and cultural districts is more interested in brick walls with ancient architectural features than walls with modern elements. In the building, the lowest TFF is the railing,

Table 2. Eye Movement Index of Street Elements in Chaozong Street Block

AOI	TFD (ms)	AFD (ms)	FN	TFF (ms)	FFD (ms)	SNA
Windows	294.50	93.42	1.86	7818.24	153.19	0.22
Doors	727.58	131.81	4.58	8082.34	137.84	1.42
Balustrade	357.21	79.92	2.33	**1339.80**	146.73	0.42
Eaves	452.75	115.64	2.75	11846.23	145.46	0.67
Brick Wall	**3005.03**	135.42	**18.69**	7885.75	123.36	**7.75**
Stainless Steel Wall	1724.30	138.00	11.66	5099.17	143.83	4.67
Shop Signs - Wall Type	260.08	84.88	1.88	12698.73	132.93	0.50
Shop Signs - Transverse Type	2214.21	127.96	14.33	2331.30	136.30	4.95
Shop Signs - Vertical Type	586.25	**140.25**	3.50	16445.62	**159.71**	1.33
Shop Signs - Suspension Type	509.21	116.67	3.08	11379.26	147.95	0.88
Road	388.46	70.08	5.16	11962.75	125.75	0.58

and the highest is the eaves. It can be seen that the first attention is more attractive to the railing, and the interest for the eaves is lower (Table 3).

Table 3. Eye Movement Index of Architectural Elements in Streets

AOI	TFD (ms)	AFD (ms)	FN	TFF (ms)	FFD (ms)	SNA
Windows	294.50	93.42	1.86	7818.24	**153.19**	0.22
Doors	727.58	131.81	4.58	8082.34	137.84	1.42
Balustrade	357.21	79.92	2.33	**1339.80**	146.73	0.42
Eaves	452.75	115.64	2.75	11846.23	145.46	0.67
Brick wall	**3005.03**	135.42	**18.69**	7885.75	123.36	**7.75**
Stainless Steel Wall	1724.30	**138.00**	11.66	5099.17	143.83	4.67

In the part of store signs, the AFD and FFD of vertical store signs are higher than other store signs, while the TFD, FN and SNA of horizontal store signs are higher than other store signs. The TFF of the horizontal store sign is the earliest, which indicates that the public is more likely to see the horizontal store sign when they first enter the block, and the total time spent on the horizontal store sign will be longer. It can be seen that the horizontal store sign is more attractive than the vertical store sign (Table 4).

Table 4. Eye Movement Index of Commercial Elements in Blocks

AOI	TFD (ms)	AFD (ms)	FN	TFF (ms)	FFD (ms)	SNA
Wall Type	260.08	84.88	1.88	12698.73	132.93	0.50
Transverse Type	**2214.21**	127.96	**14.33**	2331.30	136.30	**4.95**
Vertical Type	586.25	**140.25**	3.50	**16445.62**	**159.71**	1.33
Suspension Type	509.21	116.67	3.08	11379.26	147.95	0.88

Compare the importance of historical building elements and commercial elements in the neighborhood:

Comparing the building facades and shop signs, it is found that the commercial signs in the block are more attractive when the public enters the block space for the first time, but more attention is paid to the historical buildings in the process of block space action. This result may be due to the fact that the commercial billboards of the block are usually designed to be more prominent, with bold color matching and clear words to inform the public about the content of the store. The main key to whether a block is more attractive is whether there is an architectural style to express the atmosphere of the block within the block. It also reflects the public's attention to historical buildings (Table 5).

Table 5. Eye Movement Indexes of Architectural Elements and Commercial Elements in Blocks

AOI	TFD (ms)	AFD (ms)	FN	TFF (ms)	FFD (ms)	SNA
Architectural Element	1093.56	115.70	6.98	1339.80	141.74	2.53
Commercial Elements	892.44	117.44	5.70	2331.30	144.22	1.92

4 Discussions

4.1 Attraction Characteristics and Regeneration Strategies

Through this experiment, the elements that are more attractive to the public in the historical and cultural streets are obtained. It is found that the important factors to improve the attractiveness of the historical blocks are:

1. Architectural style: The design of historical and cultural protection buildings and ancient architectural styles has obvious visual appeal in historical and cultural districts.
2. Special structures: the design of railings or more prominent structure and shape.

3. Decorative elements: Improving the design sense of store signs on the street facade can attract more public attention.

Based on the experimental results, there are the following ways to increase the attractiveness of the blocks:

1. The protection and utilization of the original historical buildings in the historical block, the historical buildings in the block are an important carrier of historical culture, and these historical buildings give historicity and attraction to the block.
2. Create the architectural style of the block, pay attention to the unified shaping of the overall block style, and reasonably integrate the elements of historical and cultural blocks in modern architecture.
3. The design of the store's design, reflecting the characteristics of the store itself while not lacking the style elements of the block.

4.2 Limitation and Further Research Agenda

The limitations of this experiment are mainly reflected in the following aspects:

1. Limitations of eye movement data: It can only understand that these areas are more attractive to the public, and the reasons for deep preferences cannot be indicated.
2. The lack of experimental data: should be as much as possible to carry out the experiment, to reduce the error, increase the accuracy of the experiment. In the selection of experimenters, people of different types and ages should be added to better summarize the public's visual preferences.

Using eye trackers to quantitatively analyze the public's visual preferences in the space of historical and cultural blocks, the research team hopes to add more physiological detection methods in the future, not limited to visual appeal. At the same time, the public experiment requires a more realistic experience. The sub-sequent experiments can use the means of scene model construction, so that the experimenter can move freely in the scene, not limited to a specific panoramic image. In the future, more abundant means will be used to collect data, restore spatial scenes more realistically, and obtain more accurate data.

5 Conclusion

This study uses the eye tracking technology of VR eye tracker to study the visual perception preference of historical and cultural block space in real environment. Through visual preference, it is found that historical architectural elements play an important role in the space of historical and cultural blocks.

The research methods and results can be further applied in the following aspects:

1. Renew and protect the old historical and cultural blocks, enhance the protection of historical buildings in the blocks, create the historical architectural style of the whole block, and promote the vitality of the blocks.

2. Block planning designers more comprehensively understand people's visual and identity needs for the spatial experience of historical and cultural blocks, formulate targeted 'people-oriented' design strategies, and pay more attention to creating spatial feelings at the human scale to improve design and promote the inheritance and revitalization of historical and cultural space.

Fund Projects. Hunan Provincial Key Area R&D Program - Key R&D (2021WK2004), Hunan Provincial Social Science Achievement Review Committee Subjects: Study on the Mechanism of Digital Protection and Utilization of Hunan Revolutionary Heritage (XSP2023YSC065).

References

Amati, M., Ebadat, G.P., McCarthy, C., Sita, J.: How eye-catching are natural features when walking through a park? Eye-tracking responses to videos of walks. Urban For. Urban Green. **31**(1), 67–78 (2018)

Chen, C., Li, H., Lin, X., Xia, Y.: Research on the optimal design of urban street landscape based on emotional assessment. Time Build. **05**, 46–51 (2022). https://doi.org/10.13717/j.cnki.ta.2022.05.016

Chen, Y., Zhao, X.: Research on the bottom interface of streets from the perspective of pedestrians-a case study of Huaihai Road in Shanghai. City Plan. Rev. **38**(06), 24–31 (2014)

Fu, E., Wang, Y., Zhou, J., Li, X.: Eye movement analysis for restorative environmental assessment of community parks. South. Archit. **06**, 93–99 (2022)

Hamedani, Z., Solgi, E., Hine, T., Skates, H.: Revealing the relationships between luminous environment characteristics and physiological, ocular and performance measures: an experimental study. Build. Environ. **172**, 106702 (2020)

Hollander, J.B., Sussman, A., Levering, A.P., Foster-Karim, C.: Using eye-tracking to understand human responses to traditional neighborhood designs. Plan. Pract. Res. **35**(5), 485–509 (2020)

Li, X., Wang, Z., Wu, D., Tan, L., Lin, Q.: Study on the stress relief of urban riverside trails on young people based on physiological feedback. Chin. Landsc. Archit. **38**(05), 86–91 (2022). https://doi.org/10.19775/j.cla.2022.05.0086

Liu, C., Kang, J., Xie, H.: Effect of sound on visual attention in large railway stations: a case study of St. Pancras Railway Station in London. Build. Environ. **185**, 107177 (2020)

Ma, L., Xu, Y., He, S., Tang, Z., Li, T.: Research on cognitive differences of traditional village landscape based on eye movement experiment. Urban Dev. Stud. **30**(02), 86–94 (2023)

Sun, C., Yang, Y.: Visual saliency of pathfinding markers based on eye tracking-a case study of Harbin Kaide Plaza Shopping Center. Archit. J. **02**, 18–23 (2019)

Wang, F., Yan, L., Xiong, X., Wu, B.: A study on the urban memory of historical sites based on tourists' cognition: a case study of Nanluoguxiang historical site in Beijing. Acta Geogr. Sin. **67**(4), 545–556 (2012)

Wang, Q., Zhao, M., Zhao, L.: Visual analysis of landscape aesthetic evaluation based on bibliometric methods. Am. J. Civ. Eng. **9**(3), 63–69 (2021)

Xiao, J., Cao, K.: Research review, technical methods and key issues of the protection of historical blocks. Urban Plan. Forum **03**, 110–118 (2017). https://doi.org/10.16361/j.upf.201703013

Xing, Y., Leng, J.: Evaluation of public space in traditional villages based on eye tracking technology. J. Asian Archit. Build. Eng. **23**(1), 125 (2024)

Zhang, L., Li, X., Li, C., Zhang, T.: Research on visual comfort of color environment based on the eye-tracking method in subway space. J. Build. Eng. **59**, 105138 (2022)

Zhang, Y.: Study on Urban Spatial Perception Based on SD Method. Tongji University (2011)

Zhou, X.: Application of Eye Tracker in Landscape Design and Park Guide Map. Zhejiang University (2016)

Immersive AR-Assisted Assemblies
for Self-building Strategies

Teresa Han, Qi Wang, Zhiyong Dong, and Peter Búš[✉]

Institute for Future Human Habitats, Tsinghua Shenzhen International Graduate School,
Tsinghua University, Shenzhen, China
t34han@uwaterloo.ca, {wangqi24,dongzy23}@mails.tsinghua.edu.cn,
peter_bus@sz.tsinghua.edu.cn

Abstract. The research aims to develop a framework to address needs to improve design to production workflow efficiency in the context of high-density urban environments through the application of advanced digital tools. The work primarily focuses on delivering the work through augmented reality (AR) driven assembly processes.

Due to its modularity, affordability, sustainable building practices and accessibility, the research presents the self-build home as a design scheme to demonstrate such technologies. In the design proposal, an urban village rooftop intervention was proposed, adopting the open-source WikiHouse Skylark 250 modules as the design scheme for the self-build home. An AR immersive space was developed, where assembly sequence guidance was provided to aid potential self-builders to understand the building assembly process and safely, accurately and efficiently conduct on-site assemblies.

The outcomes of the research provide lasting implications to the advancement of AR-assisted production workflows, and participates in ongoing urban village regeneration efforts which supports community revitalization, increases housing opportunities and maintains its social fabric.

Keywords: Augmented reality · Self-builder · Self-assembly · AR-assisted Assembly · Urban Village China

1 Introduction

The current age of digital tools offers the opportunity to leverage its benefits to address issues of skilled labour shortage, workflow inefficiency and human errors of construction in the architecture, engineering, construction and operation (AECO) industry. This research addresses the needs to improve design to production workflows in the context of high-density urban environments through the application of augmented reality (AR) driven assembly processes.

AR lenses have the potential to play a pivotal role in on-site construction activities by providing workers with hands-free access to important information, instructions, and guidance. Popularized existing AR workflows primarily use AR as a means of onsite

© The Author(s) 2025
H. Chai et al. (Eds.): CDRF 2024, *Symbiotic Intelligence*, pp. 479–489, 2025.
https://doi.org/10.1007/978-981-96-3433-0_42

construction support by visualizing component locations or design phase visualizations (Delgado, et al., 2020; Wang, et al., 2023). Users will be prompted to identify and match real-world objects to the virtual object, and follow assembly sequence instructions to complete the production. The proposed framework aims to aid self-builders in building applications, to understand assembly sequences and safely, accurately and efficiently conduct on-site self-assemblies. In its essence, such a workflow tends to create a sustainable and educational environment for knowledge sharing and skill development.

Furthermore, the project aims to illustrate the prospective application of AR-assisted assemblies through a self-build project in the urban village setting of southern China, as a means to explore a potential opportunity for a possible strategy to revitalize urban villages.

1.1 Research Question

In order to achieve the research objectives, the following research works investigate the question: is the aid of AR technology in assembly scenarios a practical and applicable method to guide a non-expert user to the desired built scenario?

1.2 Research Methodology

The research investigates potentials for improvements in design to production workflows through a combination of theoretical and experimental approaches. The approach is divided into two phases. Phase 1 is a design phase consisting of the WikiHouse-based modular scheme application using Rhino3D as well as an immersive AR design space integration in Unity, using Microsoft Mixed Reality Toolkit 3 (MRTK3) (Microsoft, 2023). MRTK3 is an excellent resource for creating user-friendly interactions in the AR space. A sample study model will be created in Rhino, to be imported into the AR platform.

In Phase 2, the environments and models developed in Phase 1 will be used to perform a 1:10 scale assembly with the aid of Hololens 2. Hardware includes an AR headset, the Microsoft Hololens 2. A PC is used for deploying the program and debugging. All devices are connected to one common network environment. Following the self-assembly step, final conclusions and evaluation on the framework will be determined in Phase 2, based on the intuitiveness and feasibility of the workflow (Fig. 1).

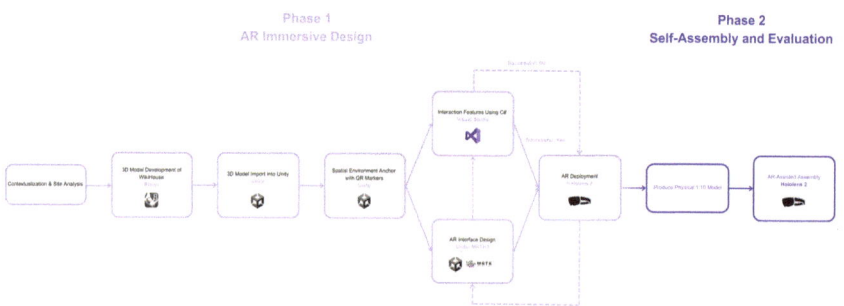

Fig. 1. AR framework development workflow

2   The Self-build Home as a Design to Production Approach for High Density Urban Environments

2.1   Urban Village Case Study

The urban village of southern China was selected to explore potentials for approaching existing site conditions and design problems. The urban village are a unique phenomenon, which pose as an opportunity to showcase the use AR-assisted assemblies by self-builders.

In the 1980s, China began unprecedented urbanization in a play to modernize cities, as such, previously underdeveloped cities experienced a series of rapid land, housing, and fiscal policy reforms. The phenomena of urban villages emerged as the rural villages surrounding cities were engulfed by the urban expansion. In present day, they occur in scattered plots all around the middle of the city, leading to a unique blend of rural and urban characteristics (Wu et al. 2022; Yuan et al. 2024). Despite their informal nature, these villages play a crucial role in providing affordable housing and diverse employment opportunities for migrant workers and low-income residents (Chen et al. 2016; Wang 2021; Wu et al. 2022).

Due to rapid urbanization efforts for megacities in China, these communities face the looming issue of urban gentrification, leading to the displacement of urban village residents and the loss of affordable housing options (Yuan et al. 2024). Urban village regeneration, referring to revitalizing urban villages while preserving their cultural heritage and social fabric, has been well recognized as having profound impacts to various social groups and urban environments. In addition, it was often found to result in improved land use efficiency and have positive effects on the housing (Lin et al. 2022). Through redevelopment projects, urban villages can be transformed into vibrant, sustainable, and inclusive neighborhoods that do not hinder the urbanization of the cities that hold them.

2.2   The Self-build Home

The self-build home (SBH) is a modern movement where individuals or groups arrange for the building of their own dwelling and, in various ways, participate in its production (Benson and Hamiduddin 2017). While they require careful planning and dedication, self-build projects offer a range of benefits and opportunities for builders in highly constrained, dense, urban environments, such as urban villages.

The SBH addresses issues regarding lack of affordable housing, particularly for low to middle-income groups in developed nations who have the means of acquiring homeownership and establishing formal settlements (Obremski and Carter 2019). Other benefits of a SBH include a greater degree of freedom regarding design and customization (Bossuyt 2021), considerable psychological benefits and social phenomenon (Bossuyt 2021; Broer and Titheridge 2010), potential cost-savings when carried out effectively (Benson and Hamiduddin 2017; Broer and Titheridge 2010).

A prime example of a successful self-build system is the WikiHouse, established by Open Systems Lab (2023). There are a wide range of benefits associated with adopting the WikiHouse-based system for a self-builder in an urban village. Firstly, the WikiHouse is particularly appealing in informal contexts, as the modular nature of the WikiHouse makes it rapidly deployable. It is also a highly accessible and inclusive building solution as it can be built without the need for specialized skills or extreme equipment (Esenarro et al. 2024). Furthermore, WikiHouse is an affordable and cost-effective housing solution freely available for anyone to use, modify, and share (Esenarro et al. 2024). Lastly, the WikiHouse promotes sustainable building practices by emphasizing the use of locally sourced, renewable materials and minimizing waste during the construction process (Esenarro, et al. 2024; Open Systems Lab 2023).

2.3 Design Proposal for an Urban Village Case Study

The paper proposes WikiHouse-based buildings on the rooftop of existing buildings in urban villages to create a micro-scale intervention, with the intent of adopting an AR-assisted assembly for a self-builder in an urban village setting. A cluster of 9 buildings are proposed to create a micro-scale intervention of a rooftop community, creating a space where old town traditions can collide with urban youth.

The proposed buildings may be adopted for both residential use and commercial use such as storefronts or public open spaces. As with previous regenerative efforts in Shenzhen, the proposal aims to preserve the original life of the community through collaging spaces, rather than replacement. The proposal assumes that a full site analysis has been performed on the buildings, and sufficient structural rework and servicing has been completed (Figs. 2 and 3).

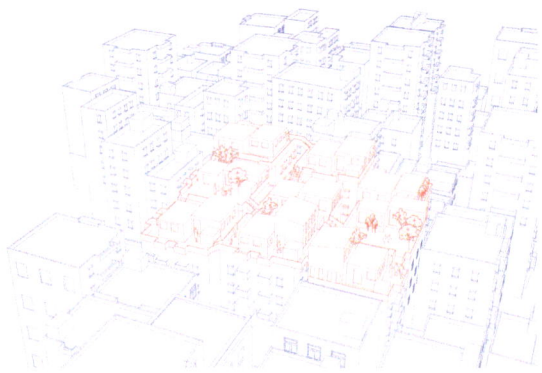

Fig. 2. Design intervention in the urban village scenario

3 AR Immersive Environment

The following section describes the details and processes to develop the AR immersive environment in Unity. The project employed the use of MRTK3, the third generation of Microsoft Mixed Reality Toolkit for Unity. It is an open-source project available for mixed reality development, containing excellent built-in interaction models and UX building blocks for the development of an AR immersive space (Microsoft 2023). Videos of the AR environment can be found on Youtube (2024). A storyboard is shown in Fig. 4 to visualize the AR self-assembly workflow.

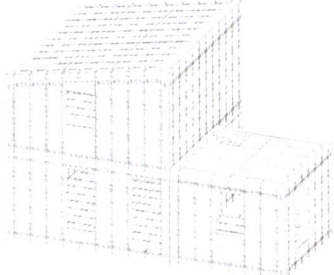

Fig. 3. Sample WikiHouse model used for AR environment

Workflow for AR Self-Assembly

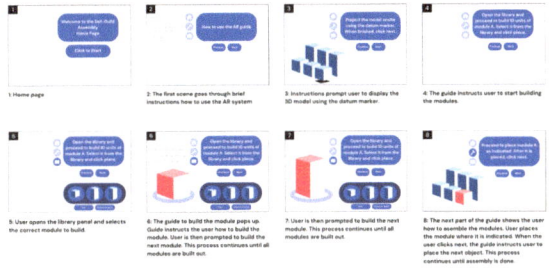

Fig. 4. AR self-assembly workflow

3.1 AR Environment Features

The following section describes the key features and processes created in the AR environment to provide the user with an intuitive and logical virtual guide.

3.1.1 Windows QR Code Tracking for Matching Virtual to Real-World Objects

Through the use of the Windows Mixed Reality QR code tracking ability (Microsoft 2024), the Hololens 2 is able to detect QR codes within view of the camera. After the QR Code is recognized and scanned, a coordinate system is established at the devices world-space location. For the developed AR environment, this technique is used as a tool for an event trigger to generate a virtual component which is superimposed onto a physical object, which aids in creating a link between virtual and real-world objects. A diagram can be seen in Fig. 5 to illustrate the process.

3.1.2 Instruction Prompts

The instruction prompts are a dialog box that guides the user through the entire AR environment, be it opening a new feature in the interface, indications to build certain WikiHouse modules etc. A welcome message is first displayed, whereafter upon button click of the "Previous" and "Next" buttons, the text will display the previous and next

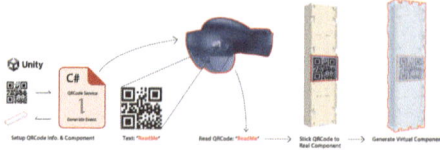

Fig. 5. QR code tracking for superimposing virtual objects onto real-world objects

instructions respectively. The instructions are called from an Excel text file. The figures below illustrate this function (Fig. 6).

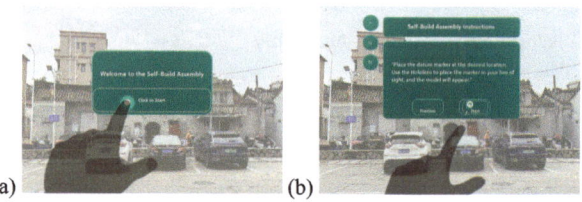

(a)   (b)

Fig. 6. Instruction prompts demonstration on Hololens 2 (a) A pop-up dialog box welcomes the user into the application (b) User clicks next to view the next line of instructions

3.1.3 Module Library

The module library is a scrollbar which holds each of the modules of the WikiHouse. It is displayed as a library of items where the user can scroll through to select their desired building module. Once the desired module is selected, it appears in front of the user in an animation. The figures below show a demonstration of the library system as described (Fig. 7).

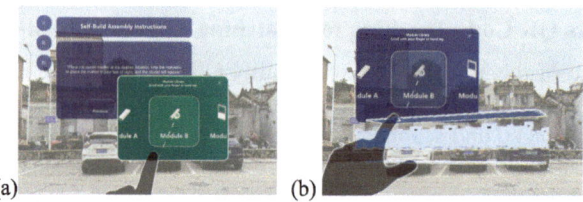

(a)   (b)

Fig. 7. Module library demonstration on Hololens 2 (a) User opens the scrollable module library (b) Once the desired module is selected, it will appear before the user in an animation

3.1.4 Assembly Sequence

The assembly sequence is a feature which walks the user step-by-step through the assembly of the WikiHouse, therefore it takes the modules and joins them together, as illustrated in Fig. 8.

Fig. 8. Proposed assembly sequence of a sample WikiHouse model

The available functions are again a "Display Previous" and "Display Next" button, which show each subsequent previous or next module in the assembly sequence respectively. Each current module to be placed is highlighted in a yellow color, making the next module that must be placed in the assembly sequence evident to the user (Fig. 9).

Fig. 9. Assembly sequence demonstration on Hololens 2 (a) User opens the assembly sequence (b) User clicks "Display Next" to view the next assembly step (c) User continues to build until the assembly is finished

3.2 Final AR-Assisted Assembly with a Scaled WikiHouse-Based Scenario

A final parts assembly was carried out using a 1:10 model of the WikiHouse-based scenario, produced using 2 mm acrylic boards and a laser cutting machine. All of the WikiHouse modules were preassembled prior to following the AR assembly sequence guidance. After the model was placed in the environment using the referenced QR marker, as seen in Fig. 10, the assembly of the WikiHouse model was carried out.

Once the AR environment was entered, the interface indicated the user to begin the first component for assembly. After each indicated component was placed, the user would click next and continue the process, as shown in Fig. 11.

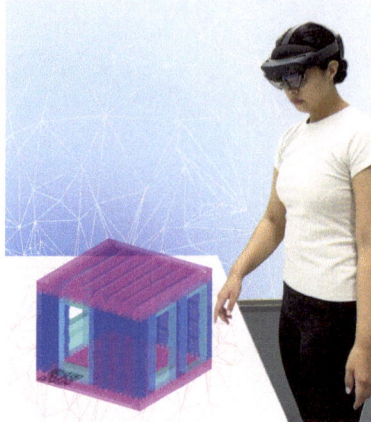

Fig. 10. Model placement in environment with QR marker

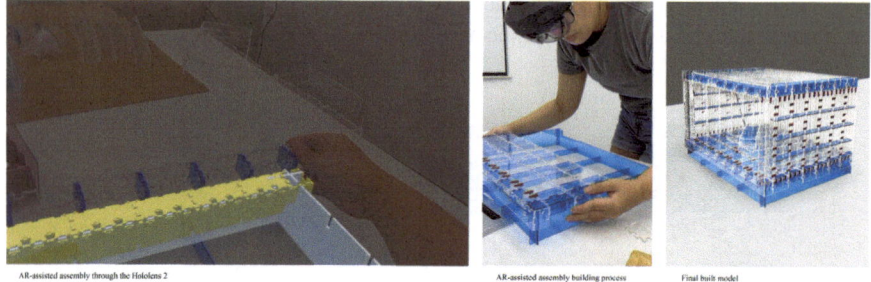

Fig. 11. AR-assisted assembly of a scaled WikiHouse with the Hololens 2

4   Evaluation, Results and Discussion

The final outcome of the research was an AR environment which provides a builder with intuitive instructions and visuals to assemble modular WikiHouse components. Through the qualitative experience of an AR- assisted assembly with a 1:10 WikiHouse sample model, it was concluded that there is indeed potential for a self-builder to learn and build with the guidance of AR technology. The qualitative observation and evaluation followed the criteria of deliverability, constructability, the level of difficulty to operate with the technology in a physical space (operability) and the overall quality of spatial experience. The logic and instructions provided by the AR environment certainly aided in the production of the WikiHouse model. Although there is identified potential in which AR technology can benefit the flow of production, further investigations can test and validate the potential through user testing and real-scale experiments and observations by non-expert builders.

Building design can be integrated with AR technology by importing the design into the environment and coding the assembly sequence. However, it can be seen from the research that the integration is limited to modular designs. The workflow at the current stage of this research is a hard-coded system, and modular components are capable of being used with AR assemblies due to their repetitive nature. Rising research of machine learning in AR provides the potential for further generative developments of building scenarios.

Furthermore, the current stage, the AR environment is majority a one-way interaction system, where the user is interacting with the AR environment. Future development involves creating a context-aware system, where the Hololens can recognize objects, or confirm their location in real-time, and thus create a communication line between the virtual design and real-world objects. This can be done through further use of ArUco markers, QR codes or AI object tracking.

Lastly, the proposal of the AR-assisted assembly must address the technological barriers that parallel the use of advanced technological tools for self-builders. A digital divide exists where the proposal of high end and technical tools seemingly contradicts the anticipated end-users: low-income, unskilled professionals. For widespread adoption of AR in self-building, particularly for low-income communities, government or local community support will be required. Furthermore, training will be required for the users prior to any assembly process, thus providing them with the skills to operate such technologies.

5 Conclusions

The following research partially addresses the needs to improve design to production workflow efficiency in the context of high-density urban environments through the application of advanced digital tools, such as AR driven assembly processes. Although the answer of the research question stated in the Sect. 1.1 is not fully answered yet, due to its modularity, affordability, sustainable building practices and accessibility, the research presents the self-build home, in particular the system based on the WikiHouse Skylark modules, as a design scheme to demonstrate such technologies and to be potentially used in 1:1 built scenario. Urban village rooftops were identified as an opportunity to showcase the use of the investigated digital tools.

An AR immersive space was developed, where the design concept was imported into the AR space. Through programming, an assembly sequence was provided too, following the intention to provide an aid to self-builders. As such, they can understand the building assembly process and safely, accurately and efficiently conduct on-site assemblies. The outcomes of the research provide lasting implications for the advancement and digitization of design to production workflows and participate in ongoing urban village regeneration efforts, which support community revitalization, increase housing opportunities, and maintain its social fabric.

Future improvements and works involve extension of a proposed mixed reality setting, where the Hololens 2 device can recognize objects and thus create a communication between the virtual design and real-world objects. This is intended to be applied in the same MRTK3-driven workflow. The next step for the research is a 1:1 assembly of the proposed WikiHouse-based scenario in the real-world setting, set up as a community-driven project adopting the AR framework presented in this paper.

The design achieved a framework which extends human capabilities and creates interdisciplinary workflows related to the AECO industry. Such research can surely provide great implications to the technological advancement and digitization of the AECO industry, and be applied in various setting from the small-scale self-building intentions to large-scale site operations.

Acknowledgements and funding information. The research presented in this paper is a part of the research project titled "高密度城市环境下的混合协同智能和人工智能驱动的建筑设计及机器人建构方法研究 (Research on hybrid collaborative intelligence and artificial intelligence-driven architectural design and robotic construction methods in high-density urban environments-HIRA)," grant n. 20231129094641002, 稳定支持 as the Stability support program, funded by 深圳市科技创新委员会 (Shenzhen Science and Technology Innovation Commission). The authors' affiliation institution, Tsinghua Shenzhen International Graduate School supports the research under the Scientific Research Start-up Funds, grant n. 002023009C, and the Shenzhen Pencheng Peacock Specific Program.

References

Benson, M., Hamiduddin, I.: Self-Build Homes, s.l. UCL (2017)

Bossuyt, D.M.: The value of self-build: understanding the aspirations and strategies of owner-builders in the Homeruskwartier, Almere. Housing Stud. **36**(5), 696–713 (2021)

Broer, S., Titheridge, H.: Eco-self-build housing communities: Are they feasible and can they lead to sustainable and low carbon lifestyles? Sustainability **2**(7), 2084–2116 (2010)

Chen, M., Liu, W., Lu, D.: Challenges and the way forward in China's new-type urbanization. Land Use Policy **55**, 334–339 (2016)

COIA: Youtube (2024). [Online]. Available at: www.youtube.com/@iFFHs-COIA

Delgado, J.M.D., Oyedele, L., Demian, P., Beach, T.: A research agenda for augmented and virtual reality in architecture, engineering and construction. Adv. Eng. Inform. **45**, 101122 (2020)

Esenarro, D., et al.: Use of digital tools (Wikihouse system) in multi-local social housing. Sustainability **16**(8), 3231 (2024)

Lin, J., Lai, Y., Chen, K., Tang, X.: What drives urban village redevelopment in china? A survey of literature based on web of science core collection database. Land **11**(4), 525 (2022)

Microsoft: Mixed Reality Toolkit 3 (2023). [Online] Available at: https://learn.microsoft.com/en-us/windows/mixed-reality/mrtk-unity/mrtk3-overview/

Microsoft: QR code tracking overview (2024). [Online] Available at: https://learn.microsoft.com/en-us/windows/mixed-reality/develop/advanced-concepts/qr-code-tracking-overview

Obremski, H., Carter, C.: Can self-build housing improve social sustainability within low-income groups? Town Plan. Rev. **90**(2), 167 (2019)

Open Systems Lab. Wikihouse (2023). [Online]. Available at: https://www.wikihouse.cc/

Wang, J., Ma, Q., Wei, X.: The application of extended reality technology in architectural design education. Buildings **13**(12), 2931 (2023)

Wang, Y.P.: Urban villages, their redevelopment and implications for inequality and integration. Urban Inequality and Segregation in Europe and China: Towards a New Dialogue, pp. 99–120 (2021)

Wu, Y., et al.: Examining the planning policies of urban villages guided by China's new-type urbanization: a case study of Hangzhou City. Int. J. Environ. Res. Public Health **19**(24), 16596 (2022)

Yuan, D., Yau, Y., Bao, H.: Urban village redevelopment in China: conflict formation and management from a neo-institutional economics perspective. Cities **145**(4), 104710 (2024)

Urban Analytics, Urban Modelling and Simulation

Food Delivery Index Assessment Based on SVI and OSM Data

Yizhou Fu[1], Zhen Yuan[2], and Yubo Liu[1(✉)]

[1] School of Architecture, State Key Laboratory of Subtropical Building Science, South China University of Technology, 381 Wushan Road, Tianhe District, Guangzhou 510641, China
liuyubo@scut.edu.cn
[2] School of Architecture, South China University of Technology, 381 Wushan Road, Tianhe District, Guangzhou 510641, China

Abstract. The food delivery industry has increasingly become an integral part of daily life for people in China. While delivery riders experience high work intensity, current urban maps lack information relevant to their fundamental needs, such as infrastructure related to safety, urban connectivity for efficiency, and environmental conditions for health. Therefore, this study focuses on the basic needs of delivery riders: safety, efficiency, and health. By referencing existing literature on bicycle index assessments, we identified three evaluation dimensions related to basic needs: infrastructure quality, connectivity, and environment. Utilizing street view images and OpenStreetMap data from Yuexiu District in Guangzhou, we extracted 14 indicators from 3 dimensions to create an urban road delivery riding index. The resulting urban delivery index map aims to assist riders in route planning, enhance map systems, and provide a foundation for future research.

Keywords: Food Delivery · Urban planning · Deep learning · Street View Image

1 Introduction

Driven by the rise of internet-based new retail models, the food delivery industry in Chinese cities has rapidly expanded. Meanwhile, the work intensity for delivery riders has significantly increased, making it increasingly difficult for them to meet their basic needs for safety, efficiency, and health [19, 34, 35]. In contrast, the primary tool for delivery riders, urban maps, has not been adapted to meet these fundamental needs.

Our literature review reveals a research gap concerning the relationship between delivery behavior and urban environments. Current research primarily focuses on bicycling, with insufficient attention given to the riding behavior of delivery riders.

Recognizing the similarities between delivery riding and bicycling, we combined existing studies on delivery rider behavior with research on bicycling environments. From the basic needs of delivery riders—safety, efficiency, and health—we identified three evaluation dimensions: infrastructure, connectivity, and environment. Focusing on Yuexiu District in Guangzhou, we utilized OpenStreetMap data, street view images (abbreviated as SVI), and DEM data to create an urban delivery index map to aid riders in route planning.

© The Author(s) 2025
H. Chai et al. (Eds.): CDRF 2024, *Symbiotic Intelligence*, pp. 493–503, 2025.
https://doi.org/10.1007/978-981-96-3433-0_43

The development of this map evaluation system represents a significant contribution to improving route decision-making, enhancing map systems, and supporting future research. This study outlines the literature review, data sources, methods, results, and finally discusses limitations and future research directions.

2 Literature Review

There are three parts of our study's literature review: studies related to the Behaviors of Delivery Riders, Riding Environment and SVI in Riding Behavior.

2.1 Studies Related to Behaviors of Delivery Riders

The high work intensity leads to aggressive riding behavior among delivery riders [35], making it crucial for navigation systems to provide comprehensive road safety information. A 2021 survey found that nearly 50% of delivery riders had experienced traffic accidents such as scratches, falls, and collisions [33]. According to a report on non-motorized bicycles [27], factors leading to these accidents can be categorized into dynamic and static factors. Existing map applications can display dynamic factors like traffic and weather conditions in real time, but static factors, such as infrastructure conditions, are not shown to delivery riders, although this information would be valuable to them.

In pursuit of delivery efficiency, riders maintain extremely tight schedules. From the existing literature, we find that riders often redistribute long-distance orders through group chats to increase their delivery efficiency [34]. One reason for this phenomenon is the varying connectivity in different urban areas; some areas are poorly connected to other parts of the city, making it difficult for riders to efficiently deliver orders. This indicates that urban connectivity parameters are of significant reference value to delivery riders.

Under high work intensity, the ability of the urban environment to meet delivery riders' health needs cannot be overlooked. A 2022 study revealed that the most urgent health education need for delivery riders, based on a survey of 240 riders, is knowledge about heatstroke prevention [19]. This reveals the potential value of the level of shade in urban environments for delivery riders.

2.2 Studies Related to Riding Environment

Through a review of domestic and international databases, we found that studies directly related to the riding environment for delivery riders are scarce, with most related research focusing on bicycling environments. Given the high similarity between electric bicycle riding and traditional bicycling, this study references primarily bicycle index studies to meet the basic needs of delivery riders.

We reviewed 28 papers on bicycle index published since 2015, categorizing 21 indicators they covered into three dimensions: infrastructure, connectivity, and environment. Considering the different characteristics of electric bicycles compared to traditional bicycles, we filtered and combined 18 indicators. Based on the previous discussion of delivery rider behavior studies, we ultimately selected 14 indicators for this study (See Sect. 3.1 for the specific screening process) (Fig. 1).

Demands Dimensions	Indicators	Publication
Safety Infrastructure	bike lane	Arellana et al. (2020); Krenn et al. (2015); Winters et al. (2016); Cain et al. (2018); Gu et al. (2018); Tran et al. (2020)
	sidewalks	Cain et al. (2018); Arellana et al. (2020)
	traffic separation	Hardinghaus et al.(2021); Krenn et al. (2015); Manton et al. (2016); Gu et al.(2018)
	curb cut	Pritchard et al. (2019); Cain et al. (2018)
	crosswalk	Cain et al. (2018); Gull' on et al. (2015); Arellana et al. (2020); Arellana et al. (2020)
	ground condition	Hardinghaus et al. (2021); Arellana et al. (2020)
	street light	Lin & Wei (2018); Gull' on et al. (2015); Hartanto et al. (2017); Arellana et al. (2020)
	traffic light	Gull' on et al. (2015)
	pavement quality	Beura & Bhuyan (2017,2018); Beura et al. (2017a,2017b,2018);Chen et al. (2017b); Lin & Wei (2018); Okon & Moreno (2019); Pritchard et al.(2019); Ryu et al. (2018); Hartanto et al. (2017)
Efficiency Connectivity	road types	Hartanto et al. (2017).
	road width	Hardinghaus et al. (2021); Bura et al. (2017a, 2017b, 2018); Beura & Bhuyan (2017, 2018); Lin & Wei (2018); Lowry et al. (2016); Okon and Moreno (2019); Pritchard et al. (2019); Ryu et al (2018); Wang et al. (2016); Cain et al. (2018); Manton et al. (2016);
	slope	Krenn et al. (2015); Hardinghaus et al. (2021); Bai et al. (2017); Chen et al. (2017b); Grigore et al. (2019); Lin & Wei (2018); Lowry et al. (2016); Ma & Dill (2017); Osama (2016); Ryu et al. (2018); Winters et al. (2016); Hartanto et al. (2017); Arellana et al.
	POI	Gu et al. (2018); Winters et al. (2016); Tran et al. (2020);
	cut-off road	Michael et al. (2021); Wang et al. (2018)
	intersection	Hardinghaus et al. (2021); Gholamialam & Matisziw (2019); Lowry et al. (2016); Ma & Dill (2017); Osama (2016); Wang et al. (2018); Gu et al. (2018); Manton et al. (2016); Winters et al. (2016); Hartanto et al. (2017); Cain et al. (2018);
	vehicles traffic flow	Bura et al. (2017a, 2017b, 2018); Beura and Bhuyan (2017, 2018); Chen et al. (2017); Grigore et al. (2019); Griswold et al. (2018); Lin & Wei (2018); Ma & Dill (2017); Okon & Moreno (2019); Osama & Sayed (2016); Pritchard et al. (2019); Ryu et al.(2018)
Health Environment	building shade	Gu et al. (2018); Hardinghaus et al. (2021); Cain et al. (2018)
	green ratio	Krenn et al. (2015); Hardinghaus et al. (2021); Gu et al. (2018); Lin & Wei (2018); Okon & Moreno (2019); Gull' on et al. (2015)
	greenery	Hartanto et al. (2017); Cain et al. (2018); Arellana et al. (2020); Tran et al. (2020)

Fig. 1. Indicators used in the post-2015 bicycle index study

2.3 Studies Related to SVI in Riding Behavior

In recent years, bicycle index research has utilized emerging technologies and data sources, such as SVI. Studies have combined SVI with computer vision technology [16] to automate indicator extraction, albeit with a limited number of indicators. Image semantic segmentation technology can be used to quantify greenery, sky landscape factors (i.e., openness), and buildings (i.e., enclosure) [12]. The integration of computer vision technology and SVI can further be used for city-scale infrastructure assessment, such as detecting and classifying traffic signals [17], and detecting accessibility issues [26]. Koichi Ito and Filip Biljecki's 2021 study [20] applied SVI and computer vision technology in 34 commonly used indicators from previous studies comprehensively assessing bicycle suitability.

In summary, SVI data can assist us in digging and identifying numerous indicators related to urban environments.

3 Methodology

We evaluated 14 indicators across three dimensions—infrastructure quality, connectivity quality, and environmental quality—in Yuexiu District, Guangzhou City, using data from SVI, OSM (OpenStreetMap), and DEM (Digital Elevation Model). Specific indicators are listed in Table 1.

Using OSM, we obtained road network data for Yuexiu District and its vicinity within 1 km. Subsequently, we generated 5045 measurement points at 100 m intervals. After excluding 14 points without SVI data and 33 points inside tunnels, we finalized 4998 effective measurement points.

The choice of Yuexiu District was based on two reasons: its dense population and frequent food delivery activities, and being an old district, its facilities may not well adapt to the demands of the new era. These reasons make Yuexiu District potentially one of the areas where the conflict between food delivery behavior and urban environment is most intense (Fig. 2).

Demands	Dimensions	Indicators		Data	Extraction	Scale
Safety	Infrastructure	presence of Sidewalks		SVI	Segmentation	1 if present 0 if not present
		presence of Crosswalk		SVI	Segmentation	1 if present 0 if not present
		presence of Curb Cut		SVI	Segmentation	1 if present 0 if not present
		Traffic Separation	presence of Road Fence	SVI	Segmentation	1 if present 0 if not present
			presence of Traffic Dividing Line	SVI	Segmentation	1 if present 0 if not present
				SVI	Calculation	0–1 (Min–Max)
		Ground Condition	presence of Catch Basin	SVI	Segmentation	1 if not present 0 if present
			presence of Manhole	SVI	Segmentation	1 if not present 0 if present
			presence of Pothole	SVI	Segmentation	1 if not present 0 if present
		presence of Street Light		SVI	Segmentation	0–1 (Min–Max)
		presence of Traffic Light		SVI	Segmentation	0–1 (Min–Max)
Efficiency	Connectivity	Type of road		OSM	Aggregation (100 m)	Primary,main road = 1 Secondary,secondary road = 0.7 Tertiary,branch road = 0.3
		Slope		DEM	Calculation	0–1 (Negative Min–Max)
		number of POI		OSM	Aggregation (500 m)	0–1 (Min–Max)
		number of Cut-off Road		OSM	Aggregation (500 m)	0–1 (Negative Min–Max)
		number of Intersection		OSM	Aggregation (500 m)	0–1 (Min–Max)
Health	Environment	Shade of the building		OSM	Aggregation (500 m,southern)	0–1 (Min–Max)
		Green ratio		SVI	Segmentation	0–1 (Min–Max)

Fig. 2. Indicators in this paper

It's important to note that many factors can affect the riding of food delivery rider. This paper focuses on indicators that do not change over time, as dynamic indicators can already be updated in real-time by existing map software, providing real time updates to food delivery riders during deliveries.

3.1 Selection of Indicators

Our 14 indicators are categorized into three dimensions: infrastructure quality, connectivity quality, and environmental quality, In terms of data sources, we extracted 9 indicators from SVI, 5 from OSM, and 1 from DEM data.

3.1.1 Infrastructure Quality

This dimension's indicators aim to assess whether the road conditions at the measurement points are suitable for safe and smooth riding by food delivery riders, comprising 8 different indicators derived from SVI analysis.

First are 2 indicators measuring the condition of sidewalks: the presence of sidewalks and curb cuts, assessing whether there are clearly curb-demarcated sidewalks within measurement points and whether these sidewalks are accessible without barriers. Food delivery riders often need to pick up orders from sidewalks, and without curb and curb cuts, it would be challenging for them to do so without disrupting urban traffic. Next are five indicators measuring road facilities: the presence of crosswalks, traffic lanes, road fences, street lights, and traffic lights. The existence of these elements is conducive to the safety of delivery workers. Conversely, the Ground Conditions indicator, which is a composite of three sub-indicators: the presence of catch basins, manholes, and potholes, affecting the safety of electric bike riding.

3.1.2 Connectivity Quality

This dimension assesses the degree of connectivity between the measurement point and other parts of the city, including 5 different indicators, four derived from OSM data and one from DEM data.

First are indicators for the number of intersections, dead-ends, and POIs. Within a 500 m radius of the measurement point. For food delivery riders, working in an area with high connectivity to surrounding parts of the city can increase work efficiency [35]. These indicators measure the quality of connectivity between the measurement point and the surrounding city. Next, the road grade and slope at the measurement point are assessed. Higher scores are assigned to points on higher-grade roads, as they indicate broader roads and better infrastructure, implying better connectivity. For slope indicators, flatter roads are more conducive to safe riding.

3.1.3 Environmental Quality

This dimension's indicators aim to assess the environmental quality at the measurement point, comprising two different indicators derived from SVI and OSM data.

First is the green view index, extracted from SVI data, measuring the extent of tree cover within the field of view at the measurement point. In Guangzhou's subtropical climate, with high temperatures and abundant sunshine, riding in shaded areas can significantly reduce the thermal load on food delivery riders. The building shade score is derived similarly, calculated from OSM data by evaluating the average height of buildings within a 50 m radius to the south of the measurement point, reflecting the quality of shade provided by urban environments.

3.2 Extraction of Indicators

The process of extracting indicator features varies by category and data source. DEM data extraction and indicator calculation are relatively straightforward, importing into ArcGIS and using the platform's tools to calculate slope. Indicators derived from OSM and SVI are detailed below.

3.2.1 OSM Data

This study obtained initial data from OpenStreetMap, then processed OSM data in ArcGIS, setting 100 m intervals for measurement points. Indicators analyzed directly from OSM data include road grade, number of intersections and dead-ends, quantity of POIs, and building shade.

Road grade was assigned based on road categories in OSM data. While Intersections and dead-ends required further processing of the road network. These were then overlaid with a 500 m buffer around the measurement points to calculate the number of intersections and dead-ends, with scores inversely related to the number of dead-ends.

POI data, primarily related to dining and retail, reflect the convenience of facilities around the measurement point. A 500 m buffer was created, and the number of POIs within each buffer was counted and then converted to a score on a scale from 0 to 1.

Building shade was mainly reflected by the average building height. Given Guangzhou's location in the northern hemisphere, only buildings to the south of the measurement point were considered for shading. we create 50 m radius buffer around each point, producing average building height data in the buffer for inclusion in the measurement point data, with scores ranged from 0 to 1.

3.2.2 SVI Data

We opted for Baidu Street View Maps for this study due to their data integrity in mainland China. After extracting a total of 5046 measurement points and their coordinates in GIS, Python code was utilized to fetch panoramic SVI data for these points from Baidu Map's Open Platform.

Considering electric bikes are not permitted to use viaduct in Guangzhou, we should collect SVI beneath viaducts where electric bikes can travel. Therefore, we collecting 7 SVIs within a 50 m buffer for every point located at viaduct. Subsequently, semantic segmentation was used to filter the images, selecting those that show the underside of the bridge for inclusion in the dataset. In total, There are 903 points in the data set on viaducts. We filtered out the SVI under viaducts from the 6403 collected images. We ended up with 4998 measurement points' panoramic SVI. For these images, we employed computer vision segmentation methods to extract features.

The In-Place Activated BatchNorm model trained on Mapillary Vistas with WideRes-Net38 and DeepLab3, developed by Bulò was chosen for its high accuracy in mean intersection over union of 53.42% [8]. This model accurately detects street-level features. Features like curb cuts, crosswalks, and traffic signals, a binary approach was used for counting. If a panoramic image detected pixels representing these objects, it scored 1; otherwise, it scored 0. A special case was the road surface condition indicator, which combines the presence of catch basins, manholes, and potholes; for each detected element in an image, it scored -0.3, with a total possible score of -1 if all three elements were detected.

For the green view index, the percentage of pixels representing vegetation in the total image area was calculated, with the specific scoring algorithm detailed in Sect. 3.3.

3.3 Construction of the Evaluation System

After obtaining statistical values for 14 indicators across three dimensions for 4998 measurement points, we calculated a comprehensive index for each point. Determining indicator weights is crucial. Methods include independent evaluation, arbitrary weighting based on expert knowledge, and equal weighting, which treats all indicators equally. Equal weighting is suitable when the relative importance of indicators is unclear or equally important. Given the lack of research on food delivery behavior and the built environment, equal weighting is appropriate.

Indicators were normalized and graded. For uneven distributions and clear groupings, the natural break method was used. All indicators were normalized to values between 0 and 1 for composite scoring. In conclusion, we have constructed a comprehensive evaluation system for urban road food delivery suitability, as summarized in Fig. 3.

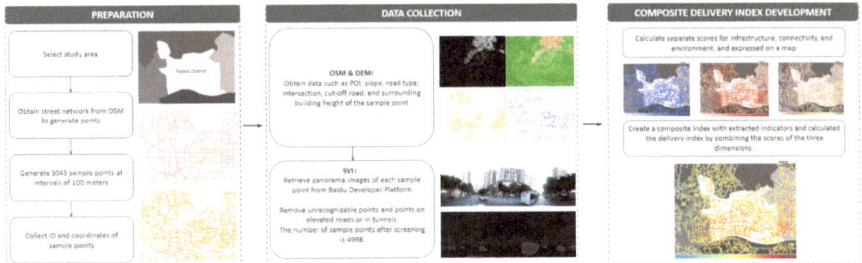

Fig. 3. Illustration of the methodology of this study

4   Results

The study focused on Yuexiu District, Guangzhou City, extracting SVI and OSM data to generate research points at 100-m intervals. This evaluation system was applied for practical analysis, and the outcomes were used to create a city-wide food delivery index map (Figs. 4 and 5).

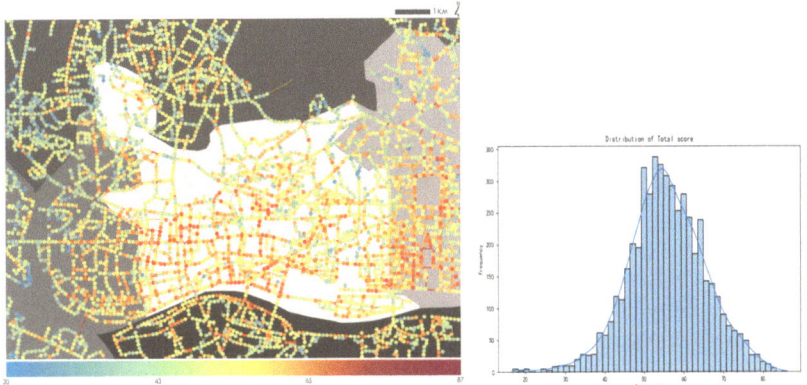

Fig. 4. Spatial and Numerical distribution of the total score in the study area

4.1   Total Food Delivery Index Score of Yuexiu District

The comprehensive Food Delivery Index score in Yuexiu District was calculated by aggregating infrastructure, connectivity, and environmental quality scores. Overall, scores varied significantly, with higher scores in the southwestern side and lower scores in the northern and eastern sides. Tianhe District to the east had higher scores, indicating better infrastructure, road connectivity, and commercial concentration. Similar scores along some roads suggested a coherent network, crucial for food delivery suitability. Measurement points were broadly distributed, with gaps in green spaces and water bodies. The histogram showed a central tendency resembling a normal distribution, with

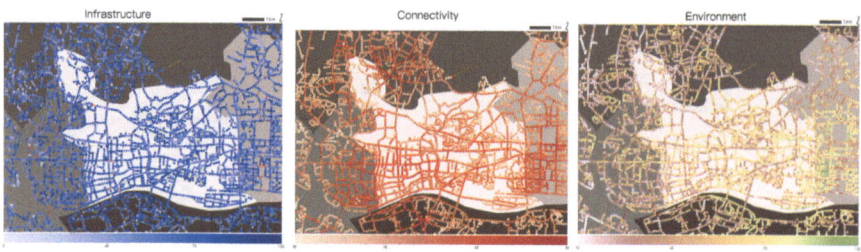

Fig. 5. Spatial distribution of the Infrastructure, Connectivity and Environment score in the study area

scores ranging from 20 to 87, most clustering between 50 and 60, and an average of 56, indicating no significant outliers. Higher-scoring roads provide better riding and delivery experiences, aiding in route planning. Lower scores were influenced by factors like traffic conditions, infrastructure, commercial facilities, and greenery. Improving these factors could enhance delivery suitability.

4.2 Score Analysis for Each Dimension

Infrastructure quality, assessed using eight SVI-derived indicators scored as 0 or 1, showed uneven distribution in Yuexiu District. Higher scores clustered around commercial centers, while industrial areas, historical sites, and old residential buildings had lower scores, likely due to lack of maintenance. The central and southern parts scored higher than the northern edges, indicating greater infrastructure investment in those areas.

Connectivity quality, assessed using five OSM and DEM-derived indicators (road type, POIs, slope, dead ends, and intersections) scored from 0 to 1, reflects road throughput and food delivery timeliness. Scores were higher centrally and lower at the edges, with the borders of Liwan and Tianhe districts scoring higher due to dense road networks. Connectivity quality decreased outward from the center, with the western and northern sides showing lower scores, possibly due to natural constraints.

Environmental quality, reflecting road shade, was measured using two indicators: building shade and tree shade. Building shade, derived from OSM data, used the average height of buildings within a 50 m semi-circle buffer south of measurement points, scored from 0 to 1. Tree shade, calculated from SVI data, used the percentage of visible vegetation, also scored from 0 to 1. Combined scores formed the overall environmental quality score.

5   Conclusion

Existing research rarely addresses the electric bike riding environment for food delivery riders. Our study, inspired by past research, uses open data sources to develop indicators measuring this environment. This study is relevant for cities in China and globally experiencing a surge in food delivery.

We assessed the Food Delivery Riding Index in Yuexiu District using 14 indicators across three dimensions. Results showed a clear central tendency, with scores mostly between 50 and 60, and few outliers. Urban core areas exhibited higher suitability due to better infrastructure and commercial density, revealing significant regional score variations.

Limitations include the use of street-level vehicle SVI, lacking open-source panoramic SVI from a non-motorized perspective, potentially causing errors. Measurement points are spaced 100 m apart, while panoramic SVI coverage may not reach 100 m, creating detection gaps. The semantic segmentation model used may make mistakes, affecting scores.

Future research could enhance data collection by using panoramic SVI from non-motorized lanes, increasing measurement density, and improving semantic segmentation models. Combining objective scores with rider surveys and interviews could refine scoring weights, achieving more accurate measurements of the food delivery riding index.

References

1. Arellana, J., Saltarín, M., Larrañaga, A.M., González, V.I., Henao, C.A.: Developing an urban bikeability index for different types of cyclists as a tool to prioritise bicycle infrastructure investments. Transp. Res. Part A Policy Pract. **139**, 310–334 (2020). https://doi.org/10.1016/j.tra.2020.07.010
2. Bai, L., Liu, P., Chan, C.Y., Li, Z.: Estimating level of service of mid-block bicycle lanes considering mixed traffic flow. Transp. Res. Part A Policy Pract. **101**, 203–217 (2017). https://doi.org/10.1016/j.tra.2017.04.031
3. Beura, S.K., Bhuyan, P.K.: Development of a bicycle level of service model for urban street segments in mid-sized cities carrying heterogeneous traffic: a functional networks approach. J. Traffic Transp. Eng. (Engl. Ed.) **4**(6), 503–521 (2017). https://doi.org/10.1016/j.jtte.2017.02.003
4. Beura, S.K., Bhuyan, P.K.: Quality of bicycle traffic management at urban road links and signalised intersections operating under mixed traffic conditions. Transp. Res. Rec. **2672**(36), 145–156 (2018). https://doi.org/10.1177/0361198118796350
5. Beura, S.K., Chellapilla, H., Bhuyan, P.K.: Urban road segment level of service based on bicycle users' perception under mixed traffic conditions. J. Mod. Transp. **25**(2), 90–105 (2017). https://doi.org/10.1007/s40534-017-0127-9
6. Beura, S.K., Kumar, N.K., Bhuyan, P.K.: Level of service for bicycle through movement at signalised intersections operating under heterogeneous traffic flow conditions. Transp. Dev. Econ. **3**(2), 21 (2017). https://doi.org/10.1007/s40890-017-0051-z
7. Beura, S.K., Manusha, V.L., Chellapilla, H., Bhuyan, P.K.: Defining bicycle levels of service criteria using Levenberg–Marquardt and self-organizing map algorithms. Transp. Dev. Econ. **4**(2), 11 (2018). https://doi.org/10.1007/s40890-018-0066-0
8. Bulò, S.R., Porzi, L., Kontschieder, P.: In-Place Activated BatchNorm for Memory-Optimized Training of DNNs. arXiv:1712.02616 [Internet]. 2018 [cited 2024 Mar 25]. Available from: http://arxiv.org/abs/1712.02616
9. Cain, K.L., Geremia, C.M., Conway, T.L., Frank, L.D., Chapman, J.E., Fox, E.H., et al.: Development and reliability of a streetscape observation instrument for international use: MAPSglobal. Int. J. Behav. Nutr. Phys. Act. **15**, 19 (2018). https://doi.org/10.1186/s12966-018-0650-z
10. Chen, C., Anderson, J.C., Wang, H., Wang, Y., Vogt, R., Hernandez, S.: How bicycle level of traffic stress correlate with reported cyclist accidents injury severities: a geospatial and

mixed logit analysis. Accid. Anal. Prev. **108**, 234–244 (2017). https://doi.org/10.1016/j.aap. 2017.09.001

11. Chen, X., Fang, X., Ye, J., Luttinen, T.: Classification criteria and application of level of service for bicycle lanes in China. Transp. Res. Rec. **2662**(1), 116–124 (2017). https://doi. org/10.3141/2662-13

12. Gong, Z., Ma, Q., Kan, C., Qi, Q.: Classifying street spaces with street view images for a spatial indicator of urban functions. Sustainability **11**, 6424 (2019). https://doi.org/10.3390/ su11226424

13. Gholamialam, A., Matisziw, T.C.: Modeling bikeability of urban systems. Geogr. Anal. **51**(1), 73–89 (2019). https://doi.org/10.1111/gean.12159

14. Grigore, E., Garrick, N., Fuhrer, R., Axhausen, I.K.W.: Bikeability in Basel. Transp. Res. Rec. **2673**(6), 607–617 (2019). https://doi.org/10.1177/0361198119839982

15. Griswold, J.B., Yu, M., Filingeri, V., Grembek, O., Walker, J.L.: A behavioral modeling approach to bicycle level of service. Transp. Res. Part A Policy Pract. **116**, 166–177 (2018). https://doi.org/10.1016/j.tra.2018.06.006

16. Gu, P., Han, Z., Cao, Z., Chen, Y., Jiang, Y.: Using open source data to measure street walkability and bikeability in China: a case of four cities. Transp. Res. Rec. **2672**(31), 63–75 (2018). https://doi.org/10.1177/0361198118758652

17. Lu, Y., Lu, J., Zhang, S., Hall, P.: Traffic signal detection and classification in street views using an attention model. Comput. Vis. Media **4**(3), 253–266 (2018). https://doi.org/10.1007/ s41095-018-0116-x

18. Hardinghaus, M., Nieland, S., Lehne, M., Weschke, J.: More than bike lanes—a multifactorial index of urban bikeability. Sustainability **13**(21), 11584 (2021). https://doi.org/10.3390/su1 32111584

19. Hu, F., Wu, Y., Zhang, H.: Analysis of health education needs and influencing factors of food delivery workers on a food delivery platform in Shanghai. Occupation Health. **38**(15), 2062–2066 (2022). https://doi.org/10.13329/j.cnki.zyyjk.2022.0405

20. Ito, K.: Assessing bikeability with street view imagery and computer vision. Transp. Res. Part C Emerg. Technol. **132**, 103371 (2021)

21. Krenn, P.J., Oja, P., Titze, S.: Development of a bikeability index to assess the bicycle-friendliness of urban environments. Open J. Civ. Eng. **5**, 451 (2015). https://doi.org/10.4236/ojce.2015.54045

22. Lin, J.J., Wei, Y.H.: Assessing area-wide bikeability: a grey analytic network process. Transp. Res. Part A Policy Pract. **113**, 381–396 (2018). https://doi.org/10.1016/j.tra.2018.04.022

23. Lowry, M.B., Furth, P., Hadden-Loh, T.: Prioritising new bicycle facilities to improve low-stress network connectivity. Transp. Res. Part A Policy Pract. **86**, 124–140 (2016). https:// doi.org/10.1016/j.tra.2016.02.003

24. Ma, L., Dill, J.: Do people's perceptions of neighborhood bikeability match "reality"? J. Transp. Land Use. **10**(1), 291–308 (2017)

25. Manton, R., Rau, H., Fahy, F., Sheahan, J., Clifford, E.: Using mental mapping to unpack perceived cycling risk. Accid. Anal. Prev. **88**, 138–149 (2016). https://doi.org/10.1016/j.aap. 2015.12.017

26. Najafizadeh, L., Froehlich, J.E.: A feasibility study of using google street view and computer vision to track the evolution of urban accessibility. In: Proceedings of the 20th International ACM SIGACCESS Conference on Computers and Accessibility. New York, NY, USA: Association for Computing Machinery, pp. 340–342 (2018). https://doi.org/10.1145/3234695.324 0999

27. Natural Resources Defense Council. Cycling Environment Risk Assessment Inside Beijing's Fourth Ring Road [Internet]. 2023 [cited 2024 Mar 25]. Available from: http://www.nrdc.cn/ Public/uploads/2023-10-25/6538da65e33b3.pdf

28. Okon, I.E., Moreno, C.A.: Bicycle Level of Service Model for the Cycloruta, Bogota Colombia. Rom. J. Transp. Infrastruct. **8**(1), 1–33 (2019). https://doi.org/10.2478/rjti-2019-0001

29. Osama, A., Sayed, T.: Evaluating the impact of bike network indicators on cyclist safety using macro-level collision prediction models. Accid. Anal. Prev. **97**, 28–37 (2016). https://doi.org/10.1016/j.aap.2016.08.010

30. Piatkowski, D.P., Marshall, W.E.: Not all prospective bicyclists are created equal: The role of attitudes, socio-demographics, and the built environment in bicycle commuting. Travel Behav. Soc. **2**(3), 166–173 (2015). https://doi.org/10.1016/j.tbs.2015.02.001

31. Pritchard, R., Frøyen, Y., Snizek, B.: Bicycle level of service for route choice—a GIS evaluation of four existing indicators with empirical data. ISPRS Int. J. Geo-Inf. **8**(5), 214 (2019). https://doi.org/10.3390/ijgi8050214

32. Ryu, S., Chen, A., Su, J., Choi, K.: Two-stage bicycle traffic assignment model. J. Transp. Eng. Part A Syst. **144**(2), 04017079 (2018). https://doi.org/10.1061/JTEPBS.0000108

33. Sun, P.: Digital labor under the "logic of algorithms": a study on food delivery riders in the platform economy. Thinking **45**(6), 50–57 (2019)

34. Sun, P., Chen, Y.: Time arbitrage" and platform labor: a study on the temporality of food delivery. Exp. Horizons **5**, 109–116 (2021)

35. Sun, P., Li, Y., Wu, J.: How the body becomes infrastructure: a study on food delivery riders in the context of platform labor. News Writ. **9**, 28–38 (2022)

36. Tran, P.T.M., Zhao, M., Yamamoto, K., Minet, L., Nguyen, T., Balasubramanian, R.: Cyclists' personal exposure to traffic-related air pollution and its influence on bikeability. Transp. Res. Part D Transp. Environ. **88**, 102563 (2020). https://doi.org/10.1016/j.trd.2020.102563

37. Wahlgren, L., Stigell, E., Schantz, P.: The active commuting route environment scale (ACRES): development and evaluation. Int. J. Behav. Nutr. Phys. Act. **7**, 58 (2010). https://doi.org/10.1186/1479-5868-7-58

38. Wang, H., Palm, M., Chen, C., Vogt, R., Wang, Y.: Does bicycle network level of traffic stress (LTS) explain bicycle travel behavior? Mixed results from an Oregon case study. J. Transp. Geogr. **57**, 8–18 (2016). https://doi.org/10.1016/j.jtrangeo.2016.08.016

39. Wang, L., Li, C., Chen, M.Z., Wang, Q.G., Tao, F.: Connectivity-based accessibility for public bicycle sharing systems. IEEE Trans. Autom. Sci. Eng. **15**(4), 1521–1532 (2018)

40. Winters, M., Teschke, K., Brauer, M., Fuller, D.: Bike Score®: associations between urban bikeability and cycling behavior in 24 cities. Int. J. Behav. Nutr. Phys. Act. **13**(1), 18 (2016)

Assisting Refined Urban Management: Building an Evaluation Framework of Data Mapping Rate Towards Digital Twin City Platform

Xinghan Chen[1], Yiping Zhang[2], and Yu Ye[1]([⊠])

[1] Department of Architecture, College of Architecture and Urban Planning, Tongji University, Shanghai, China
{2230078,yye}@tongji.edu.cn
[2] MetroDataTech Ltd., Shanghai, China
zhangyiping@metrodata.cn

Abstract. In recent years, digital twin city platforms often encounter issues such as emphasizing the physical model's accuracy over social cognition and specialized applications over a comprehensive data system, hindering the fulfilment of refined urban management's real needs. Therefore, it is essential to define the characteristics of an urban management-oriented digital twin platform and construct a detailed evaluation mechanism. This study examines the framework for evaluating mapping rates, introducing three indicators: data resolution, data freshness, and data relevance. We developed a quantifiable and replicable evaluation model to assess data completeness, update timeliness, and network correlation degree. Using Shanghai's Huamu digital twin platform as a case study, we calculated each indicator and formed a comprehensive mapping rate evaluation. This research achieves a quantitative analysis of digital twin city platforms' development quality which was previously unmeasurable. Additionally, this study aids in advancing digital twin city platforms to facilitate the development of a "bottom-up" refined urban management approach.

Keywords: digital twin city · refined urban management · data evaluation model · digital twin mapping rate

1 Background

In the context of the system reform of refined urban management [3], advances in information technology and big data offer new prospects for evolving management models [2, 7]. The digital twin city concept represents a cutting-edge urban management paradigm, merging diverse urban data to synergize digital and physical city spaces [9, 18]. This approach aims to enhance urban management, particularly in strategic decision-making for urban challenges like pollution, congestion, energy, and land use [11, 14]. In the early stages, digital twin city projects focused on data collection and visualization, utilizing municipal cloud services and technologies like 3D modelling and real-time rendering to create visual representations for urban planning and management [5, 15].

© The Author(s) 2025
H. Chai et al. (Eds.): CDRF 2024, *Symbiotic Intelligence*, pp. 504–512, 2025.
https://doi.org/10.1007/978-981-96-3433-0_44

Despite progress in developing digital twin cities, current endeavors are in their infancy and encounter two primary challenges. The first is an "overemphasis on physical precision at the expense of socio-economic digitization." Present digital twin cities prioritize replicating the physical aspects of urban environments, neglecting the complex socioeconomic activities within the city [19]. This approach often does not align with the practical needs of market entities and government regulators [13]. Moreover, the collection, purification, and integration of socioeconomic data are intrinsically more intricate, necessitating a comprehensive and methodical understanding. The second issue, "favoring scenario-specific applications over integrated data systems, " stems from a tendency in digital twin city platform construction to focus on narrow use cases, leading to fragmentation and isolation, further impeding holistic urban management [17].

These issues suggest that the construction of digital twin platforms has neglected the degree of correspondence between the physical and virtual representations of cities. As a response, our study introduces the concept of a "mapping rate" for digital twin platforms, accompanied by a systematic methodology. This framework examines critical aspects such as urban space, infrastructure, socioeconomic factors, and information dynamics, assessing their alignment and integration with their digital counterparts. Our objective is to gauge the maturity of digital twin platforms and to establish a suite of assessment tools that are practical and scalable.

2 Data Mapping Rate

Facing the demands of refined urban management, it's crucial to precisely define and scrutinize the concept of data mapping rate and identify relevant indicators (see Fig. 1). First of all, based on the nature of the digital twin, the assessment should concentrate on the detail and depth with which digital twin platforms replicate spatial, social, and economic information [1]. Accordingly, this study presents the notion of 'data resolution' to evaluate the comprehensiveness of data collection. Additionally, given the dynamic nature of data within urban management [6], the platform must ensure timely updates to fulfil the stringent demands for accuracy and efficiency in sophisticated urban management. This leads to the proposition of a 'data freshness' indicator, designed to evaluate if data is current or outdated, emphasizing the alignment of update frequencies with real-world needs. To transcend siloed approaches, data across various sectors must be integrated as per operational necessities, establishing a cohesive management network [10]. Thus, this study introduces the concept of 'data relevance,' aimed at assessing the adequacy of data interconnections across different domains in meeting practical needs.

For a quantitative assessment, the mapping rate evaluation model should be predicated on uniform and systematic urban data. Thus, investigating the data dimension structure of digital twin city platforms is essential. This study examines extensive urban data, categorized into four types. On one hand, it emphasizes data on the foundational physical elements of cities, namely "Spatial Carriers" and "Urban Infrastructure." On the other hand, in contrast to traditional projects, there is an increased emphasis on systematically collecting and integrating Socioeconomic Data, including data from "Social Entities" (e.g., residents, economic entities, governments, social organizations) and "Urban Datastreams" (e.g., human traffic, transportation flow, capital movement). These

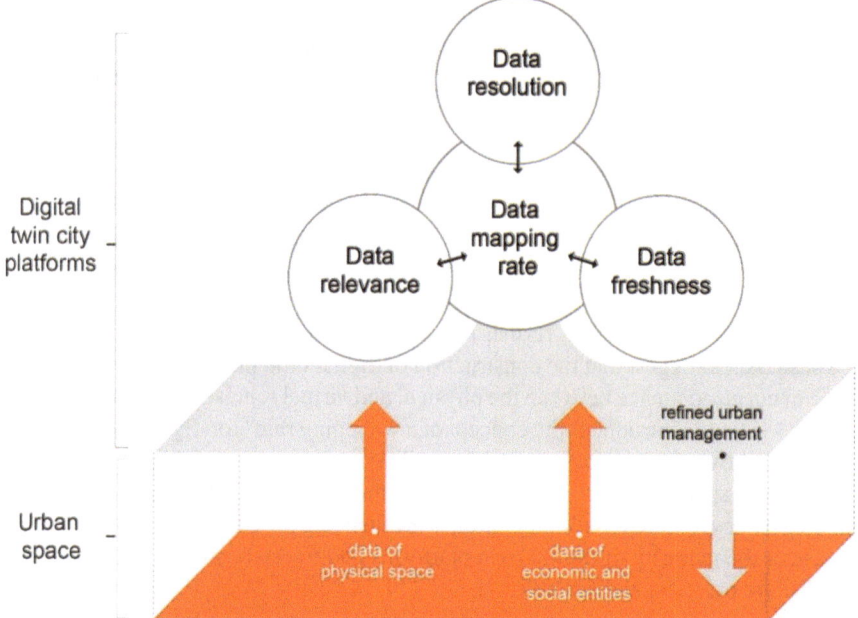

Fig. 1. Data mapping rate towards digital twin city platform

types are hierarchically categorized into "Major Categories," "Subcategories," and "Minor Categories," to capture the intricate array of data resources imperative for multi-entity and multi-structured urban management. This data system is widely adaptable for assessments across diverse digital twin cities, offering a standard benchmark to compute data mapping rates efficiently.

3   Methodology

To effectively measure data mapping rates, this study integrates methods such as the Analytic Hierarchy Process (AHP), Gaussian functions, and knowledge graphs to establish an evaluation model centered on data mapping rates. Using the Huamu digital twin platform as a case study, we conducted a detailed measurement of the previously unmeasurable data construction status for fine-grained evaluation. The socioeconomic data related to privacy were collected by the Huamu government and subsequently underwent appropriate security measures before analysis. This method will guide the sustainable development of urban digital twin platforms and precisely empower refined urban management (see Fig. 2).

3.1   Measurement of Data Resolution

Data resolution comprises systematic resolution and spatial resolution. System resolution is a measure of the completeness of data collection. We convert the coverage of data

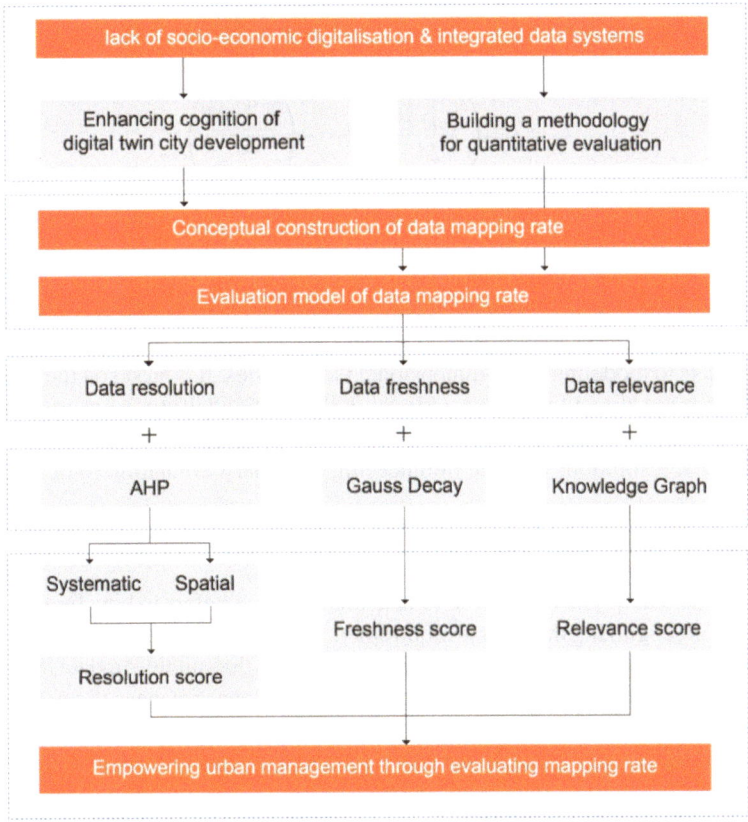

Fig. 2. Research framework

dimensions within the digital twin platform into the total area (S) of resolution units, as shown in the following formula:

$$S = \sum_i L_i \times W_i \times k_i \tag{1}$$

Here, W_i represents the type weight of the different dimensions, characterizing the differences in significance of different data types. L_i represents the depth weight of the data level, differentiating between different levels of granularity. k_i represents the data integrity in each dimension, indicating the comprehensiveness of the collection of fields in a specific dimension.

For W_i, we utilize the AHP, a quantitative method for dealing with complex decisions [16]. Weights were automatically calculated based on the significance of various data types in the analysis based on the judgments of three domain experts. For L_i, we assign exponentially increasing weights of 1, 2, and 4 to "major categories," "subcategories," and "minor categories," respectively, reflecting the rising costs of detailed data collection. For k_i, , we differentiate between "mandatory fields" (M items, constituting 60%) and "bonus fields" (N items, making up 40%) across each dimension. The field collection rate is defined as the proportion of actually collected fields to the total required fields, represented by ∂^k.

$$k_i = 0.6\left(\frac{\sum_M^k \partial^k}{M}\right) + 0.4\left(\frac{\sum_N^k \partial^k}{N}\right) \tag{2}$$

Based on this framework, attaining a k_i of 1 across all data dimensions indicates the collection of comprehensive data, equivalent to 100 points. The scoring for systematic resolution is determined by comparing this ideal total area with the actual total area of resolution units for the specific data platform.

The methodology for spatial resolution measurement is similar to systematic resolution but targets to modeling three-dimensional virtual cities. It is based on the "Technical Guidelines for the Urban Information Model (CIM) Basic Platform", ensuring its applicability to real projects. The calculation of W_i. And L_i. For spatial resolution is identical to systematical resolution, with the fferenceeing the binary evaluation (0 or 1) of k_i.

3.2 Measurement of Data Freshness

Data freshness, within the framework of data maintenance, refers to how frequently data is refreshed. Recognizing the diverse requirements for update frequencies among various data sets in urban management, it is crucial to determine the suitable update intervals for specific datasets. This study has developed a framework that balances the need and cost of updates, resulting in the establishment of five update frequency categories: "Daily," "Next-Day," "Monthly," "Semiannual," and "Annual," which signify the advised refresh cycles. The decay curves of data freshness were modelled using Gaussian functions. This assumes that data peaks in freshness upon update and decays to half at the suggested update cycle, thus determining the function forms for varying data. This method provided a formula for the decay of data freshness and its associated coefficients α_n.

$$f_n(x) = e^{\alpha_n x^2} \tag{3}$$

We performed an extensive examination of the foundational fields within the data system, leading to the development of a "Data-Field-Update Cycle" mapping table. Owing to the hierarchical tree structure of the data system, it is possible to aggregate the freshness scores from each level to determine the freshness of the immediate higher level. Consequently, the overall data freshness of the platform can be quantified using the subsequent formula:

$$X_{n-1} = \sum_i W_i \times X_n \tag{4}$$

X_n signifies the composite freshness of data across multiple dimensions at the nth level; when referring to the base-level data, X_n corresponds to the freshness of fields. Furthermore, considering the differential significance of updating diverse data types within urban management, this study introduces W_i as weighted values for each dimension using the AHP.

3.3 Measurement of Data Relevance

The evaluation of data relevance aims to ascertain the extent to which data relevant to urban management scenarios are interconnected. To measure these connections and the structure of the data network, we utilize Gephi, a software for network analysis, to apply the methods of knowledge graphs [4]. This analysis examines the correlations among data and the structural characteristics of the network, facilitating network visualization. Furthermore, a fuzzy matching algorithm is employed to compare fields and estimate the probability of their association with the same entity. This approach enables a comprehensive assessment across various service scenarios and mitigates interference caused by discrepancies in field naming.

Specifically, the construction of the data network and the calculation of data relevance are performed in Gephi. Within a complex network, different types of urban data are represented as nodes, and the field co-occurrence—determined through fuzzy matching—acts as the edge-weight connecting these nodes. The score of data relevance is quantified using the average degree [8].

$$C(i) = \sum_{i \neq j, i \neq q, j \neq q} \frac{\delta_{jq}(i)}{\delta_{jq}} \tag{5}$$

In this formula, $\delta_{jq}(i)$ represents the edge weight, which corresponds to the frequency of field co-occurrence between tables, while δ_{jq} denotes the number of nodes in the network.

4 Results

The Huamu sub-district, located in the Pudong New Area of Shanghai, is characterized by its high population density and complex management tasks, making top-down management challenging. In the context of Shanghai government's digital transformation, Huamu initiated the construction of a digital twin city platform starting from 2020. Currently, the Huamu Street sub-district has established a visualization platform along with a corresponding data collection and updating system, which is gradually being applied in grassroots management.

This study evaluated the data mapping rate of the Huamu digital twin city platform, as shown in Fig. 3. The platform achieved an overall data resolution score of 39.7. The systematic resolution score was 41.5, indicating a favorable performance. However, data collection for "Urban Datastreams" exhibited significant deficiencies. The spatial resolution score, documented at 37.1, pointed to a mediocre performance in modeling completeness and detail. The data freshness score was 34.1, which was considered moderate. Despite the relatively high freshness of "Spatial Carriers," both "Urban Datastreams" and "Urban Infrastructure" were hampered by inadequate data updating mechanisms. The data relevance score was a disappointing 13.9, with "Social Entities" and "Spatial Carriers" showing strong associations, in contrast to "Urban Infrastructure," which demonstrated the weakest connectivity with other dimensions.

For data resolution, efforts to refine the Huamu digital twin platform must prioritize the enrichment of data pertaining to "Urban Infrastructure" and "Social Entities," and pivot towards establishing effective channels for collecting "Urban Datastreams" data in the next phase. Modeling initiatives should also incorporate elements such as "Underground Spaces" and "Municipal Pipelines". For data freshness, a concentrated effort is needed to improve the recency of "Urban Datastreams" data and to implement a robust updating process for "Urban Infrastructure," with a particular focus on synchronizing data with frequent update requirements. For data relevance, it is imperative to enhance the database's completeness, encompassing more pivotal data nodes and fields. Moreover, in the process of data standardization, establishing uniform naming conventions is critical to facilitate more efficient data matching processes.

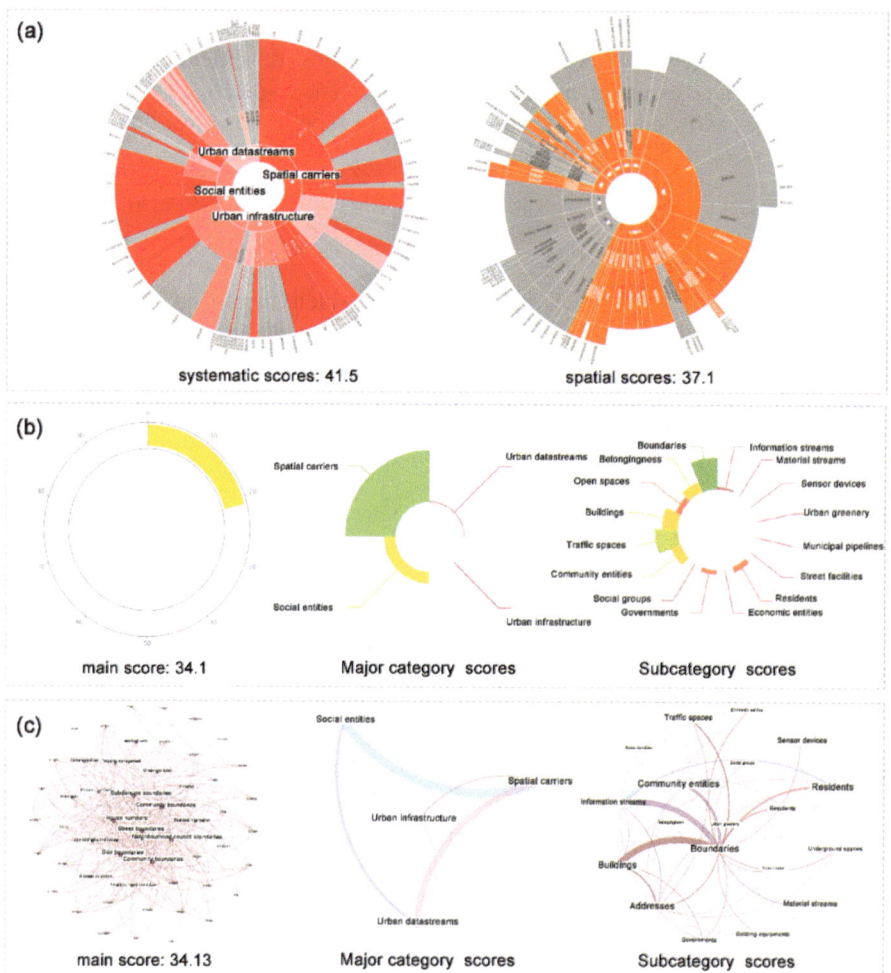

Fig. 3. Data mapping rate scores of Huamu digital twin city platform

5 Discussions and Conclusions

This study introduces a research framework aimed at evaluating the mapping rate of digital twin city platforms, fulfilling the need for refined urban management solutions. It addresses specific challenges encountered in the development of digital twin platforms by emphasizing the importance of data collection completeness, update timeliness, and network interconnectedness to define the concept of data mapping rate. Moreover, we have developed a data evaluation model focused on the data mapping rate, facilitating the previously unquantifiable assessment of digital twin platform progress. This framework, through the construction of a hierarchical data system and a combination of a series of usable and efficient methods, is widely applicable to various digital platforms. By emphasizing data freshness and relevance, it encourages relevant practices to emphasize the dynamic interrelations and effectiveness of data. Consequently, this enhances the responsiveness and adaptability of urban management, effectively addressing rapidly changing urban environments.

Furthermore, this research outlines an effective strategy for the systematization, diversification, and standardization of urban digital management practices. Unlike traditional top-down management models [12], the digital twin city management approach, predicated on the mapping rate, heralds a bottom-up system construction. Starting from the community level as the fundamental unit, it is possible to establish a higher-level digital management network, paving the way for the development of an advanced and comprehensive smart system. This evolution promises to elevate urban management and service efficiency.

However, the study acknowledges certain limitations. First, the application of the AHP introduces subjectivity in the determination of periods and weights for various indicators, although it still provides a relatively reliable benchmark for the calculation of data mapping rates. Future research may explore the use of stated preference methods for optimization. Second, the objective of measuring the mapping rate is not to indiscriminately strive for higher values. Instead, it seeks to identify an optimal range that balances management improvement and input costs, tailored to the project's specific needs. Nonetheless, given the current underdeveloped data infrastructure of many platforms, any advancement in data resolution is deemed beneficial.

References

1. Al-Sehrawy, R., Kumar, B., Watson, R.: A digital twin uses classification system for urban planning & city infrastructure management. J. Info. Technol. Constr. **26**, 832–362 (2021)
2. Allam, Z., Dhunny, Z.A.: On big data, artificial intelligence and smart cities. Cities **89**, 80–91 (2019)
3. Chen, M., Liu, W., Lu, D., Chen, H., Ye, C.: Progress of China's new-type urbanization construction since 2014: a preliminary assessment. Cities **78**, 180–193 (2018)
4. Chen, X., Jia, S., Xiang, Y.: A review: Knowledge reasoning over knowledge graph. Expert Syst. Appl. **141**, 112948 (2020)
5. Dani, A.A.H., Supangkat, S.H., Lubis, F.F., Nugraha, I.G.B.B., Kinanda, R., Rizkia, I.: Development of a smart city platform based on digital twin technology for monitoring and supporting decision-making. Sustainability. **15**(18), 14002 (2023)

6. Engin, Z., van Dijk, J., Lan, T., Longley, P.A., Treleaven, P., Batty, M., et al.: Data-driven urban management: Mapping the landscape. Journal of Urban Management. **9**(2), 140–150 (2020)

7. Hashem, I.A.T., Chang, V., Anuar, N.B., Adewole, K., Yaqoob, I., Gani, A., et al.: The role of big data in smart city. Int. J. Inf. Manage. **36**(5), 748–758 (2016)

8. Hussain, S., Muhammad, L., Yakubu, A.: Mining social media and DBpedia data using Gephi and R. J. Appl. Comp. Sci. Maths. **12**(1), 14–20 (2018)

9. Ivanov, S., Nikolskaya, K., Radchenko, G., Sokolinsky, L., Zymbler, M., (eds.): Digital twin of city: Concept overview. 2020 Global Smart Industry Conference (GloSIC). IEEE (2020)

10. Klauser, F.R., Albrechtslund, A.: From self-tracking to smart urban infrastructures: Towards an interdisciplinary research agenda on Big Data. Surveill. Soc. **12**(2), 273–286 (2014)

11. Lu, Q., Parlikad, A.K., Woodall, P., Don Ranasinghe, G., Xie, X., Liang, Z., et al.: Developing a digital twin at building and city levels: Case study of West Cambridge campus. J. Manag. Eng. **36**(3), 05020004 (2020)

12. Pissourios, I.: Policy. Top-down and bottom-up urban and regional planning: Towards a framework for the use of planning standards. European spatial research **21**(1), 83–99 (2014)

13. Rasheed, A., San, O., Kvamsdal, T.: Digital twin: Values, challenges and enablers from a modeling perspective. Ieee Access. **8**, 21980–22012 (2020)

14. Schrotter, G., Hürzeler, C.: The digital twin of the city of Zurich for urban planning. PFG–Journal of Photogrammetry, Remote Sensing and Geoinformation Science **88**(1), 99–112 (2020)

15. Shahat, E., Hyun, C.T., Yeom, C.: City digital twin potentials: a review and research agenda. Sustainability. **13**(6), 3386 (2021)

16. Vaidya, O.S., Kumar, S.: Analytic hierarchy process: an overview of applications. Eur. J. Oper. Res. **169**(1), 1–29 (2006)

17. Weil, C., Bibri, S.E., Longchamp, R., Golay, F., Alahi, A.: A Systemic Review of Urban Digital Twin Challenges, and Perspectives for Sustainable Smart Cities. Sustainable Cities and Society, 104862 (2023)

18. White, G., Zink, A., Codecá, L., Clarke, S.: A digital twin smart city for citizen feedback. Cities **110**, 103064 (2021)

19. Yossef Ravid, B., Aharon-Gutman, M.: The social digital twin: the social turn in the field of smart cities. Environment and Planning B-Urban Analytics and City Science. **50**(6), 1455–1470 (2023)

Investigating Associations Between Built Environments and Cycling Behaviour Using Street View Imagery and Strava Metro Data: A Case Study in City of Sydney, Australia

Hongming Yan[1,2], Xiaoran Huang[1,3(✉)], Jiaxin Liu[4], Sumita Ghosh[2],
and Martin Bryant[2]

[1] Faculty of Architecture and Arts, North China University of Technology, Beijing, China
`xiaoran.huang@ncut.edu.au`
[2] Faculty of Design, Architecture and Building, University of Technology Sydney, Sydney, Australia
[3] School of Design and Architecture, Swinburne University of Technology, Melbourne, Australia
[4] School of Architecture, Politecnico di Milano, Milano, Italy

Abstract. Cycling, recognized as a healthy and environmentally beneficial mode of active transport, has gained widespread acceptance and become increasingly popular. Its prevalence is profoundly influenced by the built environment, highlighting an emerging need to explore the associations between built environment factors and cycling behaviour. With the advancement of artificial intelligence, an increasing number of scholars assess the perception of the built environment using street view imagery (SVI), analysing these perceptions in conjunction with survey data. However, the usage of real-world cycling data in assessing the built environment remains limited. In our study, we explore the relationship between the built environment and cycling behaviour by correlating image segmentation analysis results of SVI from the City of Sydney with real-world cycling data from Strava Metro Data (SMD). A multivariate Poisson regression model was applied for this research. Research findings indicate a positive correlation between cycling frequency and factors such as street greenness, presence of bike lane, traffic lights, and on-street parking, while cycling frequency is negatively associated with sky openness, enclosure, street curbs, and traffic sign frame. Therefore, to build a better cycling-friendly city, urban planners and designers should focus on factors that encourage cycling and positively influence cycling behaviour. Moreover, the novel and reliable approach of integrating SVI with real-world cycling data has potential for measuring eye-level built environments in future cycling-friendly city studies.

Keywords: cycling behaviour · street view imagery · built environment · Strava Metro data

H. Chai et al. (Eds.): CDRF 2024, *Symbiotic Intelligence*, pp. 513–524, 2025.
https://doi.org/10.1007/978-981-96-3433-0_45

1 Introduction

Urban development dominated by automobiles is increasingly exacerbating climate changes. Concepts such as walkable cities (Southworth, 2005a) and 15-min living circles (Shanghai Planning and Land Resources Administration, 2016) have been proposed to encourage sustainable mobility. The objective of these concepts is not only to bring positive health benefits to individuals but also to reduce carbon footprints and mitigate global warming. Cycling, as a mode of green transportation, obtained more attention in recent years, both in developed and developing countries. It plays an important role on promoting human well-being and constructing sustainable cities, as it not only benefits on risk reduction of diseases such as diabetes, cardiovascular and certain cancers, but also alleviates urban traffic congestion and minimizes air pollution (Lee et al., 2012). Nevertheless, according to WHO (2010), there has been a significant decline in cycling rates in developing countries over the past few decades, attributed to urbanization, and rates continues to be low in developed nations. Therefore, promoting active transport such as cycling has become an essential approach when adopting green and sustainable city development agenda.

Cycling behaviour is influenced by many factors, including sociodemographic characteristics, meteorological conditions, air quality, and physical built environment is considered as one of the most essential aspects. Many scholars have reached a consensus on this issue, asserting that it can provide insights for urban design and planning policies, thereby promoting the construction of cycle-friendly cities (Lu et al., 2019). For instance, Bai et al. (2023) analysed the impact of built environment factors on cycling behaviour in urban greenways, revealing that street-level greening are positively correlated with cycling frequency. Yang et al. (2019) demonstrated that the presence of cycling infrastructure is a related built environment factor affecting cycling behaviour. While many studies have explored the impact of the built environment on cycling behaviour, comprehensive research examining the relationship from the perspective of street-scale environmental factors is scarce. Most studies either explore the impact of one or a few street environmental factors on cycling behaviour or use large-scale factors such as vegetation cover, land use, and architectural design (Bai et al., 2023). In this study, 23 street environment features are extracted from SVI to comprehensively explore the associations between these features and cycling behaviour. We aim to answer the following research question: what are the associations between built environment characteristics and cycling behaviour? How can we further improve cycling frequency and experience in the urban renewal process based on this information?

2 Literature Review

2.1 Street View Images in Urban Cycling Research

Urban cycling research has increasingly been accorded paramount attention among the public and governments. Many studies are dedicated to developing urban bikeability scores to assess the construction of cycling-friendly cities, while pyramids of studies focus on exploring the relationship between the built environment and cycling behaviour (Koh and Wong, 2013; Tran et al., 2020). The use of Geographic Information System

(GIS), in conjunction with census data and government open-source geographic information such as land-use data and Normalized Difference Vegetation Index (NDVI), has become prevalent in pinpointing objective measures of built environment (Yang et al., 2019). However, the use of such meso-scale data leads to limitations in the scale of research. Recent advancements in computer vision technologies, coupled with the extensive increase in the spatial coverage of SVI, have gradually contributed to the field in urban mobility (Ito and Biljecki, 2021). In these studies, only a few environmental factors derived from SVI are considered, often combined with other meso-scale data for cycling assessments. For instance, Tran et al. (2020) used only three out of twelve environmental factors derived from SVI, and although Gu et al. (2018) relied solely on SVI for data source, which only examined three environmental factors from SVI for consideration with urban bikeability. Most previous research has combined perceptual data collected from interviews or questionnaires with built environment factors to explore the relationship between cycling behaviour and environmental factors, as they are widely used methods to gather human's perceptual data (Yang et al., 2019). However, it is notable that the disadvantages of interviews and questionnaires, such as time-consuming, labor-intensive, and the limitations on small sample sizes, can adversely affect the research outcomes.

2.2   Objective and Subjective Measures

Numerous studies have delved into urban walkability and bikeability by analysing the objective factors of the built environment (Winters et al., 2016). Objective factors refer to the permanently present elements within the static street environment. The status quo of street infrastructure has been utilized in many previous studies and is considered as a significant objective factor in exploring cycling behaviour (Arellana et al., 2020; Cain et al., 2018). Cain et al. (2018) revealed that the quality of built environment, such as the presence of bicycle facilities and sidewalks, significantly impacts human's physical activity behaviour. Arellana et al. (2020) considered nine factors related to street facility construction, including the presence of bike lanes, sidewalks, curbs, trees, and others, to develop an index for assessing urban bikeability. Additionally, subjective factors, typically referring to individuals' perceptions of their environment, also play a crucial role in the study of urban mobility. Most research integrates subjective with objective factors to investigate cycling behaviour (Koh and Wong, 2013). As mentioned in the previous section, this non-observation data is usually collected through time-consuming interviews and questionnaires. However, in the study by Ito and Biljecki (2021), they demonstrated that data obtained solely from the SVI could be utilized to comprehensively assess urban bikeability from both subjective and objective perspectives. Moreover, urban greenness, spatial enclosure, and street congestion are three common subjective factors obtainable from SVI, which have been employed in numerous studies to evaluate human's perceptions of streets (Gu et al., 2018; Tran et al., 2020; Zhou et al., 2019).

2.3 Summary

In summary, there are notable gaps in urban cycling behaviour research that merit emphasis. Firstly, current research primarily focuses on correlational analysis using data collected from interviews and questionnaires with data from the SVI, with less exploration given to the integration of real-time cycling data. However, real-time cycling datasets, such as bike-sharing datasets and the Strava Metro Dataset (SMD), which contain more detailed data on bicycle usage frequency can enhance our understanding of built environment factors that influence cycling (Griffin and Jiao, 2015; Sun et al., 2017). Secondly, the majority of data derived from SVI represents only a minor portion of the environmental factors selected for previous studies, suggesting that the use of SVI is not yet exhaustive. Thirdly, Australia, exhibits a vibrant cycling culture with a sizable population of cyclists, complemented by its robust infrastructure and 'sunny climate'. While a few studies have utilized the Strava data to analyze urban cycling behavior in Australia, such research combining the active transport data with SVI to investigate the influence of built environment factors on cycling behavior in the country is relatively rare. Therefore, this pilot study aims to validate the built environment factors affecting cycling behaviour in the central area of Sydney, Australia, by utilizing real-time cycling frequency data from SMD and a more comprehensive set of subjective and objective environmental factors obtained solely from SVI. The study further explored whether different built environment factors impact cyclist behaviour for different cycling purposes, by comparing the associations on commute cyclist and those on leisure cyclist.

3 Methods and Materials

3.1 Study Area

The city of Sydney is located in the central core area of the Greater Sydney. According to Australia Bureau of Statistics (2022), This city encompasses an area of 2506.6 hectares and has a population of 224,331. As the CBD area of Sydney, nearly half of the top 500 companies have their main offices, and the headquarters for almost 80% of the international and national banks operating in Australia. The City of Sydney with the most complex road network and built environment is the busiest city in Australia. As the first SMD-based cycling research, the City of Sydney was selected for this pilot study.

3.2 Methodology

Firstly, this study used SVI as the only data source to evaluate 23 factors under two categories. Cyclist perception includes greenness, water view ratio, sky openness, and enclosure. Street infrastructure as objective measures consists of 19 indicators including bike lanes, bike rack, rail track, on-street parking, sidewalk, crosswalk, curb cuts, traffic lights, traffic sign, traffic sign frame, street lights, junction box, surveillance, potholes, manhole & catch basin, signage, banner, street amenities and utility pole & pole. At the meanwhile, for the SMD, missing and aberrant data were eliminated. Subsequently, we processed the cycling activity data into an average monthly cycling frequency, as we are unable to depict the distinct monthly scenarios and monthly variations using SVI. Finally, a Multivariate Poisson Regression model was applied to explore the associations between built environments and cycling behaviour (Fig. 1).

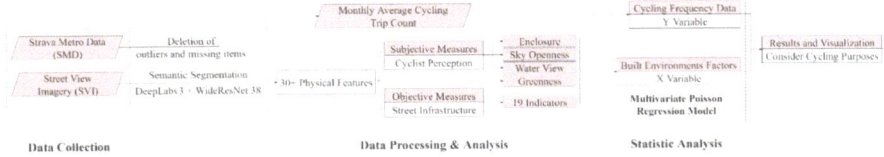

Fig. 1. Research framework and methodology

3.3 Data Collection

3.3.1 Street View Images Data

This study relies on the API of Google Street View (GSV) to retrieve information on SVI, as its widespread reach and superior quality across Australia. 11,618 SVI sampling points were generated with 50-m spacing along the street network in City of Sydney (data of street network were obtained from OpenStreetMap). For each location point, we could obtain unique panorama images by using GSV API. However, significant image distortion of panorama images can impact the segmentation results (Yu et al., 2019). Therefore, we collected images (640 × 640 pixels) with four headings of 0°, 90°, 180°, and 270° for each sample point and then merge them to mitigate this limitation. In this study, we collected 46,472 images from four directions and joint the images with 11,618 panorama images in total. After filtering out some irrelevant images, 9,301 panoramic street view images were subjected to image segmentation.

3.3.2 Strava Metro Data

The data of cycling frequency was obtained from Strava Metro Data (SMD). Strava (San Franciso, CA, USA) comprises both web-based and app-based formats to allow users to track their daily activities including, running, walking, cycling, etc. GPS enabled smartphone collect location points, time data, demographic data, and purposes of cycling that are subsequently aggregated as Big Data within SMD.

Currently, students and researchers worldwide have access to request cycling data from Strava free of charge. We have applied for and acquired the monthly cycling data of the year 2023 in the City of Sydney. This research contains aggregated and de-identified data from Strava Metro. Based on the descriptive analysis of the dataset (Table 1), there are total 24,502 road segments with 71,143,145 trips in the SMD in 2023. 66% of cycling journeys are undertaken for leisure purposes, while 34% are motivated by commuting needs. The majority of cycling activities are concentrated in the morning and evening, accounting for 58% and 23% of total journeys, while midday and overnight cycling trips together comprise only 19%. From a demographic perspective, 86% of cycling activity participants are male, with the primary age groups involved being those between 35 to 54 years old and 18 to 34 years old. In addition, after excluding cases lacking complete data for all twelve months, we conducted a temporal analysis (Fig. 2). It was noted that cycling activities progressively increased from January to March. A notable decline from April to July, marking the period with the lowest frequency of cycling activities within the year. This was followed by a steady resurgence in cycling frequency, despite a slight reduction observed in November.

Table 1. Descriptive Analysis of Strava cyclists data

Strava Metro Data of City of Syndey 2023				
Total Segment Count			24,502	
Total Trip Count			71,143,145	
Commute Trip Count			24,391,075	(34%)
Leisure Trip Count			46,792,780	(66%)
Trip Count by Time	Morning	Midday	Evening	Overnight
	58%	17%	23%	3%
Cycling Participation by Gender	Male	Female	Unspecified	
	86%	12%	2%	
Cycling Participation by Age	18-34	35-54	55-64	65+
	32%	51%	14%	4%

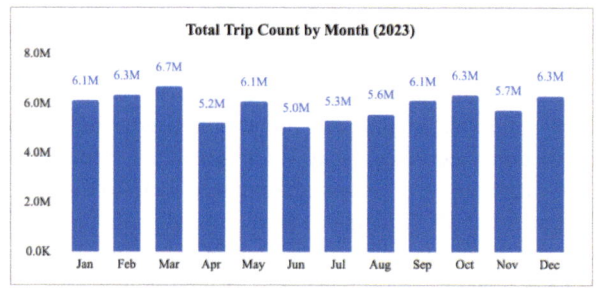

Fig. 2. Shows the total trip count by month

4   Data Processing and Analysis

4.1   Information Extraction from SVI

Deeplabv3 model pre-trained with ADE20K dataset is widely used by urban researcher for image semantic segmentation, as it can segment 150 categories, the most labels up to now (Gong et al., 2023). Nevertheless, Mapillary Vistas dataset lies in its global coverage and diversity of the trained images, including street images from Australia. Moreover, it specializes in segmenting labels of street infrastructure, such as bike lanes, curbs, and potholes, which are elements not achievable by ADE20K segmentation. Consequently, for the task of segmentation, we have chosen the In-Place Activated BatchNorm (IPAB) model, which has been trained on the Mapillary Vistas dataset using WideResNet38 and DeepLabv3, developed by Bulo et al. (2018). Several cyclist perception indicators such as eye-level greenness, water view factor, and sky openness can be quantified by the proportions of vegetation, water, and sky in the image. Moreover, the measurement of enclosure can be calculated by using the ratio of vertical objects to horizontal features and a higher value of enclosure denotes a more compact spatial configuration (Bai et al., 2023; Zhou et al., 2019a). In terms of street infrastructure indicators, we utilised binary values to quantify them, as calculating the pixels of street infrastructure elements like streetlights, bike lanes, and street amenities is not meaningful, scoring 1 if present in the image and 0 if absent (Ito and Biljecki, 2021).

4.2 Cycling Frequency Data and Spatial Process

Firstly, to mitigate the impact of the temporal dimension on the results, we utilized the monthly average trip count per segment as the data for cycling frequency. For instance, some segments lack cycling frequency data for all 12 months; hence, using the total number of trips as the dependent variable would be inaccurate. Secondly, it is necessary to correlate the cycling frequency data with the image segmentation data for statistical analysis. Spatially, the monthly average cycling frequencies obtained from SMD are based on data from 30,332 road segments, while the results of image segmentation are based on data from 9,301 street view points. We employed a Near analysis tool in ArcGIS Map to spatially link the data from these two layers, resulting in 7,775 corresponding data points. Finally, after excluding outliers and eliminating missing values, over 6,204 data points were used for the final statistical analysis.

4.3 Statistical Analysis

After completing the data processing, we initially employed scatterplot analysis for a preliminary investigation of several variables, aiming to better understand the characteristics of the data and the relationships between each variable. Variables of street infrastructure were all binary data. Thus, it is not meaningful to analysis their correlation. Figure 3 illustrated that there are no significant linear relationships either among the selected independent variables or between the independent and dependent variables.

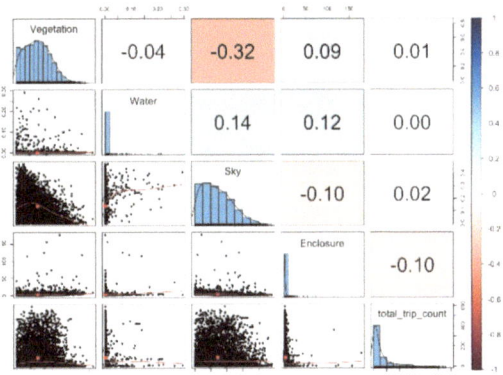

Fig. 3. Correlation matrix of variables

Furthermore, a correlation matrix (Fig. 3) based on Pearson's correlation coefficient was computed, which further substantiated the absence of multicollinearity among the independent variables. According to previous studies (Bai et al., 2023; Wang et al., 2020), this research utilized a multivariate Poisson regression model to explore the impact of built environment factors on cycling frequency, given that cycling frequency is a count variable. Moreover, the cycling frequencies on commute and leisure individually tested within two separate regression models to investigate variations in cycling purposes.

5 Results and Discussion

5.1 Spatial Distribution of Cycling Frequency

Figure 4 demonstrates the spatial distribution information of different cycling purposes. From an overall perspective on cycling frequency distribution, it is evident that the majority of higher frequency cycling routes are located on primary roads, while secondary and tertiary roads exhibit lower cycling frequencies. This could suggest that cyclists prefer to select primary roads for their wider lanes and better street infrastructure. In comparison between cycling frequencies for commuting purposes and leisure activities, leisure cycling is more frequent, aligning with the data (Table 1) showing that leisure constitutes the main purpose of cycling activities (66%). Furthermore, in terms of spatial distribution differences, cycling for commuting purposes is more frequent in central urban areas, particularly near public transport stations. Although there are some overlapping hotspots for both purposes, streets with higher cycling frequencies for leisure purposes are predominantly located around urban green spaces and waterfront areas.

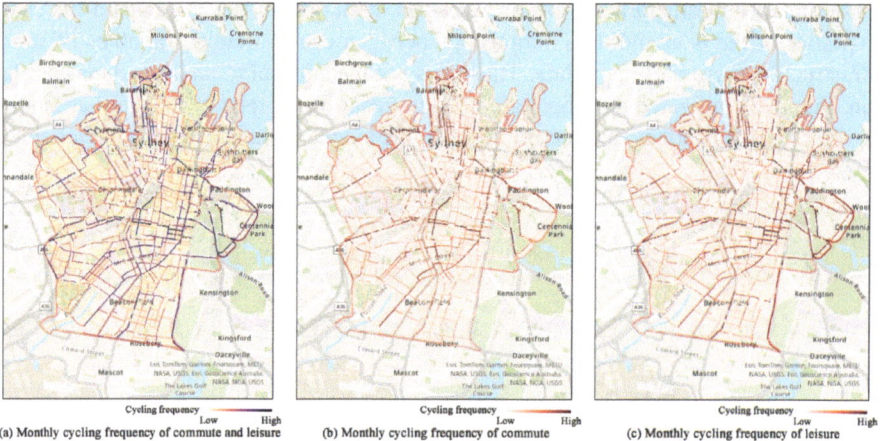

Fig. 4. Spatial distribution of cycling frequency (a) cycling frequency of both commute and leisure cyclist (b) commute cycling frequency and (c) leisure cycling frequency

5.2 Spatial Distribution of Cycling Perception Indicators

Figure 5 illustrates the spatial distribution of four cycling perception indicators including vegetation view (VV), sky openness (SO), water view (WV) and enclosure (ENS) based on image segmentation analysis. The central area of the City of Sydney shows lower VV and SO, which can possibly be attributed to the area being the center of Sydney's skyscrapers. The high-density development results in reduced levels of greenery and limited sky exposure. Moreover, the distribution of VV and SO appears to be similar, suggesting a potential linear relationship between these two factors, consistent with findings in the correlation matrix (Fig. 3). Regarding water view, it is distinctly noticeable that areas with high waterfront visibility predominantly lie near coastal zones and around water features within green spaces. Highly enclosed areas are primarily found in the city's central area. As you move away from the city centre, the sense of enclosure tends to lessen, a pattern that may be related to the density of urban development.

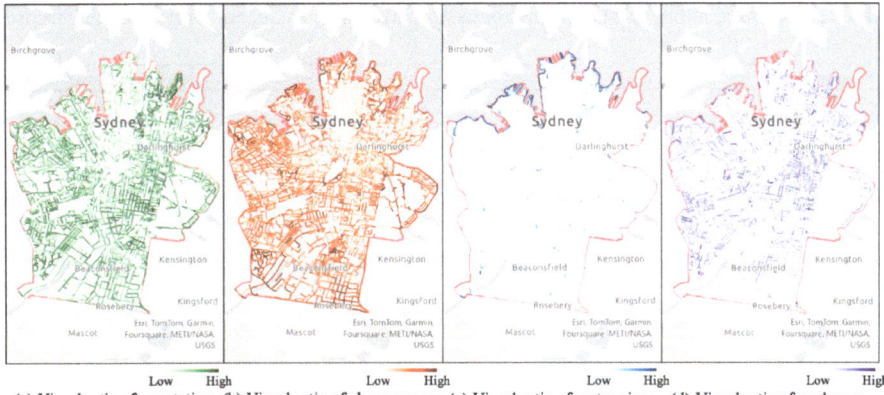

Fig. 5. Spatial distribution of four different cycling perception indicators (a) vegetation view (b) sky openness (c) water view and (d) enclosure

5.3 Correlation Between Built Environments and Cycling Behaviour

We applied multivariate Poisson regression analysis in R Studio to explore the relationship between built environment factors and cycling behaviour (Table 2). For perception indicators, our results suggest that the visibility of vegetation positively influenced cycling frequencies regardless of the rider's purpose. Previous studies have also found that exposure to greenness was positively associated with the frequency of cycling (Bai et al., 2023; Wang et al., 2020). Views of water were strongly positively correlated with leisure cycling in the City of Sydney, while they had a weak negative correlation with cycling for commuting purposes. This finding indicates that water bodies are more attractive leisure cyclists, who prefer to cycle around waterfronts. A clear sky was consistently associated with lower frequencies of cycling. This phenomenon could be due to the fact that a wide-open sky often results in higher temperatures from greater sun exposure, whereas cyclists may prefer to ride in cooler conditions (Meng et al., 2016). The enclosure aspect showed a negative relationship in cycling frequency, indicating that streets that are more enclosed may deter cycling. In terms of street infrastructure, the presence of bike lanes and traffic lights was positively correlated with cycling frequencies, particularly commute cyclists, as these features tend to make the streets safer. Cyclists are more likely to use bike lanes because such infrastructure can enhance their perceived safety (Rita et al., 2023). However, features such as curb cuts and large traffic sign frames had a negative effect on cycling frequency, although less significant. This can be explained by the fact that large sign frames often exist on highways or expressways, which restrict cycling activities. In addition, our regression results also indicate that advertisements in signage and banner were positively associated with cycling frequencies. The abundance of street advertisements may serve as an indicator of an urban street's vibrancy, which could potentially attract a greater number of cyclists.

Table 2. The associations of 24 built environments and cycling frequency with different purposes

Model predictor	Model 1: Total cycling frequency	Model 2: Commute cycling frequency	Model 3: Leisure cycling frequency
	Coef. (p-value)	Coef. (p-value)	Coef. (p-value)
Cycling perception indicators			
Vegetation	0.009 ***	0.006 ***	0.010 ***
Water	0.042 ***	-0.067 *	0.061 ***
Sky	-0.014 ***	-0.018 ***	-0.012 ***
Enclosure	-0.647 ***	-0.577 ***	-0.685 ***
Street infrastructure indicators (presence of ...)			
Bike.Lane.True	0.135 **	0.259 ***	0.073
Bike.Rack.True	0.142 *	0.146 .	0.142 *
Rail.Track.True	0.125 .	0.200 **	0.085
On.Street.Parking.True	0.199 ***	0.201 ***	0.200 ***
Sidewalk.True	0.095	0.060	0.104
Crosswalk.True	0.065	0.090 .	0.052
Curb.Cut.True	-0.097 **	-0.103 *	-0.096 **
Traffic.Light.True	0.350 ***	0.360 ***	0.347 ***
Traffic.Sign.True	-0.020	-0.002	-0.028
Traffic.Sign.Frame.True	-0.346 *	-0.499 **	-0.287 *
Street.Light.True	0.042	0.065	0.034
Junction.Box.True	-0.006	0.003	-0.011
Surveillance.True	-0.172	-0.177	-0.172
Pothole.True	-0.051	-0.151	-0.003
Manhole.Catch.Basin.Ture	-0.041	0.006	-0.064 .
Signage.Ads.True	0.271 ***	0.275 ***	0.274 ***
Banner.Ads.True	0.117 *	0.152 *	0.099 .
Street.Amenities.True	-0.071 *	-0.067 .	-0.073 *
Utility.Pole.True	0.420 *	0.774 **	0.298 .

Coef. = Coefficient; p-value: 0 '***' 0.001 '**' 0.01 '*' 0.05 '.' 0.1 ' ' 1

6   Conclusion and Future Research

This study is a pilot study to explore the association between built environments and cycling behaviour in Australia by using real-time cycling frequency data and street-view data. The findings indicate that vegetation showed a significant positive correlation with cycling frequency both on commute purpose and leisure purpose. In contrast, sky openness and enclosure were negatively associated with cycling behaviour. Furthermore, street infrastructure indicators such as presence of bike lane, on-street parking and traffic lights were three key factors have positive influence on cycling behaviour in that they increase frequencies, while the presence of traffic sign frame and curbs showed negative impacts on it. Therefore, urban planners and policymakers should pay more attention to these key subjective and objective environmental factors to promote a better cycle-friendly cities.

Several limitations needed to be presented. First, our segmentation utilized pre-trained IPAB models on old Mapillary Vistas dataset, which have a limitation on categories (66 labels) extracted from images. Second, some limitations on the cycling data such as the absence of personal and economic data about cyclists restricts the range of our statistical evaluation. Third, as a pilot study of City of Sydney, future investigation could explore the comparison among other Australian cities. In future research, the new Mapillary Vistas dataset, featuring 124 labels capable of capturing more comprehensive street information, along with a more powerful model like One-former, can be utilized for segmentation tasks. Additionally, this study did not include several variables, such as urban density, road width, and land-use data. Future studies should consider incorporating these variables. Furthermore, spatial heterogeneity and the non-linear relationship between built environments and cycling behaviour should also be taken into account in future research.

References

Arellana, J., Saltarín, M., Larrañaga, A.M., González, V.I., Henao, C.A.: Developing an urban bikeability index for different types of cyclists as a tool to prioritise bicycle infrastructure investments. Transp. Res. Part Policy Pract. **139**, 310–334 (2020). https://doi.org/10.1016/j.tra.2020.07.010

Bai, Y., Bai, Y., Wang, R., Yang, T., Song, X., Bai, B.: Exploring associations between the built environment and cycling behaviour around urban greenways from a human-scale perspective. Land **12**, 619 (2023). https://doi.org/10.3390/land12030619

Bulo, S.R., Porzi, L., Kontschieder, P.: In-place Activated BatchNorm for Memory-Optimized Training of DNNs. In: 2018 IEEE/CVF Conference on Computer Vision and Pattern Recognition. Presented at the 2018 IEEE/CVF Conference on Computer Vision and Pattern Recognition (CVPR), pp. 5639–5647. IEEE, Salt Lake City, UT (2018). https://doi.org/10.1109/CVPR.2018.00591

Cain, K.L., et al.: Development and reliability of a streetscape observation instrument for international use: MAPS-global. Int. J. Behav. Nutr. Phys. Act. **15**, 19 (2018). https://doi.org/10.1186/s12966-018-0650-z

Gong, W., Huang, X., White, M., Langenheim, N.: Walkability perceptions and gender differences in urban fringe new towns: a case study of shanghai. Land **12**, 1339 (2023). https://doi.org/10.3390/land12071339

Griffin, G.P., Jiao, J.: Where does bicycling for health happen? Analysing volunteered geographic information through place and plexus. J. Transp. Health **2**, 238–247 (2015). https://doi.org/10.1016/j.jth.2014.12.001

Gu, P., Han, Z., Cao, Z., Chen, Y., Jiang, Y.: Using Open Source Data to Measure Street Walkability and Bikeability in China: A Case of Four Cities. Transp. Res. Rec. **2672**, 63–75 (2018). https://doi.org/10.1177/0361198118758652

Ito, K., Biljecki, F.: Assessing bikeability with street view imagery and computer vision. Transp. Res. Part C Emerg. Technol. **132**, 103371 (2021). https://doi.org/10.1016/j.trc.2021.103371

Koh, P.P., Wong, Y.D.: Influence of infrastructural compatibility factors on walking and cycling route choices. J. Environ. Psychol. **36**, 202–213 (2013). https://doi.org/10.1016/j.jenvp.2013.08.001

Lee, I.-M., Shiroma, E.J., Lobelo, F., Puska, P., Blair, S.N., Katzmarzyk, P.T.: Effect of physical inactivity on major non-communicable diseases worldwide: an analysis of burden of disease and life expectancy. The Lancet **380**, 219–229 (2012). https://doi.org/10.1016/S0140-6736(12)61031-9

Lu, Y., Yang, Y., Sun, G., Gou, Z.: Associations between overhead-view and eye-level urban greenness and cycling behaviors. Cities **88**, 10–18 (2019). https://doi.org/10.1016/j.cities.2019.01.003

Meng, M., Zhang, J., Wong, Y.D., Au, P.H.: Effect of weather conditions and weather forecast on cycling travel behavior in Singapore. Int. J. Sustain. Transp. **10**, 773–780 (2016). https://doi.org/10.1080/15568318.2016.1149646

Rita, L., Peliteiro, M., Bostan, T.-C., Tamagusko, T., Ferreira, A.: Using deep learning and google street view imagery to assess and improve cyclist safety in London. Sustainability **15**, 10270 (2023). https://doi.org/10.3390/su151310270

Shanghai Urban Planning and Land Resources Administration. Shanghai Planning Guidance of 15-Minute Community-life Circle (2016). https://hd.ghzyj.sh.gov.cn/zcfg/ghss/201609/P020160902620858362165.pdf

Southworth, M.: Designing the Walkable City. J. Urban Plan. Dev. **131**, 246–257 (2005). https://doi.org/10.1061/(ASCE)0733-9488(2005)131:4(246)

Sun, Y., Mobasheri, A.: Utilizing crowdsourced data for studies of cycling and air pollution exposure: a case study using strava data. Int. J. Environ. Res. Public Health **14**, 274 (2017). https://doi.org/10.3390/ijerph14030274

Tran, P.T.M., Zhao, M., Yamamoto, K., Minet, L., Nguyen, T., Balasubramanian, R.: Cyclists' personal exposure to traffic-related air pollution and its influence on bikeability. Transp. Res. Part Transp. Environ. **88**, 102563 (2020). https://doi.org/10.1016/j.trd.2020.102563

Wang, R., Lu, Y., Wu, X., Liu, Y., Yao, Y.: Relationship between eye-level greenness and cycling frequency around metro stations in Shenzhen, China: A big data approach. Sustain. Cities Soc. **59**, 102201 (2020). https://doi.org/10.1016/j.scs.2020.102201

Winters, M., Teschke, K., Brauer, M., Fuller, D.: Bike Score®: Associations between urban bike-ability and cycling behavior in 24 cities. Int. J. Behav. Nutr. Phys. Act. **13**, 18 (2016). https://doi.org/10.1186/s12966-016-0339-0

World Health Organization: Global recommendations on physical activity for health. Recomm. Mond. Sur Act. Phys. Pour Santé 58 (2010)

Yang, Y., Wu, X., Zhou, P., Gou, Z., Lu, Y.: Towards a cycling-friendly city: An updated review of the associations between built environment and cycling behaviors (2007–2017). J. Transp. Health **14**, 100613 (2019). https://doi.org/10.1016/j.jth.2019.100613

Zhou, H., He, S., Cai, Y., Wang, M., Su, S.: Social inequalities in neighborhood visual walkability: Using street view imagery and deep learning technologies to facilitate healthy city planning. Sustain. Cities Soc. **50**, 101605 (2019). https://doi.org/10.1016/j.scs.2019.101605

Visual Typology: A Numerical Taxonomy of Urban Spaces Using Isovist Analysis

Chengxuan Li[✉]

Architectural Association School of Architecture, 36 Bedford Sq., London WC1B 3ES, UK
Chengxuan.Li@aaschool.ac.uk, CL2749@cornell.edu

Abstract. Urban spaces possess diverse visual qualities that significantly impact comfort, aesthetics, and navigation. This paper introduces a novel approach towards classifying urban spaces based on their visual characteristics through isovist analysis. An isovist is the polygon representing the visible areas from a given vantage point. The geometrical attributes of the isovist polygon enables a quantitative measure of visual qualities in the urban setting. However, the potential for classifying urban spaces based on the geometrical attributes of isovist polygons remains largely untapped. This paper presents a methodology to systematically categorise urban spaces using isovists and their geometrical attributes. By aggregating ten dimensions of geometrical attributes through a Gaussian Mixture Model (GMM) clustering analysis, this workflow produces a classifier that categorises urban spaces into 10 distinct spatial types, each possessing unique visual and spatial characteristics. This method successfully captures intrinsic spatial typologies across diverse urban contexts and can reflect the values embedded in urban design schemes. By facilitating meaningful and discussions in urban planning and design, this research contributes to a deeper and numerical understanding of the spatial and visual aspects of urban design. Further research avenues include the extension of this methodology to 3D analysis and refining tessellation algorithms for improved computational efficiency and accuracy.

Keywords: Urban Spaces · Visual Qualities · Isovist · Cluster Analysis · Gaussian Mixture Model (GMM)

1 Introduction

Historically, architects and urban designers have developed a strong interest in the visual qualities of urban spaces [1–4]. The optical perception of urban spaces directly influences aspects of comfort, aesthetics, and navigation. To systematically quantify the visual qualities of a specified location in space, the concept of the isovist was introduced by Benedikt [5] as 'the set of all points visible from a given vantage point in space'. The initial application of this measure featured only with the qualitative assessment of the visual quality of specific point(s) within a given space; yet later this quantitative model has been extensively used by Space Syntax [6, 7] and in complexity theories to understand cities and their morphology [8, 9].

© The Author(s) 2025
H. Chai et al. (Eds.): CDRF 2024, *Symbiotic Intelligence*, pp. 525–535, 2025.
https://doi.org/10.1007/978-981-96-3433-0_46

Isovist analysis is a highly suggestive way of measuring and analysing the visual qualities of spatial configurations, and the generalisation of isovists to map the overall visual qualities of a given space necessitates the development of a field of isovist points inside a given space, distributed uniformly and contiguously to map out the visual field conditions within this spatial system of a building or an urban condition [10]. The isovist field enables larger-scale analysis and comparison of visibility conditions across all spaces within a given context or spatial condition.

To date, the isovist field is predominantly used to quantitatively assess the spatial qualities of a smaller scale similar to an architectural interior [10]. The discussion of their results centres around the relationship between numerical performances and visual ground truths. However, the lack of research into the usage of isovist fields as a way of numerically classifying urban spaces into taxonomical categories or clusters, is a valuable opportunity truly missed. Research into the numerical classification of urban spaces and urban forms concentrates on the morphological aspects of the built-up fabric [11–15], while focusing little on the visual, perceptual qualities of the urban spaces; admittedly, [12] operates on the level of street segments and pathways as the basic units of classification, taking into account several pedestrians' perspectives, its inability to address the visual and optical qualities of urban spaces is epitomic of this gap. Given the astounding diversity of cities and the apparently difference in visual and perceptual qualities of all types of public and semi-public spaces within cities across contexts, there is a true demand for a highly view-based methodological framework for the classification of urban spaces into categorical types of distinct visual qualities, thus facilitating meaningful comparisons and discussions of different urban spaces and urban forms.

This paper therefore presents the development of an innovative way of making use of isovists and isovist fields to categorise urban spaces based on the geometrical attributes of the isovist polygon at a given location. By the creation of an isovist field based on a grid of sample points in the urban space upon which isovist analysis is carried out, each sample point is aggregated with ten dimensions of geometrical attributes that describe each isovist numerically. All points sampled from isovist fields for each urban condition are concatenated for clustering using Gaussian Mixture Model (GMM), yielding 10 categorical types. Through a visual inspection, clusters are interpreted in view of their patterns in connection with the original spatial implications of the corresponding urban design scheme.

2 Generating the Isovist and Its Geometrical Attributes

The generation of individual isovist polygons follow the 'radiate' [5] or ray casting method. For each given sample point, rays emit in all directions, evenly spaced upon specific angular intervals. A ray would terminate where it intersects with an obstacle (e.g. solid walls of buildings), and will be 'occluded' where no obstacle is encountered within a specific distance or max vista limit defined in prior [5]. The collection of all end points of each ray are labelled accordingly as 'occlusion' or 'intersection', and are arranged in a sequential order to form the vertices of a polygon.

For each sample point, isovist analysis results in a closed polygon denoting the area visible from the point. Extending Benedikt [5], the geometrical attributes are calculated for each resultant polygon, as listed below. In each of the equations, x_i, y_i refers to the x- and y-components of the i-th member of the array of vectors from the sample point to the corresponding member in the array of vertices comprising the polygon.

Area. Area refers to the area measure of the set of all points visible from a given sample point. In a typical cartesian coordinate system, the area is defined as

$$A_v = \sum_{i=1}^{n} \frac{1}{2} |x_{i-1} \cdot y_i - x_i \cdot y_{i-1}|$$

Average Radial. Average Radial (Average Radii) refers to the arithmetic mean of the view lengths (radials) in all directions from a given sample point, defined as

$$Q_v = \frac{1}{n} \cdot \sum_{i=1}^{n} \sqrt{x_i^2 + y_i^2}$$

Perimeter. Perimeter refers to the perimeter, i.e. the summation of the lengths of all edges of the isovist polygon. Perimeter is defined as

$$P_v = \sum_{i=1}^{n} \sqrt{(x_i - x_{i-1})^2 + (y_i - y_{i-1})^2}$$

Closed Perimeter. Closed perimeter refers to the summation of the lengths of all closed edges (i.e. solid walls) of the isovist polygon. A closed edge is a straight line connecting two consecutive vertices of the isovist polygon that are both labelled as 'intersection'. Closed Perimeter is expressed as

$$U_v = \sum_{v_i \in V_{\text{intersection}}}^{n} \frac{\sqrt{(x_i - x_{i-1})^2 + (y_i - y_{i-1})^2} + \sqrt{(x_i - x_{i+1})^2 + (y_i - y_{i+1})^2}}{2}$$

where v_i represents the i-th member of the array of vertices; $V_{\text{intersection}}$ denotes the set of all vertices labelled as 'intersection'.

Compactness. Compactness is the isoperimetric quotient of the isovist polygon, defined as

$$C_v = \frac{4\pi A_v}{P_v^2}$$

Drift. Drift refers to the distance between the sample point and the area centroid of the isovist polygon, defined as

$$D_v = \sqrt{(x_m - x_v)^2 + (y_m - y_v)^2}$$

where x_m, y_m are x- and y- coordinates of the area centroid and x_v, y_v are x- and y- coordinates of the sample point.

Occlusivity. Occlusivity is the portion of the rays occluded, defined as

$$O_v = 1 - \frac{U_v}{P_v}$$

Variance. Variance refers to the mean squared difference of all radial lengths of an isovist polygon, given as

$$T_v = \frac{1}{n} \sum_{i=1}^{n} \left(\sqrt{x_i^2 + y_i^2} - Q_v \right)^2$$

Skewness. Skewness refers to the mean of the cube of deviation between all radial lengths and the mean radial lengths of an isovist polygon, given as

$$S_v = \frac{1}{n \cdot T_v^{3/2}} \sum_{i=1}^{n} \left(\sqrt{x_i^2 + y_i^2} - Q_v \right)^3$$

Vista. Vista refers to the maximum view length possible at the sample point, defined as

$$H_v = \sup \left(\left\{ \sqrt{x_i^2 + y_i^2} | i \in 1, n \right\} \right)$$

Amongst the attributes introduced above, Area and Average Radius are representative of the overall average visibility in all directions from a sample point. However, the order of rays in each direction (angle) of ray emission affects the shape of the polygon and thus the Area, yet if the ray lengths are fixed, the Average Radius remains unchanged. This marks the fundamental difference between the two measures. Perimeter and Compactness are more sensitive to the overall shape of the isovist polygon and could describe the distribution of visibility, e.g. homogeneous in all directions, or heterogeneous with strong directionality. Occlusivity and Closed Parameter measures are sensitive to the proportion of directions where views extend beyond the given range of visibility, and could thus be used to account for urban axial connections such as boulevards and monumental avenues. Vista is representative of the maximum visibility length, while Variance and Skewness are statistically important to account for the homogeneity/heterogeneity of views in different directions as well as the degree of symmetricity or asymmetricity.

To compute these geometrical attributes, a piece of software has been developed in a grasshopper environment in Rhinoceros 3D in C#/.NET Framework. An alternative version has also been made available in python, using t4gpd [16] package, and shapely [17] geometries.

3   Sampling Isovist Fields

To sample all spaces within an urban condition, a means of tessellation is necessary to arrange the sample points to which isovist polygons will be generated and geometrical attributes will be aggregated [8, 18]. Without delving into advanced adaptive tessellation methods which could be the opportunity for future research [10], this paper refers to the method of simple, pixel-based uniform grid [8]. This tessellation allows for the enumeration of all spaces equally, and is necessary for the preparation of a detailed taxonomy that encapsulates different spatial and visual qualities across the space (Figs. 1 and 2).

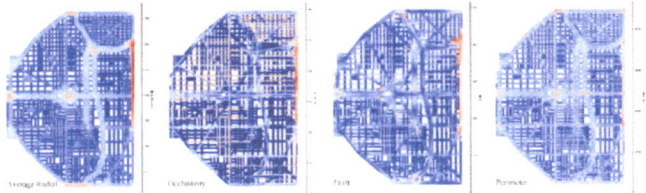

Fig. 1. A selected area of Burnham's Chicago plan showing values of several attributes

Fig. 2. Scatter matrix showing the interrelations between different geometrical attributes of the isovist polygons of each point in the isovist field of Burnham's Chicago plan

4 A Methodological Foundation for the Classification of Urban Spaces

The aim of classification relates to the need to reduce complexities and to identify patterns. Existing literature on the classification of urban spaces [11, 12, 19–23] outlines a general workflow for the classification of urban spaces. In this research, isovist polygons are mapped among all urban public spaces based on 19 typical figure-ground plans from a hybrid of as-found urban conditions, unrealised urban design schemes, and historical urban environments based on historical maps. The data source is a set of figure-ground vector drawings, appropriated from [24]. Upon the geometrical attributes computed from these polygons, a cluster analysis is carried out (Fig. 3).

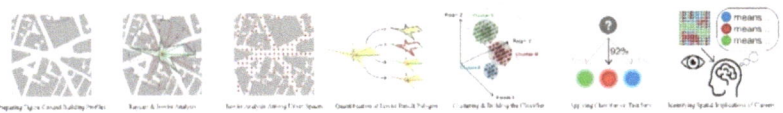

Fig. 3. Workflow for the classification of urban spaces based on isovist analysis

The most suitable algorithm for cluster analysis is Gaussian Mixture Model (GMM). It is a probabilistic derivative of K-Means algorithm [25], and has been tested under similar tasks of urban space classification in [26]. The optimal number of clusters is determined by the local minimum of the Bayesian Information Criterion (BIC). This cluster analysis gives 10 clusters and upon a visual inspection of the clustering result, clustering results appear well-defined and able to reflect the patterned nature of urban forms, consistent with the intuitive knowledge of urban morphology.

5   Results and Discussion

The cluster analysis gives 10 types. Upon a visual inspection, the interpretive description for each type is given in the table below (Fig. 4 and Table 1):

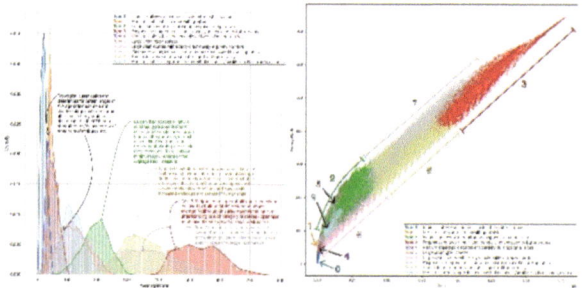

Fig. 4. Interpretation of the clusters' preferences for certain Average Radius and Area values

Table 1. Classification result and intuitive interpretation of clusters

Type	Description
Type 0	Interstitial alleyways and small scale urban public spaces
Type 1	Medium scale public spaces with great vista
Type 2	Natural surfaces or large urban open spaces with great vista
Type 3	Peripheries or large unbuilt land, generally not enclosed by built-up fabric
Type 4	Medium scale public spaces enclosed by buildings on all sides
Type 5	Large urban open spaces
Type 6	Large open spaces with axiality/directionality, e.g. view corridors
Type 7	Peripheries or large unbuilt land with one side enclosed by built-up fabric
Type 8	Threshold transition between unbuilt land and built urbanity
Type 9	Medium scale urban public spaces with directionality/axiality, e.g. Boulevards, avenues

GMM produces clusters each possessing a certain 'preference' over a range of values of each geometrical attribute. For example, by examining the distribution of Average Radius measure of the points under each cluster, it is clearly distinguishable that spaces

Fig. 5. Typological classification of urban spaces in Burnham's Chicago plan, showing 10 different categories of urban spaces

of Type 3 (peripheries or large unbuilt land, generally not enclosed by built-up fabric) favour larger values of Average Radius whilst spaces of Type 0 (interstitial alleyways and small scale urban public spaces) favour the least Average Radius value. Such interpretive analysis of the numerical result confirms the visual inspection and intuitive interpretation of the meaning of each cluster (Figs. 5 and 6).

Fig. 6. Typological classification of urban spaces in Colin Rowe's urban design for the Roma Interrotta project; Le Corbusier's masterplan for Antwerp; Vienna general plan for ringstrasse; and historical plan of Karlsruhe

In terms of the spatial distribution of the clusters, it is observed that the hierarchical arrangement of urban spaces in the Roma Interrotta project of Rowe [4] confirms the intuitive interpretation of meanings of clusters. The entirely different spatial configuration of

Vienna ringstrasse masterplan confirms the success of the classifier in decerning spaces with great visibility and increased vista, differentiated from the interstitial alleyways and granular street blocks in the inner city (Figs. 7 and 8).

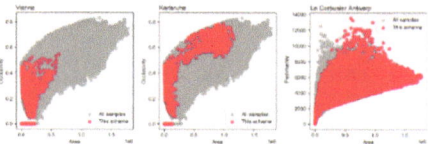

Fig. 7. Scatter plot showing the comparison between visual qualities of each sampled point within each scheme and that of the entire concatenated dataset of all samples within 19 sites

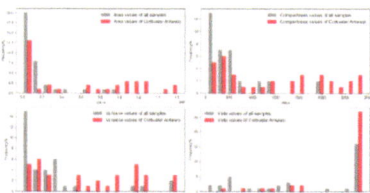

Fig. 8. Histograms showing the distribution of Area, Compactness, Variance, and Vista values across all sample points in Le Corbusier's scheme and in all schemes surveyed

Le Corbusier's Antwerp plan presents new spatial types totally alien to the other schemes surveyed, confirming the initial difference in design intentions in favour of a car-enabled estate with low ground space intensity [27]. The direct numerical consequence regarding visual qualities relate to higher Area values due to increased visibility and the scarcity of human-scaled public spaces; higher Compactness signify more homogeneous and uniform visual experiences in all directions; increased vista is also indicative of both characters.

6   Applications

Within the test set, it is possible to map out how the urban spaces within each input urban condition breaks down into different types of spaces (Figs. 9 and 10).

Meanwhile, the robustness of the classifier also enables an automatic spatial classification of any given urban condition. For example, by mapping out the distribution of Type 0 spaces (interstitial alleyways and small scale urban public spaces), it is possible to examine the correlation between the abundance of human-scale urban spaces and urban density (measured by Ground Space Index or GSI [27, 28], a measure of ground area that buildings occupy in relation to the total area of the land). By establishing an analytical repository of western European cities of similar (although slightly different) bioclimatic and social contexts, this comparison is suggestive of the numerical interpretation of the spatial implications of different urban design tactics, from vehicle-oriented developments to historic centres of European cities. Meanwhile, the comparison between

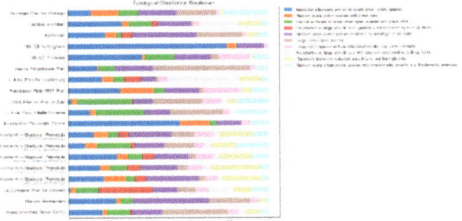

Fig. 9. Typological breakdown of 19 schemes surveyed, showing the proportion of each cluster of space involved in each urban condition

Fig. 10. Comparing urban layouts according to GSI and abundance of Type 0 spaces (left); catalogue of urban conditions of similar GSI but different frequency of Type 0 spaces (right)

cities of similar GSI values yet different in the abundance of Type 0 spaces provokes the awareness of how different European cities would tackle the century-old yet profoundly important question of urban density. For example, one could draw the conclusion that for a more efficient, vehicle enabled layout, Barcelona's might be a good lesson to be learnt; for more human-scale layouts favouring pedestrian access, the historic city centres of Marseilles, Rome, Paris, The Hague etc. are epitomic cases for urban planners and designers. The coincidence of such observations with textbook conclusions again confirms the validity of this approach, whilst this method is also suggestive of undiscovered spatial patterns and enables new aspects of urban spatial analysis.

7 Conclusion

7.1 Summary of Key Findings

Isovist analysis is a useful tool to quantify visual qualities of the built environment, both at the scale of a building and at the scale of urban developments and public spaces. This paper presents a method to classify urban spaces based on visual qualities, measured by the geometrical attributes of the resultant polygon of isovist analysis. Using a Gaussian Mixture Model, this paper proposes an analytical workflow that extracts the inherent patterns of spatial characterisations embedded in the 19 urban schemes used for classification and produces an urban space classifier capable of classifying urban spaces into 10 basic spatial types. This methodology proves successful in capturing the intrinsic spatial typology within differing urban conditions and distinct urban contexts and could reflect the values and visions embedded in certain urban design schemes.

7.2 Further Research

Currently, this workflow exists in 2D and could only be applied to vector-based figure-ground urban plans. However, without significant modifications to the fundamentals of the workflow, it could anticipate an augmentation of scale of operation through the incorporation of 3D geometries and 3D isovist analytics, as outlined in [29]. Additionally, as the current parameter that controls the generation of the grid of the isovist field is a trade-off between efficiency and accuracy, smarter and context-adaptive tessellation algorithms could be applied to the creation process of the isovist field to reduce computational load and maximise accuracy [10]. Meanwhile, adding spatial lagging [11] which allows for each sample point to be context-sensitive and aggregate characteristics from neighbouring sample points would also be a feasible strategy to address spatial continuity in data and avoid outliners due to unforeseen errors in computation.

Reference

1. Sitte, C.: The art of building cities city building according to its artistic fundamentals. Reinhold Publishing Corporation, New York (2013)
2. Lynch, K.: The image of the city. M.I.T. Press, Cambridge, Mass (2008)
3. Krier, R.: Urban space. 5. Academy Editions, London (1991)
4. Peterson, S., Row, K.: Urban design tactics. Archit. Des. **49**(3–4), 76–81 (1979)
5. Benedikt, M.L.: To take hold of space: isovists and isovist fields. Environ. Plann. B. Plann. Des. **6**(1), 47–65 (1979)
6. Hillier, B., Hanson, J.: The Social Logic of Space. Cambridge University Press, New York (1989)
7. Hillier, B.: What are cities for? And how does this relate to their spatial form? The Journal of Space Syntax. **6**(2), 199–212 (2016)
8. Batty, M.: Exploring isovist fields: space and shape in architectural and urban morphology. Environ. Plann. B. Plann. Des. **28**(1), 123–150 (2001)
9. Batty, M.: Cities and complexity: understanding cities with cellular automata, agent-based models, and fractals. 1. paperback ed., p. 565. MIT, Cambridge, Mass. London (2007)
10. Conroy Dalton, R., Dalton, N.S., McElhinney, S., Mavros, P.: Isovist in a Grid: Benefits and limitations. Proceedings 13th International Space Syntax Symposium, p. 17 (2022)
11. Fleischmann, M., Feliciotti, A., Romice, O., Porta, S.: Morphological tessellation as a way of partitioning space: Improving consistency in urban morphology at the plot scale. Comput. Environ. Urban Syst. **80**, 101441 (2020)
12. Usui, H., Asami, Y.: Size distribution of urban blocks in the Tokyo Metropolitan Region: estimation by urban block density and road width on the basis of normative plane tessellation. Int. J. Geogr. Inf. Sci. **32**(1), 120–139 (2018)
13. Basaraner, M., Cetinkaya, S.: Performance of shape indices and classification schemes for characterising perceptual shape complexity of building footprints in GIS. Int. J. Geogr. Inf. Sci. **31**(10), 1952–1977 (2017)
14. Biljecki, F., Chow, Y.S.: Global building morphology indicators. Comput. Environ. Urban Syst. **95**, 101809 (2022)
15. Urquizo, J., Calderón, C., James, P.: Metrics of urban morphology and their impact on energy consumption: A case study in the United Kingdom. Energy Res. Soc. Sci. **32**, 193–206 (2017)
16. Leduc, T.: t4gpd [Software] (2024). https://doi.org/10.5281/zenodo.5771916
17. Gillies, S., et al.: Shapely [Software] (2024). https://doi.org/10.5281/zenodo.5597138

18. Turner, A., Penn, A.: Making isovists syntactic: isovist integration analysis. The 2nd International Symposium on Space Syntax (1999)
19. Schirmer, P.M., Axhausen, K.W.: A multiscale classification of urban morphology. Journal of Transport and Land Use [Internet]. [cited 2024 March 29] **9**(1) (2015). https://www.jtlu.org/index.php/jtlu/article/view/667
20. Arribas-Bel, D., Fleischmann, M.: Spatial Signatures - Understanding (urban) spaces through form and function. Habitat Int. **128**, 102641 (2022)
21. Alexiou, A., Singleton, A., Longley, P.A.: A classification of multidimensional open data for urban morphology. Built Environment. **42**(3), 382–395 (2016)
22. Steiniger, S., Lange, T., Burghardt, D., Weibel, R.: An approach for the classification of urban building structures based on discriminant analysis techniques. Trans. GIS **12**(1), 31–59 (2008)
23. Kropf, K.: Aspects of urban form. Urban Morphology. **13**(2), 105–120 (2009)
24. Graves, C.P.: The genealogy of cities, p. 367. Kent State Univ. Press, Kent, Ohio (2009)
25. Reynolds, D.: Gaussian Mixture Models. In: Li, S.Z., Jain, A., (eds.) Encyclopedia of Biometrics [Internet], p. 659–63. Springer US, Boston, MA (2009). [cited 2024 Mar 26]. https://doi.org/10.1007/978-0-387-73003-5_196
26. Fleischmann, M., Feliciotti, A., Romice, O., Porta, S.: Methodological foundation of a numerical taxonomy of urban form. Environment and Planning B: Urban Analytics and City Science. **49**(4), 1283–1299 (2022)
27. Berghauser Pont, M., Haupt, P.: Space, density and urban form. Delft University of Technology, Netherlands (2009)
28. Berghauser Pont, M., Haupt, P.: Spacematrix: Space, Density and Urban Form [Internet]. revised. TU Delft OPEN Publishing [cited 2023 Oct 9] (2023). https://books.open.tudelft.nl/home/catalog/book/38
29. Kim, G., Kim, A., Kim, Y.: A new 3D space syntax metric based on 3D isovist capture in urban space using remote sensing technology. Comput. Environ. Urban Syst. **74**, 74–87 (2019)

Multi-agent Simulation-Based Urban Waterfront Public Space Quality Comprehensive Measurement Indexes

Chunxia Yang[1], Ming Zhan[1(✉)], and Ziying Yao[2]

[1] Tongji University, 1239 Siping Road, Shanghai, China
945581064@qq.com
[2] Shanghai Urban Planning and Design Research Institute, 331 Tongren Road, Shanghai, China

Abstract. From the perspective of behavior, the study combines multi-agent simulation and measurement index system for the waterfront area. Connecting with the waterfront space characteristics, the study combines spatial, behavioral and correlation measurement indexes to form the measurement dimensions of access-efficiency, stay-comfort and water-friendliness. Based on literature research and behavior simulation output characteristics, the study initially constructs an index system for measuring the quality of public space in urban waterfront, with a total of 18 measuring indexes. 6 typical waterfront sections along the Huangpu River are selected as samples and the indexes of each period are output through the behavior simulation model based on the Anylogic. Through the extraction of the mean value, maximum value and minimum value, an urban waterfront public space measurement index system suitable for the Huangpu River in Shanghai is finally formed. By measuring different samples and normalizing the results, it is found that stay-comfort is the most important index, while access-efficiency and water-friendliness are of less importance. The study proposes a comprehensive measurement index system from behavioral needs, which forms a more accurate and intelligent index for analyzing waterfront areas in three dimensions. In addition, based on user's recreational behavior, a dynamic measurement method is developed to provide a more efficient quantitative reference for the identification and optimization of weakness in urban waterfront.

Keywords: waterfront public space · behavior simulation · multi-agent · measurement index

1 Research Background and Research Path

To promote scientific and quantitative development of the digital city, it's important to quantitatively analyze the quality of urban public space, and construct effective spatial measurement and optimization standards. Waterfront area is an important part to spark public vitality, but traditional planning and empirical design are disconnected from actual usage. Scientific evaluation and measurement methods are urgently needed to increase the quality of public space.

H. Chai et al. (Eds.): CDRF 2024, *Symbiotic Intelligence*, pp. 536–548, 2025.
https://doi.org/10.1007/978-981-96-3433-0_47

Traditional measurement methods of public space mainly rely on on-site environmental auditing, which relies on site survey and subjective evaluation of users. In the face of large-scale measurement, it is difficult to accurately grasp the relationship between space and behavior. The development of multi-source new data and new technical tools has provided support for large-scale, objective and efficient measurement of urban space, forming a new off-site environmental audit method. For example, Wang and Ma (2020) evaluated the vitality of the waterfront public space of the Huangpu River in Shanghai through multi-source data. However, research based on multi-source data can only obtain current and past data and analyze the existing site conditions, lacking of detailed collection of individual data. Since 2003, behavior simulation technology has been introduced. It expands the dynamic analysis of time dimension, can more accurately simulate and analyze the interaction between space and behavior, and allows dynamic measurement and analysis of public space.

The existing research on public space measurement indexes can be divided into spatial perspective and behavioral perspective. The measurement from the perspective of space focuses on the physical environment characteristics of the public space itself. The measurement dimension mainly refers to the 3D dimension of density, diversity and design proposed by Cervero and Kockelman (1997), as well as the 5D dimension formed by the introduction of destination accessibility and traffic distance by Ewing and Cervero (2008). The measurement indexes are mainly static indexes such as the area, scale and quantity of space elements. With the concern for humanism, the research from the perspective of behavior has been paid more attention, and the measurement of dynamic behavior has been adopted. The measurement indexes such as fluctuation index and revisiting rate have emerged. At present, the waterfront measurement research is mainly static spatial measurement. It is necessary to explore the measurement dimensions and indexes for the unique characteristics of the waterfront. For the leisure space, the dimensions of traffic, residence, and vitality, as well as several indexes of space and behavior, static or dynamic, can be roughly sorted out, providing a basis for the establishment of a more comprehensive waterfront public space comprehensive measurement index system.

With the help of multi-agent simulation technology, this research proposes a comprehensive measurement index system of waterfront public space quality based on behavior needs, and forms a dynamic measurement method according to the rules of users' recreational behavior. This study selects six waterfront public space sections from north to south on both sides of the Huangpu River as the research objects for in-depth investigation and measurement. Six typical time periods (7: 00–9: 00, 9: 00–11: 00, 11: 00–13: 00, 13: 00–15: 00,15: 00–17: 00 and 17: 00–19: 00)are selected for the survey (Fig. 1). The study can be divided into 3 parts: the measurement dimension and indexes establishment, simulation model construction and index output and measurement results analysis (Fig. 2).

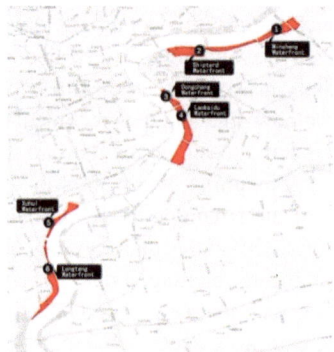

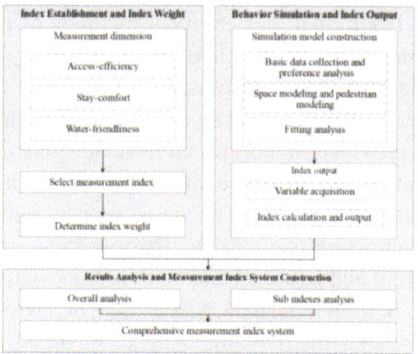

Fig. 1. Six sites along the Huangpu River (left)

Fig. 2. Research framework (right)

2 Index Establishment and Index Weight

2.1 Establishment of Measurement Dimensions and Indexes

The determination of measurement dimensions and indexes should not only conform to waterfront behavior characteristics, but also combine the characteristics of simulation output. This study establishes three measurement dimensions of access-efficiency, stay-comfort and water-friendliness, and divides 18 indexes into three categories: inherited, improved and newly added.

The dimension of access-efficiency includes the measurement of external accessibility and internal connectivity of waterfront public space. The stay behavior of people reflects the perfection of facility service level and whether landscape elements can promote the production of various behaviors, reflecting the quality of waterfront public space. The hydrophilic behavior of the crowd reflects whether the configuration and organization of the unique elements of the waterfront such as the base surface and the shoreline are reasonable. The hydrophilic convenience is the unique measurement dimension of the waterfront different from other urban spaces, which can be divided into two levels: the overall and the coastal. In the dimension of water-friendliness, the accessibility of sight, the density of shoreline and the fluctuation index of crowd in the study of waterfront are used to measure the overall and coastal hydrophilicity. The density of vertical water flow and the hydrophilicity index of shoreline are added to reflect the influence of base depth and elevation on hydrophilicity respectively (Table 1).

2.2 Weight of Measurement Index

20 experts in the field of architecture, planning and landscape are invited to conduct a questionnaire survey to compare the scores of various measurement indexes according to different measurement dimensions. The weight of each index of urban waterfront public space for the comprehensive measurement can be obtained by establishing the hierarchical structure model through the analytic hierarchy process. The weight of each index is shown in Table 2.

Table 1. Measurement indexes of 3 dimensions and calculation formulas of newly added indexes

Dimension		Index	Type	Index Connotation
Access-efficiency	External accessibility	walking detour coefficient	improved	The ratio of the actual walking distance of pedestrians to the ideal straight-line distance
		Walking time	inherited	Pedestrian walking time
	Internal connectivity	Pedestrian density	inherited	Ratio of pedestrian number to accessible area
		Space utilization	improved	The ratio of the space area used by the crowd to the total area of the site
		Base-plane connection coefficient	added	The ratio of the actual walking distance to the ideal straight-line distance when the base plane is converted
				$R = \dfrac{\sum_{i=1}^{N}\sqrt{(M_i - M_{i+1})^2 + (N_i - N_{i+1})^2}}{\sqrt{(M_0 - M_e)^2 + (N_0 - N_e)^2}}$ (M_i, N_i)——Instantaneous coordinate of particle (M_0, N_0)——coordinates of the starting point of the particle conversion base plane (M_e, N_e)——Coordinates when the particle completes the base plane conversion
Stay comfort	Group stay	Stay capacity	inherited	Number of people staying longer than three minutes
		Stay rate	inherited	Ratio of the number of people staying to the total number of people in the site
		Stay surface density	inherited	The ratio of the amount of residence to the area of residence
		Attraction point utilization	added	The ratio of the number of people staying in the influence area of the attraction point to the total number of people.
				$D_c = \dfrac{Q_c}{Q} * 100\%$ Q_c——the number of attracted particles in the area Q — — Total number of slow-moving arriving particles in the area
		Attraction point average visit rate	added	Average number of visits to the point of attraction
				$D_q = \dfrac{Q_{cs}}{q} * 100\%$ Q_{cs}——Total times of being attracted in the area q — — Number of attraction points in the area
	Individual stay	Type of stay behavior	improved	Number of types of stay behaviors
		Stay time per capita	inherited	Average of individual stay times
		Times of stays per capita	inherited	The average number of times an individual stays during the entire waterfront behavior
Water-friendliness	Overall hydrophilicity	Sight reaching water rate	improved	Percentage of the number of particles that see the waterfront
		Flow density perpendicular to water	added	The ratio between the number of people in the space perpendicular to the water and the depth of the waterfront interface
				$D_v = \sum \dfrac{Q_l}{S_l / L_l} * 100\%$ Q_l——Number of particles per base level S_l / L_l——Depth of each base level
	Coastal hydrophilicity	Shoreline line density	inherited	The ratio of the number of people staying on various shorelines to the overall length of the shoreline
		Hydrophilicity of shoreline at different elevations	added	The ratio of the number of people staying on shorelines at different elevations to the height difference of the section
				$V_h = \sum \dfrac{Q_l}{H_l}$ Q_l——Number of particles per base level H_l — — Height difference between each base level
		Shoreline population fluctuation index	improved	The ratio of the maximum to the minimum number of shoreline populations

According to the spatio-temporal relationship of index measurement, it can be divided into static index and dynamic index. The former reflects a certain situation in a space range at a specific time point, which represents the instantaneous static situation in public space. The latter reflects the quantitative characteristics of a certain spatial range in a specific period of time.

Table 2. Weight of comprehensive measurement index of urban waterfront public space quality

Target Layer	Index Layer 1	Weight	Index Layer 2	Weight	Index Layer 3	Weight	Type
Comprehensive Measurement of Public Space Quality of Urban Waterfront	X Access-efficiency	0.2143	Xa-External accessibility	0.1014	Xa1-Walking detour coefficient	0.0486	D
					Xa2-Walking time	0.0528	D
			Xb-Internal connectivity	0.1128	Xb1-Pedestrian density	0.0296	S
					Xb2-Space utilization	0.0413	S
					Xb3-Base-plane connection coefficient	0.0419	D
	Y Stay-comfort	0.5354	Ya-Group stay	0.3970	Ya1-stay capacity	0.0761	D
					Ya2-stay rate	0.1072	D
					Ya3-stay surface density	0.0522	D
					Ya4-Attraction point utilization	0.0763	S
					Ya5-Attraction point average visit rate	0.0851	D
			Yb-Individualstay	0.1384	Yb1-stay behavior type	0.0704	D
					Yb2-stay time per capita	0.0422	D
					Yb3-Times of stays per capita	0.0258	D
	Z Water-friendliness	0.2503	Za-Overall hydrophilic	0.1053	Za1-Sight reaching water rate	0.0817	S
					Za2-Flow density perpendicular to water	0.0236	S
			Zb-Coastal hydrophilicity	0.1450	Zb1-Shoreline line density	0.0407	S
					Zb2-Hydrophilicity of shoreline at different elevations	0.0632	S
					Zb3-Shoreline Population Fluctuation Index	0.0411	D

Note: D represents dynamic measurement index, S represents static measurement index.

3 Behavior Simulation and Index Output

3.1 Construction of Behavior Simulation Model

Based on the Anylogic platform, the behavior simulation model of 6 samples is constructed. The modeling process can be divided into following 3 parts:

Basic Data Collection and Preference Analysis

Basic data can be divided into spatial element data and crowd behavior data. This study classifies the spatial elements of the waterfront area into 4 categories of base level, shoreline form, buildings, facilities and their subcategories. Using the method of field survey to collect the information. Crowd behaviors are divided into 5 categories of viewing, leisure, sports, entertainment, consumption and their subcategories. By means of mapping, counting and questionnaire, the information of crowd characteristics, behavior types, locations and duration are obtained.

The distribution of pedestrian in waterfront shows the probability of pedestrian choosing different space elements to carry out behaviors is different. The attraction weight of each space element expresses the relationship between spatial elements and crowd behaviors, which is an interactive parameter that can be identified by the model platform.Taking the behavior data analysis of Xuhui Waterfront from 15:00 to 17:00 as an example, the corresponding relationship between some spatial elements and behavior types is summarized in Table 3.

Table 3. Spatial-behavioral analysis data example

Spatial Elements		Behavior Type	Number of People	Attraction Weight
Base level	Shaded Pathway	viewing	7	2.0173%
		leisure	102	29.3948%
		entertainment	1	0.2882%
		sports	1	0.2882%
		consumption	7	2.0173%
	Accessible Grass	leisure	16	4.6110%
		entertainment type	2	0.5764%
		consumption	132	38.0403%
	Square	viewing	8	2.3055%
		leisure	67	19.3084%
		sports	1	0.2882%
Facilities configuration	Formal seat	leisure	1	0.2882%

Space Modeling and Pedestrian Modeling

Firstly, the base map is imported into the Anylogic platform and various elements are translated based on the spatial markup module of the Anylogic to build a simulation environment. As static agents, the spatial module has three basic attributes: service radius, pedestrian capacity and attraction weight. For the interaction between the spatial elements and pedestrian, it is necessary to quantify the space marking module to obtain various variables required for index output.

In order to fully consider the characteristic differences of people, each pedestrian is simulated as a unique agent particle with its own attributes. Besides the three initial attributes of gender, age and behavior type, the pedestrian agent has four basic attributes: vision range, planned recreation time, basic speed and element perception radius. Variable of Anylogic is used to express the dynamic attributes, such as moving speed, moving direction and moving time, which may change with the running of the model. To count the state of pedestrians, variable extraction (coordinates, velocities, stay time, walking distance, stay times, etc.) of particles is required. Then, the behavior activity chain is constructed to express the path selection and decision-making process of pedestrian in the waterfront. The agent particles can synthesize various attractive and repulsive forces in the simulation environment at each step of the continuous process and realize the simulation of waterfront random behavior activities.

Fitting Analysis

The preliminary model simulation results are compared with the site survey results and the element attraction is adjusted through multiple simulations to make the simulation results highly consistent with the actual situation. On the basis of qualitative fitting, quantitative fitting analysis is added to improve the accuracy of behavior simulation model. SPSS Statistics is used to analyze the bivariate correlation between the simulated data and the measured data of the attraction points in each period. The Pearson coefficients in each period are all greater than 0.6, and the significance levels are far less than 0.01, which proved that the model is effective. The simulation results can be presented as density map (Fig. 3).

Fig. 3. Simulation result of Xuhui Waterfront in 7:00–9:00 and 15:00–17:00

3.2 Measurement Index Output

The dynamic simulation process of the model is operated to obtain the initial variables of index calculation, including spatial characteristics, pedestrian statistics and pedestrian dynamic parameters (Table 4). Then, according to the measurement index calculation formula, the operation module of Anylogic are used to convert the initial variables into index measurement results. After completing all index calculations, each index will be exported to an Excel table.

Table 4. Variable acquisition

Variable Type	Variable Name	Variable Acquisition Method
Spatial characteristics	Site area	Obtained from the environment module area calculation function
	Shoreline length	Obtained by the environment module length calculation function
	Section length	Obtained by the area and length calculation function of the environment module
	Height of base surface	Obtained by the environment module height calculation function
Pedestrian statistics	Number of tourists	Set statistics area and call function of the number of pedestrians in the area
	Sight condition	Set sight reaching water area and call area pedestrian quantity function;
Dynamic parameters	Location parameter	Call the coordinate function to calculate the pedestrian position
	Moving time	Call the time function to record the start time
	State of motion	Get the current velocity of the particle, <0.1 m/s is considered as stay state
	Length of stay	Call the time function to record the start time in accordance with the stay state
	Times of stops	Record the number of times agent particles stay

4   Results Analysis and Measurement Index System Construction

4.1   Overall Analysis

Through the analysis of the results, it is found that the comprehensive measurement results of each site are higher and lower during periods of 15: 00–17: 00 and 7: 00–9: 00. In order to understand the relationship between the quality of public space in each site, this study selects the mean value, peak hours results and trough hours results for comparison and analysis after data normalization (Table 5). From 7:00 to 9:00, there are few people in the sites and the activities are not fully carried out, so the comprehensive quality is mainly based on the mean value and peak hours value. From the perspective of each measurement dimension, the access-efficiency, stay comfort and water-friendliness of most sites are better in the peak hours, but worse in the trough hours. Combined with the parallel comparison of the mean value and peak hours value throughout the day, the sites can be divided into high-quality sections, including the Shipyard Waterfront, Longteng Waterfront and Dongchang Waterfront. The lower quality sections include Xuhui Waterfront, Laobaidu Waterfront and Minsheng Waterfront. Targeted optimization can be carried out according to the measurement results of each site.

Table 5. Comprehensive and multi-dimensional measurement results

Measure Dimension		Dongchang	Xuhui	Laobaidu	Long Teng	Minsheng	Shipyard
Overall score		41/45/36	30/43/16	33/40/26	46/54/29	37/37/36	50/48/43
Score ranking		3/3/3	6/4/6	5/5/4	2/1/4	4/6/2	1/2/1
Access-efficiency		7/7/5	7/7/6	9/11/8	14/12/10	15/9/10	15/10/11
Sub-items	External accessibility	1/2/0	2/2/2	4/4/4	8/8/8	8/8/8	8/8/9
	Internal connectivity	6/6/5	5/6/4	5/7/4	6/4/3	7/2/2	7/2/2
Stay-comfort		29/25/26	16/25/7	20/23/14	20/26/13	16/24/20	26/28/25
Sub-items	Group stay	25/17/23	10/17/5	16/20/10	17/22/10	11/17/12	20/19/20
	Individual stay	4/9/3	6/9/2	4/3/4	4/4/3	5/7/7	7/9/6
Water-friendliness		6/10/5	6/10/3	5/5/3	11/16/5	5/4/6	8/10/6
Sub-items	Overall	4/1/4	1/1/1	2/3/3	7/8/4	2/1/3	6/6/6
	Coastal	2/9/1	5/9/1	2/3/0	4/8/2	3/3/4	3/4/1

Note: mean value/ peak hours results/ trough hours results

4.2 Sub Indexes Analysis

Based on the measurement results in 6 sites, the distribution rule of each sub index in different sites and different periods is summarized, and the spatial quality differences of samples are compared (Fig. 4). Pedestrian density has both time and site rules, which increases gradually with time and reaches the peak at 15: 00–17: 00. The indexes of space utilization, stay capacity, stay surface density, stay time per capita, flow density perpendicular to water, shoreline line density and hydrophilicity of shoreline at different elevations show obvious time regularity. The space utilization, stay capacity, stay surface density, flow density perpendicular to water, shoreline line density and hydrophilicity of shoreline increase gradually with time and reach the peak at 15: 00–17: 00, with great differences among sites. Walking detour coefficient, walking time, base-plane connection coefficient and stay behavior type show different site rules. Walking detour coefficient and walking time are related to the size of the neighborhood around the site, the density of the road network, the layout of entrances and exits, and the road network structure. There are obvious differences among sites in the indexes of stay rate, attraction point utilization, times of stays per capita, per capita stay times, sight reaching water rate and shoreline population fluctuation index, and there is no obvious rule.

4.3 Comprehensive Measurement Index System

Combined with the AHP index weight calculation results based on the expert question-naire, a comprehensive quality measurement index system suitable for the waterfront

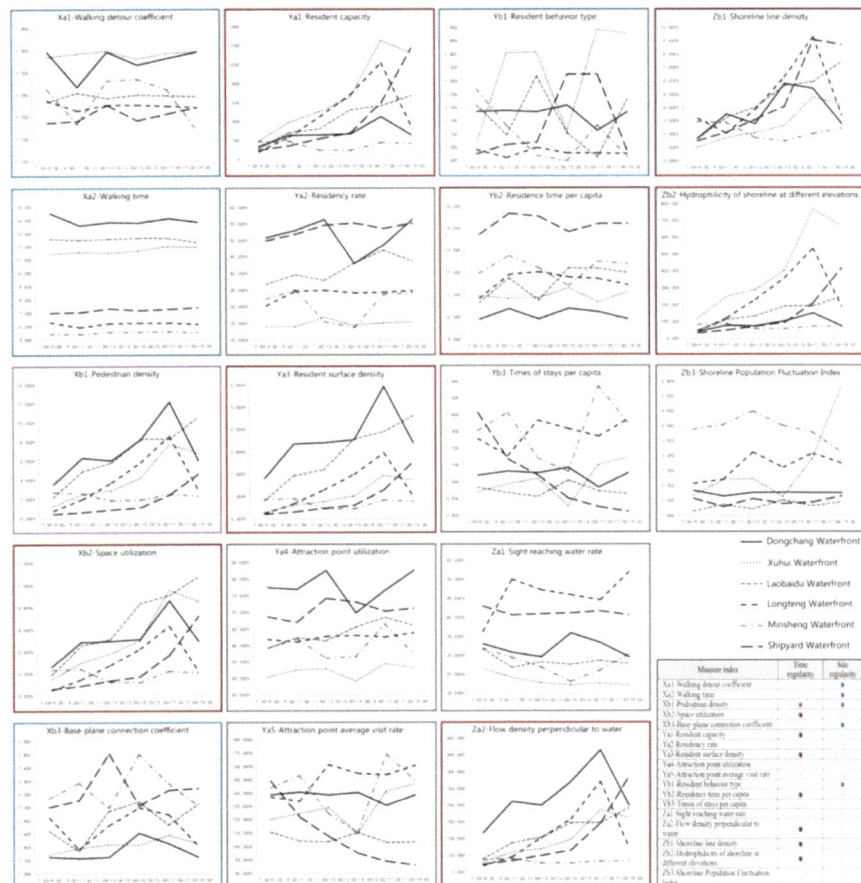

Fig. 4. Summary of time and site rules of sub indexes

public space of Shanghai Huangpu River is formed (Table 6). Among them, the index weight is listed, and the contribution of each index to the three dimensions and the overall quality of waterfront public space can be known. At the same time, the reference mean value, maximum value and minimum value of each sample measurement index are also listed, which can provide a basis for the quality measurement of these samples or other similar projects in the future.

Table 6. Measurement index system

Measure Dimension		Measure Index	Index Weight	Reference Mean Value	Reference Maximum Value	Reference Minimum Value
X Access-efficiency	Xa-External accessibility	Xa1-Walking detour coefficient	0.0486	1.263	1.347	1.169
		Xa2-Walking time	0.0528	9.517	14.537	5.404
	Xb-Internal connectivity	Xb1-Pedestrian density	0.0296	4.563%	0.431%	12.198%
		Xb2-Space utilization	0.0413	1.695%	0.051%	10.785%
		Xb3-Base-plane connection coefficient	0.0419	1.841	2.811	1.230
Y Stay-comfort	Ya-Group stay	Ya1-Stay capacity	0.0761	434	82	1255
		Ya2-Stay rate	0.1072	39.484%	23.762%	56.399%
		Ya3-Stay surface density	0.0522	1.863%	0.215%	5.935%
		Ya4-Attraction point utilization	0.0763	64.590%	49.155%	82.646%
		Ya5-Attraction point average visit rate	0.0851	129.612%	86.124%	168.996%
	Yb-Individual stay	Yb1-Stay behavior type	0.0704	1.164	1.000	1.492
		Yb2-Stay time per capita	0.0422	11.928	9.915	14.691
		Yb3-Times of stays per capita	0.0258	1.120	0.824	1.572
Z Water-friendliness	Za-Overall hydrophilic	Za1-Sight reaching water rate	0.0817	33.796%	22.158%	52.043%
		Za2-Flow density perpendicular to water	0.0236	122.492	19.759	365.310

(*continued*)

Table 6. (*continued*)

Measure Dimension		Measure Index	Index Weight	Reference Mean Value	Reference Maximum Value	Reference Minimum Value
	Zb-Coastal hydrophilicity	Zb1-Shoreline line density	0.0407	4.499%	1.024%	17.723%
		Zb2-Hydrophilicity of shoreline at different elevations	0.0632	207.688	34.430	766.308
		Zb3-Shoreline population fluctuation index	0.0411	2.399	1.156	5.345

5   Conclusion

Based on the multi-agent behavior simulation technology, this study measures six waterfront areas on both sides of the Huangpu River in Shanghai, and finds that the dimensions of stay-comfort has a more important impact on the quality of public space in urban waterfront areas than the dimensions of access-efficiency and water-friendliness. At the same time, the dynamic measurement based on behavior simulation has obtained the distribution rules of each sub index in different sites and different time periods. Through the extraction of mean value, maximum value and minimum value, an urban waterfront public space measurement index system suitable for the Huangpu River in Shanghai is finally formed. The reference values of measurement results can assist in the diagnosis and renovation of site problems, and the dynamic simulation of key indexes can also assist in the flow hierarchical warning and control of waeterfront areas.

References

Wang, W.Q., Ma, X.J.: Vitality assessment of waterfront public space based on multi-source data: a case study of the Huangpu river waterfront. Urban Planning Forum **255**, 48–56 (2020)

Cervero, R., Kockelman, K.: Travel demand and 3Ds: density, diversity, and design. Transp. Res. Rec. **2**(3), 199–219 (1997)

Ewing, R.H.: Characteristics, causes, and effects of sprawl: a literature review. In: Marzluff, J.M., Shulenberger, E., Endlicher, W., et al. (eds.) Urban Ecology. Springer, Boston (2008)

Author Index